Advances in Computational Fluid Dynamics

Advances in Computational Fluid Dynamics delves into the emergent ways that engineers are utilizing computer simulations to enhance efficiency, reduce costs, and innovate across aerospace, automotive, energy, and biomedical engineering fields. It provides the most recent tools and strategies for improving prediction accuracy, design, and optimization.

Highlighting the practical uses of computational fluid dynamics (CFD) in solving real-world engineering issues, the book covers a wide range of physical problems from turbulence modeling and high-performance computing to the integration of machine learning and multiphysics simulation. It includes case studies in aerodynamic designs, energy conversion processes, and cooling systems and examines AI integration and machine learning techniques.

The book will interest researchers, upper-level undergraduate, and graduate engineering students studying practical applications of CFD.

Advances in Computational Fluid Dynamics

Edited by
Basma Souayeh and Katta Ramesh

CRC Press
Taylor & Francis Group
Boca Raton London New York

CRC Press is an imprint of the
Taylor & Francis Group, an **informa** business

Designed cover image: Shutterstock

First edition published 2026
by CRC Press
2385 NW Executive Center Drive, Suite 320, Boca Raton FL 33431

and by CRC Press
4 Park Square, Milton Park, Abingdon, Oxon, OX14 4RN

CRC Press is an imprint of Taylor & Francis Group, LLC

ISBN: 978-1-041-00819-4 (hbk)
ISBN: 978-1-041-00820-0 (pbk)
ISBN: 978-1-003-61182-0 (ebk)

DOI: 10.1201/9781003611820

Typeset in Times
by Newgen Publishing UK

Contents

Part IV Bio-Convective and Reactive Flow Dynamics

Part V Microchannels and Solar Applications

Part VI Specialized Flow Configurations

About the Editors

Basma Souayeh is an associate professor at King Faisal University, Saudi Arabia and a researcher in the Laboratory of Fluid Mechanics and Heat Transfer (URPF Lab) at the University of Tunis El Manar, Tunisia. She has more than 12 years of experience in education and research fields with more than 140 peer-reviewed international papers published in ISI journals in physics. She is a co-editor of *Mathematical Modelling of Fluid Dynamics and Heat Transfer* (CRC Press, 2023). She also serves as a reviewer for several international ISI scientific journals and has been both a speaker and a session chair at several national and international conferences. Dr. Souayeh's research interests include computational physics, mechanical engineering, thermal engineering, energy, computational fluid mechanics, numerical simulation, numerical modeling, CFD simulation, finite volume method, finite difference method, engineering thermodynamics, energy engineering, heat and mass transfer, nanofluids, CFD coding, and fluid flow and mechanics.

Katta Ramesh is an associate professor and deputy dean (internationalization) in the School of Mathematical Sciences at Sunway University, Malaysia. He holds the position of adjunct professor at esteemed institutions including Chandigarh University, India; Lovely Professional University, India; Graphic Era (Deemed to be University), India; and Symbiosis Institute of Technology, India. With over a decade of teaching and research experience, Dr. Ramesh earned his Ph.D. in applied mathematics, specializing in the modeling of physiological flows, from Visvesvaraya National Institute of Technology, Nagpur, India, in 2016. Dr. Ramesh has contributed 100 articles to reputable international journals, edited one book, four book chapters, and two conference proceedings. He is a lifetime member of the Indian Society of Theoretical and Applied Mechanics, Indian Science Congress, Indian Mathematical Society, and Ramanujan Mathematical Society.

Contributors

Kareem Akhtar
University of Engineering and Technology
Pakistan

Mir Waqas Alam
Department of Physics
King Faissal University
Saudi Arabia

Liaqat Ali
School of Sciences
Chaoyang University
China

Ali F. Ali Fadiel
Department of Mechanical Engineering Technology
Higher Institute for Sciences and Technology
Libya

Aishwarya Anand
Department of Studies in Mathematics
Davangere University
India

Abdul Jabbar Ansari
Department of Mechanical Engineering
Bansal Institute of Engineering and Technology
India

Retna Apsari
Department of Physics
Universitas Airlangga
Indonesia

Suresh Babu Baluguri
Department of Mathematics
Sreyas Institute of Engineering and Technology
India

Veeranna Bhoomika
Department of Studies in Mathematics
Davangere University
India

Vasavi Cheruku
Department of Mathematics
Sreyas Institute of Engineering and Technology
India

M. Revathi Devi
Department of Mathematics
Chennai Institute of Technology
India

Monaem Elmnifi
Department of Mechanical Engineering Technology
Belgorod State Technological University named after. V.G. Shukhov
Russia

Ragulkumar Elumalai
Centre for Computational Modeling
Chennai Institute of Technology
India

Jayaprada Evani
Department of Mathematics
Excellencia College
India

Srinivas Gosukonda
Department of Mathematics
Geethanjali College of Engineering and Technology
India

Ikram Ul Haq
School of Mechanical Engineering
Tianjin University
China

Waqar Ul Hassan
School of Mechanical Engineering
Tianjin University
China

Perumandla Karunakar
Department of Mathematics
VIT-AP University
India

Panchappa Mallikarjuna Kavana
Department of Studies in Mathematics
Davangere University

Ikhyun Kim
Department of Mechanical Engineering
Keimyung University
South Korea

Pardeep Kumar
School of Basic & Applied Sciences
K.R. Mangalam University
India

R. Naveen Kumar
Department of Mathematics
Amrita School of Engineering
India

B. Madhusudhana Rao
Department of Information Technology
University of Technology and Applied Sciences
Oman

Ulavathi Shettar Mahabaleshwar
Department of Studies in Mathematics
Davangere University
India

Abhilash Mahantheshappa
Department of Studies in Mathematics
Davangere University
India

Hasan Shakir Majdi
Department of Chemical Engineering and Petroleum Industries
Al-Mustaqbal University College
Iraq

Akshay Kumar Maurya
Department of Mechanical Engineering
Bansal Institute of Engineering and Technology
India

Ragavi Moorthy
Department of Mathematics
Vellore Institute of Technology
India

Murugan Muthtamilselvan
Department of Mathematics
Bharathiar University
India

R. Nanda Kumar
Department of Artificial Intelligence and Data Science
Anand Institute of Higher Technology
India

A.J.D. Nanthakumar
Vehicle Dynamics Laboratory
SRM Institute of Science and Technology
India

Bumireddy Neelakanta Reddy
Department of Mathematics
Sri Venkateswara Degree College
India

Kandavkovi Mallikarjuna Nihaal
Department of Studies in Mathematics
Davangere University

Mallikarjun Basavaraju Patil
Department of Studies and Research in Mathematics
Tumkur University
India

Sambath Paulsamy
Department of Mathematics
SRM Institute of Science and Technology
India

Francis Peter
Department of Mathematics
Easwari Engineering College
India

Tamalapakula Poornima
Department of Mathematics
Vellore Institute of Technology
India

K. Prabakaran
Department of Mathematics
SRM Institute of Science and Technology
India

Dhanapal Prakash
Department of Mathematics
SRM Institute of Science and Technology
India

Suresh Prithvi
Department of Studies and Research in Mathematics
Tumkur University
India

Elumalai Ragupathi
Department of Mathematics
SRM Institute of Science and Technology
India

D.R. Sasi Rekha
Department of Mathematics
Dayananda Sagar College of Engineering
India

Sudha Mahanthesh Sachhin
Department of Studies in Mathematics
Davangere University
India

Gadhigeppa Myacher Sachin
Department of Studies in Mathematics
Davangere University
India

Bellagubbala Bala Sai Sankar
Department of Mathematics
VIT-AP University
India

Shravanakumari Shanthaveerappa Mangala
Department of Studies in Mathematics
Davangere University

Oleg G. Shvydkyi
Independent Researcher
Ukraine

Narsu Sivakumar
Department of Mathematics
SRM Institute of Science and Technology
India

Varshegaa Sivanarayana
Department of Mathematics
SRM Institute of Science and Technology
India

Raju Snehashree
Department of Studies and Research in Mathematics
Tumkur University
India

Basma Souayeh
Department of Physics
King Faisal University
Saudi Arabia

Pandikunta Sreenivasulu
Department of Mathematics
Sri Venkateswara College of Engineering
India

Tanu Suraddihalli Murugesh
Department of Studies in Mathematics
Davangere University
India

Muhammad Anas Wazir
Ghulam Ishaq Khan Institute of Engineering Sciences & Technology
Pakistan

Acknowledgments

This work was supported by the Deanship of Scientific Research, Vice Presidency for Graduate Studies and Scientific Research, King Faisal University, Saudi Arabia (Grant No. 251772).

Funding
This work was supported by the Deanship of Scientific Research, Vice Presidency for Graduate Studies and Scientific Research, King Faisal University, Saudi Arabia (Grant No. 251772).

Basma Souayeh
Affiliation: Department of Physics, College of Science, King Faisal University, PO Box 400, Al-Ahsa, 31982, Saudi Arabia.
Email: bsouayeh@kfu.edu.sa

Part I

Analytical and Theoretical Fluid Dynamics

1 Steady Navier–Stokes Solutions in Rotating Disk Boundary Layers

Oleh G. Shvydkyi

1.1 INTRODUCTION

The motion of compressible fluid entrained by a rotating disk of finite radius is schematically shown in Figure 1.1.

The disk is located in the plane $z = 0$ of cylindrical polar coordinates and rotates around the z axis with a constant angular velocity Ω. The liquid is considered from the side of the disk where $z > 0$. The boundary conditions for the problem are as follows:

$$\text{in the plane } z = 0 \text{ we have } v_r = 0;\ v_\varphi = \Omega r;\ v_z = 0. \tag{1.1}$$

In other words, the fluid adheres to the surface of disk. We assumed in our problems that the radius of the disk is arbitrarily equal to unity. Although we impose no boundary conditions at $r = \infty$, and $z = H$, where H is an immersion depth of the disk, the equation of tunnel mathematics allows the correct solution of the problem to be calculated at some distance from the edge of the disk.

This chapter provides the solution of a problem when the rotating disk has a finite radius (arbitrarily equal to unity). Like the solution for incompressible fluid [1], this solution applies only to the volume of the fluid enclosed in a cylinder of approximately unit radius with height z_l (where z_l is the vicinity of the z-coordinate, where the Navier–Stokes equation works and the fluid moves in the horizontal mainly radial direction on Figure 1.1).

The solution implies obtaining analytical expressions (formulae) for three components of velocity, density and pressure (as well as temperature for hypersonic case) in cylindrical polar coordinates. Numerical methods are not applicable. The formulae for the velocity components are derived from the relations of tunnel mathematics since the latter are satisfied by all vector fields (including those obeying the Navier–Stokes equations). Using relations of tunnel mathematics, we inevitably obtain unnecessary solutions and must cut them off by means of a physical analysis of the task. The Navier–Stokes equation is then used to verify the obtained solutions and calculate the pressure (as well as temperature for the hypersonic case).

The theory of turbulent incompressible and compressible fluids is in whole semi-empirical, and has some extent of untransparency due to complexity of the Navier–Stokes equations. The current state of affairs in this field can be found in the recent

DOI: 10.1201/9781003611820-2

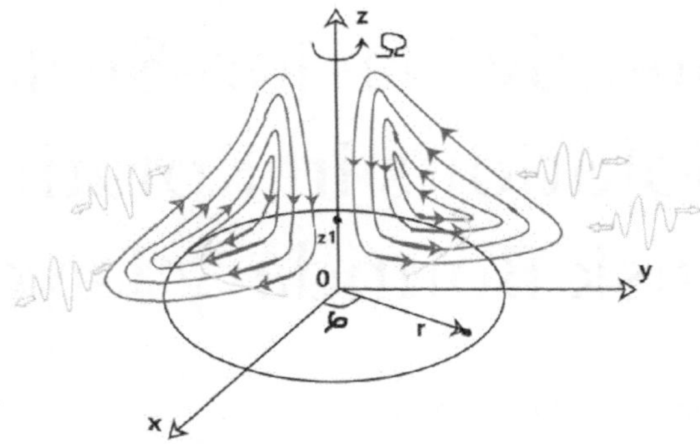

FIGURE 1.1 Streamlines of compressible fluid entrained by a rotating disk of finite radius in the planes parallel to the z axis.

treatise by Weyburne [2]. For the subsonic case, when we can neglect the increase in temperature, the results for compressible and incompressible fluids are almost similar. For the hypersonic case of the compressible fluid, the thickness of the boundary layer is to be increased with temperature [3–5]. Besides, for gases, such as air, which have a Prandtl number less than 1, the thermal boundary layer is thicker than the velocity (pressure) boundary layer [6]. The currently used analytical formulae for velocity components in the turbulent boundary layer over the rotating disk can be found in [7]. The study of the turbulent boundary layer by numerical methods can be found in [8], and by analytical methods in [9]. The application of tensor tunnel mathematics to model the Karman solution for a laminar boundary layer over a rotating disk is given in [10].

1.2 THEORY

The solution of the problem for a compressible fluid entrained by a rotating disk of finite radius will be symmetric with respect to the φ-coordinate. Figure 1.1 schematically shows fluid streamlines in the planes parallel to the z axis.

If tunnel mathematics equations are applied to the components of the vector velocity field in the Cartesian coordinate system, they appear as follows[11]:

$$\frac{\partial u}{\partial x}+\frac{wxy}{(x^2+y^2)^{3/2}}=\frac{\partial v}{\partial y}-\frac{wxy}{(x^2+y^2)^{3/2}}=-\frac{2y}{\sqrt{x^2+y^2}}\frac{\partial u}{\partial z}+\frac{\partial w}{\partial z}. \tag{1.2}$$

$$\frac{\partial v}{\partial x}+\frac{wy^2}{(x^2+y^2)^{3/2}}=-\frac{\partial u}{\partial y}+\frac{wx^2}{(x^2+y^2)^{3/2}}=-\frac{2y}{\sqrt{x^2+y^2}}\frac{\partial v}{\partial z}. \tag{1.3}$$

$$\frac{\partial w}{\partial x}=-i\frac{\partial w}{\partial y}=-\frac{\partial u}{\partial z}-i\frac{\partial v}{\partial z}. \tag{1.4}$$

The components of the velocity field in expressions (1.2)–(1.4) are set as follows:

$$V_x = u;\ V_y = v;\ V_z = \mathrm{Re}w, \tag{1.5}$$

where Rew is a real part of function w; moreover, the functions u, v, w are the components of the function of the spatial complex variable [11]:

$$P(L) = u(x, y, z) + iv(x, y, z) + fw(x, y, z). \tag{1.6}$$

Now the fluid is considered as compressible:

$$\nabla \cdot V = \mathrm{div}V = \frac{\partial u}{\partial x} + \frac{\partial v}{\partial y} + \frac{\partial \mathrm{Re}w}{\partial z} \neq 0. \tag{1.7}$$

Again we suppose in (1.2) and (1.4) that

$$\frac{\partial u}{\partial z} = 0. \tag{1.8}$$

The function w is an analytic function of the variables x and y and, therefore, a harmonic function on the xy plane. But now we assume that

$$\frac{\partial w}{\partial z} \neq 0. \tag{1.9}$$

Any vector fields including those obeying the Navier–Stokes equations satisfy relations (1.2)–(1.4).

The steady Navier–Stokes equation for compressible fluid has such form [12, 13]:

$$\rho\left(\vec{V}\mathrm{grad}\right)\vec{V} = -\,\mathrm{grad}P + \mu\Delta\vec{V} + \left(\zeta + \frac{\mu}{3}\right)\mathrm{grad\ div}\vec{V}, \tag{1.10}$$

where $\vec{V}$ is the velocity vector;

P – pressure in fluid;
μ – coefficient of dynamic viscosity;
ρ – fluid density;
ζ – second viscosity.

The second viscosity ζ is of the order of dynamic viscosity μ. Then taking into account the axial symmetry of the Navier–Stokes equations in cylindrical polar coordinates [14] we obtain from (1.10)

$$\rho\left(v_r\frac{\partial v_r}{\partial r} + v_z\frac{\partial v_r}{\partial z} - \frac{v_\varphi^2}{r}\right)$$

$$=-\frac{\partial P}{\partial r}+\mu\left(\frac{1}{r}\frac{\partial}{\partial r}\left(r\frac{\partial v_r}{\partial r}\right)+\frac{\partial^2 v_r}{\partial z^2}-\frac{v_r}{r^2}+\frac{1}{3}\frac{\partial}{\partial r}\left(\frac{1}{r}\frac{\partial(rv_r)}{\partial r}+\frac{\partial v_z}{\partial z}\right)\right); \quad (1.11)$$

$$\rho\left(v_r\frac{\partial v_\varphi}{\partial r}+v_z\frac{\partial v_\varphi}{\partial z}+\frac{v_r v_\varphi}{r}\right)=\mu\left(\frac{1}{r}\frac{\partial}{\partial r}\left(r\frac{\partial v_\varphi}{\partial r}\right)+\frac{\partial^2 v_\varphi}{\partial z^2}-\frac{v_\varphi}{r^2}\right); \quad (1.12)$$

$$\rho\left(v_r\frac{\partial v_z}{\partial r}+v_z\frac{\partial v_z}{\partial z}\right)$$

$$=-\rho g-\frac{\partial P}{\partial z}+\mu\left(\frac{1}{r}\frac{\partial}{\partial r}\left(r\frac{\partial v_z}{\partial r}\right)+\frac{\partial^2 v_z}{\partial z^2}+\frac{1}{3}\frac{\partial}{\partial z}\left(\frac{1}{r}\frac{\partial(rv_r)}{\partial r}+\frac{\partial v_z}{\partial z}\right)\right); \quad (1.13)$$

where g is the gravitational acceleration.

For compressible fluid the density ρ is already the variable quantity, which is to be determined from the equation of continuity:

$$\rho \operatorname{div}\vec{V}+\vec{V}\operatorname{grad}\rho=0. \quad (1.14)$$

Taking into account (1.8) we can obtain from equations (1.2) such a relation for divergence of the velocity field:

$$\nabla\cdot\vec{V}=\operatorname{div}\vec{V}=\frac{\partial u}{\partial x}+\frac{\partial v}{\partial y}+\frac{\partial \mathrm{Re}w}{\partial z}=3\frac{\partial \mathrm{Re}w}{\partial z}. \quad (1.15)$$

Comparing (1.15) with last terms in the right-hand side of (1.11) and (1.13) we obtain the following auxiliary relation:

$$\frac{1}{r}\frac{\partial(rv_r)}{\partial r}+\frac{\partial \mathrm{Re}w}{\partial z}=3\frac{\partial \mathrm{Re}w}{\partial z},$$

whence

$$\frac{1}{r}\frac{\partial(rv_r)}{\partial r}=2\frac{\partial \mathrm{Re}w}{\partial z}. \quad (1.16)$$

Now we show function w in the following form:

$$w=\Omega C_0\alpha(z)\left(\log\frac{r}{r_0}+i\varphi\right), \quad (1.17)$$

where C_0 and r_0 are the constants; $[r_0]=m$; in order to obtain $V_z=\mathrm{Re}w<0$ we should impose $r_0>1\ m$; and $\alpha(z)>0$, where $\alpha(z)$ is a function of the z-coordinate.

We can at once find an expression for radial velocity v_r from (1.16) and (1.17):

$$v_r = \Omega C_0 \frac{\partial \alpha}{\partial z} r \left(\log \frac{r}{r_0} - \frac{1}{2} \right); \tag{1.18}$$

where we put the constant of integration equal to zero.

Taking into account (1.15) and (1.17) we can rewrite the equation of continuity (1.14) in such form:

$$3\rho \Omega C_0 \frac{\partial \alpha}{\partial z} \log \frac{r}{r_0} = -\left(v_r \frac{\partial \rho}{\partial r} + v_z \frac{\partial \rho}{\partial z} \right) \tag{1.19}$$

For subsonic and hypersonic cases, we can put product $v_z \frac{\partial \rho}{\partial z}$ equal to zero since in the former case the thickness of the boundary layer is small enough, and in the latter case we assume that $\frac{\partial \rho}{\partial z} \ll \frac{\partial \rho}{\partial r}$ and the magnitude of centrifugal forces is very large. After differentiating (1.17) we obtain such relations:

$$\frac{\partial w}{\partial x} = \frac{\Omega C_0 \alpha \left(x - iy \right)}{x^2 + y^2}; \tag{1.20}$$

$$\frac{\partial w}{\partial y} = \frac{\partial v}{\partial z} = \frac{\Omega C_0 \alpha \left(y + ix \right)}{x^2 + y^2} = \frac{\Omega C_0 \alpha f^*}{\sqrt{x^2 + y^2}} = \frac{\Omega C_0 \alpha \left(f + 2 \sin \varphi \right)}{r}. \tag{1.21}$$

We used in expression (1.21) the following relations of tunnel mathematics for the operator f and its conjugate operator f^* [11]:

$$f = ie^{i\varphi} = -\sin \varphi + i \cos \varphi = \frac{-y + ix}{\sqrt{x^2 + y^2}}. \tag{1.22}$$

$$f^* = ie^{-i\varphi} = \sin \varphi + i \cos \varphi = \frac{y + ix}{\sqrt{x^2 + y^2}}. \tag{1.23}$$

$$f \times f^* = -1. \tag{1.24}$$

$$f^* = f + 2 \sin \varphi. \tag{1.25}$$

Special attention should be paid to the operator f since the factor will be used for building the spatial coordinate z in the equations on the plane (which corresponds to the relation (1.6)). We can see from equation (1.21) that this factor is $\frac{\alpha}{r}$ (we omitted the constants).

Generally speaking, the complex function w must be selected in such a manner that its real part corresponds to the pattern of occurring physical phenomena (i.e. to the projection V_z of the velocity field in Figure 1.1), and the integrating of relations (1.2) and (1.3) does not yield to integrals that could not be solved in quadratures.

Now we proceed to transform the relations (1.2) and (1.3) to cylindrical polar coordinates [14]; and taking into account expressions (1.17) and (1.21) we obtain

$$\frac{\partial v_r}{\partial r}-\frac{\partial v_\varphi}{\partial r}\tan\varphi+\frac{v_r}{r}+\frac{v_\varphi}{r}\cot\varphi+\frac{w\sin\varphi\cos\varphi}{r}$$

$$=\frac{\partial v_r}{\partial r}+\frac{\partial v_\varphi}{\partial r}\cot\varphi+\frac{v_r}{r}-\frac{v_\varphi}{r}\tan\varphi-\frac{w\sin\varphi\cos\varphi}{r}=\Omega C_0\frac{\partial\alpha}{\partial z}\log\frac{r}{r_0}; \quad (1.26)$$

$$\frac{\partial v_r}{\partial r}\tan\varphi+\frac{\partial v_\varphi}{\partial r}-\frac{v_r}{r}\cot\varphi+\frac{v_\varphi}{r}+\frac{w\sin^2\varphi}{r}$$

$$=-\frac{\partial v_r}{\partial r}\cot\varphi+\frac{\partial v_\varphi}{\partial r}+\frac{v_r}{r}\tan\varphi+\frac{v_\varphi}{r}+\frac{w\cos^2\varphi}{r}=\frac{-2(\sin\varphi)\,\Omega C_0\alpha\left(f+2\sin\varphi\right)}{r}; \quad (1.27)$$

Deriving equations (1.26) and (1.27), the well-known formulas were used for the transition from Cartesian velocity projections to cylindrical polar [14] ones:

$$u=v_r\cos\varphi-v_\varphi\sin\varphi; \quad (1.28)$$

$$v=v_r\sin\varphi+v_\varphi\cos\varphi. \quad (1.29)$$

Since in the region under consideration in Figure 1.1 the fluid moves horizontally (mainly radially) we can put $w = 0$ in (1.26) and (1.27). By integrating these equations under such conditions we obtain two solutions for both of the projections of the velocity v_φ and v_r. The solutions that correspond to the physical pattern of the phenomena of this problem (particularly to the boundary conditions (1.1)) are as follows:

$$v_\varphi=\Omega r; \quad (1.30)$$

$$v_r=\frac{1}{2}\Omega C_0\frac{\partial\alpha}{\partial z}r\left(\log\frac{r}{r_0}-\frac{1}{2}\right). \quad (1.31)$$

Expressions (1.30) and (1.31) are the solutions of equations (1.26).

Solution (1.31) coincides with (1.18) as far as constant factor.

1.3 RESULTS AND DISCUSSIONS

To extend solutions (1.30) and (1.31) into space, we use the factor at the operator f in equation (1.27):

$$z=-\frac{\alpha}{r}, \quad (1.32)$$

We assumed $\sin\varphi$ in (1.27) as a constant.

1.3.1 Extension for Azimuthal Velocity

By means of (1.32) the extension of (1.30) into space is fulfilled very simply:

$$v_\varphi = \frac{\Omega r}{\dfrac{z}{z_{0\varphi}} + 1}, \tag{1.33}$$

where $\alpha = -1$ since (1.30) does not contain α; $z_{0\varphi}$ is a normalization factor, its dimension is meters. We added unity in the denominator in order to satisfy the boundary conditions (1.1).

1.3.2 Extension for Radial Velocity

For the projection v_r in (1.31) we obtain the following expression:

$$v_r = -\Omega C_0 r \left(\frac{z}{\delta}\right)^2 \left(\log\frac{rz}{r_0\delta} - \frac{1}{2}\right); \tag{1.34}$$

Deriving (1.34) we put in (1.31) and (1.32) $\alpha = \frac{z^2}{\delta}$; δ is the thickness of the boundary layer since we seek a solution in its area.

1.3.3 Finding the Auxiliary Relations

Using the true Navier–Stokes equations we can at once verify the solutions (1.33) and (1.34) that have been expanded into space by means of the apparatus of tunnel mathematics. The simplest way to do this is to make use of equation (1.12) since it does not contain the pressure P. Substituting solutions (1.33) and (1.34), and the real part of relation (1.17) with $\alpha = \frac{z^2}{\delta}$ as well as their corresponding derivative into equation (1.12) we conclude the following important auxiliary relation for the z-coordinate in the area of boundary layer $\left(\frac{z}{\delta} \sim 1\right)$:

$$\rho v_r \left(2 + \frac{1}{1 + \dfrac{z_{0\varphi}}{\delta}}\right) = \frac{2\mu r}{\delta^2 \left(1 + \dfrac{z_{0\varphi}}{\delta}\right)^2}. \tag{1.35}$$

If $z_{0\varphi} \gg \delta$ as it is for laminar boundary layer then we obtain

$$z_{0\varphi}\sqrt{\frac{\mu r}{\rho v_r}} = \text{const} \tag{1.36}$$

If $z_{0\varphi} \ll \delta$ as it is for the turbulent boundary layer then we obtain

$$\delta = \sqrt{\frac{2\mu r}{3\rho v_r}} \tag{1.37}$$

equation (1.37) is a dynamic relation; recall that ρ is a variable quantity.

Formula (1.12), where the pressure is absent due to axial symmetry, can be used for the calculation of friction force acting in a fluid, includingacting per unit disk surface in a direction perpendicular to its radius. Having performed the appropriate calculations with (1.33), we obtain the following formula for the friction force:

$$F_{\text{fr}} = \mu \frac{\partial v_\varphi}{\partial z} = -\frac{\mu \Omega r}{z_{0\varphi}\left(\frac{z}{z_{0\varphi}} + 1\right)^2}. \tag{1.38}$$

Directing z to zero in expression (1.38) we obtain the relation for the friction force acting on a unit surface of the disk in the direction perpendicular to its radius:

$$F_{\text{fr}} = -\frac{\mu \Omega r}{z_{0\varphi}}. \tag{1.39}$$

Thus we conclude from (1.39) that constant $z_{0\varphi}$ must be determined experimentally.

1.3.4 Calculation of Pressure and Other Parameters of the Boundary Layer using the Navier–Stokes Equations

1.3.4.1 Calculation of Density for Subsonic and Hypersonic Cases

Using (1.19), (1.34), and taking into account that $v_z \frac{\partial \rho}{\partial z} = 0$ we find an expression for the density of compressible fluid:

$$\rho = \rho_0 \left[\frac{r}{r_0} \left(\log \frac{rz}{r_0 \delta} - \frac{1}{2} \right)^{-\left(\log \frac{z}{\delta} - \frac{1}{2} \right)} \right]^{\frac{6\delta}{z}}, \tag{1.40}$$

where $\rho_0 = 1 \frac{kg}{m^3}$ is the mean value of density for gases at normal temperature, $r_0 = 1.65$ m.

Neglecting the term $\left(-\frac{1}{2}\right)$ in brackets in (1.40) we can rewrite the expression for density in terms of radial velocity:

$$\rho = \rho_0 \left(-\frac{v_r}{\Omega C_0} \left(\frac{\delta}{z} \right)^2 \right)^{\frac{6\delta}{z}} \left(r_0^{\frac{\delta}{z}} \log \frac{rz}{r_0 \delta} \right)^6 \tag{1.41}$$

Whence we can arrive at the following approximate relation for radial velocity in the area of the boundary layer:

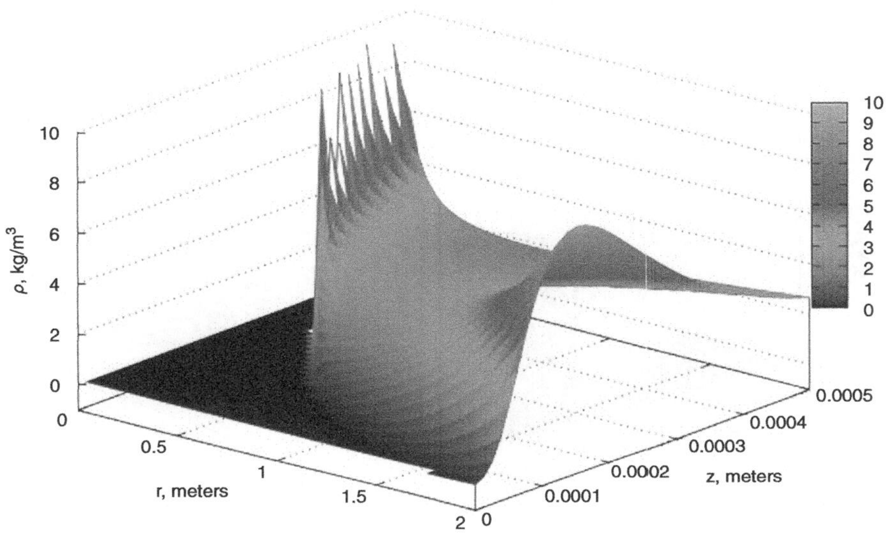

FIGURE 1.2 Density of compressible fluid entrained by a rotating disk in the area of the boundary layer for laminar streaming (thickness of the boundary layer $\delta = 0.00025$ m).

$$v_r \sim \rho^{z/6\delta}, \tag{1.42}$$

where $\frac{z}{\delta} \sim 1$ in the area of the boundary layer.

Relation (1.42) is very similar to the $\frac{1}{7}$ *th power velocity distribution law* of velocity in the area of the boundary layer from the theory of turbulent motion [13, 7]:

$$u \sim \mathrm{Re}^{1/7} \sim \rho^{1/7}, \tag{1.43}$$

where Re is the Reynolds number.

The approximate plots for the density from (1.40) are shown in Figures 1.2 and 1.3 for laminar and turbulent flows, respectively. We recall that the radius of the disk is equal to unity.

In order to correctly depict values of (1.40) in the area of the boundary layer, on the left we have shifted the values of the r-coordinate by unity and the values of the z-coordinate by δ.

On both plots the complex interaction pattern of the main stream and compression and rarefaction simple cylindrical waves propagating in fluid is clearly tracked. Also on both plots in the vicinity of the origin of the frame of references, the dead zones occur. For turbulent flow, the dead zone is smaller.

1.3.4.2 Subsonic Case

Equation (1.11) is amenable to integration if we assume that on most of the disk the function $\log\frac{rz}{r_0\delta}$ and especially its high powers change slowly enough, so we can

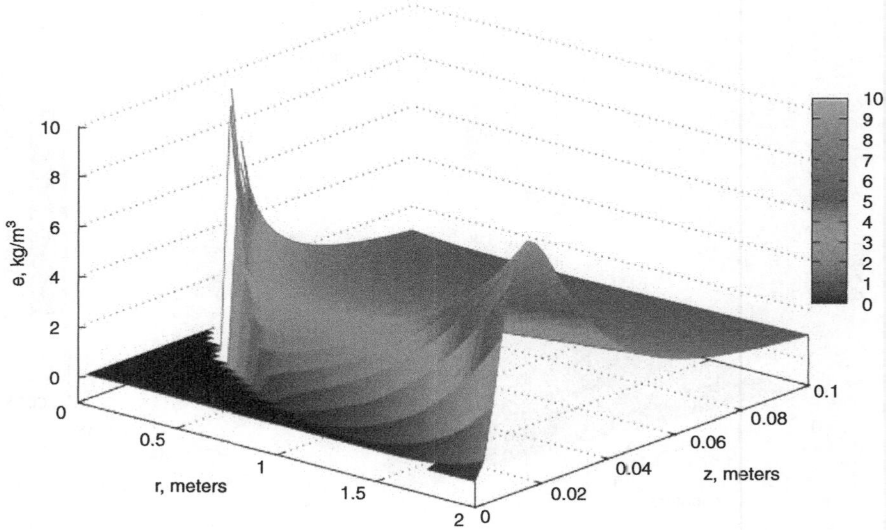

FIGURE 1.3 Density of compressible fluid entrained by a rotating disk in the area of the boundary layer for turbulent streaming (thickness of the boundary layer $\delta = 0.025$ m).

consider them asconstants. Also we assume that in the area of the boundary layer the relation $\frac{z}{\delta} \sim 1$ holds. In order to simplify the integration, we will neglect the term $\left(-\frac{1}{2}\right)$ in brackets in (1.40) and (1.34).

At these conditions, the approximate formula for pressure in the area of the boundary layer appears as follows:

$$P = -\frac{\rho_0 \Omega^2 r^{\left(\frac{6\delta}{z}+2\right)}}{\left(\frac{6\delta}{z}+2\right) r_0^{\frac{6\delta}{z}}} \left(C_0^2 \left(\frac{z}{\delta}\right)^3 \left(\log\frac{rz}{r_0\delta}\right)^2 \left(\frac{z}{\delta} - 1 + \frac{\log\frac{rz}{r_0\delta}}{\left(\frac{6\delta}{z}+2\right)\left(\log\frac{rz}{r_0\delta}+1\right)}\right) - \frac{1}{\left(\frac{z}{z_{0\varphi}}+1\right)^2} \right)$$
$$+\mu\Omega C_0 \frac{z}{\delta} \left(\frac{z}{\delta} \left(-\frac{1}{2}\left(\log\frac{rz}{r_0\delta}+2\right)^2 - \left(\frac{r}{z}\right)^2 \left(\log\frac{rz}{r_0\delta}+1\right) + \frac{1}{2}\left(\log\frac{rz}{r_0\delta}\right)^2 \right) + 2\log\frac{r}{r_0} \right). \tag{1.44}$$

The corresponding derivatives were calculated using formulas (1.17), (1.33), and (1.34). Generally speaking, we must still add to the right-hand side of formula (1.44) the hydrostatic pressure $\rho g (H - z)$, where H is an immersion depth of the disk.

In Figure 1.4, the plot corresponding to formula (1.44) in the region of the boundary layer is shown with the following values of the constants:

$$C_0 = 0.01;$$

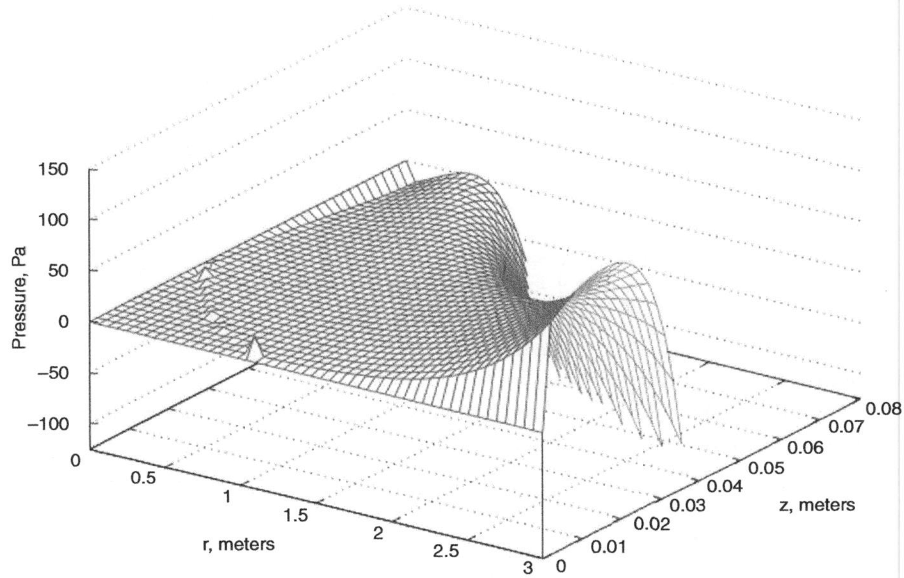

FIGURE 1.4 Plot corresponding to formula (1.44) (turbulent motion in the area of the boundary layer, $R_r = \frac{v_r r}{v} > 3 \cdot 10^5$).

$$\delta = 0.025\ m;$$

$$r_0 = 6\ m;$$

$$z_{0\varphi} = 0.001\ m;$$

$$\Omega = 600\ \frac{\text{rad}}{s};$$

$$\rho_0 = 1\ \frac{kg}{m^3}$$

$$\mu = 0.00001\ \text{Pa sec.}$$

Such values of constants correspond to the Reynolds number is to be lower in order for the viscosity to have an appreciable effect in the area of the boundary layer.

We have again shifted, on the left, the values of the r-coordinate by unity and the values of the z-coordinate by δ in inertia terms in (1.44). We recall that the radius of the disk is equal to 1 meter.

The motion of compressible fluid continues over significantly larger distances from the edge of the disk than that of incompressible fluid, and it is difficult to determine where the compressible fluid is at rest. This pattern occurs due to the lower density of the compressible fluid and complex interaction of the main stream and the compression and rarefaction simple cylindrical waves propagating in the subsonic fluid.

Integrating (1.13) we obtain the following expression for pressure:

$$P = -\frac{\rho_0 (\Omega C_0)^2 r^{\frac{6\delta}{z}}}{r_0^{\frac{6\delta}{z}}} \left(\frac{z^4}{\delta^2} \left(\log \frac{rz}{r_0 \delta} \right) \left(-\frac{1}{5}\frac{z}{\delta} + \frac{1}{2} \log \frac{rz}{r_0 \delta} \right) - \frac{g(H-z)}{(\Omega C_0)^2} \right) + 4\mu \Omega C_0 \frac{z}{\delta} \log \frac{r}{r_0}. \tag{1.45}$$

We omit constants of integration.

Comparing (1.44) and (1.45) and baring in mind that we can take C_0 as an arbitrarily small quantity we find a useful relation:

$$\frac{r^2}{\left(\frac{6\delta}{z}+2\right)\left(\frac{z}{z_{0\varphi}}+1\right)^2} = \frac{gH}{\Omega^2}, \tag{1.46}$$

where H is an immersion depth of the disk; g is the gravitational acceleration.

From (1.46) we easily find

$$\frac{z}{z_{0\varphi}} + 1 \approx \frac{r\Omega}{\sqrt{8gH}}. \tag{1.47}$$

We see from (1.47) that the thickness of the boundary layer $\left(\frac{z}{\delta} \sim 1\right)$ for compressible fluid increases proportionally to the angular velocity Ω at the same r-coordinate. The theory ([13, 15]) says that the turbulent boundary layer on rotating streamline bodies grows rapidly with the rotation parameter $\lambda = \frac{\Omega R}{U_m}$, where U_m is the axial reference velocity; R is the largest radius of the body. So we see the full compliance of our result with theory. It is surprising that we obtained this result just not taking into account a rise of temperature. Such a result is very different from the case of incompressible fluid, where the thickness of the boundary layer decreases inversely proportionally to the square root of Ω [12, 13].

1.3.4.3 Hypersonic Case (Mach Number *M* for Azimuthal Velocity is about 9; Adiabatic Wall)

For a hypersonic compressible flow, the viscosity μ must be regarded as dependent on the space coordinate, because μ varies considerably with temperature, and the changes in velocity and pressure together with the heat due to friction bring about considerable temperature variations.

So, in the Cartesian coordinate system, the Navier–Stokes equations for a compressible flow in a gravitational field look like this [13]:

$$\rho\frac{Du}{Dt}=-\frac{\partial P}{\partial x}+\frac{\partial}{\partial x}\left[\mu\left(2\frac{\partial u}{\partial x}-\frac{2}{3}\mathrm{div}\vec{V}\right)\right]+\frac{\partial}{\partial y}\left[\mu\left(\frac{\partial u}{\partial y}+\frac{\partial v}{\partial x}\right)\right]$$
$$+\frac{\partial}{\partial z}\left[\mu\left(\frac{\partial w}{\partial x}+\frac{\partial u}{\partial z}\right)\right];$$

$$\rho\frac{Dv}{Dt}=-\frac{\partial P}{\partial y}+\frac{\partial}{\partial y}\left[\mu\left(2\frac{\partial v}{\partial y}-\frac{2}{3}\mathrm{div}\vec{V}\right)\right]+\frac{\partial}{\partial z}\left[\mu\left(\frac{\partial v}{\partial z}+\frac{\partial w}{\partial y}\right)\right]$$
$$+\frac{\partial}{\partial x}\left[\mu\left(\frac{\partial u}{\partial y}+\frac{\partial v}{\partial x}\right)\right];$$

$$\rho\frac{Dw}{Dt}=-\rho g-\frac{\partial P}{\partial z}+\frac{\partial}{\partial z}\left[\mu\left(2\frac{\partial w}{\partial z}-\frac{2}{3}\mathrm{div}\vec{V}\right)\right]+\frac{\partial}{\partial x}\left[\mu\left(\frac{\partial w}{\partial x}+\frac{\partial u}{\partial z}\right)\right]$$
$$+\frac{\partial}{\partial y}\left[\mu\left(\frac{\partial v}{\partial z}+\frac{\partial w}{\partial y}\right)\right]. \tag{1.48}$$

The equation for temperature variations has such form [13]:

$$\rho c_p\frac{dT}{dt}=\frac{dP}{dt}+\frac{\partial}{\partial x}\left(k\frac{\partial T}{\partial x}\right)+\frac{\partial}{\partial y}\left(k\frac{\partial T}{\partial y}\right)+\frac{\partial}{\partial z}\left(k\frac{\partial T}{\partial z}\right)+\mu\phi; \tag{1.49}$$

where $\frac{D}{Dt}$, $\frac{d}{dt}$ are the substantial derivatives; T is a temperature; c_p [J/kg K] represents the specific heat at constant pressure per unit mass (in general c_p depends on temperature, but we will assume its value as a constant equal to 1200 J/kg K (air at 1500 K)); k [J/m sec K] is a thermal conductivity; Φ [1/sec^2] represents the dissipation function:

$$\phi=2\left(\left(\frac{\partial u}{\partial x}\right)^2+\left(\frac{\partial v}{\partial y}\right)^2+\left(\frac{\partial w}{\partial z}\right)^2\right)+\left(\frac{\partial u}{\partial y}+\frac{\partial v}{\partial x}\right)^2+\left(\frac{\partial v}{\partial z}+\frac{\partial w}{\partial y}\right)^2+\left(\frac{\partial w}{\partial x}+\frac{\partial u}{\partial z}\right)^2$$
$$-\frac{2}{3}\left(\frac{\partial u}{\partial x}+\frac{\partial v}{\partial y}+\frac{\partial w}{\partial z}\right)^2. \tag{1.50}$$

In order to obtain a physically admissible result, we need to assume that both viscosity μ and thermal conductivity k for the hypersonic case are proportional at least to the ninth power of temperature:

$$\mu=\mu_0\left(\frac{T}{T_0}\right)^9; \tag{1.51}$$

$$k=k_0\left(\frac{T}{T_0}\right)^9; \tag{1.52}$$

where $\mu_0 = 0.00001$ Pa sec; $k_0 = 0.0242 \frac{J}{m\sec K}$; $T_0 = 300$ K (room temperature). We introduced (1.52) in order to eliminate from considerations the Prandtl's number.

Now we proceed to the treatment of equations (1.48)–(1.52) by means relations of tunnel mathematics (1.2)–(1.4). During this treatment, we will use such a rule: if a term containing the φ-coordinate is added to a non-containing φ-coordinate term, then the former will be neglected; if a term containing the φ-coordinate is multiplied by a non-containing φ-coordinate term, then the former will be treated as a constant. For instance, after such a treatment the dissipation function from (1.50) appears simply enough as follows:

$$\Phi = \left(\frac{v_z}{r}\right)^2 \tag{1.53}$$

$$\text{where } v_z = \Omega C_0 \frac{z^2}{\delta} \log \frac{r}{r_0}. \tag{1.54}$$

The equations (1.48) acquires such a view in vector form:

$$\rho\left(\vec{V}\text{grad}\right)\vec{V} = \vec{F}_{\text{ext}} - \text{grad}P + \mu\left(\Delta\vec{V} + \frac{1}{3}\text{grad div}\vec{V}\right) + \frac{\mu_0 v_z}{T_0^9 r}\left(\text{div}\vec{T}\right)\vec{l}, \tag{1.55}$$

where $\vec{F}_{\text{ext}}$ represents external force fields; and we have introduced the vector field of temperatures for reasons of mathematical convenience. This field is defined in the following manner in cylindrical coordinates:

$$\vec{T} = T^9\left(A\vec{e_r} + B\vec{e_\varphi} + C\vec{k}\right) \tag{1.56}$$

where A, B, C are the constants.

Vector $\vec{l}$ is an ordinary vector:

$$\vec{l} = \left(A_1\vec{e_r} + B_1\vec{e_\varphi} + C_1\vec{k}\right) \tag{1.57}$$

where A_l, B_l, C_l are the constants.

So, the Navier–Stokes equations (1.48) in cylindrical polar coordinates assume such a form for the hypersonic case:

$$\begin{aligned}
&\rho\left(v_r\frac{\partial v_r}{\partial r} + v_z\frac{\partial v_r}{\partial z} - \frac{v_\varphi^2}{r}\right) \\
&= -\frac{\partial P}{\partial r} + \mu_0\left(\frac{T}{T_0}\right)^9\left[\frac{1}{r}\frac{\partial}{\partial r}\left(r\frac{\partial v_r}{\partial r}\right) + \frac{\partial^2 v_r}{\partial z^2} - \frac{v_r}{r^2} + \frac{1}{3}\frac{\partial}{\partial r}\left(\frac{1}{r}\frac{\partial(rv_r)}{\partial r} + \frac{\partial v_z}{\partial z}\right)\right] \\
&\quad + \frac{A_1\mu_0 v_z}{T_0^9 r}\left[\frac{A}{r}\frac{\partial(rT^9)}{\partial r} + C\frac{\partial T^9}{\partial z}\right]
\end{aligned} \tag{1.58}$$

$$\rho\left(v_r\frac{\partial v_\varphi}{\partial r}+v_z\frac{\partial v_\varphi}{\partial z}+\frac{v_r v_\varphi}{r}\right)=\mu_0\left(\frac{T}{T_0}\right)^9\left(\frac{1}{r}\frac{\partial}{\partial r}\left(r\frac{\partial v_\varphi}{\partial r}\right)+\frac{\partial^2 v_\varphi}{\partial z^2}-\frac{v_\varphi}{r^2}\right)$$
$$+\frac{B_1\mu_0 v_z}{T_0^9 r}\left(\frac{A}{r}\frac{\partial\left(rT^9\right)}{\partial r}+C\frac{\partial T^9}{\partial z}\right). \tag{1.59}$$

$$\rho\left(v_r\frac{\partial v_z}{\partial r}+v_z\frac{\partial v_z}{\partial z}\right)$$

$$=-\rho g-\frac{\partial P}{\partial z}+\mu_0\left(\frac{T}{T_0}\right)^9\left(\frac{1}{r}\frac{\partial}{\partial r}\left(r\frac{\partial v_z}{\partial r}\right)+\frac{\partial^2 v_z}{\partial z^2}+\frac{1}{3}\frac{\partial}{\partial z}\left(\frac{1}{r}\frac{\partial(rv_r)}{\partial r}+\frac{\partial v_z}{\partial z}\right)\right)$$
$$+\frac{C_1\mu_0 v_z}{T_0^9 r}\left(\frac{A}{r}\frac{\partial\left(rT^9\right)}{\partial r}+C\frac{\partial T^9}{\partial z}\right); \tag{1.60}$$

Accordingly, equation (1.49) for temperature variations acquires such a form:

$$\rho c_p\left(v_r\frac{\partial T}{\partial r}+v_z\frac{\partial T}{\partial z}\right)=v_r\frac{\partial P}{\partial r}+v_z\frac{\partial P}{\partial z}+k_0\left(\frac{T}{T_0}\right)^9\left[\frac{1}{r}\frac{\partial}{\partial r}\left(r\frac{\partial T}{\partial r}\right)+\frac{\partial^2 T}{\partial z^2}\right]$$
$$+\frac{9k_0T^8}{T_0^9}\left[\left(\frac{\partial T}{\partial r}\right)^2+\left(\frac{\partial T}{\partial z}\right)^2\right]+\mu_0\left(\frac{T}{T_0}\right)^9\Phi. \tag{1.61}$$

Flow in the hypersonic case is always turbulent, so we cannot neglect by function w in (1.26) anymore. Solving (1.26) with w we obtain the following expressions for the velocity projections v_φ and v_r:

$$v_\varphi=\frac{\Omega r}{\frac{z}{z_\mu}+1}-C_\varphi\Omega C_0\frac{z^2}{\sqrt{\delta\delta_T}}\left(\log\frac{rz}{r_0z_\mu}+1\right)=\frac{\Omega r}{\frac{z}{z_\mu}+1}-C_\varphi\frac{v_z}{\log\frac{r}{r_0}}\left(\log\frac{rz}{r_0z_\mu}+1\right); \tag{1.62}$$

where we used (1.54); also we changed the notation for normalization factor $z_{0\varphi}$ by z_μ because it is connected with friction forces, which act in fluid; $C_\varphi=-1$; δ_T is the thickness of thermal boundary layer. We will seek a solution of (1.58)–(1.61) in the vicinity of the z-coordinate where the approximate relation $\frac{z}{\sqrt{\delta\delta_T}}1$ holds. It is easily seen from (1.62) that the non-slip condition is preserved.

$$v_r=-\Omega r\frac{z}{\sqrt{\delta\delta_T}}\left(C_0\frac{z}{\sqrt{\delta\delta_T}}\left(\log\frac{rz}{r_0\sqrt{\delta\delta_T}}-\frac{1}{2}\right)-C_r\right); \tag{1.63}$$

where C_r is a constant. We will assume for the hypersonic case that $C_r = 0.05$, and $C_0 = 0.1$.

In Figures 1.5 and 1.6 the plots corresponding to (1.62) and (1.63) in the region of the boundary layer are shown:

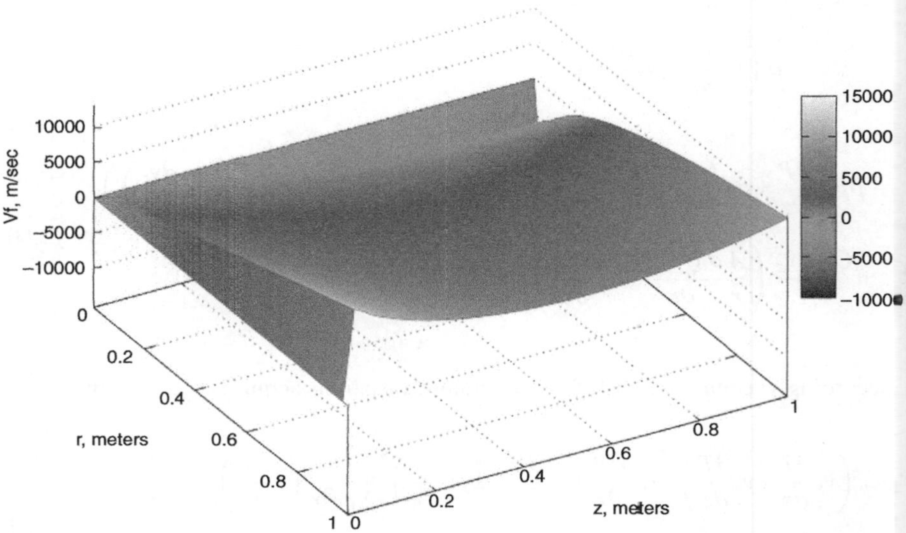

FIGURE 1.5 Plot corresponding to formula (1.62) for hypersonic azimuthal velocity (the thickness of velocity boundary layer $\delta = 0.25$ m).

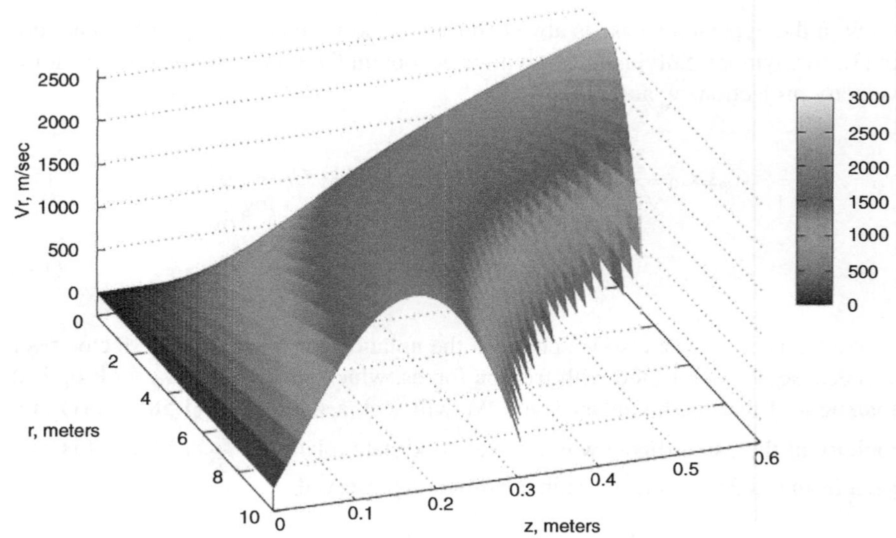

FIGURE 1.6 Plot corresponding to formula (1.63) for supersonic radial velocity (the thickness of the velocity boundary layer $\delta = 0.25$ m).We recall that the radius of the disk is equal to 1 meter only.

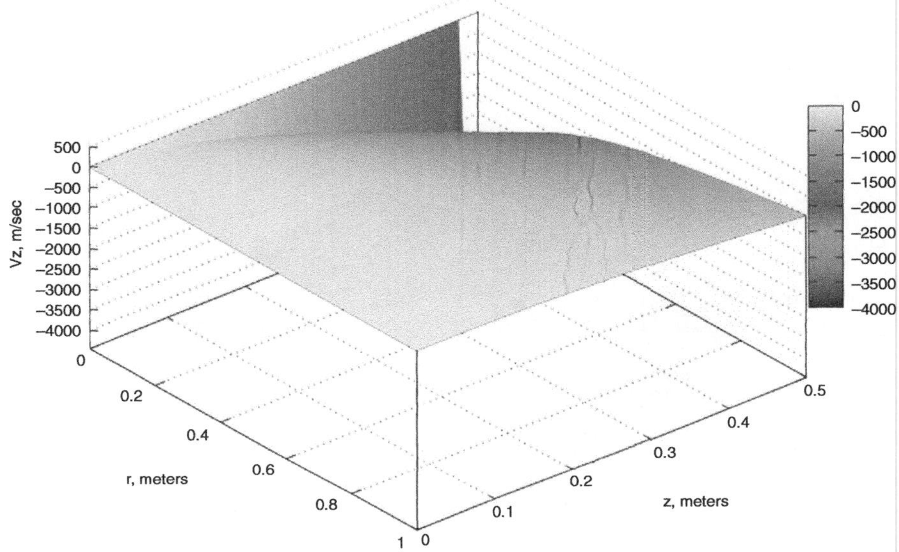

FIGURE 1.7 Plot corresponding to formula (1.54) for the transonic axial velocity (the thickness of the velocity boundary layer $\delta = 0.25$ m).

In Figure 1.7 the plot corresponding to (1.54) for axial velocity v_z in the region of the boundary layer is shown:

Accordingly, we should rewrite the expression (1.40) for the density of gas. Solving (1.19) with (1.63) we have

$$\rho = \rho_0 \left[\frac{1}{e} \left(\frac{rz}{r_0\sqrt{\delta\delta_T}} \right)^2 \left(C_0 \frac{z}{\sqrt{\delta\delta_T}} \left(\log \frac{rz}{r_0\sqrt{\delta\delta_T}} - \frac{1}{2} \right) - r \right)^{\left(1 + \frac{2r}{C_0} \cdot \frac{\sqrt{\delta\delta_T}}{z}\right)} \right]^{\frac{3\sqrt{\delta\delta_T}}{z}} \tag{1.64}$$

where e is the exponent equal to 2.71. In Figure 1.8 the plot corresponding to (1.64) in the region of the boundary layer is shown.

In order to correctly depict values of (1.64) in the area of the boundary layer we have made the transformation of the coordinate:

$$z \to \left(10z + \sqrt{\delta\delta_T}\right); \; r \to (r + r_0) \tag{1.65}$$

Probably, (1.65) is a consequence of neglecting $\frac{\partial p}{\partial z}$ in (1.19).

1.3.4.3.1 Calculation of Temperature Distribution

In principle, we have a sufficient number of equations in order to find the temperature and pressure distributions. So, we have no reason to use the state equation of gas.

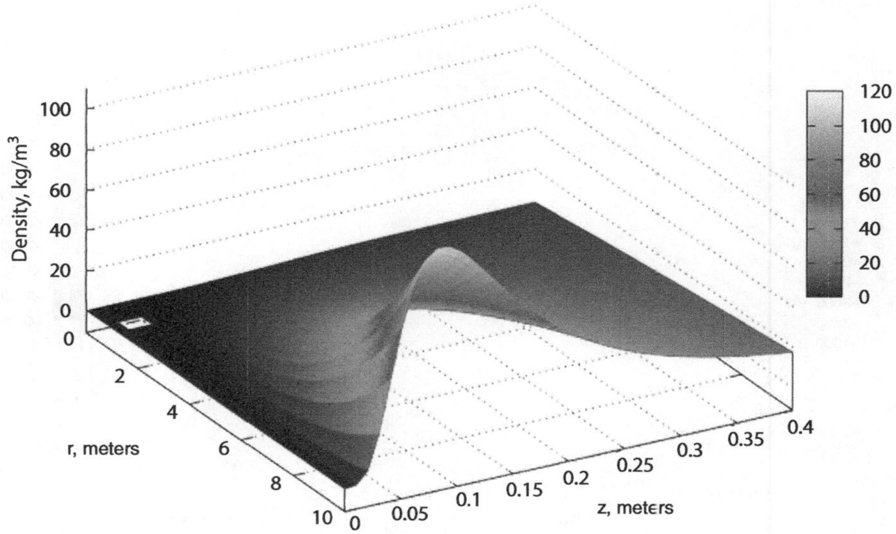

FIGURE 1.8 Plot corresponding to formula (1.64) for the density of gas (the thickness of the velocity boundary layer $\delta = 0.25$ m).

It is very convenient for hypersonic flows of fluids because we cannot use the state equation of perfect gas in this case.

In order to find the temperature distribution from (1.61) we change derivatives $\frac{\partial T}{\partial r}$ and $\frac{\partial^2 T}{\partial r^2}$ by $\frac{\partial T}{\partial z}$ and $\frac{\partial^2 T}{\partial z^2}$ respectively and afterwards carry out the integration of (1.61) by dz. It is very convenient because all velocities in (1.62), (1.63), and (1.54) as well as density in (1.64) are independent on z in the area of the boundary layer (the relations $\frac{z}{\sqrt{\delta \delta_T}} 1$ and $\frac{z}{z_\mu} 1$ hold).

From (1.59) we find such auxiliary relations:

$$\frac{B_1 A \mu_0 v_z}{T_0^9 r}\left(\frac{1}{r}\frac{\partial\left(rT^9\right)}{\partial r}+\frac{\partial T^9}{\partial z}\right)=\rho\frac{2v_r v_\varphi}{r}-\mu_0\left(\frac{T}{T_0}\right)^9\frac{2\Omega r}{z_\mu^2\left(\frac{z}{z_\mu}+1\right)^3} \quad (1.66)$$

where we assumed that $A = C$ and $v_\varphi \gg v_z$.

$$\frac{\partial T}{\partial r}=\frac{r}{9B_1 A v_z}\left(\frac{1}{T^8}\cdot\frac{\rho T_0^9}{\mu_0}\cdot\frac{2v_r v_\varphi}{r}-T\cdot\frac{2\Omega r}{z_\mu^2\left(\frac{z}{z_\mu}+1\right)^3}\right)-\frac{1}{9}\frac{T}{r}-\frac{\partial T}{\partial z} \quad (1.67)$$

$$\frac{\partial^2 T}{\partial r^2}=\frac{1}{9B_1Av_z}\left(\left(1-\frac{1}{\log\frac{r}{r_0}}\right)+r\left(\frac{\partial}{\partial r}-\frac{\partial}{\partial z}\right)\right)\times\left(\frac{1}{T^8}\cdot\frac{\rho T_0^9}{\mu_0}\cdot\frac{2v_rv_\varphi}{r}-T\cdot\frac{2\Omega r}{z_\mu^2\left(\frac{z}{z_\mu}+1\right)^3}\right)$$
$$+\frac{10}{81}\frac{T}{r^2}+\frac{2}{9}\cdot\frac{1}{r}\cdot\frac{\partial T}{\partial z}+\frac{\partial^2 T}{\partial z^2} \tag{1.68}$$

$$\left(\frac{\partial T}{\partial r}\right)^2+\left(\frac{\partial T}{\partial z}\right)^2=\left(\frac{r}{9B_1Av_z}\left(\frac{1}{T^8}\cdot\frac{\rho T_0^9}{\mu_0}\cdot\frac{2v_rv_\varphi}{r}-T\cdot\frac{2\Omega r}{z_\mu^2\left(\frac{z}{z_\mu}+1\right)^3}\right)-\frac{1}{9}\frac{T}{r}\right)^2$$
$$-\frac{2r}{9B_1Av_z}\left(\frac{1}{T^8}\cdot\frac{\rho T_0^9}{\mu_0}\cdot\frac{2v_rv_\varphi}{r}-T\cdot\frac{2\Omega r}{z_\mu^2\left(\frac{z}{z_\mu}+1\right)^3}\right)\cdot\frac{\partial T}{\partial z} \tag{1.69}$$
$$+\frac{2}{9}\cdot\frac{T}{r}\cdot\frac{\partial T}{\partial z}+2\left(\frac{\partial T}{\partial z}\right)^2$$

Using (1.66)–(1.69), (1.58) and (1.60) we collect all terms in (1.61) and arrive to such a relation:

$$\frac{\partial T}{\partial z}\left(c_p\left(v_z-v_r\right)+2c_dv_r\right)=-v_r\left(v_r\frac{\partial v_r}{\partial r}-\frac{v_\varphi^2}{r}\right)-v_z\left(v_r\frac{\partial v_z}{\partial r}+g\right)+$$
$$\frac{2v_rv_\varphi}{B_1r}\left(A_1v_r+C_1v_z\right)+\frac{1}{T^8}\cdot\frac{2\rho T_0^9}{9\mu_0B_1}\frac{v_rv_\varphi}{Av_z}\left(v_r\left(c_d-c_p\right)\right) \tag{1.70}$$
$$+T\cdot\left(\frac{v_r}{r}\left(c_d\left(\frac{8}{9}-\frac{1}{log\frac{r}{r_0}}\right)+\frac{c_p}{9}\right)+\frac{2\Omega r^2v_r}{9B_1Av_zz_\mu^2\left(\frac{z}{z_\mu}+1\right)^3}\left(c_p-9c_d\right)\right)$$

where $c_d=\frac{2k_0v_\varphi}{9\mu_0B_1Av_z}$; we denote this quantity as a dynamic specific heat in contrast to the specific heat at constant pressure c_p. We omitted in (1.70) all terms containing high powers of temperature (eighth and higher) because their contribution to the final result during integration by dz from 0 meter to approximately 1 meter is negligible (fortunately, the thicknesses of the velocity boundary layer and the thermal boundary layer are less than unity; for instance, we assume that $\delta = 0.25\ m$, and $\delta_T = 0.5\ m$).

Now we impose on (1.70) the boundary condition of the adiabatic wall:

$$\left(\frac{\partial T}{\partial z}\right)_{z=0}=0 \tag{1.71}$$

We recall that at $z=0$, the boundary condition (1.1) is preserved.

Applying (1.71) to (1.70) we obtain such important relations:

$$\frac{2\rho T_0^9}{9\mu_0 B_1}\frac{v_r v_\varphi}{Av_z} = T^9 \cdot \frac{18\Omega r^2}{9B_1 Av_z z_\mu^2}. \tag{1.72}$$

$$\left(\frac{T}{T_0}\right)^9 = \frac{\rho v_r v_\varphi z_\mu^2}{9\mu_0 \Omega r^2} \tag{1.73}$$

We will apply (1.73) for the calculation of a pressure distribution lower.

Substituting (1.72) into (1.70) we obtain such a differential equation for temperature in the hypersonic case:

$$\alpha \frac{\partial T}{\partial z} = \beta + \gamma T, \tag{1.74}$$

where α, β, γ are independent on z in the area of the boundary layer:

$$\alpha = c_p\left(v_z - v_r\right) + 2c_d v_r \tag{1.75}$$

$$\beta = -v_r\left(v_r \frac{\partial v_r}{\partial r} - \frac{v_\varphi^2}{r}\right) - v_z\left(v_r \frac{\partial v_z}{\partial r} + g\right) + \frac{2v_r v_\varphi}{B_1 r}\left(A_1 v_r + C_1 v_z\right) \tag{1.76}$$

$$\gamma = v_r\left[\frac{2\Omega r^2}{9B_1 Av_z z_\mu^2}\left(9\left(c_d - c_p\right) + \frac{1}{\left(\frac{z}{z_\mu}+1\right)^3}\left(c_p - 9c_d\right)\right) + \frac{1}{r}\left(c_d\left(\frac{8}{9} - \frac{\Omega C_0 z^2}{v_z\sqrt{\delta\delta_T}}\right) + \frac{c_p}{9}\right)\right] \tag{1.77}$$

The solution of (1.74) is

$$T = -\frac{\beta}{\gamma} + \left(T_w + \frac{\beta}{\gamma}\right)\exp\left\{\frac{\gamma}{\alpha}z\right\} \tag{1.78}$$

where T_w is the temperature of the wall, i.e. the temperature of the disk surface. We will assume such an expression for T_w (by the type of radial velocity profile in (1.63))

$$T_w = (-120K)\left(\frac{r}{1}\right)^3\left(\log\frac{r}{r_0} - \frac{1}{2}\right) + 300K \tag{1.79}$$

In Figures 1.9 and 1.10 the plots corresponding to (1.78) in the region of the boundary layer are shown.

For depicting Figures 1.9 and 1.10 such values of constants were used:

$$\mu_0 = 0.00001 \text{ Pa sec};$$

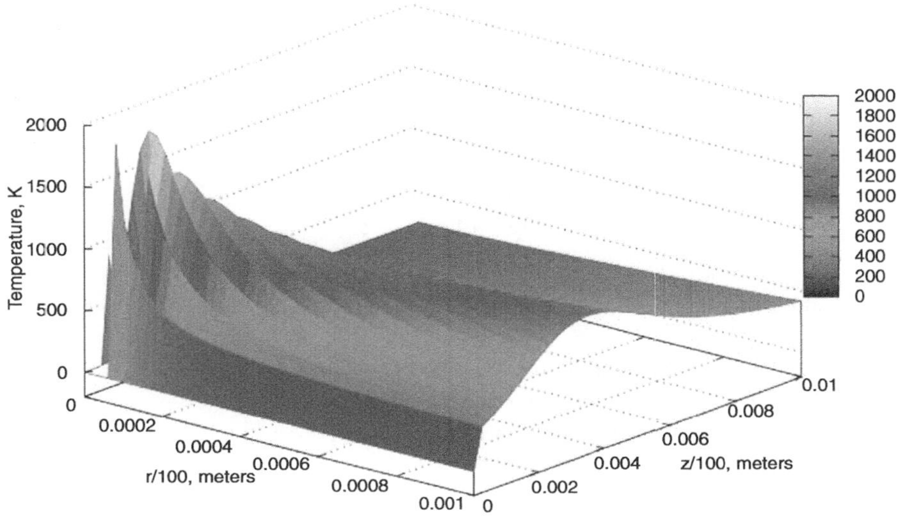

FIGURE 1.9 Temperature distribution in the area of the thermal boundary layer for the hypersonic case ($\delta_T = 0.5$ m; the values of the *r*- and *z*-coordinates are reduced by 100 times).

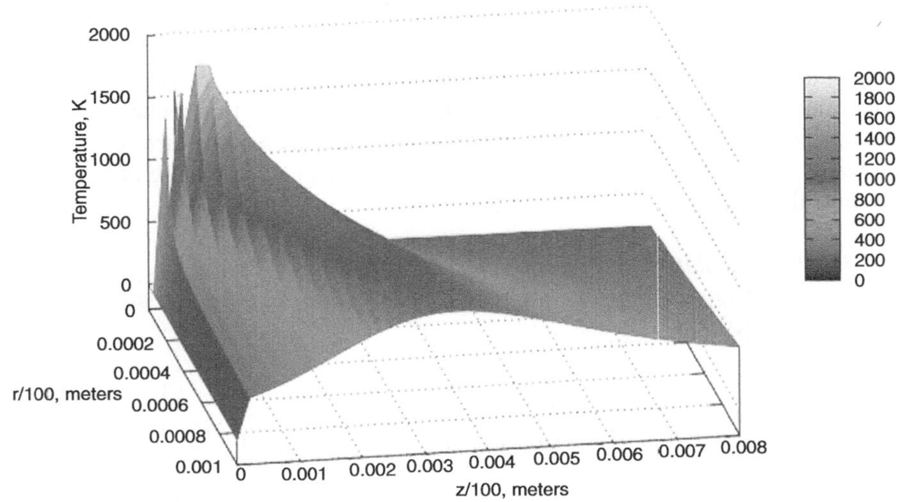

FIGURE 1.10 Temperature distribution in the area of the thermal boundary layer for the hypersonic case ($\delta_T = 0.5$ m; the values of the *r*- and *z*-coordinates are reduced by 100 times; the preceding plot is turned about the temperature axis for convenience).

$$k_0 = 0.0242 \frac{J}{m \sec K}$$

$$c_p = 1200 \frac{J}{\text{kg}K}, \left(\text{for} T = 1500 \text{ K}\right);$$

$$\Omega = 14000 \frac{\text{rad}}{\text{sec}};$$

Max. values of velocities in the area of the thermal boundary layer:

$v_\varphi = 7000$ m/sec, (hypersonic for $T = 1500$ K, $M_\varphi \approx 9$);
$v_r = 1500$ m/sec, (supersonic for $T = 1500$ K, $M_r \approx 2$);
$v_z = -700$ m/sec, (transonic for $T = 1500$ K);
$\delta = 0.25\ m$, (velocity boundary layer);
$\delta_T = 0.5\ m$, (thermal boundary layer);
$z_\mu = 0.35\ m$;
$r_0 = 16.5\ m$;
$C_0 = 0.1$;
$C_r = 0.05$;
$C_\varphi = -1$;
$A = A_1 = B_1 = C_1 = 0.5$.

It is seen from Figures 1.9 and 1.10 that the hottest place (the temperature center) in hypersonic gas over the rotating disk tends to the origin of the coordinate system where method of tunnel mathematics is not very accurate. This tendency must be verified experimentally. As the z-coordinate increases, the temperature center moves outwards to the edge of the disk. It looks like this is a peculiar feature of hypersonic gas over the rotating disk.

1.3.4.3.2 Calculation of Pressure Distribution

Substituting (1.66) and (1.73) into (1.58), we obtain the approximate expression for pressure distribution:

$$P = \rho \left[\frac{1}{\left(\frac{6\sqrt{\delta \delta_T}}{z} + 2 \right)} \left(-2v_r^2 + \left(v_r + v_\varphi \right)^2 \right) - \frac{4C_0 z_\mu^2}{81} \left(\frac{z}{\sqrt{\delta \delta_T}} \right)^3 \frac{v_r v_\varphi}{r^2} \right], \quad (1.80)$$

where ρ is taken from (1.64). Generally speaking, we must still add to the right-hand side of formula (1.80) the hydrostatic pressure $\rho g \left(H - z\right)$, where H is an immersion depth of the disk.

In order to integrate (1.58) we first rewrite (1.64) as

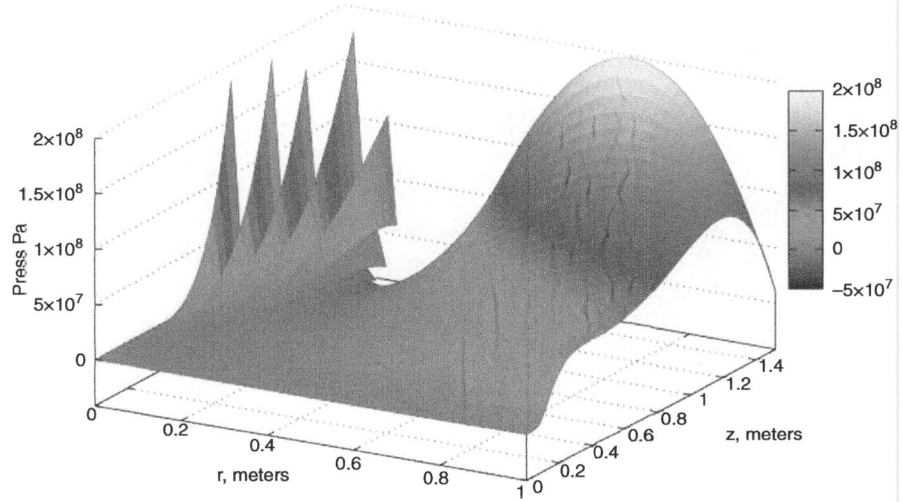

FIGURE 1.11 Pressure distribution in the area of the thermal boundary layer and beyond for the hypersonic case ($\delta = 0.25$ m; $\delta_T = 0.5$ m; radius of the disk is 1 meter).

$$\rho \approx \rho_0 \left(\frac{1}{e(r_0\Omega)^2} \right)^{\frac{3\sqrt{\delta\delta_T}}{z}} \cdot v_r^{\frac{6\sqrt{\delta\delta_T}}{z}}, \tag{1.81}$$

and after we pass from integration by dr to integration by dv_r with the aid of approximate relation

$$dr \approx \frac{r dv_r}{v_r} \tag{1.82}$$

All integrals in (1.58) are recurrent, so we easily find (1.80). In Figures 1.11–1.13 the plots corresponding to (1.80) in the region of the boundary layer are shown.

It is seen from Figures 1.11 and 1.12 that the pressure profile for the hypersonic case has two prominent peaks. The first peak corresponds to $z = \delta = 0.25\ m$, where δ is the thickness of the velocity boundary layer. The second peak is far outside the thermal boundary layer, where the method of tunnel mathematics is not very accurate. As the r-coordinate increases, this peak gradually disappears (Figure 1.13).

Integrating (1.60) in a similar way, we find such a relation:

$$P = \rho \left[-\frac{v_z v_r}{r \log \frac{r}{r_0}} \cdot z + g(H - z) + \frac{2 v_r v_\varphi}{r} \cdot z \right] \tag{1.83}$$

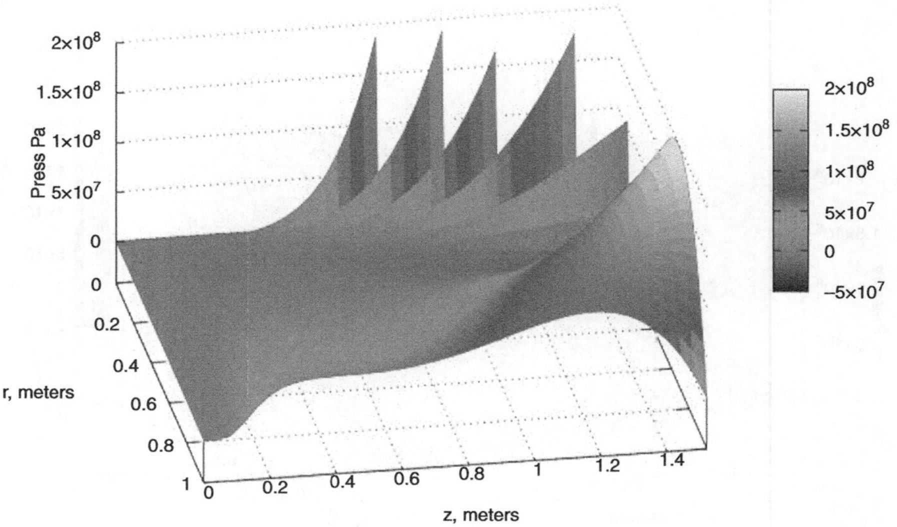

FIGURE 1.12 Pressure distribution in the area of the thermal boundary layer and beyond for the hypersonic case ($\delta = 0.25$ m; $\delta_T = 0.5$ m; radius of the disk is 1 meter). The preceding plot is turned about the pressure axis for convenience.

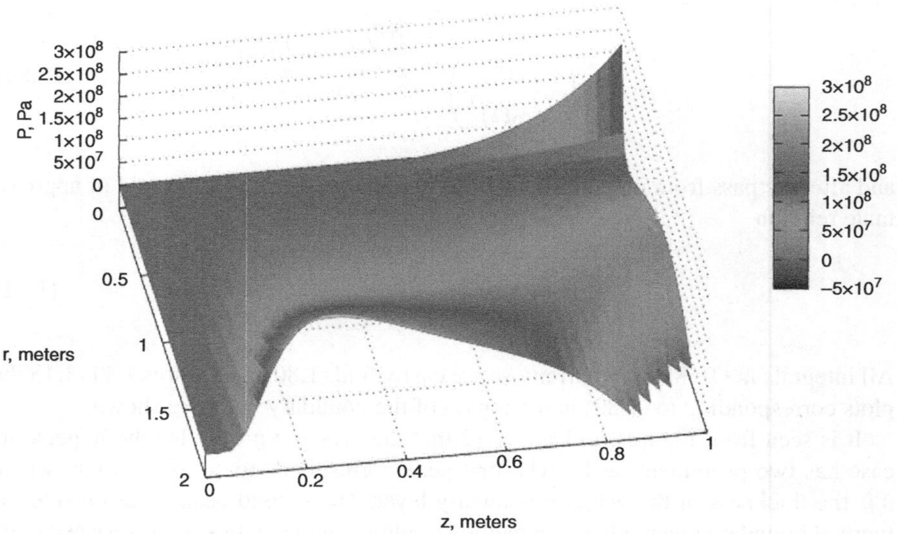

FIGURE 1.13 Pressure distribution in the area of the thermal boundary layer and beyond for the hypersonic case as the r-coordinate increases ($\delta = 0.25$ m; $\delta_T = 0.5$ m; radius of the disk is 1 meter).

Equating (1.80) and (1.83) in the area of boundary layer, i.e. at $z=\sqrt{\delta\delta_T}$, and taking into account that ratio $\frac{v_\varphi}{v_r}$ increases as the velocity of hypersonic gas increases, we obtain a very important relation for the thickness of the boundary layer:

$$\sqrt{\delta\delta_T}\approx\frac{v_\varphi}{v_r}\cdot\frac{r}{16}-z_\mu^2\cdot\frac{2C_0}{81r} \tag{1.84}$$

We see from (1.84) that both the velocity boundary layer and the thermal boundary layer cease to depend directly from angular velocity Ω and are determined by ratio $\frac{v_\varphi}{v_r}$ and normalization factor $z_\mu=z_{0\varphi}$ from (1.39).

1.4 CONCLUSIONS

The solution of the Navier–Stokes equations using the tunnel mathematics apparatus is simple and elegant, and also requires a good mathematical training and a deep physical analysis of the problem. This method does not require the use of specific software and can be used for the primary analysis of hydrodynamic problems. The results obtained using this method for a compressible fluid entrained by a disk of finite radius in the area of the boundary layer for subsonic and hypersonic flows corresponds to the theory of compressible fluids.

DECLARATION OF INTERESTS

The authors report no conflict of interest.

DATA AVAILABILITY

The data that supports the findings of this study are available within the article.

REFERENCES

[1] Shvydkyi, O. G. (2024). "Analytical solution of the steady Navier-Stokes equation for an incompressible fluid entrained by a rotating disk of finite radius in the area of boundary layer." *Annals of Mathematcial and Physics*,7(3), 305–313. https://doi.org/10.17352/amp.000137

[2] David Weyburne (2021). Aspects of Bo undary Layer Theory by David Weyburne ISBN.

[3] K. G. Winter, J. C. Rotta and K. G. Smith. (1968). "Studies of the turbulent boundary layer on a waisted body of revolution in subsonic and supersonic flow." https://reports.aerade.cranfield.ac.uk/handle/1826.2/2902

[4] D. A. Spence. (1959). "The growth of compressible turbulent boundary layers on isothermal and adiabatic walls." https://reports.aerade.cranfield.ac.uk/handle/1826.2/3760

[5] Coles, D. E. (1962). *The Turbulent Boundary Layer in a Compressible Fluid.* Santa Monica, CA: RAND Corporation. www.rand.org/pubs/reports/R403.html

[6] Sivells, J. C. and R. G. Payne. (1959). "A method of calculating turbulent-boundary-layer growth at hypersonic mach numbers." *Arnold Engineering Development Center* 7, AEDC-TR-59-3.

[7] Henrik Alfredsson, P., Kato, K. and Lingwood, R. J. (January 2024). "Flows over rotating disks and cones." *Annual Review of Fluid Mechanics*, 56, 45–68. https://doi.org/10.1146/annurev-fluid-121021-043651

[8] Appelquist, E., Schlatter, P., Alfredsson, P. H. and Lingwood, R. J. (2018). "Turbulence in the rotating-disk boundary layer investigated through direct numerical simulations." *European Journal of Mechanics B/Fluids*, 70, 6–18. https://doi.org/10.1016/j.euromechflu.2018.01.008

[9] Zhang, F. C., Xie, J. H., Chen, S. X. and Zheng, X. (2024). "Robust relation of streamwise velocity autocorrelation in atmospheric surface layers based on an autoregressive moving average model." *Journal of Fluid Mechanics*, 981, A20. https://doi.org/10.1017/jfm.2024.85

[10] Shvydkyi, O. G. (2024). "Sharp cone-broad cone-disk: Analytical solutions in the tunnel mathematics space to the steady Navier-Stokes equations in the area of boundary layer for incompressible symmetric flows entrained by these rotating bodies." *Acceleron Aerospace Journal*, 3(6), 624–647. https://doi.org/10.61359/11.2106-2464

[11] Shvydkyi, O. G. (2023). "Application of tunnel mathematics for solving the steady Lamé and Navier-Stokes equations." *Innovation in Science and Technology*, 2(4), 1–14. https://doi.org/10.56397/IST.2023.07.01

[12] Landau, L. D. and E. M. Lifshitz. (1987). "Fluid mechanics." *Course of Theoretical Physics*, 6, 45, 162.

[13] Schlichting, H. (1979). *Boundary-Layer Theory* (7th ed., pp. 265–271, 335–336, 600, 695). New York, USA: McGraw Hill.

[14] Soler, T. and Han, J.-Y. (2024). "Orthogonal curvilinear coordinates [and more] for Engineers and Physical Scientists" (NEW BOOK, September 2024, Printed by Excel Book Writing). https://orthogonal-curvilinear-coordinates-book-ts.com/

[15] Parr, O. (1963). "Untersuchungen der dreidimensionalen Grenzschicht an rotierenden Drehkörpern bei axialer Anströmung." *Ingenieur-Archiv*, 32, 393–413. https://doi.org/10.1007/BF00531788

2 MHD Maxwell Fluid with Slip over Porous Stretching Sheets

Mallikarjun Basavaraju Patil, Suresh Prithvi, and Basma Souayeh

2.1 INTRODUCTION

The boundary layer is the thin layer in the vicinity of the bottom fluid, the boundary layer plays a vital role in polymer extraction, thinning of metallic foils, crude oil extraction, producing thin plastic films, nuclear power plants, cooling the sheets, air-lift, and many more. Thus study of the boundary layer is crucial. The fundamental principles of the MHD state where current flows through conducting fluids when a magnetic field is present, causes the fluid to undergo forces that alter the magnetic field. The Navier–Stokes equations, which deal with fluid dynamics, and the Maxwell equations, which deal with electromagnetism, are the foundation of the MHD equation set. Maxwell fluid is a type of fluid that anticipates the relaxation time impact on the fluid flow, which cannot be done by other fluids. Since the Maxwell fluid is viscoelastic, the impact of viscosity and elasticity on the boundary layer is studied. Since magnetic fields have a significant influence on fluid motion, we will be using both magnetic fields and Maxwell fluids in our study. As fluid motion involves the no-slip condition, let us examine the Thomson–Troian slip in the current study of fluid motion.

Sakiadis [1, 2] was the pioneer in studying the fluid flow across a continuous surface, a continuous flat surface has greater drag than a flat finite surface. In their analysis, they allowed stretching sheets under a dusty ternary nanofluid, Mahabaleshwar *et al.* [3] discovered that increased inverse Darcy number enhanced the velocity profile for the stretching sheet and the effect of the ternary nanofluid was better than that of the dusty fluid. The fluid's velocity was discovered to be lowered by the magnetic field perpendicular to the stretched sheet by Mahesh *et al.* [4] when they analyzed the nanofluid flow over an inclined magnetic field. When Vishalakshi *et al.* [5] examined the Rivlin–Ericksen flow on a stretching sheet, they discovered that when the visco-elastic parameter increased, the concentration also decreased and the fluid velocity was reduced by a uniform magnetic field. The influence of MHD UCM fluid over a stretching sheet is discussed by Mamaloukas *et al.* [6] and found that the increase in elastic parameter decreased the velocity, the obtained governing equations were solved using the RK4 method. When Mahabaleshwar *et al.* [7] investigated the Newtonian flow across a porous stretching or shrinking sheet under slip conditions, they discovered that the Navier slip parameter caused a decrease in both the axial and

DOI: 10.1201/9781003611820-3

transverse velocity. The entropy generation of the MHD Casson CNT-based nanofluid flow across a wedge was examined by Umadevi and Mallikarjun [8]. After applying the HAM to solve the resulting equations, it turned out that the entropy generation increased as the magnetic parameter and wedge angle increased. Viscous dissipation and internal energy for both uniform prescribed heat flux and temperature have been studied by Mohammadein and Gorla [9].

Asaithambi [10] has used the finite difference approach to convert the semi-infinite interval to a finite interval, thus solving the Falkner–Skan problem. Visco elastic fluid flow across a magnetic field-influenced stretching sheet has been discussed by Siddheshwar and Mahabaleshwar [11]. Umadevi and Mallikarjun [12] analyzed the influence of the Dufour-sorbet and melting process of Eyring–Powell fluid on a wedge and found that the larger values of Dufour and soret parameters increased the concentration and temperature, and the velocity declines for growing melting parameter values. The second-grade fluid across a porous plate has been highlighted by Hayaat and Khan [13]. When aligned MHD micropolar fluid was applied over a wedge with slip conditions, Mallikarjun and Umadevi [14] examined the effects. They discovered that the temperature rose and that the velocity profile was suppressed by the velocity slip parameter. Singh *et al.* [15] in their study of MHD flow across a porous stretching sheet studied mass transpiration. The flow of a ternary nanofluid across a stretching sheet under Thomson–Troian slip conditions, where the sheet is assumed to be stretched vertically, was examined by Li *et al.* [16]. The computed equations were solved through the RKF-45 method and it was found that the slip parameter raised the temperature while decreasing the fluid's velocity. Additionally, the silver suspension particles increased the fluid's temperature but decreased its velocity.

Sekhar *et al.* [17] numerically investigated the viscoelastic MHD fluid flow in the presence of temperature and velocity slip along a stretching sheet to study the heat radiations and observed that the increased Brownian parameter reduced the heat transfer rate while being accelerated by the thermophoresis parameter. The effect of thermal radiation upon the Maxwell MHD fluid over the stretching sheet was examined by Aliakbar *et al.* [18] who noted that for increasing values of elastic number, heat transfer capacity increases and heat transfers from the sheet to the fluid due to magnetic parameter. Lie symmetry transformation has been implemented for computing non-Fourier heat flux with quadratic convection by Bhatti *et al.* [19]. In addition to chemical reaction catalysis, Razaq *et al.* have examined the convective condition and Cattaneo–Christov heat flow of ternary nanofluid [20]. The impact of upper convected Maxwell fluid flow on the moving plate has been studied by Sadeghy *et al.* [21]. Nadeem *et al.* [22] deliberated the Maxwell fluid flow over a nanomaterial MHD moving surface. Magnetic dipole in Maxwell fluid flow past a stretching sheet has been debated by Kumar *et al.* [23].

The homotopy analysis method is not yet used to solve differential equations with Thomson–Troian slip boundary condition so here we allow the flow of Maxwell fluid over a porous stretching sheet and analyze the outcomes for Thomson–Troian velocity slip condition given by [4], in the presence of inclined magnetic field considering Brownian and thermophoresis parameters. The so-obtained PDEs are converted to

ODEs through some similarity transformations and solved via homotopy analysis for the first time. The chapter is organized as follows: the mathematical model and formulation are given in section 2, solution of the problem through the homotopy analysis method is presented in section 3, the outcomes of the study in the form of results and discussion are outlined in section 4 along with the graphical and tabular form, the conclusion of the study is given in section 5 and references can be found in the final section.

2.2 MATHEMATICAL MODEL AND FORMULATION

Under the Thomson tunnel slip condition, consider an MHD Maxwell fluid flowing across a porous linearly stretched sheet $u = ax$. Figure 2.1 illustrates the application of an inclined magnetic field with a strength of B_0 at an angle of τ.

The following definition of the governing boundary-layer equations is based on the aforementioned assumptions [18, 17]:

$$\frac{\partial u}{\partial x}+\frac{\partial u}{\partial y}=0, \tag{2.1}$$

$$u\frac{\partial u}{\partial x}+v\frac{\partial u}{\partial y}=-\lambda\left(u^2\frac{\partial^2 u}{\partial x^2}+v^2\frac{\partial^2 u}{\partial y^2}+2uv\frac{\partial^2 u}{\partial x\partial y}\right)+v\frac{\partial^2 u}{\partial y^2}-\frac{\sigma\sin^2(\tau)B_0^2}{\rho}u-\frac{vu}{k_p}, \tag{2.2}$$

$$\frac{\partial T}{\partial y}v+\frac{\partial T}{\partial x}u=\tau^*(\frac{D_T}{D_\infty}\left(\frac{\partial T}{\partial y})^2+D_B\frac{\partial C}{\partial y}\frac{\partial T}{\partial y}\right)+\alpha\frac{\partial^2 T}{\partial y^2}-\frac{1}{(\rho C_p)}\frac{\partial q_r}{\partial y}+(T-T_\infty)\frac{Q_0}{(\rho C_p)}, \tag{2.3}$$

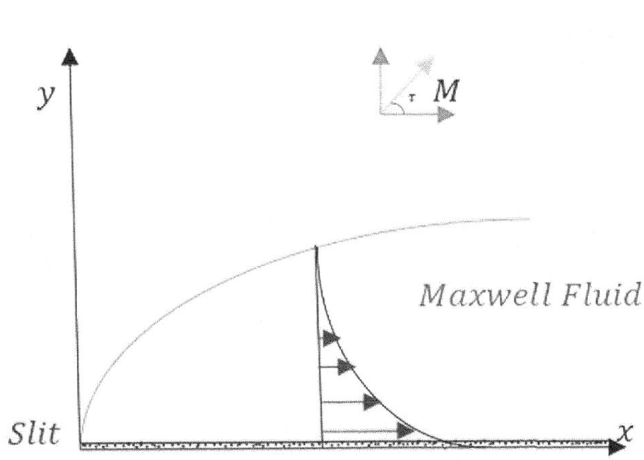

FIGURE 2.1 Geometry of Maxwell fluid over porous stretching sheet under Thomson–Troian slip.

$$v\frac{\partial C}{\partial y}+u\frac{\partial C}{\partial x}=\frac{\partial^2 T}{\partial y^2}\frac{D_T}{D_\infty}+\frac{\partial^2 C}{\partial y^2}D_B \tag{2.4}$$

corresponding boundary constraints [16]

$$u=u_w+u_1(1-u_2\frac{\partial u}{\partial y})^{-0.5}\frac{\partial u}{\partial y},\qquad v=v_w,\qquad C=C_w,\qquad T=T_w,\quad \text{for}\quad y\to 0, \tag{2.5}$$

$$T\to T_\infty,\qquad C\to C_\infty,\qquad u\to 0,\qquad \text{as}\quad y\to\infty, \tag{2.6}$$

Here,

λ – relaxation time
ν – kinematic viscosity
ρ – density of fluid
σ – electrical conductivity
τ – angle of inclinatiion
τ^* – heat capacity
α – thermal diffusivity
B_0 – magnetic field strength
k_p – permeability of porous medium
D_B – Brownian diffusion coefficient
D_T – thermophoresis diffusion coefficient
C_p – specific heat at constant pressure
q_r – radiative heat flux
Q_0 – heat production or absorption coefficient
u_w – sheet velocity
u_1, u_2 – slip parameters
T_w – wall temperature
C_w – all concentration
T_∞ – temperature away from the wall
C_∞ – concentration away from the wall

We use the following commonalities to non-dimensionalize and condense the governing equations to ODE.

$$u=axf',\qquad v=-(a\nu)^{1/2}f,\qquad \eta=\sqrt{\frac{a}{\nu}}y,\qquad \phi=\frac{C-C_\infty}{C_w-C_\infty},\qquad \theta=\frac{T-T_\infty}{T_w-T_\infty} \tag{2.7}$$

The subsequent differential equations can be obtained by substituting Eq. (2.7) into Eqs. (2.2), (2.3), and (2.4).

$$f'^2 - ff'' + \beta\left(f^2 f''' - 2ff'f''\right) - f''' + \left(Da^{-1} + M\sin^2(\tau)\right)f' = 0 \quad (2.8)$$

$$\frac{1}{Pr}\left(\frac{4R}{3} + 1\right)\theta'' + Nb\phi'\theta' + f\theta' + Nt\theta'^2 + Q\theta = 0 \quad (2.9)$$

$$\phi'' + PrLef\phi' + \frac{Nt}{Nb}\theta'' = 0 \quad (2.10)$$

Accordingly, the boundary constraints diminish to

$$\begin{gathered} f'(0) = \frac{f''(0)}{\sqrt{1 - \gamma_2 f''(0)}}\gamma_1 + 1, \\ f(0) = s, \quad \phi(0) = 1, \quad f'(\infty) = 0, \\ \theta(0) = 1, \quad \phi(\infty) = 0, \quad \theta(\infty) = 0 \end{gathered} \quad (2.11)$$

where magnetic parameter $M = \frac{\sigma B_0^2}{\rho a}$, elasticity number $\beta = a\lambda$, inverse Darcy number $Da^{-1} = \frac{\nu}{k_p a}$, Prandl number $Pr = \frac{\nu \rho C_p}{k}$, Brownian parameter $Nb = \frac{\tau^* D_B (C_w - C_\infty)}{\nu}$, thermophoresis parameter $Nt = \frac{\tau^* D_T (T_w - T_\infty)}{T_\infty \nu}$, radiation parameter $R = \frac{16\sigma^* T_\infty^3}{3kk^*}$, heat sink/source parameter $Q = \frac{Q_0}{a\rho C_p}$, Lewis number $\mathrm{Le} = \frac{\nu}{D_B}$.

The Sherwood number, Nusselt number, and local skin friction coefficient are described as

$$\mathrm{Sh} = \frac{xq_m}{D_B(C_w - C_\infty)}$$

$$\mathrm{Nu} = \frac{xq_w}{k(T_w - T_\infty)}$$

$$C_f = \frac{\tau_w}{\rho u_\infty^2}$$

where τ_w, q_w, and q_m are wall shear stress, wall heat gradient, and wall mass gradient.

After similarity transformation, we get

$$\sqrt{1/\mathrm{Re}}\mathrm{Nu} = -\left(1 + \frac{4R}{3}\right)\theta'(0)$$

$$\sqrt{\mathrm{Re}}C_f = (1 + \beta)f''(0)$$

$$\sqrt{1/\mathrm{Re}}\mathrm{Sh} = -\phi'(0)$$

2.3 HOMOTOPY ANALYSIS METHOD

Shijun Liao created the HAM analytical method in 1992, which made it easier to solve coupled and nonlinear ODEs by getting series solutions. Hayat *et al.* [13] state that the initial approximations and matching linear operators for an analytical study on HAM could be chosen as

$$f(\Gamma) = s + r\left(1 - e^{-\Gamma}\right) \qquad \phi(\Gamma) = e^{-\Gamma} \qquad \theta(\Gamma) = e^{-\Gamma} \tag{2.12}$$

where r is the root of the equation $1 - r - \frac{\gamma_1 r}{\sqrt{1+\gamma_2 r}} = 0$.

Let us define the Auxiliary equation a

$$L_f = \frac{\partial^3 f}{\partial \Gamma^3} - \frac{\partial f}{\partial \Gamma}, \qquad L_\theta = \frac{\partial^2 \theta}{\partial \Gamma^2} - \theta, \qquad L_\phi = \frac{\partial^2 \phi}{\partial \Gamma^2} - \phi \tag{2.13}$$

having the property

$$L_f\left[p_1 + p_2 e^{\Gamma} + p_3 e^{-\Gamma}\right] = 0, \qquad L_\theta\left[p_4 e^{\Gamma} + p_5 e^{-\Gamma}\right] = 0, \qquad L_\phi\left[p_6 e^{\Gamma} + p_7 e^{-\Gamma}\right] = 0 \tag{2.14}$$

here p_1 to p_7 are arbitrary constants.

The $0th$ deformation equation is

$$L_f\left[\hat{f}(\Gamma,\delta) - f_o(\Gamma)\right](1-\delta) = H_f N_f\left[\hat{f}(\Gamma,\delta)\right]\delta \tag{2.15}$$

$$L_\theta\left[\hat{\theta}(\Gamma,\delta) - \theta_o(\Gamma)\right](-\delta+1) = H_\theta \delta N_\theta\left[\hat{\phi}(\Gamma,\delta), \hat{f}(\Gamma,\delta), \hat{\theta}(\Gamma,P)\right] \tag{2.16}$$

$$L_\phi\left[\hat{\phi}(\Gamma,\delta) - \phi_o(\Gamma)\right](-\delta+1) = H_\phi \delta N_\phi\left[\hat{\phi}(\Gamma,\delta), \hat{f}(\Gamma,\delta), \hat{\theta}(\Gamma,\delta)\right] \tag{2.17}$$

H_θ, H_f, and H_ϕ – auxiliary parameters, $\delta \in [0,1]$ – embedding parameter, N_θ, N_f, and N_ϕ – nonlinear operators as given as below

$$N_f\left[\hat{f}(\Gamma,\delta)\right] = \beta\left(-2\hat{f}(\Gamma,\delta)\frac{\partial \hat{f}(\Gamma,\delta)}{\partial \Gamma}\frac{\partial^2 \hat{f}(\Gamma,\delta)}{\partial \Gamma^2} + \hat{f}^2(\Gamma,\delta)\frac{\partial^3 \hat{f}(\Gamma,\delta)}{\partial \Gamma^3}\right) - \frac{\partial^3 \hat{f}(\Gamma,\delta)}{\partial \Gamma^3}$$
$$- \hat{f}(\Gamma,\delta)\frac{\partial^2 \hat{f}(\Gamma,\delta)}{\partial \Gamma^2} + \frac{\partial \hat{f}(\Gamma,\delta)}{\partial \Gamma}^2 + \left(Da^{-1} + M\sin^2(\tau)\right)\frac{\partial \hat{f}(\Gamma,\delta)}{\partial \Gamma} \tag{2.18}$$

$$N_\theta\left[\hat{f}(\Gamma,\delta), \hat{\phi}(\Gamma,\delta), \hat{\theta}(\Gamma,\delta)\right] = \frac{1}{Pr}\left(1 + \frac{4R}{3}\right)\frac{\partial^2 \hat{\theta}(\Gamma,\delta)}{\partial \Gamma^2} + \hat{f}(\Gamma,\delta)\frac{\partial \hat{\theta}(\Gamma,\delta)}{\partial \Gamma} + Q\hat{\theta}(\Gamma,\delta)$$
$$+ Nb\frac{\partial \hat{\theta}(\Gamma,\delta)}{\partial \Gamma}\frac{\partial \hat{\phi}(\Gamma,\delta)}{\partial \Gamma} + Nt\left(\frac{\partial \hat{\theta}(\Gamma,\delta)}{\partial \Gamma}\right)^2 \tag{2.19}$$

$$N_{\phi}\left[\hat{f}(\Gamma,\delta),\hat{\phi}(\Gamma,\delta),\hat{\theta}(\Gamma,\delta)\right]=\frac{\partial^{2}\hat{\phi}(\Gamma,\delta)}{\partial\Gamma^{2}}++PrLe\hat{f}(\Gamma,\delta)\,\hat{\phi}(\Gamma,\delta)+\frac{Nt}{Nb}\frac{\partial^{2}\hat{\theta}(\Gamma,\delta)}{\partial\Gamma^{2}} \tag{2.20}$$

Corresponding boundary constraints are

$$\begin{aligned}
&\frac{\partial\hat{f}(0,\delta)}{\partial\Gamma}=1+\frac{\gamma_1\dfrac{\partial^{2}\hat{f}(0,\delta)}{\partial\Gamma^{2}}}{\sqrt{1-\gamma_2\dfrac{\partial^{2}\hat{f}(0,\delta)}{\partial\Gamma^{2}}}},\quad \hat{f}(0,\delta)=s,\\
&\hat{\phi}(0,\delta)=1,\quad \hat{\theta}(\infty,\delta)=0,\quad \hat{\theta}(0,\delta)=1,\\
&\hat{\phi}(\infty,\delta)=0,\quad \frac{\partial\hat{f}(\infty,\delta)}{\partial\Gamma}=0
\end{aligned} \tag{2.21}$$

as δ varies from 0 to 1 we get the required solution as

$$\begin{aligned}
&\hat{\theta}(\Gamma,1)=\theta(\Gamma),\quad \hat{\theta}(\Gamma,0)=\theta_o(\Gamma)\\
&\hat{\phi}(\Gamma,1)=\phi(\Gamma),\quad \hat{\phi}(\Gamma,0)=\phi_o(\Gamma)\\
&\hat{f}(\Gamma,1)=f(\Gamma),\quad \hat{f}(\Gamma;0)=f_o(\Gamma)
\end{aligned} \tag{2.22}$$

The Taylor series expansion of $\hat{\phi}(\Gamma,\delta)$, $\hat{\theta}(\Gamma,\delta)$,, $\hat{f}(\Gamma;\delta)$ about δ is

$$\begin{aligned}
&\hat{\theta}(\Gamma,\delta)=\theta_o(\Gamma)+\sum_{n=1}^{n=\infty}\delta^{n}\theta_n(\Gamma),\\
&\hat{f}(\Gamma,\delta)=f_o(\Gamma)+\sum_{n=1}^{n=\infty}\delta^{n}f_n(\Gamma),\\
&\hat{\phi}(\Gamma,\delta)=\phi_o(\Gamma)+\sum_{n=1}^{n=\infty}\delta^{n}\phi_n(\Gamma)
\end{aligned} \tag{2.23}$$

here $f_n(\Gamma)=\frac{1}{n!}\frac{\partial^{n}f(\Gamma,\delta)}{\partial\delta^{n}}|_{\delta=0}$, $\theta_n(\Gamma)=\frac{1}{n!}\frac{\partial^{n}\theta(\Gamma,\delta)}{\partial\delta^{n}}|_{\delta=0}$ and $\phi_n(\Gamma)=\frac{1}{n!}\frac{\partial^{n}\phi(\Gamma,\delta)}{\partial\delta^{n}}|_{\delta=0}$ for $\delta=1$ the series solution converges as below

$$\begin{aligned}
&f(\Gamma)=f_o(\Gamma)+\sum_{n=1}^{n=\infty}\delta^{n}f_n(\Gamma)\\
&\theta(\Gamma)=\theta_o(\Gamma)+\sum_{n=1}^{n=\infty}\delta^{n}\theta_n(\Gamma)\\
&\phi(\Gamma)=\phi_o(\Gamma)+\sum_{n=1}^{n=\infty}\delta^{n}\phi_n(\Gamma)
\end{aligned} \tag{2.24}$$

the nth order deformation equations are as follows:

$$L_f\left[f_n(\Gamma)-f_{n-1}(\Gamma)\,a_n\right]=R_{f,n}H_f,$$
$$L_\theta\left[\theta_n(\Gamma)-\theta_{n-1}(\Gamma)\,a_n\right]=R_{\theta,n}H_\theta,\tag{2.25}$$
$$L_\phi\left[\phi_n(\Gamma)-\phi_{n-1}(\Gamma)\,a_n\right]=R_{\phi,n}H_\phi,$$

where

$$a_n=\begin{cases}0, & n>1,\\ 1, & n\leq 1,\end{cases}\tag{2.26}$$

$$R_{f,n}=f'''_{m-1}(\Gamma)+\sum_{k=0}^{m-1}\Gamma\Big[f_{-1-k+m}(\Gamma)f_k(\Gamma)+f_{-1-k+m}(\Gamma)f_k(\Gamma)$$
$$-f_{-1-k+m}(\Gamma)f'_k(\Gamma)-f_{-1-k+m}(\Gamma)f_k(\Gamma)$$
$$+\beta f_{-1-k+m}(\Gamma)\sum_{l=0}^{k}\left(2f_{k-1}(\Gamma)f'_l-f_{k-1}(\Gamma)f'''_l(\Gamma)\right)\Big]\tag{2.27}$$
$$+\left(Da^{-1}+M\sin^2[\tau]\right)f_{m-1}$$

$$R_{\theta,n}=\frac{1}{Pr}\left(1+\frac{4R}{3}\right)\theta'_{m-1}(\Gamma)+\sum_{k=0}^{m-1}\Gamma\Big[f_{-1-k+m}(\Gamma)\,\theta'_k(\Gamma)+Nt\,\theta'_{-1-k+m}(\Gamma)\,\theta'_k(\Gamma)$$
$$+Nb\,\theta'_{-1-k+m}(\Gamma)\,\phi'_k(\Gamma)\Big]+Q\,\theta_{m-1}(\Gamma)\tag{2.28}$$

$$R_{\phi,n}=\phi''_{m-1}(\Gamma)+\mathrm{PrLe}\sum_{k=0}^{m-1}f_{-1-k+m}(\Gamma)\,\phi_k+\tfrac{Nt}{Nb}\theta''_{m-1}(\Gamma)\tag{2.29}$$

corresponding boundary constraints

$$f_n(0)=f_n(0)=\theta_n(0)=\phi_n(0)=f_n(\infty)=\theta_n(\infty)=\phi_n(\infty)=0\tag{2.30}$$

The general solutions are of the form:

$$f_n(\Gamma)=p_1+p_2e^{\Gamma}+p_3e^{-\Gamma}+f'_n(\Gamma)\tag{2.31}$$

$$\theta_n(\Gamma)=p_4e^{\Gamma}+p_5e^{-\Gamma}+\theta'_n(\Gamma)\tag{2.32}$$

$$\phi_n(\Gamma)=p_6e^{\Gamma}+p_7e^{-\Gamma}+\phi'_n(\Gamma)\tag{2.33}$$

2.4 RESULT VALIDATION

The governing nonlinear partial differential equations from (2.1) to (2.6) were converted to ordinary differential equations through similarity variables and thus the obtained equations (2.8)–(2.11) were solved with the help of the homotopy analysis method. The obtained solutions are presented in the form of graphs and the impact of various parameters like velocity slip, magnetic, elasticity, porous, thermophoresis, Brownian, Lewis, angle of inclination, and radiation are discussed below. The outcomes are plotted with $s = 0.1$, $\mathrm{Pr} = 2$, $R = 0.1$, $\mathrm{Nb} = 0.5$, $\mathrm{Nt} = 0.3$, $Q = -0.1$, $\mathrm{Le} = 2$, $Da^{-1} = 0.5$, $\beta = 0.5$, $M = 0.5$, $\tau = \pi/4$.

As shown in Figure 2.2, a rise in the velocity slip parameter lowers the fluid's axial velocity owing to the motion of the fluid and the sheet. As illustrated in Figure 2.4, a rise in the magnetic field diminishes the fluid's axial velocity since the Lorentz force leads the fluid's axial velocity to drop when it is applied against its motion. Furthermore, the angle at which the magnetic field is applied also changes the way the Lorentz force affects the fluid's velocity, as can be seen in Figure 2.3, an increase in angle from 0 to $\pi/2$ results in a reduction in the fluid's axial velocity. As shown in Figure 2.5, increasing the permeability of the stretching sheet reduces the fluid's axial velocity by expanding the region through which the fluid flows. Figure 2.6 illustrates how the Maxwell parameter reduces the fluid's axial velocity. As seen in

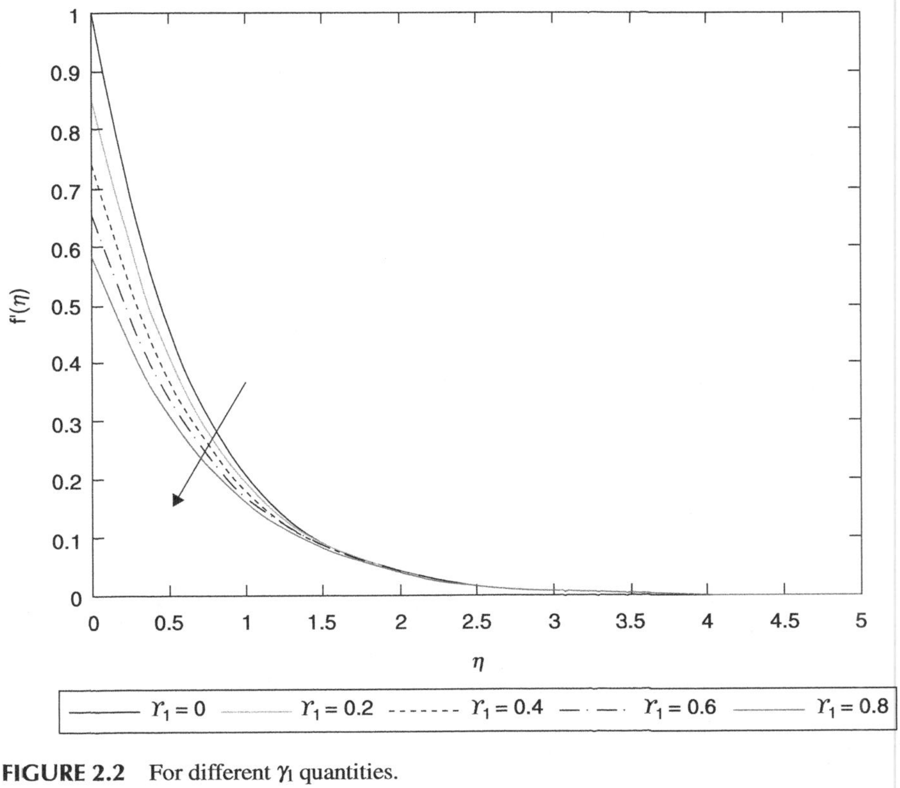

FIGURE 2.2 For different γ_1 quantities.

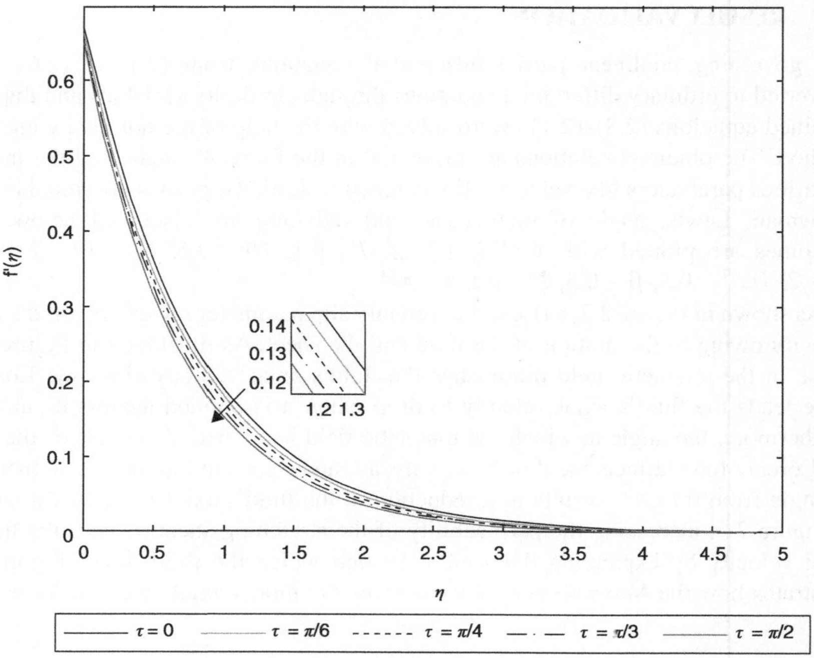

FIGURE 2.3 $f'(\eta)$ for different τ quantities.

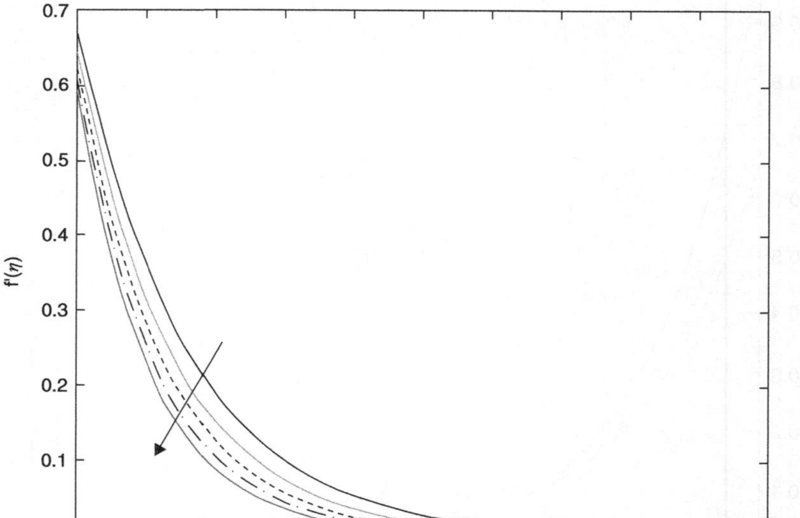

FIGURE 2.4 $f'(\eta)$ for different M quantities.

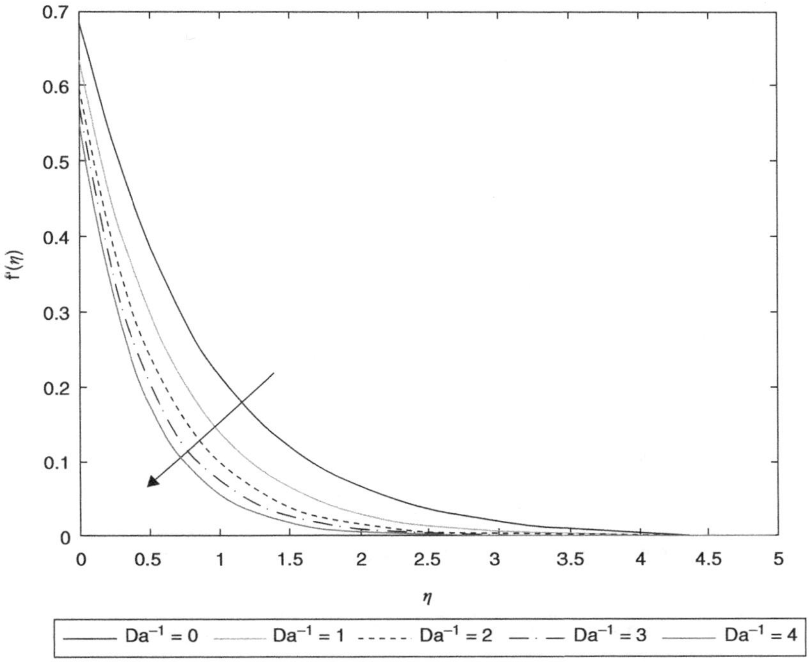

FIGURE 2.5 $f'(\eta)$ for different Da^{-1} quantities.

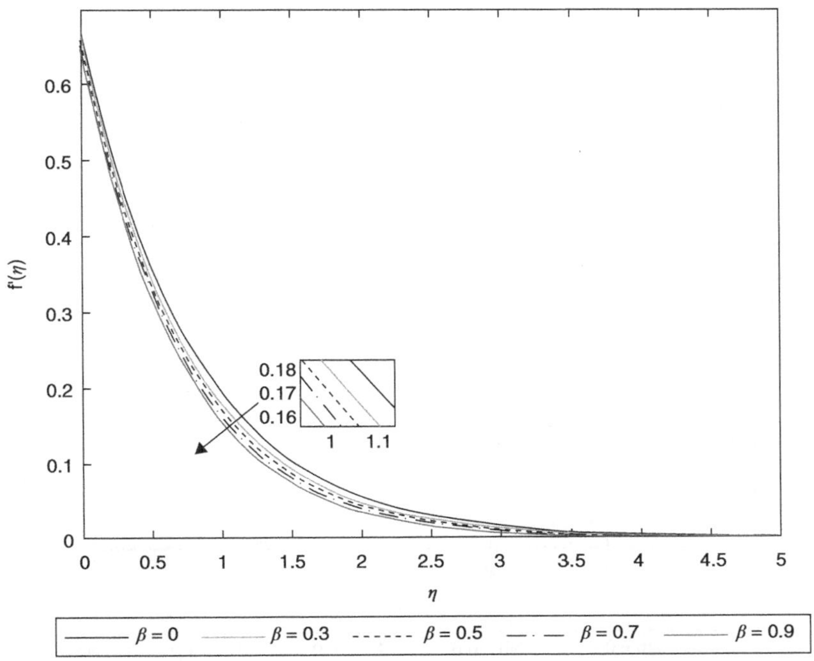

FIGURE 2.6 $f'(\eta)$ for different β quantities.

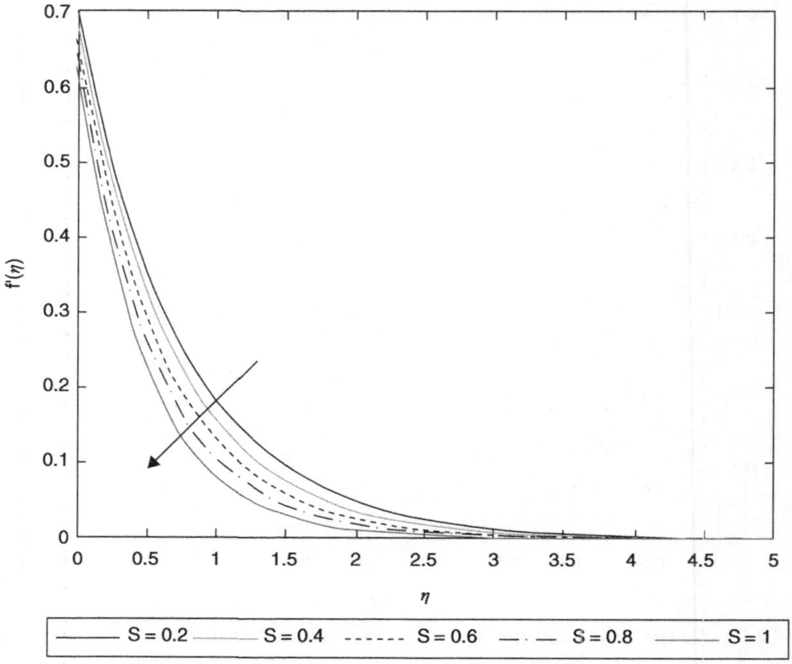

FIGURE 2.7 $f'(\eta)$ for different S quantities.

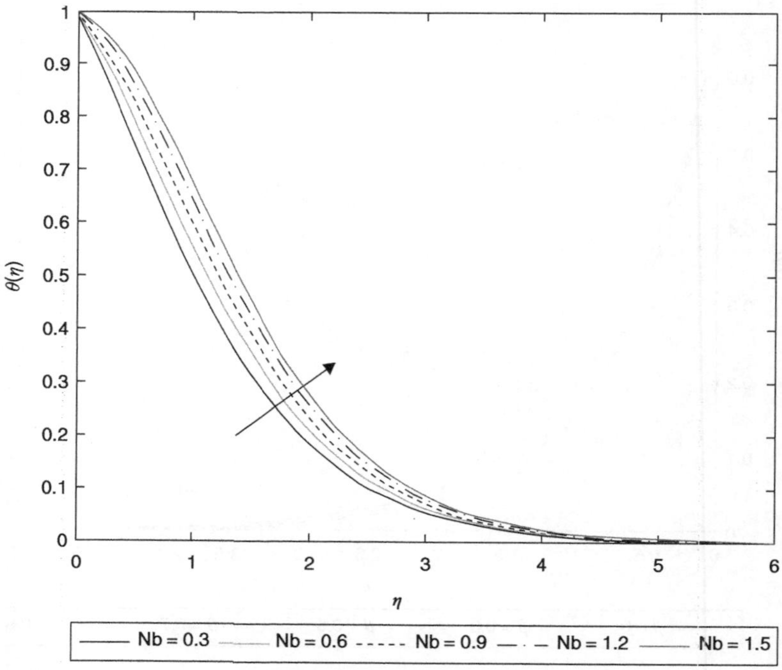

FIGURE 2.8 $\theta(\eta)$ for different Nb quantities.

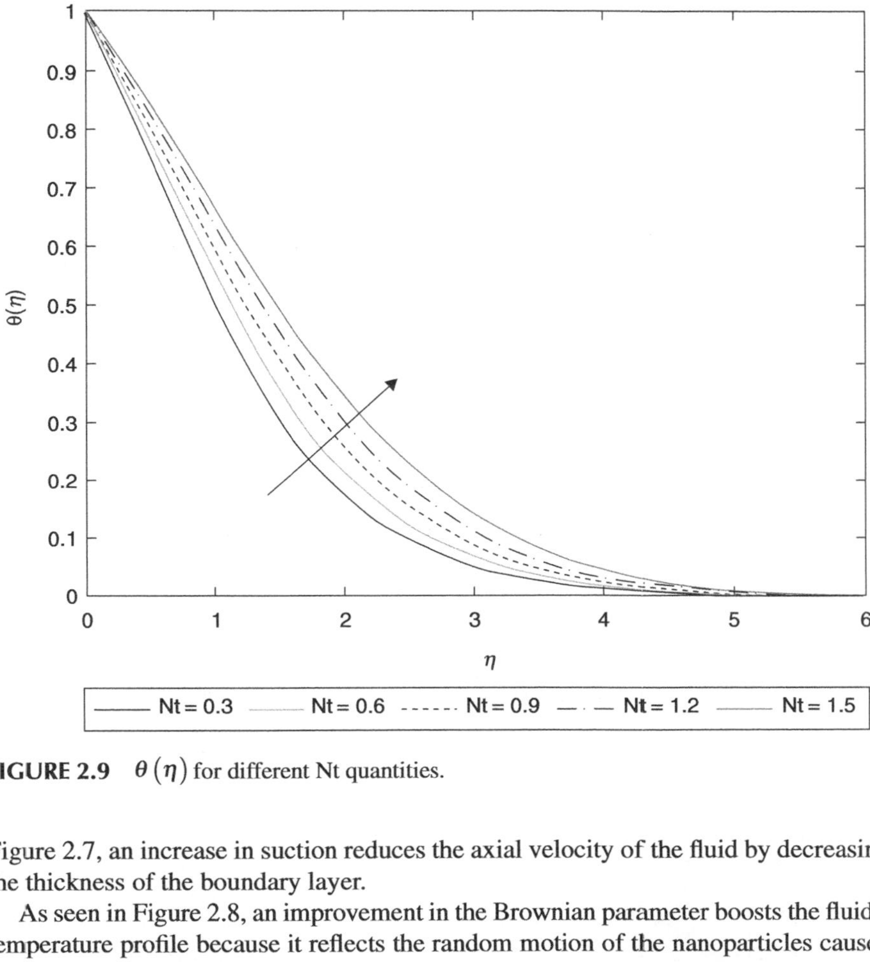

FIGURE 2.9 $\theta(\eta)$ for different Nt quantities.

Figure 2.7, an increase in suction reduces the axial velocity of the fluid by decreasing the thickness of the boundary layer.

As seen in Figure 2.8, an improvement in the Brownian parameter boosts the fluid's temperature profile because it reflects the random motion of the nanoparticles caused by collisions between fluid molecules, thereby accelerating the rate of temperature transmission. As observed in Figure 2.9, an increase in the value of the thermophoresis parameter improves the fluid's temperature profile because this parameter represents how nanoparticles migrate as a result of a temperature gradient inside the fluid. As seen in Figure 2.10, a higher heat sink parameter reduces the fluid's internal temperature profile because heat sink characteristics drain heat away from the fluid, lowering its temperature. The temperature gradient of the fluid varies as the radiation parameter increases, as seen in Figure 2.11. The rate of transfer of heat rises as a result of the sheet and fluid's increased relative movement. Figure 2.14 illustrates how increasing the flow rate parameter causes the temperature of the fluid gradient to rise.

As seen in Figure 2.12, the fluid's concentration drops as the Brownian parameter rises because of its random motion. As seen in Figure 2.13, a rise in Lewis number lowers the fluid's concentration profile and therefore reflects a drop in the fluid's mass diffusivity. Figure 2.15 illustrates how an increase in the velocity slip variable enhances the concentration within the boundary layer and, in turn, the concentration profile within the fluid.

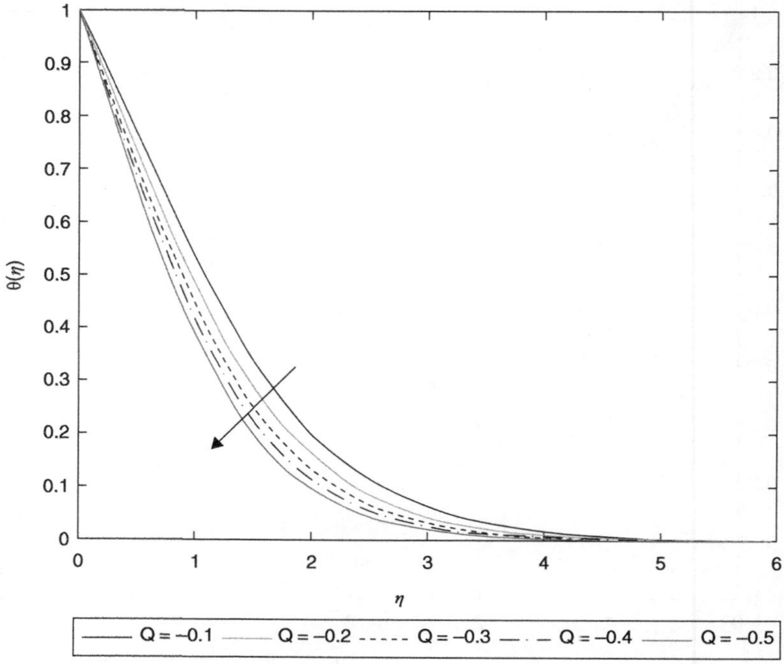

FIGURE 2.10 $\theta(\eta)$ for different Q quantities.

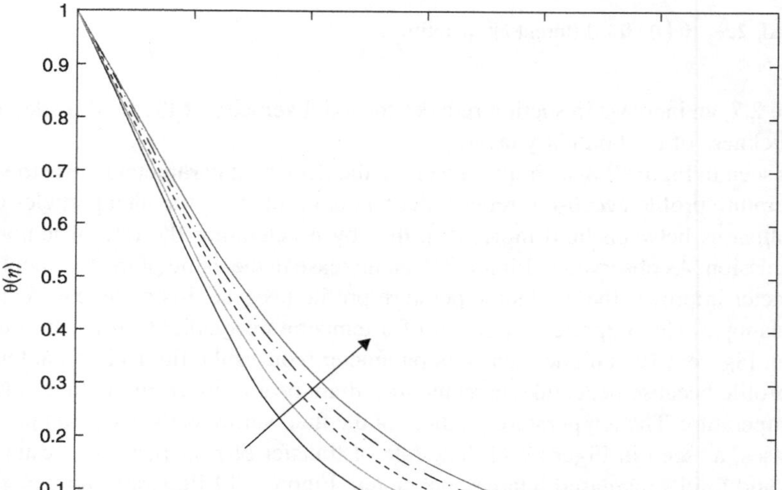

FIGURE 2.11 $\theta(\eta)$ for different R quantities.

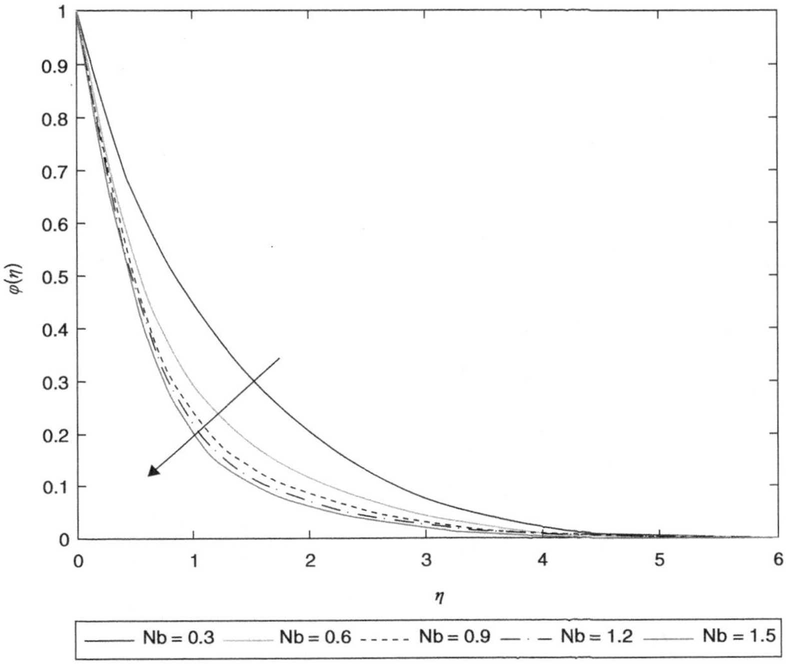

FIGURE 2.12 $\phi(\eta)$ for different Nb quantities.

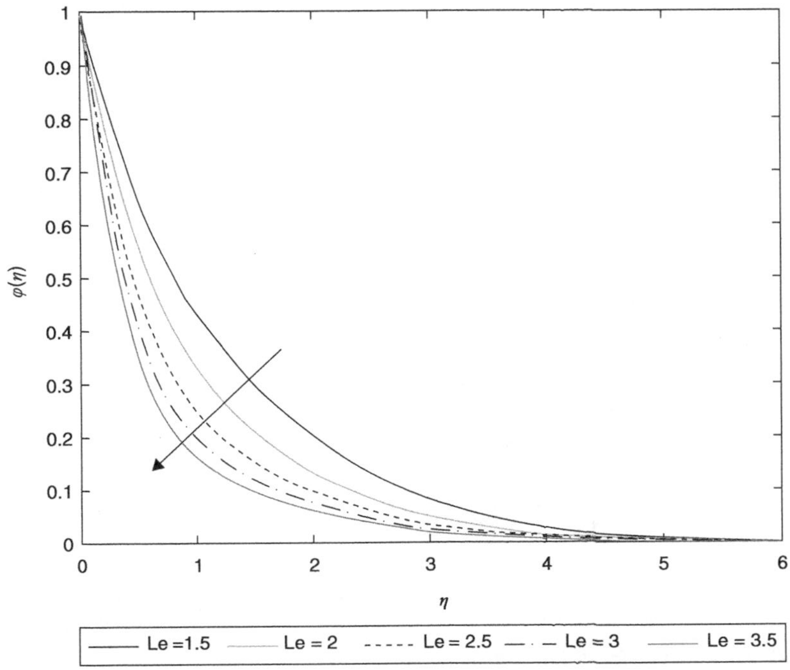

FIGURE 2.13 $\phi(\eta)$ for different Le quantities.

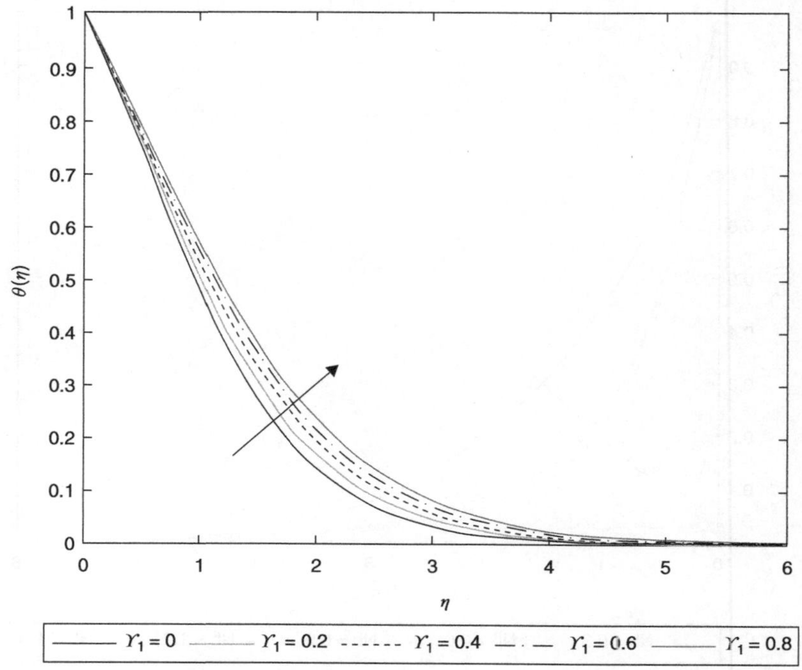

FIGURE 2.14 $\theta(\eta)$ for different γ_1 quantities.

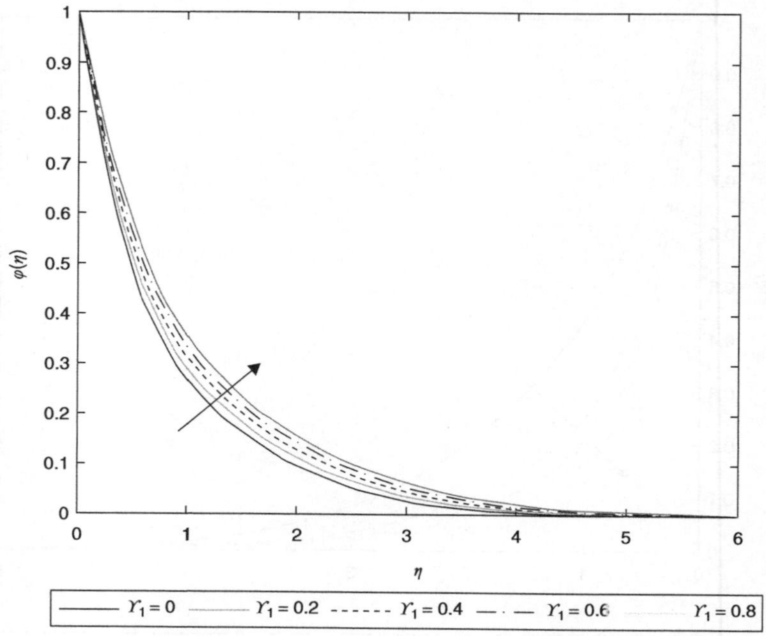

FIGURE 2.15 $\phi(\eta)$ for different γ_1 quantities.

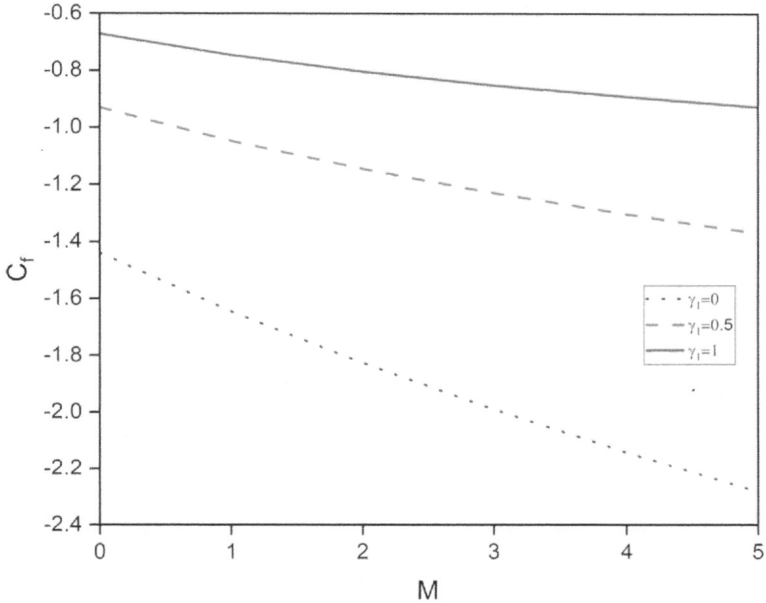

FIGURE 2.16 The local skin friction coefficient with respect to M for different values γ_1.

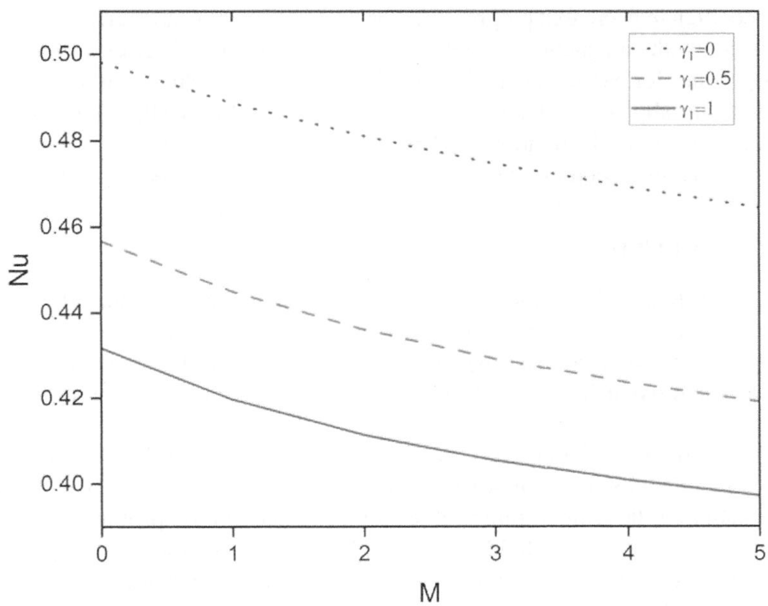

FIGURE 2.17 Nusselt number with respect to M for different values γ_1.

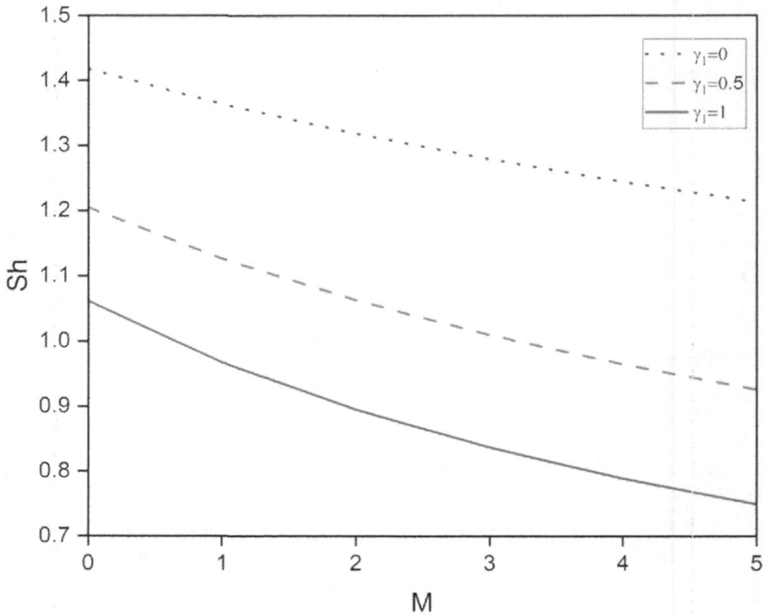

FIGURE 2.18 Sherwood number with respect to M for different values γ_1.

Figure 2.16 illustrates why the local skin friction coefficient changes about the magnetic parameter for different values for the velocity slip parameter. The local skin friction coefficient decreases progressively as the velocity slip parameter increases. The link connecting the magnetic parameter and the Nusselt number across a variety of velocity slip values for parameters is shown in Figure 2.17. The Nusselt number decreases as the velocity slip parameter rises. For a variety of velocity slip parameter values, Figure 2.18 illustrates the relationship between the Sherwood number and the magnetic parameter. The Sherwood number decreases when the velocity slip parameter rises.

2.5 CONCLUSION

The Maxwell fluid is allowed to flow over a porous stretching under the influence of an inclined magnetic field with Thomson–Troian slip condition, which is analyzed through the homotopy analysis method that provides more accurate results. The outcomes are plotted via graphs and listed in Table 2.1. The following observations are made through our study.

- The Thomson–Troian slip parameter plays a crucial role in decreasing the axial velocity in comparison with no-slip, also increased slip parameter increased the temperature profile and the concentration profile.
- A stronger magnetic field results in a lower axial velocity.
- Porosity is another parameter that affects the axial velocity. Here we can observe that increased porosity decreased the axial velocity.
- Increased elastic number decreased the Sherwood and Nusselt number and numerically decreased the skin friction coefficient.

TABLE 2.1
Comparision of the Nusselt Number with Previous Research [17]

Pr	Sekhar *et al.*	Present Method
0.72	0.463255	0.46325
1	0.581992	0.58199
3	1.165213	1.16521

Note: When $s=0$, $R=0$, $\mathrm{Nb}=0$, $\mathrm{Nt}=0$, $Q=0$, $\beta=0$, $M=0$, $\tau=0$, $\mathrm{Da}^{-1}=0$, $\gamma_1=0$, $\gamma_2=0$

REFERENCES

[1] Sakiadis, B. C. (1961). Boundary-layer behavior on continuous solid surfaces: I. Boundary-layer equations for two-dimensional and axisymmetric flow. *AIChE Journal*, 7(1), 26–28. https://doi.org/10.1002/aic.690070108

[2] Sakiadis, B. C. (1961b). Boundary-layer behavior on continuous solid surfaces: II. The boundary layer on a continuous flat surface. *AIChE Journal*, 7(2), 221–225. https://doi.org/10.1002/aic.690070211

[3] Mahabaleshwar, U., Maranna, T., Pérez, L., Bognár, G., and Oztop, H. (2023). An impact of radiation on laminar flow of dusty ternary nanofluid over porous stretching/shrinking sheet with mass transpiration. *Results in Engineering*, 18, 101227. https://doi.org/10.1016/j.rineng.2023.101227

[4] Mahesh, R., Mahabaleshwar, U., Kumar, P. V., Öztop, H. F., and Abu-Hamdeh, N. (2023). Impact of radiation on the MHD couple stress hybrid nanofluid flow over a porous sheet with viscous dissipation. *Results in Engineering*, 17, 100905. https://doi.org/10.1016/j.rineng.2023.100905

[5] Vishalakshi, A. B., Mahabaleshwar, U. S., and Sheikhnejad, Y. (2022). Impact of MHD and mass transpiration on Rivlin–Ericksen liquid flow over a stretching sheet in a porous media with thermal communication. *Transport in Porous Media*, 142(1–2), 353–381. https://doi.org/10.1007/s11242-022-01756-w [6] Mamaloukas, C., Abel, M. S., Tawade, J. V., and Mahabaleshwar, U. S. (2011). *On effects of a transverse magnetic field on a UCM fluid over a stretching sheet*. ResearchGate. www.researchgate.net/publication/267472399

[7] Mahabaleshwar, U., Nagaraju, K., Sheremet, M., Baleanu, D., and Lorenzini, E. (2019). Mass transpiration on Newtonian flow over a porous stretching/shrinking sheet with slip. *Chinese Journal of Physics*, 63, 130–137. https://doi.org/10.1016/j.cjph.2019.11.016

[8] Umadevi, K. B., and Mallikarjun, B. P. (2022). Entropy generation on magnetohydrodynamic laminar boundary-layer Casson CNT-based nanofluid flow across a wedge by using the homotopy analysis method. *Heat Transfer*, 51(7), 6340–6364. https://doi.org/10.1002/htj.22594

[9] Mohammadein, A., and Gorla, R. S. R. (2001). Heat transfer in a micropolar fluid over a stretching sheet with viscous dissipation and internal heat generation. *International Journal of Numerical Methods for Heat and Amp Fluid Flow*, 11(1), 50–58. https://doi.org/10.1108/09615530110364088

[10] Asaithambi, A. (2003). A second-order finite-difference method for the Falkner–Skan equation. *Applied Mathematics and Computation*, 156(3), 779–786. https://doi.org/10.1016/j.amc.2003.06.020

[11] Siddheshwar, P., and Mahabaleswar, U. (2004). Effects of radiation and heat source on MHD flow of a viscoelastic liquid and heat transfer over a stretching sheet. *International Journal of Non-Linear Mechanics*, 40(6), 807–820. https://doi.org/10.1016/j.ijnonlinmec.2004.04.006

[12] Umadevi, K. B., and Mallikarjun, B. P. (2022a). Homotopy analysis method for an influence of Dufour-Soret and melting process on magnetohydrodynamic boundary-layer flow towards a wedge in an eyring powell fluid. *Indian Journal of Science and Technology*, 15(16), 700–711. https://doi.org/10.17485/ijst/v15i16.2354

[13] Hayat, T., and Khan, M. (2005). Homotopy solutions for a generalized second-grade fluid past a porous plate. *Nonlinear Dynamics*, 42(4), 395–405. https://doi.org/10.1007/s11071-005-7346-z

[14] Mallikarjun, B. P., and Umadevi, K. B. (2023). An aligned magnetic field impact on a micropolar fluid boundary-layer flow across awedge subjected to slip effects: Homotopy analysis method. *Special Topics and Reviews in Porous Media an International Journal*, 14(5), 67–78. https://doi.org/10.1615/specialtopicsrevporousmedia.2023047134

[15] Singh, J., Mahabaleshwar, U. S., and Bognár, G. (2019). Mass transpiration in non-linear MHD flow due to porous stretching sheet. *Scientific Reports*, 9(1). https://doi.org/10.1038/s41598-019-52597-5

[16] Li, S., Puneeth, V., Saeed, A. M., Singhal, A., Al-Yarimi, F. A. M., Khan, M. I., and Eldin, S. M. (2023). Analysis of the Thomson and Troian velocity slip for the flow of a ternary nanofluid past a stretching sheet. *Scientific Reports*, 13(1), 2340. https://doi.org/10.1038/s41598-023-29485-0

[17] Sekhar, P. R., Sreedhar, S., Ibrahim, S. M., Kumar, P. V., and Omprakash, B. (2023). Numerical investigation of heat radiation on MHD viscoelastic nanofluid flow over a stretching sheet with heat source and slip conditions. *International Journal on Interactive Design and Manufacturing (IJIDeM)*, 18(5), 2991–3000. https://doi.org/10.1007/s12008-023-01407-4

[18] Aliakbar, V., Alizadeh-Pahlavan, A., and Sadeghy, K. (2007). The influence of thermal radiation on MHD flow of Maxwellian fluids above stretching sheets. *Communications in Nonlinear Science and Numerical Simulation*, 14(3), 779–794. https://doi.org/10.1016/j.cnsns.2007.12.003

[19] Bhatti, M. M., Shahid, A., Sarris, I. E., and Bég, O. A. (2022). Spectral relaxation computation of Maxwell fluid flow from a stretching surface with quadratic convection and non-Fourier heat flux using Lie symmetry transformations. *International Journal of Modern Physics B*, 37(9). https://doi.org/10.1142/s0217979223500820

[20] Razaq, A., Hayat, T., Khan, S. A., and Momani, S. (2023). ATSS model based upon applications of Cattaneo-Christov thermal analysis for entropy-optimized ternary nanomaterial flow with homogeneous-heterogeneous chemical reactions. *Alexandria Engineering Journal*, 79, 390–401. https://doi.org/10.1016/j.aej.2023.08.013

[21] Sadeghy, K., Najafi, A., and Saffaripour, M. (2005). Sakiadis flow of an upper-convected Maxwell fluid. *International Journal of Non-Linear Mechanics*, 40(9), 1220–1228. https://doi.org/10.1016/j.ijnonlinmec.2005.05.006

[22] Nadeem, S., Akhtar, S., and Abbas, N. (2020). Heat transfer of Maxwell base fluid flow of nanomaterial with MHD over a vertical moving surface. *Alexandria Engineering Journal*, 59(3), 1847–1856. https://doi.org/10.1016/j.aej.2020.05.008

[23] Kumar, R. N., Jyothi, A., Alhumade, H., Gowda, R. P., Alam, M. M., Ahmad, I., Gorji and Prasannakumara, B. (2021). Impact of magnetic dipole on thermophoretic particle deposition in the flow of Maxwell fluid over a stretching sheet. *Journal of Molecular Liquids,* 334, 116494. https://doi.org/10.1016/j.molliq.2021.116494

3 Couple Stress Effects in Magnetic Hybrid Nanofluid Flows

Sudha Mahanthesh Sachhin, Gadhigeppa Myacher Sachin, Ulavathi Shettar Mahabaleshwar, and Basma Souayeh

3.1 INTRODUCTION

Ternary nanofluid flow involves the movement of a fluid mixture consisting of a base fluid and two types of nanoparticles. This combination enhances the thermophysical properties of the fluid. Ternary nanofluids are designed to provide better performance than single or binary nanofluids by optimizing the composition of nanoparticles to suit specific applications. These fluids are extensively used in heat transfer systems, including cooling in electronic devices, solar energy collectors, and industrial heat exchangers. In industrial processes, ternary nanofluids improve lubrication, machining, and cutting efficiency. Additionally, they enhance renewable energy technologies like solar and geothermal systems and are used in wastewater treatment and pollution control. The importance of ternary nanofluid flow lies in its ability to improve energy efficiency, reduce operating costs, and support sustainability goals. Their adaptability to different systems makes them critical for advancing technologies in energy, healthcare, and environmental protection. Souayeh and Ramesh [1] studied the ternary nanofluid suspensions over the stretching surface with the influence of buoyancy forces. Priyadharshini *et al.* [2] explored the synthesis of ternary nanoparticles mixed in water movement across an expanding surface. Varatharaj *et al.* [3] examined the heat exchange effect on nanofluid flow with boundary slip conditions across porous medium. Akshatha *et al.* [4] examined the couple stress impact on sodium alginate suspensions flow via a permeable surface with heat radiation. Ouyang *et al.* [5] explored the stability analysis for Maxwell fluid flow with the impact of viscous dissipation.

The presence of heat sources or sinks in fluid flow refers to the addition or removal of heat within the fluid as it moves through a system. Heat sources supply thermal energy, while heat sinks absorb it, influencing the fluid's temperature and flow behavior. This interaction can significantly alter the thermal and hydrodynamic properties of the fluid. In engineering systems, heat sources and sinks are crucial in controlling the temperature distribution, pressure drop, and flow dynamics in heat exchangers, cooling systems, and reactors. In electronic cooling, heat sinks help dissipate heat,

DOI: 10.1201/9781003611820-4

preventing overheating of components. Proper control of these thermal interactions ensures better performance in industrial, automotive, and energy systems. Moreover, understanding their impact helps in designing more effective cooling and heating strategies. Mishra *et al.* [6] examined the convective movement of viscoelastic fluid flow over porous plate. Deng *et al.* [7] explored the numerical approach to study the heat source impact on natural convective cavity. Tsai *et al.* [8] explored the heat source impact unsteady expanding surface via quiescent liquid medium. Li *et al.* [9] studied the volumetric heat source impact on fluid movement with buoyancy forces.

Thermal radiation on fluid flow refers to the transfer of heat through electromagnetic waves from the fluid or surrounding surfaces. This form of heat transfer occurs without direct contact, affecting the temperature distribution and dynamics of the flow. It is significant in systems where temperature differences lead to radiation emission, impacting both the fluid and its surroundings. In applications like combustion chambers, furnaces, and cooling systems, thermal radiation plays a critical role in heat exchange. It also influences processes in space environments where conduction and convection are limited. Radiative heat transfer is crucial in the design of solar collectors, where it aids in absorbing solar energy. The importance of thermal radiation in fluid flow is seen in optimizing energy efficiency, controlling temperature, and enhancing the performance of heat-sensitive systems. Understanding its effects is essential in industries, such as aerospace, electronics, and energy, where managing heat is vital for system reliability and longevity. Kumar *et al.* [10] examined the thermal radiation impact on nanofluid movement across vertical surface. Turkyilmazoglu [11] explored the thermal radiation influence on convective magnetic fluid movement via porous disk. Misra and Sinha [12] studied the magnetic field and radiation impact on unsteady blood movement in capillaries. Li *et al.* [13] studied the theoretical analysis of unsteady fluid movement across an expanding surface.

The fluid flow within porous media is influenced by the geometry, size, and distribution of the pores, affecting the permeability and resistance to flow. This behavior is crucial in many natural and engineered systems. Applications of porous media include groundwater flow, oil and gas reservoir simulations, filtration systems, and soil–water interactions in agriculture. In engineering, they are used in heat exchangers, fuel cells, and catalytic reactors where fluid flows through porous materials to enhance performance. Understanding these interactions is essential for designing efficient systems in environmental management, energy production, and industrial applications. Eid and Nafe [14] examined heat transfer analysis on the hybrid nanofluid movement across porous medium. Wang *et al.* [15] studied the heat sink and forced convection impact on nanomaterials with porous materials. Krishna *et al.* [16] explored the radiation influence on magnetic fluid movement across vertical accelerated surface. Waini *et al.* and Mahabaleshwar *et al.* [17, 18] studied the radiation impact on hybrid nanofluid movement along a stretched surface.

Inspired by the aforementioned studies, the authors examined the influence of couple stress on magnetic hybrid nanofluid movement across expanding permeable surface with heat transfer, considered molybdenum disulfide and graphene oxide nanoparticles dispersed in ethylene glycol to create a hybrid nanofluid, governing equations are calculated by using similarity variables, thermal radiation and heat

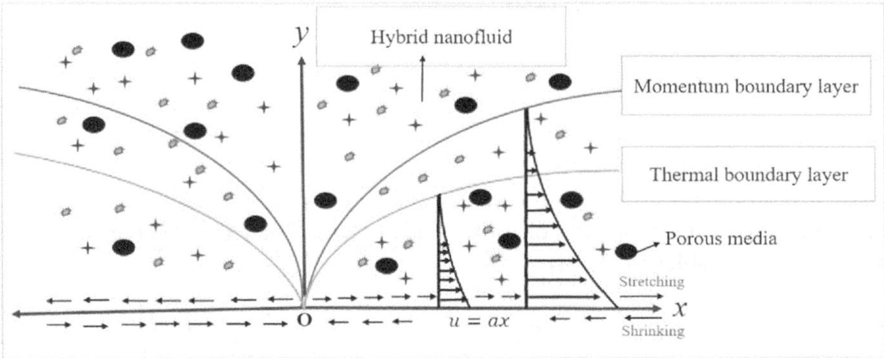

FIGURE 3.1 Schematic diagram of the fluid flow.

TABLE 3.1
Thermophysical Properties [4, 5, 14, 18]

Properties	$C_2H_6O_2$	MoS_2	GO
σ	1.07×10^{-4}	2.09×10^{-4}	4.57×10^{-6}
$\rho(kgm^{-3})$	1111	5.06×10^{3}	1800
$C_p(JkgK^{-1})$	2430	397.21	717
$\kappa(WmK^{-1})$	0.253	34.5	5000

source included in energy equation and solved by using hypergeometric series solution method, the current work has many uses in industries like polymer processing, metalworking, and glass manufacturing, among others. Figure 3.1 represents the schematic diagram; Table 3.1 represents the thermophysical properties.

3.2 MATHEMATICAL FORMULATION AND SOLUTIONS [4, 5, 14, 18]

$$\frac{\partial u}{\partial x}+\frac{\partial v}{\partial y}=0, \tag{3.1}$$

$$u\frac{\partial u}{\partial x}+v\frac{\partial u}{\partial y}=\frac{\mu_{eff}}{\rho_{hnf}}\frac{\partial^2 u}{\partial y^2}-\frac{\eta_0}{\rho_{hnf}}\frac{\partial^4 u}{\partial y^4}-\frac{\sigma_{hnf}}{\rho_{hnf}}B_0^2\sin^2(\tau)\,u-\frac{v_{hnf}}{K}u, \tag{3.2}$$

$$u\frac{\partial T}{\partial x}+v\frac{\partial T}{\partial y}=\frac{\kappa_{hnf}}{(\rho C_p)_{hnf}}\frac{\partial^2 T}{\partial y^2}-\frac{1}{(\rho C_p)_{hnf}}\frac{\partial q_r}{\partial y}+\frac{Q_0(T-T_\infty)}{(\rho C_p)_{hnf}}, \tag{3.3}$$

Boundary conditions [4, 5, 14, 18]

$$\begin{aligned} &u=ax, \quad v=v_w \quad -\kappa\frac{\partial T}{\partial Y}=h(T_w-T), \quad \text{as} \quad y\to 0,\\ &u=0, \qquad\qquad T=T_\infty, \qquad\qquad \text{at} \quad y\to\infty, \end{aligned} \tag{3.4}$$

where u and v are the component of velocity, ν denotes dynamic viscosity,κ denotes thermal conductivity, σ denotes electrical conductivity, and T denotes temperature.

Similarity variables [4, 5, 14]:

$$u=axf_\eta(\eta), \quad v=-\sqrt{a\nu}f(\eta), \quad \eta=y\sqrt{\frac{a}{\nu}}, \quad \theta(\eta)=\frac{T-T_\infty}{T_w-T_\infty}. \tag{3.5}$$

The radiation heat flux solved by utilizing Rosseland's approach [4, 5, 14].

$$q_r=-\frac{4\sigma^*}{3k^*}\frac{\partial T^4}{\partial y}, \tag{3.6}$$

where k^* denotes absorption parameter and σ^* denotes Stephen Boltzmann constant, expand T^4 with Taylor series and ignore the high order terms to get

$$T^4\cong 4T_\infty^3T-3T_\infty^4, \tag{3.7}$$

the radiation effect in Eq. (3.3) is obtained as

$$\frac{\partial q_r}{\partial y}=-\frac{16\sigma^*T_\infty^3}{3k^*}\frac{\partial^2 T}{\partial y^2} \tag{3.8}$$

adding Eq. (3.7) and (3.8) in Eq. (3.3) to obtain

$$u\frac{\partial T}{\partial x}+v\frac{\partial T}{\partial y}=\left(\frac{\kappa_{hnf}}{(\rho C_p)_{hnf}}+\frac{1}{(\rho C_p)_{hnf}}\frac{16\sigma^*T_\infty^3}{3k^*}\right)\frac{\partial^2 T}{\partial y^2}+\frac{Q_0(T-T_\infty)}{(\rho C_p)_{hnf}}, \tag{3.9}$$

Eqs. (3.2) and (3.3) are solved as

$$\begin{aligned} &\lambda f_{\eta\eta\eta}(\eta)+A_2\left[f_{\eta\eta}(\eta)f(\eta)-(f_\eta(\eta))^2\right]-Cf_{\eta\eta\eta\eta\eta}(\eta)\\ &\quad-\left[A_3M\sin^2(\tau)+A_1Da^{-1}\right]f_\eta(\eta)=0 \end{aligned} \tag{3.10}$$

$$(A_4+Nr)\theta_{\eta\eta}(\eta)+A_5\Pr\theta_\eta(\eta)f(\eta)+Ni\Pr\theta(\eta)=0. \tag{3.11}$$

where

$M=\frac{\sigma_f B_0^2}{\rho_f a}$ is the magnetic parameter,

$Da^{-1}=\frac{\nu_f}{Ka}$ is the inverse Darcy number.

$\lambda=\frac{\mu_{eff}}{\mu_f}$ is the viscosity ratio.

Pr denotes the Prandtl number,

$A_1=\frac{\mu_{hnf}}{\mu_f},\quad A_2=\frac{\rho_{hnf}}{\rho_f},\quad A_3=\frac{\sigma_{hnf}}{\sigma_f},\quad A_4=\frac{\kappa_{hnf}}{\kappa_f},\quad A_5=\frac{(\rho C_p)_{hnf}}{(\rho C_p)_f}$. are constants.

Modified B. Cs are calculated as [4, 5, 14]

$$\begin{aligned} &f(\eta)=S, \quad f_\eta(\eta)=\alpha, \quad \theta_\eta(\eta)=-Bi(1-\theta(\eta)), && \text{as} \quad \eta\to 0,\\ &f_\eta(\eta)=0, \quad \theta(\eta)=0, && \text{at} \quad \eta\to\infty \end{aligned} \tag{3.12}$$

$$\left.\begin{aligned} &\frac{\rho_{hnf}}{\rho_f}=(1-\phi_2)\left(1-\phi_1+\phi_1\left(\frac{\rho_{s1}}{\rho_f}\right)\right)+\phi_2\left(\frac{\rho_{s2}}{\rho_f}\right)\\ &\frac{\sigma_{hnf}}{\sigma_f}=\frac{\sigma_{s_2}+2\sigma_{bf}+2\phi_2(\sigma_{s_2}-\sigma_f)}{\sigma_{s_2}+2\sigma_{bf}-\phi_2(\sigma_{s_2}-\sigma_f)}\\ &\text{where}\\ &\frac{\sigma_{bf}}{\sigma_f}=\frac{\sigma_{s_1}+2\sigma_f+2\phi_1(\sigma_{s_1}-\sigma_f)}{\sigma_{s_1}+2\sigma_f-\phi_1(\sigma_{s_1}-\sigma_f)}\\ &\frac{\mu_{hnf}}{\mu_f}=\frac{1}{(1-\phi_1)^{2.5}(1-\phi_2)^{2.5}}\\ &\frac{(\rho C_p)_{hnf}}{(\rho C_p)_f}=(1-\phi_2)\left(1-\phi_1+\phi_1\left(\frac{(\rho C_P)_{s_1}}{(\rho C_P)_f}\right)\right)+\phi_2\left(\frac{(\rho C_P)_{s_2}}{(\rho C_P)_f}\right)\\ &\frac{\kappa_{hnf}}{\kappa_f}=\frac{\kappa_{s_2}+2\kappa_{bf}+2\phi_2(\kappa_{s_2}-\kappa_f)}{\kappa_{s_2}+2\kappa_{bf}-\phi_2(\kappa_{s_2}-\kappa_f)}\\ &\text{where}\\ &\frac{\kappa_{bf}}{\kappa_f}=\frac{\kappa_{s_1}+2\kappa_f+2\phi_1(\kappa_{s_1}-\kappa_f)}{\kappa_{s_1}+2\kappa_f-\phi_1(\kappa_{s_1}-\kappa_f)} \end{aligned}\right\} \tag{3.13}$$

3.3 ANALYTICAL SOLUTION FOR THE MOMENTUM EQUATION

Here to obtain the exact solution for Eq. (3.10) with B. Cs (3.12) we consider [4, 18, 14].

$$f(\eta)=S+\frac{\alpha}{\beta}(1-e^{-\beta\eta}) \tag{3.14}$$

Differentiate Eq. (14) to get

$$f_\eta(\eta)=\alpha e^{-\beta\eta},\qquad f_{\eta\eta}(\eta)=-\alpha\beta e^{-\beta\eta},\quad f_{\eta\eta\eta}(\eta)=\alpha\beta^2 e^{-\beta\eta}. \tag{3.15}$$

Substitute Eqs. (3.14) and (3.15) in (3.10), we get the solution β.

$$\lambda\beta^2-C\beta^4-A_2S\beta-A_2\alpha-A_3M\sin^2(\tau)-A_1Da^{-1}=0 \tag{3.16}$$

$$\beta_1=\frac{1}{2}\sqrt{\frac{2\lambda}{3C}+\frac{\chi_1}{3C\chi_2}+\frac{\chi_2}{3\times 2^{1/3}C}}-\frac{1}{2}\sqrt{\frac{4\lambda}{3C}-\frac{\chi_1}{3C\chi_2}-\frac{\chi_2}{3\times 2^{1/3}C}}$$
$$-\frac{2SA_2}{C\sqrt{\frac{2\lambda}{3C}+\frac{\chi_1}{3C\chi_2}+\frac{\chi_2}{3\times 2^{1/3}C}}},$$

$$\beta_2=\frac{1}{2}\sqrt{\frac{2\lambda}{3C}+\frac{\chi_1}{3C\chi_2}+\frac{\chi_2}{3\times 2^{1/3}C}}+\frac{1}{2}\sqrt{\frac{4\lambda}{3C}-\frac{\chi_1}{3C\chi_2}-\frac{\chi_2}{3\times 2^{1/3}C}}$$
$$-\frac{2SA_2}{C\sqrt{\frac{2\lambda}{3C}+\frac{\chi_1}{3C\chi_2}+\frac{\chi_2}{3\times 2^{1/3}C}}}$$

$$\beta_3=-\frac{1}{2}\sqrt{\frac{2\lambda}{3C}+\frac{\chi_1}{3C\chi_2}+\frac{\chi_2}{3\times 2^{1/3}C}}-\frac{1}{2}\sqrt{\frac{4\lambda}{3C}-\frac{\chi_1}{3C\chi_2}-\frac{\chi_2}{3\times 2^{1/3}C}+\frac{2SA_2}{C\sqrt{\frac{2\lambda}{3C}+\frac{\chi_1}{3C\chi_2}+\frac{\chi_2}{3\times 2^{1/3}C}}}}$$

$$\beta_4=-\frac{1}{2}\sqrt{\frac{2\lambda}{3C}+\frac{\chi_1}{3C\chi_2}+\frac{\chi_2}{3\times 2^{1/3}C}}+\frac{1}{2}\sqrt{\frac{4\lambda}{3C}-\frac{\chi_1}{3C\chi_2}-\frac{\chi_2}{3\times 2^{1/3}C}}$$
$$+\frac{2SA_2}{C\sqrt{\frac{2\lambda}{3C}+\frac{\chi_1}{3C\chi_2}+\frac{\chi_2}{3\times 2^{1/3}C}}}$$

where

$$\chi_1=2^{1/3}\left(12CDa^{-1}A_1+\lambda^2+12CA_3M\sin^2(\tau)\right)$$

$$\chi_2=\left[\begin{array}{l}-2\lambda^3+27CS^2A_2^2+72C\lambda\left(Da^{-1}A_1+A_3M\sin^2(\tau)\right)+\\ \sqrt{\begin{array}{l}-4\left(12CDa^{-1}A_1+\lambda^2+12CA_3M\sin^2(\tau)\right)^3+\\ \left(-2\lambda^3+27CS^2A_2^2+72C\lambda\left(Da^{-1}A_1+A_3M\sin^2(\tau)\right)^2\right)\end{array}}\end{array}\right]^{1/3} \tag{3.17}$$

3.4 ANALYTICAL SOLUTION FOR THE TEMPERATURE EQUATION

To get the exact solution for Eq. (3.11), consider a new variable by Eq. (3.11) and B. Cs (3.12), which gives

$$t\frac{d^2\theta}{d\theta^2}+(1-m+nt)\frac{d\theta}{dt}+L\theta=0 \tag{3.18}$$

where

$$m=\frac{A_5S\Pr}{(A_4+Nr)\,\beta}+\frac{A_5\Pr\alpha}{\beta^2\,(A_4+Nr)},n=\frac{A_5\Pr\alpha}{(A_4+Nr)},L=\frac{Ni\Pr}{\beta^2(A_4+Nr)}.$$

Additional boundary conditions are

$$\frac{\Pr}{\beta}\frac{d\theta}{dt}\left(\frac{\Pr}{\beta^2}\right)=Bi\left(1-\theta\left(\frac{\Pr}{\beta^2}\right)\right),\text{and } \theta(0)=0, \tag{3.19}$$

to solve Eq. (3.18) by power series solution $\theta(t)=\sum_{r=0}^{\infty}a_rt^{r+k}$ and utilizing the Frobenius method to obtain an analytical solution as

$$\theta(t)=\frac{\frac{Bi}{A_4}\left(e^{-\beta\eta}\right)^{\frac{k_1+k_2}{2}}H\left[\frac{k_1+k_2}{2},k_2+1,-npe^{-\beta\eta}\right]}{\left(\beta\frac{k_1+k_2}{2}\right)H\left[\frac{k_1+k_2}{2},k_2+1,-np\right]} \\ -\frac{np}{k_2+1}H\left[\frac{k_1+k_2}{2}+1,k_2+2,-np\right]+\frac{L}{A_4}H\left[\frac{k_1+k_2}{2},k_2+1,-np\right] \tag{3.20}$$

where

$$L=\frac{\Pr Ni}{\beta^2(A_4+Nr)},\quad k_1=m,\quad k_2=\sqrt{m^2-4L},\quad m=\frac{A_5S\Pr}{(A_4+Nr)\quad\beta}+\frac{A_5\Pr d}{\beta^2(A_4+Nr)},$$
$$n=\frac{A_5\Pr d}{A_4+Nr},$$

3.5 RESULT AND DISCUSSION

The present study examines the influence of couple stress on magnetic hybrid nanofluid movement across expanding permeable surface with heat transfer, considered molybdenum disulfide and graphene oxide nanoparticles dispersed in ethylene glycol to create hybrid nanofluid, governing equations are calculated by using similarity variables, thermal radiation and heat source are included in the energy equation and solved by using the hypergeometric series solution method. The effect of several physical parameters on fluid flow through graphs is studied in the below section.

Figures 3.2(a) and (b) represent the Brinkman ratio effect on axial velocity for both expanding and shrinking boundary, increasing the Brinkman ratio decays the velocity of the fluid movement. Physically, viscous forces penetrate deeper into the porous structure, creating smoother transitions between the porous and free-flow regions.

Figures 3.3(a) and (b) represent the magnetic parameter effect on axial momentum for both stretching and shrinking boundary, rising the magnetic term decays the velocity of the fluid movement. Physically, the presence of a magnetic field gives a rapid deduction of momentum, especially near the fluid boundary, the Lorentz force resists fluid flow, opposing the fluid's motion, increasing the magnetic field's strength leads to a decay in fluid momentum.

Figures 3.4 (a) and (b) portrays the mass transpiration parameter effect on axial momentum for both stretching and shrinking surface, increasing the mass transpiration parameter reduces the velocity. Physically, mass suction gradually reduces fluid momentum, particularly near the plate surface. As the suction term increases, the velocity profile decreases, making the velocity boundary layer thinner. The velocity moves away from the channel center line towards the plate, with the most significant velocity reduction occurring very close to the fluid boundary.

Figures 3.5(a) and (b) represent the Biot number parameter effect on temperature for both expanding and shrinking boundary, increasing the Biot number increases the temperature of the fluid flow. Physically, the Biot number significantly influences fluid temperature by determining thermal gradient characteristics. For small Biot numbers, the temperature remains nearly uniform, while larger Biot numbers create significant non-uniform temperature distributions. As the Biot number increases, temperature profiles become more complex, with enhanced thermal gradients and variations across the fluid flow.

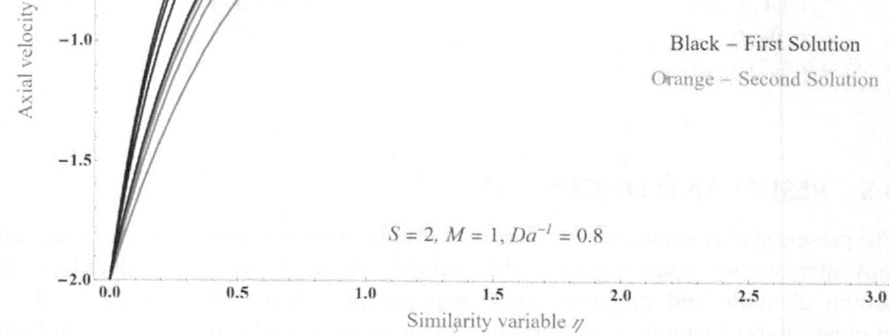

(a)

FIGURE 3.2(a) Brinkman parameter effect on axial velocity for $\alpha < 0$.

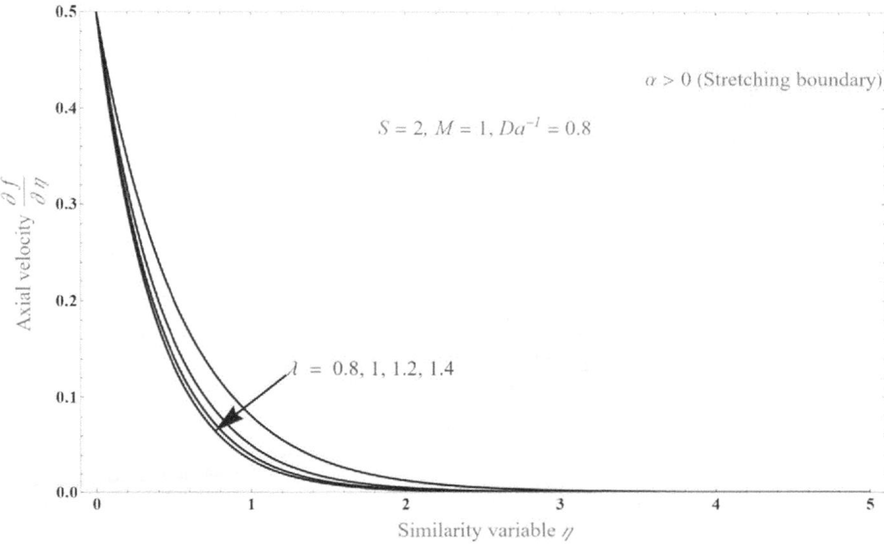

(b)

FIGURE 3.2(b) Brinkman term effect on axial momentum for $\alpha > 0$.

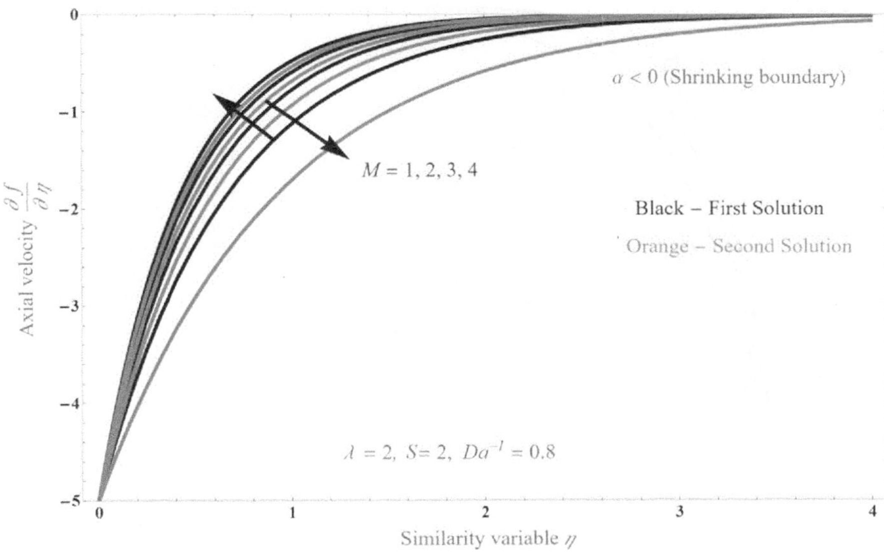

(a)

FIGURE 3.3(a) Magnetic term effect on axial velocity for $\alpha < 0$.

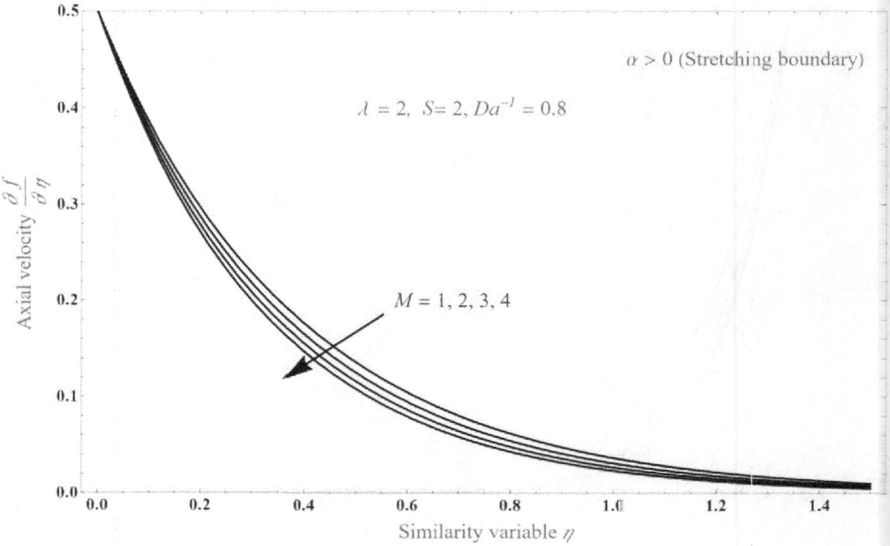

(b)

FIGURE 3.3(b) Magnetic term effect on axial velocity for $\alpha > 0$.

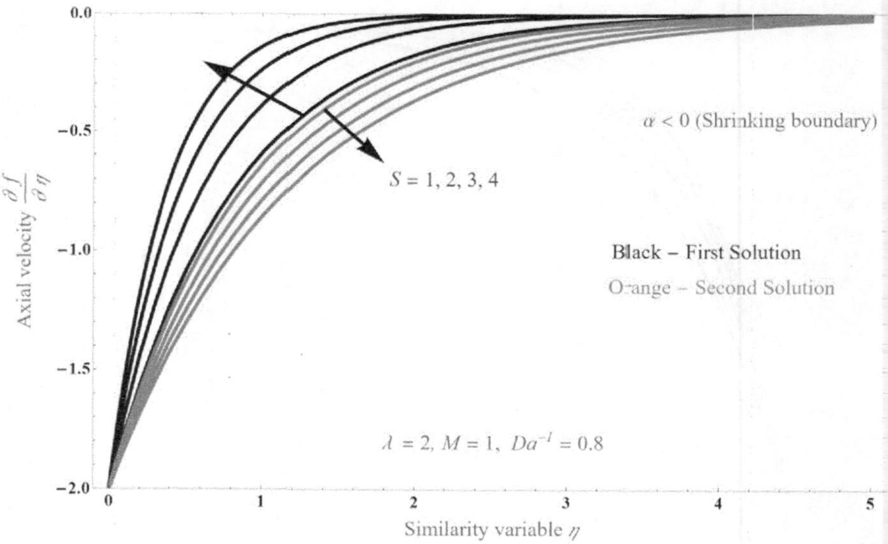

(a)

FIGURE 3.4(a) Mass suction parameter effect on axial velocity for $\alpha < 0$.

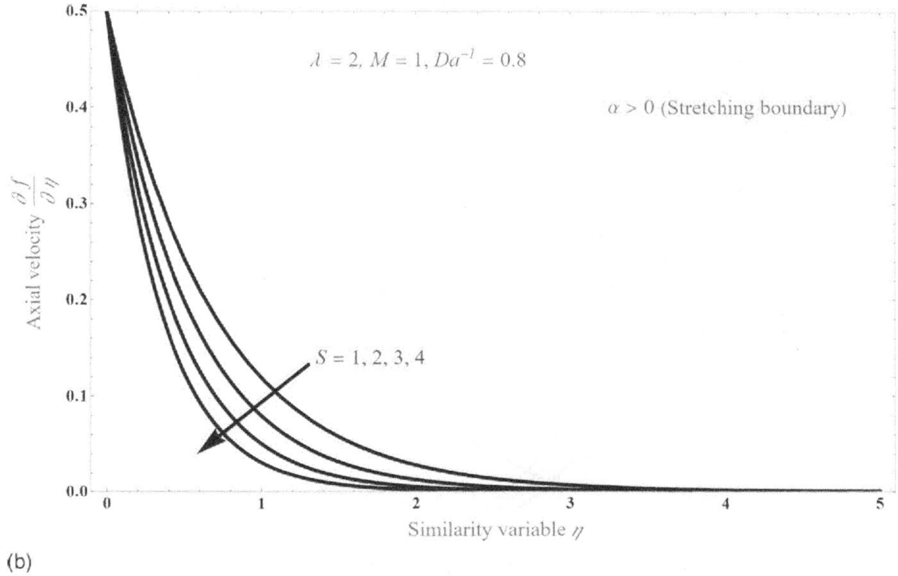

FIGURE 3.4(b) Mass suction parameter effect on axial velocity for $\alpha > 0$.

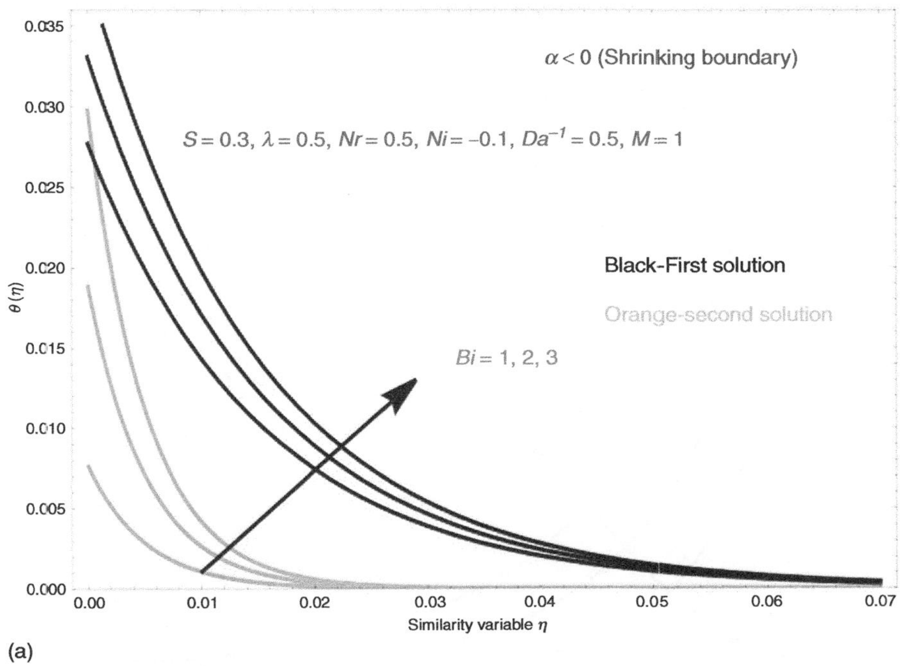

FIGURE 3.5(a) Biot number effect on temperature for $\alpha < 0$.

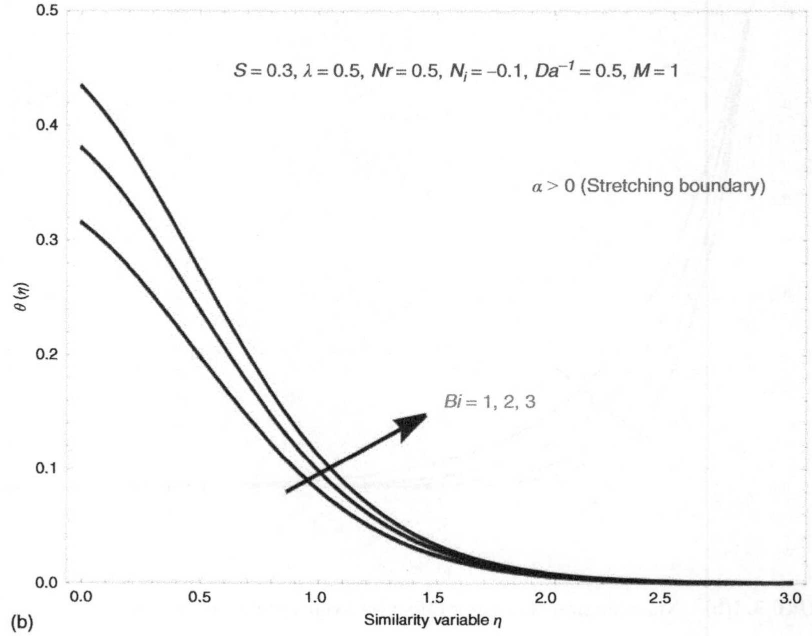

(b)

FIGURE 3.5(b) Biot number effect on temperature for $\alpha > 0$.

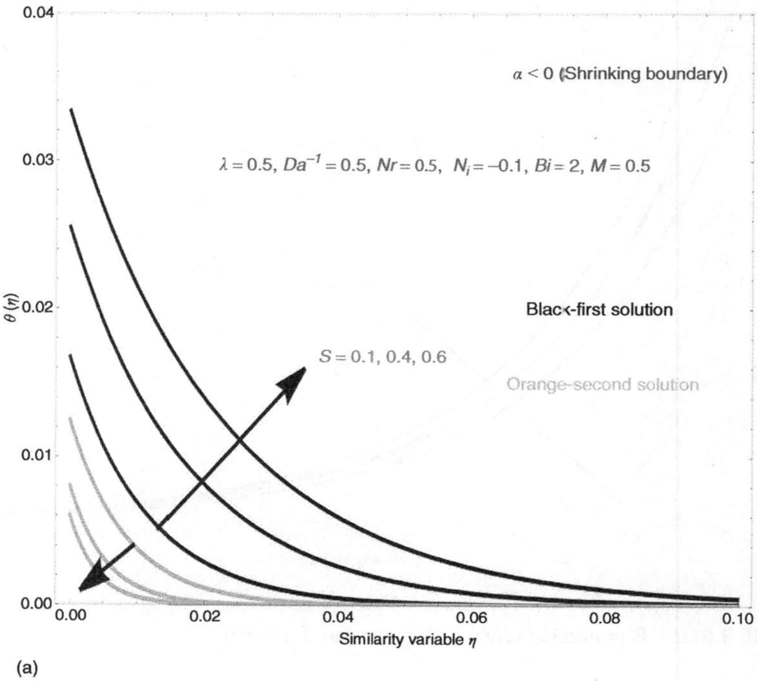

(a)

FIGURE 3.6(a) Mass transpiration effect on temperature for $\alpha < 0$.

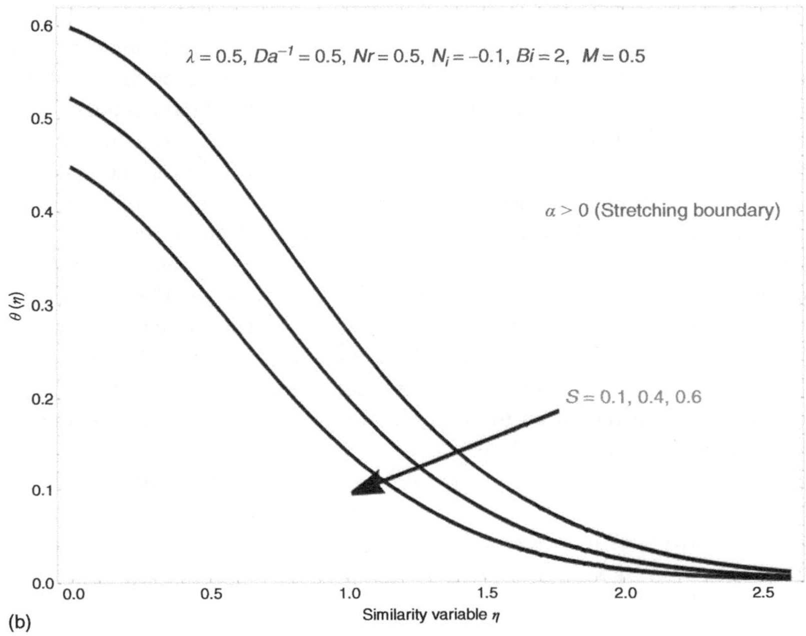

FIGURE 3.6(b) Mass transpiration effect on temperature for $\alpha > 0$.

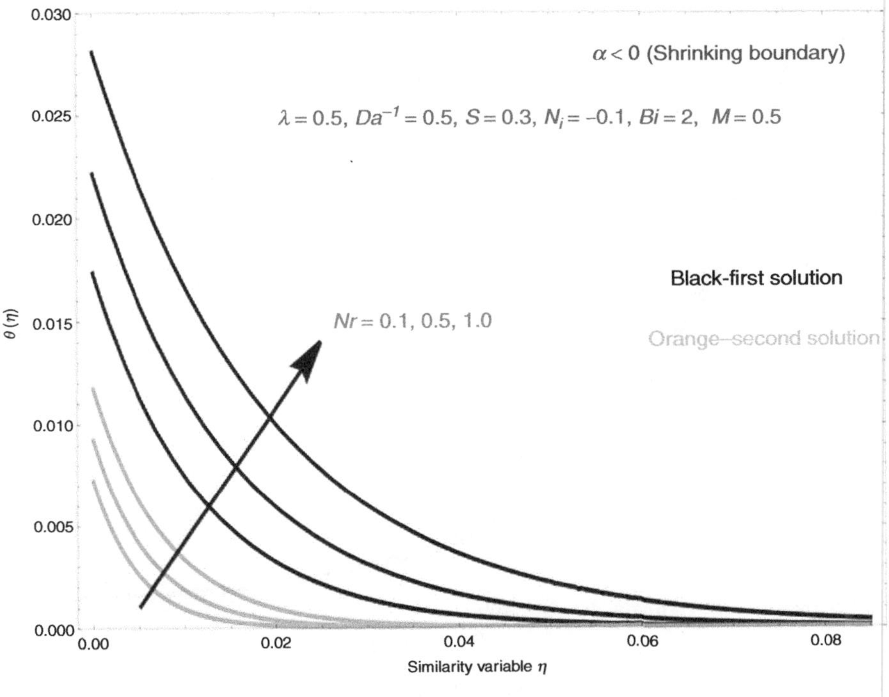

FIGURE 3.7 Thermal radiation effect on temperature.

Figures 3.6(a) and (b) portrays the mass suction parameter effect on temperature for both expanding and shrinking surface, increasing the mass transpiration increases the temperature on shrinking boundary and decreases the stretching boundary of the fluid flow. Physically, mass suction and injection significantly alter fluid temperature profiles by modifying the thermal boundary layer. When suction increases, fluid temperature decreases as ambient fluid is brought closer to the surface, reducing thermal boundary layer thickness. Conversely, injection increases fluid temperature by expanding the thermal boundary layer and introducing warmer fluid towards the surface.

Figure 3.7 represents the radiation term effect on temperature, increasing the thermal radiation increases the energy of the fluid movement. Physically, radiation significantly influences fluid temperature by modifying heat transfer characteristics. As the radiation parameter increases, fluid temperature generally rises, with the energy boundary layer thickness decreasing correspondingly. The radiation impact creates an equivalent increased thermal diffusivity, leading to complex temperature profile changes across the fluid flow.

Figure 3.8 represents the magnetic parameter effect on temperature, rising the magnetic field, increases the energy of the fluid movement. Physically, the magnetic field significantly influences fluid temperature by modifying heat transfer characteristics. As the magnetic field strength rises, the temperature profile generally rises, with thermal boundary layer thickness changing correspondingly. Thermomagnetic

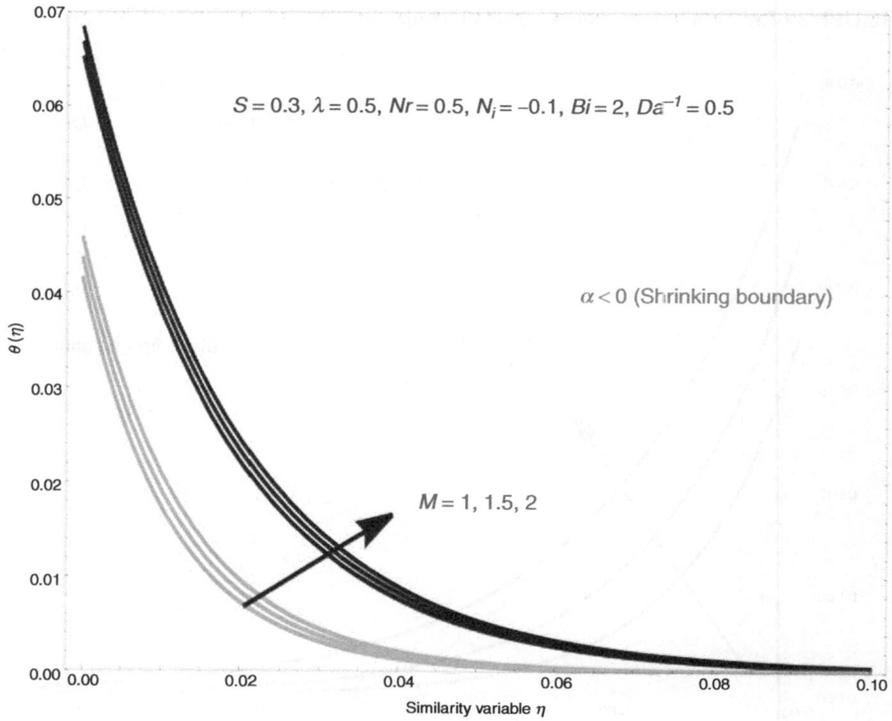

FIGURE 3.8 Magnetic parameter effect on temperature.

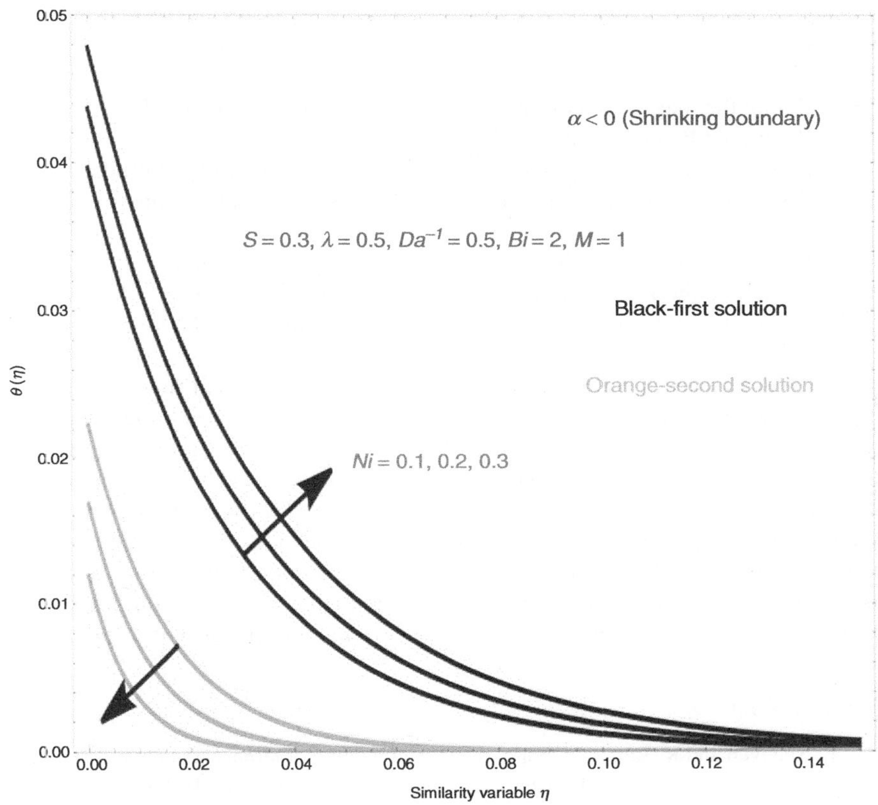

FIGURE 3.9 Heat source/sink term effect on temperature.

convection enables heat flow control, allowing directed heat dissipation along magnetic field patterns.

Figure 3.9 represents the heat source parameter effect on temperature, rising the heat source term, increases the energy of the fluid. Physically, the heat source significantly influences fluid temperature by generating additional thermal energy within the flow field. As the heat source term increases, the energy profile rises, with the fluid experiencing enhanced thermal distribution and increased velocity. The heat source effectively augments temperature by creating more heat energy, leading to a non-uniform temperature gradient across the fluid flow.

3.6 CONCLUSION

The present study examines the influence of couple stress on magnetic hybrid nanofluid movement across expanding permeable surface with heat transfer, considered molybdenum disulfide and graphene oxide nanoparticles dispersed in ethylene glycol to create hybrid nanofluid, governing equations are calculated by using similarity variables, thermal radiation and heat source included in energy equation and solved by using the hypergeometric series solution method, results of the present study are as follows:

- Rising the magnetic term decreases the momentum of the fluid.
- Rising the Brinkman term reduces the velocity of the fluid.
- Rising the mass transpiration term reduces the velocity of the fluid.
- Rising the Biot number parameter increases the temperature of the fluid movement.
- Rising the mass transpiration parameter rises the temperature.
- Rising the thermal radiation term increases the temperature.
- Rising the heat source term increases the temperature of the fluid flow.

The study is limited by assumptions of steady, two-dimensional flow and constant properties, which may not accurately represent practical scenarios involving time-dependent or three-dimensional flows. The impacts of thermal radiation, chemical reactions, and non-uniform magnetic fields are not included, potentially limiting the scope of the results. Future research could address these gaps by considering unsteady and three-dimensional models, variable magnetic field effects, and additional thermophysical phenomena. Experimental validation and advanced computational techniques, such as machine learning, could also be explored to improve the applicability and accuracy of hybrid nanofluid flow models in real-world applications.

REFERENCES

[1] B. Souayeh and K. Ramesh, "Numerical scrutinization of ternary nanofluid flow over an exponentially stretching sheet with gyrotactic microorganisms," *Mathematics*, vol. 11, no. 4, Art. no. 4, Jan. 2023, doi: 10.3390/math11040981.

[2] P. Priyadharshini, M. V. Archana, N. A. Shah, and M. H. Alshehri, "Ternary hybrid nanofluid flow emerging on a symmetrically stretching sheet optimization with machine learning prediction scheme," *Symmetry*, vol. 15, no. 6, Art. no. 6, Jun. 2023, doi: 10.3390/sym15061225.

[3] K. Varatharaj, R. Tamizharasi, and K. Vajravelu, "Ternary hybrid nanofluid flow and heat transfer at a permeable stretching sheet with slip boundary conditions," *Eur. Phys. J. Spec. Top.*, Aug. 2024, doi: 10.1140/epjs/s11734-024-01295-z.

[4] H. D. Akshatha, S. M. Sachhin, U. S. Mahabaleshwar, R. K. Lodhi, and K. Ramesh, "Dynamics of sodium alginate-based ternary nanofluid flow over a stretching sheet with Al2O3, SiO2, and TiO2 nanoparticles," *Multiscale Multidiscip. Model. Exp. Des.*, vol. 8, no. 1, p. 32, Nov. 2024, doi: 10.1007/s41939-024-00591-4.

[5] Y. Ouyang, M. F. Md Basir, K. Naganthran, and I. Pop, "Dual solutions in Maxwell ternary nanofluid flow with viscous dissipation and velocity slip past a stretching/shrinking sheet," *Alex. Eng. J.*, vol. 105, pp. 437–448, Oct. 2024, doi: 10.1016/j.aej.2024.07.093.

[6] S. R. Mishra, G. C. Dash, and M. Acharya, "Mass and heat transfer effect on MHD flow of a visco-elastic fluid through porous medium with oscillatory suction and heat source," *Int. J. Heat Mass Transf.*, vol. 57, no. 2, pp. 433–438, Feb. 2013, doi: 10.1016/j.ijheatmasstransfer.2012.10.053.

[7] Q.-H. Deng, "Fluid flow and heat transfer characteristics of natural convection in square cavities due to discrete source–sink pairs," *Int. J. Heat Mass Transf.*, vol. 51, no. 25, pp. 5949–5957, Dec. 2008, doi: 10.1016/j.ijheatmasstransfer.2008.04.062.

[8] R. Tsai, K. H. Huang, and J. S. Huang, "Flow and heat transfer over an unsteady stretching surface with non-uniform heat source," *Int. Commun. Heat Mass Transf.*, vol. 35, no. 10, pp. 1340–1343, Dec. 2008, doi: 10.1016/j.icheatmasstransfer.2008.07.001.

[9] Y. Li, Y.-H. Feng, X.-X. Zhang, and C.-S. Wu, "An improved simulation of heat transfer and fluid flow in plasma arc welding with modified heat source model," *Int. J. Therm. Sci.*, vol. 64, pp. 93–104, Feb. 2013, doi: 10.1016/j.ijthermalsci.2012.08.007.

[10] M. A. Kumar, Y. D. Reddy, V. S. Rao, and B. S. Goud, "Thermal radiation impact on MHD heat transfer natural convective nano fluid flow over an impulsively started vertical plate," *Case Stud. Therm. Eng.*, vol. 24, p. 100826, Apr. 2021, doi: 10.1016/j.csite.2020.100826.

[11] M. Turkyilmazoglu, "Thermal radiation effects on the time-dependent MHD permeable flow having variable viscosity," *Int. J. Therm. Sci.*, vol. 50, no. 1, pp. 88–96, Jan. 2011, doi: 10.1016/j.ijthermalsci.2010.08.016.

[12] J. C. Misra and A. Sinha, "Effect of thermal radiation on MHD flow of blood and heat transfer in a permeable capillary in stretching motion," *Heat Mass Transf.*, vol. 49, no. 5, pp. 617–628, May 2013, doi: 10.1007/s00231-012-1107-6.

[13] B.-W. Li, W. Wang, and J.-K. Zhang, "Combined effects of magnetic field and thermal radiation on fluid flow and heat transfer of mixed convection in a vertical cylindrical annulus," *J. Heat Transf.*, vol. 138, no. 6, pp. 062501, Mar. 2016, doi: 10.1115/1.4032609.

[14] M. R. Eid and M. A. Nafe, "Thermal conductivity variation and heat generation effects on magneto-hybrid nanofluid flow in a porous medium with slip condition," *Waves Random Complex Media*, vol. 32, no. 3, pp. 1103–1127, May 2022, doi: 10.1080/17455030.2020.1810365.

[15] J. Wang *et al.*, "Simulation of hybrid nanofluid flow within a microchannel heat sink considering porous media analyzing CPU stability," *J. Pet. Sci. Eng.*, vol. 208, p. 109734, Jan. 2022, doi: 10.1016/j.petrol.2021.109734.

[16] M. V. Krishna, N. A. Ahammad, and A. J. Chamkha, "Radiative MHD flow of Casson hybrid nanofluid over an infinite exponentially accelerated vertical porous surface," *Case Stud. Therm. Eng.*, vol. 27, p. 101229, Oct. 2021, doi: 10.1016/j.csite.2021.101229.

[17] I. Waini, A. Ishak, T. Groşan, and I. Pop, "Mixed convection of a hybrid nanofluid flow along a vertical surface embedded in a porous medium," *Int. Commun. Heat Mass Transf.*, vol. 114, p. 104565, May 2020, doi: 10.1016/j.icheatmasstransfer.2020.104565.

[18] U. S. Mahabaleshwar, S. M. Sachhin, L. M. Pérez, and H. F. Oztop, "An impact of inclined MHD on biviscosity Bingham hybrid nanofluid flow over porous stretching/shrinking sheet with heat transfer," *Journal of Molecular Liquids*, vol. 398, p. 124244, Mar. 2024, doi: 10.1016/j.molliq.2024.124244.

4 Jeffrey Fluid Flow with Heat and Mass Transfer

Sudha Mahanthesh Sachhin, Gadhigeppa Myacher Sachin, Ulavathi Shettar Mahabaleshwar, Aishwarya Anand, Tanu Suraddihalli Murugesh, and Basma Souayeh

4.1 INTRODUCTION

A stretching surface refers to a continuously extending boundary, which is essential for optimizing manufacturing processes, enhancing heat transfer efficiency, and improving product quality. They are widely used in aerodynamic studies, biomedical engineering, and cooling systems. Mathematical modeling of stretching surfaces helps analyze boundary layer behavior, ensuring precise control in various technological applications. Cortell [1] studied the heat transfer and temperature conditions on viscous fluid flow over effect on a nonlinear expanding surface. Dutta *et al.* [2] examined the Prandtl number and surface heat flux influence on the incompressible viscous fluid flow over stretching of the surface. Chakrabarti and Gupta [3] studied the magnetic field influence on the fluid flow across an expanding surface. Makinde and Aziz [4] examined the Brownian motion impact on nanofluid movement across a linearly expanding sheet.

A heat source/sink refers to a region in a system where heat is either generated (source) or absorbed (sink), affecting temperature distribution. It plays a vital role in thermal management, influencing energy efficiency and system stability. Heat sources are found in combustion engines, electronic devices, and nuclear reactors. Proper analysis of heat sources and sinks is crucial in designing efficient heat exchangers, insulation materials, and industrial processes. Their study helps to improve thermal performance in aerospace, manufacturing, and biomedical applications, ensuring optimal temperature regulation. Bataller [5] examined the heat generation impact on second grade fluid movement across an isothermal expanding sheet. Jalili *et al.* [6] examined the impact of heat source on non-Newtonian fluid movement across an expanding surface. Abel and Mahesha [7] explored the heat transfer impact on visco-elastic fluid flow across a flat surface with magnetic field. Deng [8] examined the numerical analysis of the heat source effect on natural convective fluid flow inside a square cavity.

 DOI: 10.1201/9781003611820-5

A Jeffery fluid behavior is crucial in optimizing processes like extrusion, lubrication, and biomedical fluid transport. It is widely applied in blood flow analysis, food processing, and chemical industries, where accurate viscosity and elasticity control are needed. Researchers use Jeffery fluid models to enhance efficiency in engineering applications involving complex fluids. Dadhich *et al.* [9] examined the mass transfer radiated Jeffery fluid movement across linear expand of the surface. Nagler [10] studied the wall friction impact on non-Newtonian fluid flow inside a wedge using the Jaffery-Hamel model. Ellahi *et al.* [11] examined the Hall current effect on Jeffery peristaltic movement of fluid across a rectangular duct. Jena *et al.* [12] studied the magnetic field impact on Jeffrey fluid flow across porous media using the hypergeometric approach.

The Schmidt number plays a crucial role in fluid dynamics and mass transfer studies, particularly in predicting how substances mix in a fluid. A high Schmidt number indicates slower diffusion, which is important in chemical engineering, environmental studies, and combustion processes. It helps in designing efficient mixing equipment, analyzing pollutant dispersion, and optimizing industrial gas–liquid interactions. The Schmidt number is widely used in heat exchangers, biomedical engineering, and aerospace applications to improve mass transfer efficiency and process control. Hanazaki *et al.* [13] explored the Schmidt number influence on fluid flow across a vertically moving sphere. Scalo *et al.* [14] examined the Schmidt number impact on fluid flow with mass transfer and wall shear interface. The authors of [15–23] examined the heat transfer influence on fluid flow across various boundaries.

4.2 MATHEMATICAL FORMULATION

The rheological equations of Jeffrey fluid are [9, 10, 11]

$$\tau = -\rho I + S$$

$$S = \frac{\mu}{1+\Gamma}\left[M_1 + \Gamma_1\left(\frac{\partial M_1}{\partial t} + V.\nabla\right)M_1\right]$$

where μ is the dynamic viscosity, Γ and Γ_1 are the material parameters, τ is the Cauchy stress tensor, and the Rivlin–Ericksen tensor $M_1 = (\nabla V) + (\nabla V)^1$.

The governing equations are [2, 6, 9, 12]

$$\frac{\partial u}{\partial x} + \frac{\partial v}{\partial y} = 0 \tag{4.1}$$

$$u\frac{\partial u}{\partial x} + v\frac{\partial v}{\partial y} = \frac{\nu}{1+\Gamma}\left[\frac{\partial^2 u}{\partial y^2} + \Gamma_1\left(u\frac{\partial^3 u}{\partial x \partial y^2} + v\frac{\partial^3 u}{\partial y^3} - \frac{\partial u}{\partial x}\frac{\partial^2 u}{\partial y^2} + \frac{\partial u}{\partial y}\frac{\partial^2 u}{\partial x \partial y}\right)\right], \tag{4.2}$$

$$\rho Cp\left(u\frac{\partial T}{\partial x} + v\frac{\partial T}{\partial y}\right) = \kappa\frac{\partial^2 T}{\partial y^2} + Q(T - T_\infty), \tag{4.3}$$

$$u\frac{\partial C}{\partial x}+\frac{\partial C}{\partial y}=D_b\frac{\partial^2 C}{\partial y^2}, \tag{4.4}$$

The boundary conditions are as follows [2, 6, 9, 12]:

$$\begin{aligned} &u=U_w(x)=ax, \quad v=0, &&\text{at} \quad y=0,\\ &u\to 0, &&\text{as} \quad y\to\infty, \end{aligned} \tag{4.5}$$

$$\begin{aligned} &T=T_w=T_\infty+B_1\left(\frac{x}{l}\right)^n, \quad C=C_w=C_\infty+B_2\left(\frac{x}{l}\right)^n &&\text{at} \quad y=0,\\ &T\to T_\infty, \qquad C\to C_\infty, &&\text{as} \quad y\to\infty. \end{aligned} \tag{4.6}$$

Similarity variables [2, 6, 9, 12] are

$$\begin{aligned} &\eta=\sqrt{\frac{a}{v}}y, \qquad u=axf'(\eta), \qquad v=-\sqrt{av}f(\eta),\\ &\theta(\eta)=\frac{T-T_\infty}{T_w-T_\infty}, \qquad \varphi(\eta)=\frac{C-C_\infty}{C_w-C_\infty}. \end{aligned} \tag{4.7}$$

The modified governing equations are as follows:

$$f_{\eta\eta\eta}+(1+\Gamma)\left(ff_{\eta\eta}-f_\eta{}^2\right)+\beta\left(f_{\eta\eta}{}^2-ff_{\eta\eta\eta\eta}\right)=0, \tag{4.8}$$

$$\theta_{\eta\eta}+\Pr\left(f\theta_\eta-n\theta f_\eta+\gamma\theta\right)=0, \tag{4.9}$$

$$\varphi_{\eta\eta}+Sc(f\varphi_\eta-n\varphi f_\eta)=0 \tag{4.10}$$

Modified boundary conditions are

$$\begin{aligned} &f(0)=0, \quad f'(0)=1, \quad \theta(0)=1, \quad \varphi(0)=1,\\ &f'(\infty)\to 0, \quad \theta(\infty)\to 0, \qquad \varphi(\infty)\to 0. \end{aligned} \tag{4.11}$$

Non-dimensional parameters are

$$\beta=\Gamma_1 a, \quad \Pr=\frac{\mu Cp}{\kappa}, \quad Sc=\frac{v}{D_b}, \quad \gamma=\frac{Q}{a\rho Cp}. \tag{4.12}$$

4.3 ANALYTICAL SOLUTIONS

The olution for the momentum equation by boundary conditions is defined as

$$f(\eta) = \frac{1-\exp(-\alpha\eta)}{\alpha}, \qquad f_\eta(\eta) = \exp(-\alpha\eta), \tag{4.13}$$

where

$$\alpha = \pm\sqrt{\frac{1+\Gamma}{1+\beta}}. \tag{4.14}$$

The analytical solution of Eqs. (4.2) are given as

$$u = ax\exp(-\alpha\eta)\ , v = -\sqrt{av}\frac{1-\exp(-\alpha\eta)}{\alpha}, \tag{4.15}$$

using the above equation temperature and concentration equation defined as

$$\theta_{\eta\eta} + \Pr\left(\frac{1-\exp(-\alpha\eta)}{\alpha}\right)\theta_\eta - n\Pr\theta\exp(-\alpha\eta) + \gamma\theta = 0, \tag{4.16}$$

$$\varphi_{\eta\eta} + Sc\left(\frac{1-\exp(-\alpha\eta)}{\alpha}\right)\varphi_\eta - nSc\,\varphi\exp(-\alpha\eta) = 0. \tag{4.17}$$

To obtain the solution of Eqs. (4.16) and (4.17), respectively, we introduce new variables ζ and ξ as follows:

$$\zeta = \frac{\Pr}{\alpha^2}\exp(-\alpha\eta), \xi = \frac{Sc}{\alpha^2}\exp(-\alpha\eta) \tag{4.18}$$

Adding Eq. (4.18) into Eqs. (4.16) and (4.17), we get

$$\zeta\theta_{\eta\eta}(\zeta) + \left(1 - \frac{\Pr}{\alpha^2} + \zeta\right)\theta_\eta(\zeta) - \left(n - \frac{\gamma}{\alpha^2\zeta}\right)\theta(\zeta) = 0, \tag{4.19}$$

$$\xi\varphi_{\eta\eta}(\zeta) + \left(1 - \frac{Sc}{\alpha^2} + \xi\right)\varphi_\eta(\xi) - n\varphi(\xi) = 0. \tag{4.20}$$

Modified B. Cs are

$$\begin{aligned} \theta\left(\zeta = \frac{\Pr}{\alpha^2}\right) &= 1, \qquad \theta(\zeta = 0) = 0,\\ \varphi\left(\xi = \frac{Sc}{\alpha^2}\right) &= 1, \qquad \varphi(\xi = 0) = 0. \end{aligned} \tag{4.21}$$

The analytical solutions for temperature and concentration are

$$\begin{aligned} \theta(\zeta) &= \left(\frac{\alpha^2}{\Pr}\zeta\right)^{K_1+K_2} \frac{{}_1H_1\left[K_1+K_2-n, 2K_1+1, -\zeta\right]}{{}_1H_1\left[K_1+K_2-n, 2K_1+1, -\dfrac{\Pr}{\alpha^2}\right]}, \\ \varphi(\xi) &= \left(\frac{\alpha^2}{\Pr}\xi\right)^{\frac{Sc}{\alpha^2}} \frac{{}_1H_1\left[\dfrac{Sc}{\alpha^2}-n, 2\dfrac{Sc}{\alpha^2}+1, -\xi\right]}{{}_1H_1\left[\dfrac{Sc}{\alpha^2}-n, 2\dfrac{Sc}{\alpha^2}+1, -\dfrac{Sc}{\alpha^2}\right]}, \end{aligned} \tag{4.22}$$

with $K_1 = \frac{\Pr}{\alpha^2}$, $K_2 = \frac{\sqrt{\Pr^2-4\alpha^2\gamma}}{2\alpha^2}$ and ${}^1H_1(a,b,z)$ is Kummer's function written as

$$_1H_1(a,b,z) = 1 + \sum_{\eta=1}^{\infty}\left(\frac{(a)_m}{(b)_m}\frac{z^m}{n!}\right), \tag{4.23}$$

Here (a_n) and (b_n) are Pochhammer's symbols defined by

$$\begin{aligned} (a)_m &= a(a+1)(a+2)\ldots(a+m-1), \\ (b)_m &= b(b+1)(b+2)\ldots(b+m-1). \end{aligned} \tag{4.24}$$

In terms of η, it is written as

$$\begin{aligned} \theta(\eta) &= \exp\left[-\alpha(K_1+K_2)\eta\right] \times \frac{{}_1H_1\left[K_1+K_2-n, 2K_1+1, -\dfrac{\Pr}{\alpha^2}\exp(-\alpha\eta)\right]}{{}_1H_1\left[K_1+K_2-n, 2K_1+1, -\dfrac{\Pr}{\alpha^2}\right]}, \\ \varphi(\eta) &= \exp\left(\frac{Sc}{\alpha^2}\right) \frac{{}_1H_1\left[\dfrac{Sc}{\alpha^2}-n, 2\dfrac{Sc}{\alpha^2}+1, -\dfrac{Sc}{\alpha^2}\exp(-\alpha\eta)\right]}{{}_1H_1\left[\dfrac{Sc}{\alpha^2}-n, 2\dfrac{Sc}{\alpha^2}+1, -\dfrac{Sc}{\alpha^2}\right]}. \end{aligned} \tag{4.25}$$

4.4 RESULT AND DISCUSSION

The impact of a heat source and Schmidt number on Jeffrey fluid movement over an expanding surface is crucial in controlling heat and mass transfer in industrial and engineering applications. Heat sources influence temperature distribution, enhancing or reducing thermal efficiency, while the Schmidt number determines mass diffusion rates. Governing PDEs are calculated to ODEs via similarity terms, the analytical hypergeometric approach and Kummer's function are utilized to solve the temperature and concentration equations, Figure 4.1 represents the schematic diagram, several parameters, such as Schmidt number, are examined through graphs in the below section.

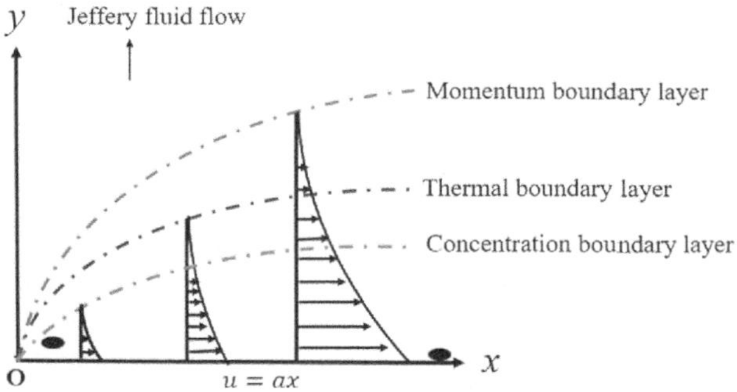

FIGURE 4.1 Schematic diagram.

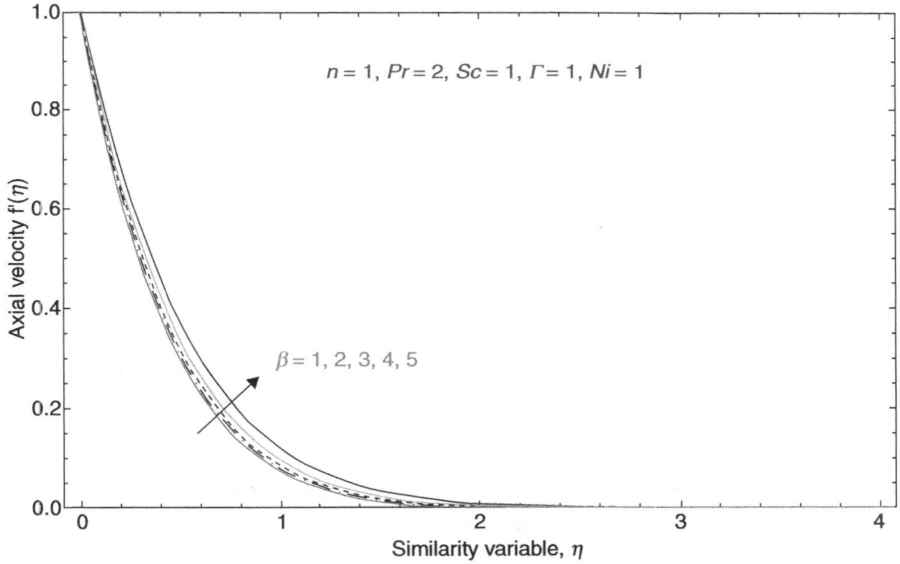

FIGURE 4.2 β influence on axial velocity.

Figure 4.2 represents the β influence on the $f'(\eta)$, increasing the β increases the velocity, β is a dimensionless term that quantifies the fluid's elasticity by comparing the relaxation time of a material to the characteristic time of a process. Its influence on velocity in fluid flow is significant, Deborah number indicates stronger elastic effects, causing resistance to deformation and enhancing the velocity near the stretching surface.

Figure 4.3 portrays the impact of Γ on velocity, enhancing the Γ decays the velocity of the flow. Γ significantly affects the velocity of non-Newtonian fluids, such as Jeffrey fluid, over a stretching sheet. When this ratio increases, the fluid's elasticity dominates, leading to higher resistance to deformation, which reduces the velocity near the surface.

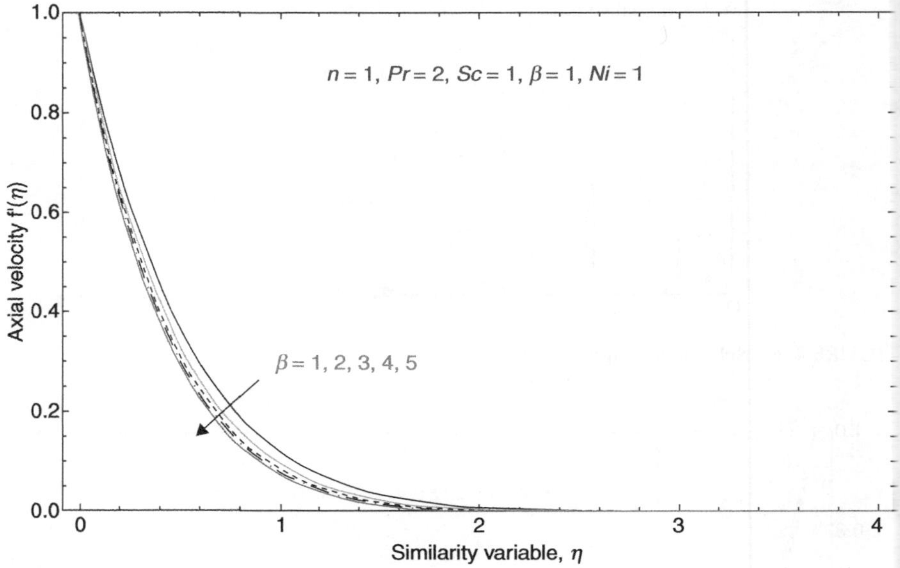

FIGURE 4.3 Γ influence on velocity.

Figure 4.4 represents the *Sc* influence on $\varphi(\eta)$, enhancing the *Sc* decays the concentration. The *Sc* directly influences the $\varphi(\eta)$ profile in mass transfer processes by determining the relative strength of momentum diffusivity to mass diffusivity, a lower Schmidt number enhances mass diffusion, resulting in a more uniform concentration distribution. This effect is critical in chemical engineering, environmental studies, and biomedical applications, where controlling mass transfer efficiency is essential for processes like pollutant dispersion, drug delivery, and industrial mixing Figures 4.5(a) and (b) represents the impact of term *n* on both temperature and concentration profile, enhancing the parameter *n* reduces both $\varphi(\eta)$ and the temperature profile.

Figure 4.6 portrays the impact of the heat source on the $\theta(\eta)$ profile, enhancing the heat source term increases the $\theta(\eta)$ of the fluid flow. A heat source significantly affects the temperature distribution in a fluid flow by supplying thermal energy, which increases the overall temperature of the system. A stronger heat source leads to higher thermal conductivity, enhancing heat transfer and creating a steeper temperature gradient. This can accelerate fluid motion due to thermal buoyancy effects. Conversely, a weaker heat source results in slower heat diffusion, maintaining a lower temperature. Understanding this influence is crucial in industrial applications like cooling systems, polymer processing, and energy generation, where precise thermal control is essential for efficiency and performance optimization.

Figure 4.7 represents the effect of Prandtl number on $\theta(\eta)$, enhancing the Prandtl number parameter decays $\theta(\eta)$, a lower Prandtl number enhances thermal diffusivity,

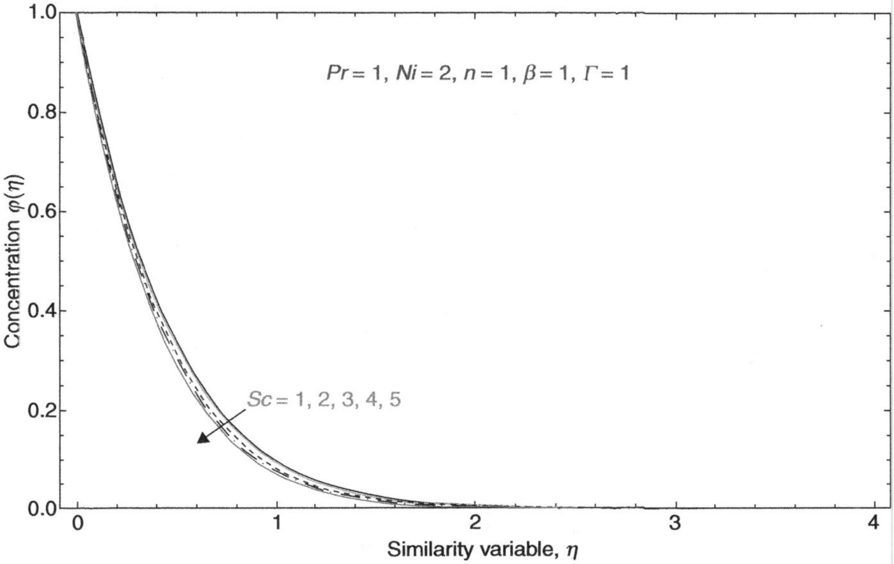

FIGURE 4.4 Schmidt number influence on concentration.

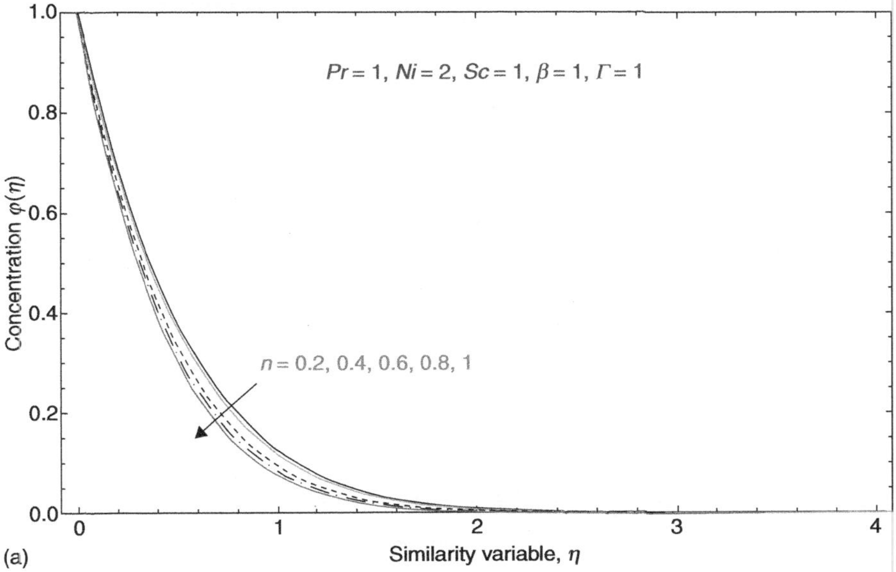

FIGURE 4.5(a) Parameter n influence on concentration.

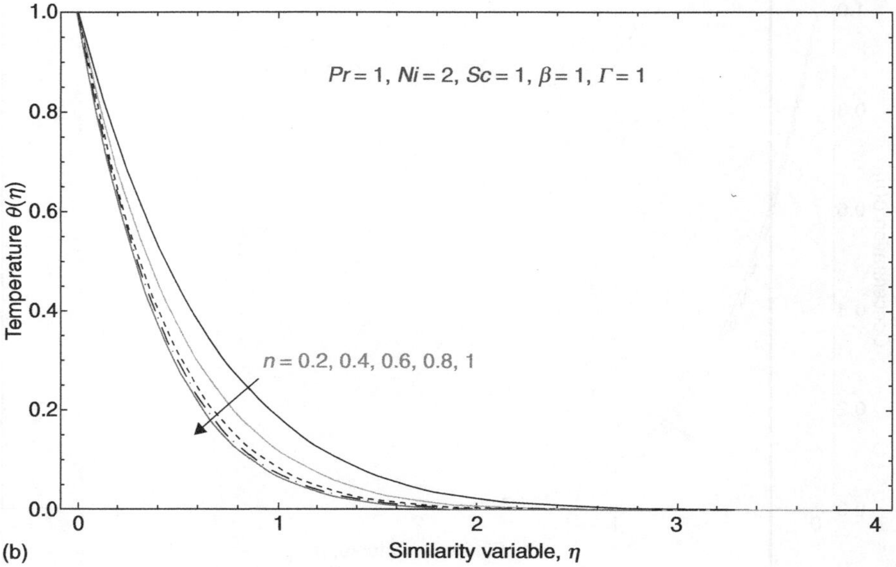

FIGURE 4.5(b) Parameter n influence on temperature.

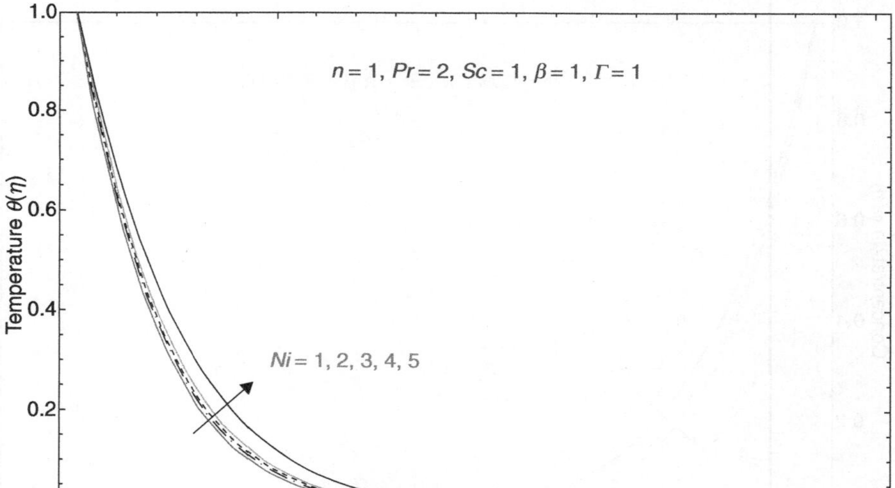

FIGURE 4.6 Heat source impact on $\theta(\eta)$ of the fluid flow.

allowing heat to spread more efficiently and leading to a more uniform temperature distribution. This influence is vital in heat exchangers, aerodynamic heating, and industrial cooling systems, where controlling thermal diffusion is essential for performance and efficiency.

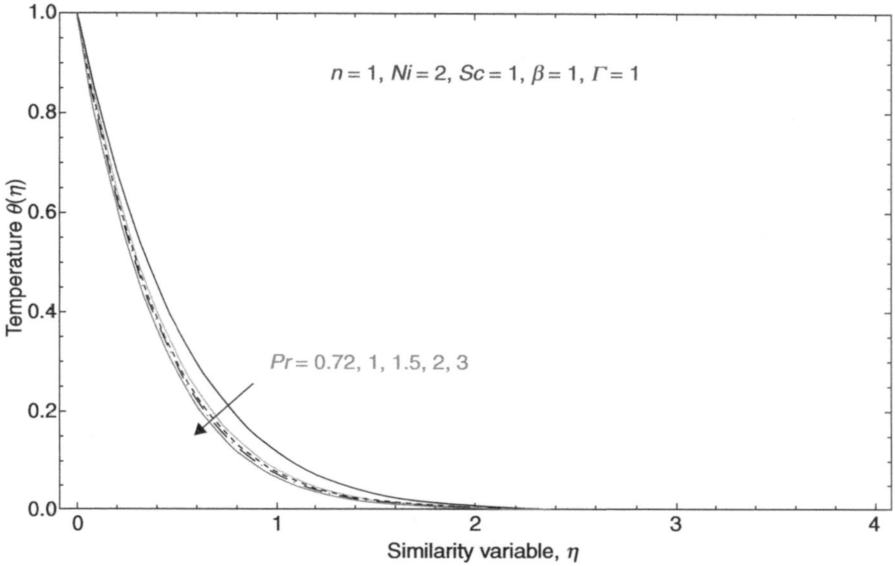

FIGURE 4.7 Prandtl number impact on $\theta\left(\eta\right)$.

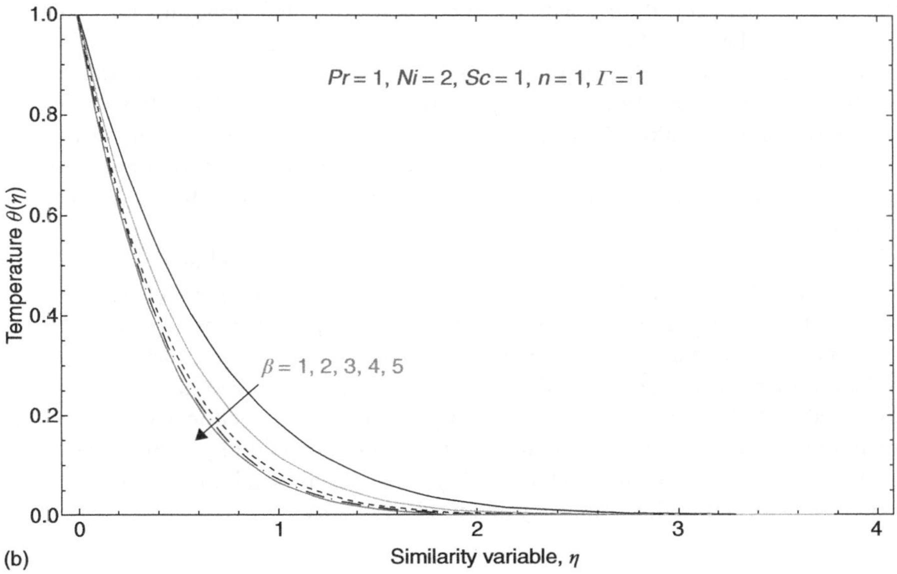

FIGURE 4.8 Deborah number impact on $\theta\left(\eta\right)$.

Figure 4.8 portrays the Deborah number influence on temperature of the fluid flow, enhancing the Deborah number decays $\theta\left(\eta\right)$. The Deborah number influences the temperature distribution in viscoelastic fluid flows by determining

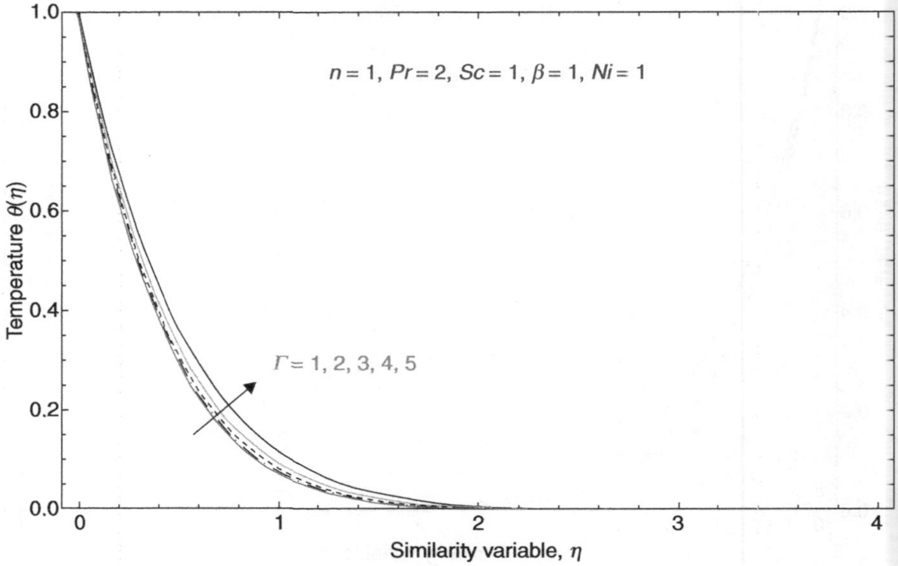

FIGURE 4.9 Γ impact on temperature of the fluid flow.

the balance between fluid elasticity and the applied deformation rate. A higher Deborah number signifies greater fluid elasticity, which resists deformation and reduces heat transfer efficiency, leading to increased temperature retention within the fluid. Conversely, a lower Deborah number allows the fluid to behave more like a Newtonian fluid, enhancing thermal diffusion and lowering the temperature gradient. This effect is crucial in polymer processing, biomedical fluid applications, and industrial heat management, where controlling thermal behavior is essential for process efficiency and stability.

Figure 4.9 represents the Γ impact on temperature of the fluid flow, enhancing the Γ increases the temperature of the fluid flow. The Γ significantly affects the temperature distribution in viscoelastic fluid flows by altering heat transfer characteristics. A higher ratio enhances fluid elasticity, increasing resistance to deformation and reducing thermal diffusivity, leading to higher temperature retention. Conversely, a lower ratio weakens elastic effects, allowing better heat conduction and resulting in a more uniform temperature distribution. This factor is crucial in polymer processing, biofluid dynamics, and industrial thermal systems, where precise temperature control is essential for improving efficiency and product quality.

4.5 CONCLUSION

The impact of a heat source and Schmidt number on Jeffrey fluid movement over an expanding surface is crucial in controlling heat and mass transfer in industrial and engineering applications. Heat sources influence temperature distribution, enhancing or reducing thermal efficiency, while the Schmidt number determines mass diffusion rates. Governing PDEs are calculated to ODEs via similarity terms, the analytical hypergeometric approach and Kummer's function are utilized to solve the temperature and concentration equations. Several parameters, such as Schmidt number, have been examined. The results of the current work reveal the following:

- Enhancing the heat source term increases the temperature.
- Increasing the Prandtl number decays the temperature.
- Increasing the Deborah number increases the momentum and decays the temperature.
- Enhancing the Γ parameter increasess the temperature and decays the velocity.

This study is limited to steady, laminar conditions with constant liquid properties, which may not fully capture real-world complexities, such as turbulence, unsteadiness, and variable thermal conductivity. Additionally, external effects like magnetic fields and chemical reactions are not considered, restricting broader applicability. Future research can extend the analysis to unsteady and turbulent flows, incorporate variable liquid properties, and explore the impact of external forces like radiation and magnetic fields, enhancing its relevance in biomedical, industrial, and aerospace applications.

REFERENCES

[1] R. Cortell, "Viscous flow and heat transfer over a nonlinearly stretching sheet," *Appl. Math. Comput.*, vol. 184, no. 2, pp. 864–873, Jan. 2007, doi: 10.1016/j.amc.2006.06.077.

[2] B. K. Dutta, P. Roy, and A. S. Gupta, "Temperature field in flow over a stretching sheet with uniform heat flux," *Int. Commun. Heat Mass Transf.*, vol. 12, no. 1, pp. 89–94, Jan. 1985, doi: 10.1016/0735-1933(85)90010-7.

[3] A. Chakrabarti and A. S. Gupta, "Hydromagnetic flow and heat transfer over a stretching sheet," *Q. Appl. Math.*, vol. 37, no. 1, pp. 73–78, Apr. 1979, doi: 10.1090/qam/99636.

[4] O. D. Makinde and A. Aziz, "Boundary layer flow of a nanofluid past a stretching sheet with a convective boundary condition," *Int. J. Therm. Sci.*, vol. 50, no. 7, pp. 1326–1332, Jul. 2011, doi: 10.1016/j.ijthermalsci.2011.02.019.

[5] R. C. Bataller, "Viscoelastic fluid flow and heat transfer over a stretching sheet under the effects of a non-uniform heat source, viscous dissipation and thermal radiation," *Int. J. Heat Mass Transf.*, vol. 50, no. 15, pp. 3152–3162, Jul. 2007, doi: 10.1016/j.ijheatmasstransfer.2007.01.003.

[6] P. Jalili, S. M. Sharif Mousavi, B. Jalili, P. Pasha, and D. D. Ganji, "Thermal evaluation of MHD Jeffrey fluid flow in the presence of a heat source and chemical reaction," *Int. J. Mod. Phys. B*, vol. 38, no. 08, p. 2450113, Mar. 2024, doi: 10.1142/S0217979224501133.

[7] M. S. Abel and N. Mahesha, "Heat transfer in MHD viscoelastic fluid flow over a stretching sheet with variable thermal conductivity, non-uniform heat source and radiation," *Appl. Math. Model.*, vol. 32, no. 10, pp. 1965–1983, Oct. 2008, doi: 10.1016/j.apm.2007.06.038.

[8] Q.-H. Deng, "Fluid flow and heat transfer characteristics of natural convection in square cavities due to discrete source–sink pairs," *Int. J. Heat Mass Transf.*, vol. 51, no. 25, pp. 5949–5957, Dec. 2008, doi: 10.1016/j.ijheatmasstransfer.2008.04.062.

[9] Y. Dadhich, R. Jain, A. R. Kaladgi, M. Alwetaishi, A. Afzal, and C. A. Saleel, "Thermally radiated jeffery fluid flow with nanoparticles over a surface of varying

thickness in the influence of heat source," *Case Stud. Therm. Eng.*, vol. 28, p. 101549, Dec. 2021, doi: 10.1016/j.csite.2021.101549.

[10] J. Nagler, "Jeffery-Hamel flow of non-Newtonian fluid with nonlinear viscosity and wall friction," *Appl. Math. Mech.*, vol. 38, no. 6, pp. 815–830, Jun. 2017, doi: 10.1007/s10483-017-2206-8.

[11] R. Ellahi, M. M. Bhatti, and I. Pop, "Effects of hall and ion slip on MHD peristaltic flow of Jeffrey fluid in a non-uniform rectangular duct," *Int. J. Numer. Methods Heat And Fluid Flow*, vol. 26, no. 6, pp. 1802–1820, Aug. 2016, doi: 10.1108/HFF-02-2015-0045.

[12] S. Jena, S. R. Mishra, and G. C. Dash, "Chemical reaction effect on MHD Jeffery fluid flow over a stretching sheet through porous media with heat generation/absorption," *Int. J. Appl. Comput. Math.*, vol. 3, no. 2, pp. 1225–1238, Jun. 2017, doi: 10.1007/s40819-016-0173-8.

[13] H. Hanazaki, K. Konishi, and T. Okamura, "Schmidt-number effects on the flow past a sphere moving vertically in a stratified diffusive fluid," *Phys. Fluids*, vol. 21, no. 2, p. 026602, Feb. 2009, doi: 10.1063/1.3075953.

[14] C. Scalo, U. Piomelli, and L. Boegman, "High-Schmidt-number mass transport mechanisms from a turbulent flow to absorbing sediments," *Phys. Fluids*, vol. 24, no. 8, p. 085103, Aug. 2012, doi: 10.1063/1.4739064.

[15] P. Ma, X. Li, and D. N. Ku, "Heat and mass transfer in a separated flow region for high Prandtl and Schmidt numbers under pulsatile conditions," *Int. J. Heat Mass Transf.*, vol. 37, no. 17, pp. 2723–2736, Nov. 1994, doi: 10.1016/0017-9310(94)90389-1.

[16] C. P. Koutsou, S. G. Yiantsios, and A. J. Karabelas, "A numerical and experimental study of mass transfer in spacer-filled channels: Effects of spacer geometrical characteristics and Schmidt number," *J. Membr. Sci.*, vol. 326, no. 1, pp. 234–251, Jan. 2009, doi: 10.1016/j.memsci.2008.10.007.

[17] G. Bognár, "On similarity solutions of boundary layer problems with upstream moving wall in non-Newtonian power-law fluids," *IMA J. Appl. Math.*, vol. 77, no. 4, pp. 546–562, Jun. 2012, doi: 10.1093/imamat/hxr033.

[18] S. A. Lone, F. Ali, A. Saeed, and G. Bognár, "Irreversibility analysis with hybrid cross nanofluid of stagnation point and radiative flow (TiO2+CuO) based on engine oil past a stretchable sheet," *Heliyon*, vol. 9, no. 4, p. e15056, Apr. 2023, doi: 10.1016/j.heliyon.2023.e15056.

[19] I. F. Barna, G. Bognár, M. Guedda, L. Mátyás, and K. Hriczó, "Analytic self-similar solutions of the Kardar-Parisi-Zhang interface growing equation with various noise terms," *Math. Model. Anal.*, vol. 25, no. 2, pp. 241–256, Mar. 2020, doi: 10.3846/mma.2020.10459.

[20] M. Klazly and G. Bognár, "A novel empirical equation for the effective viscosity of nanofluids based on theoretical and empirical results," *Int. Commun. Heat Mass Transf.*, vol. 135, p. 106054, Jun. 2022, doi: 10.1016/j.icheatmasstransfer.2022.106054.

[21] G. Bognar, "Ferrofluid flow in magnetic field above stretching sheet with suction and injection," *Math. Model. Anal.*, vol. 25, no. 3, pp. 461–472, Jul. 2020, Accessed: Feb 20, 2025. [Online]. Available: www.academia.edu/121605678/Ferrofluid_Flow_in_Magnetic_Field_Above_Stretching_Sheet_with_Suction_and_Injection.

[22] S. M. Sachhin, U. S. Mahabaleshwar, and A. Chan, "Effect of slip and thermal gradient on micropolar nano suspension flow across a moving hydrogen fuel-cell membrane," *Int J. Hydrog. Energy*, vol. 63, pp. 59–81, Apr. 2024, doi: 10.1016/j.ijhydene.2024.02.332.

[23] S. M. Sachhin, U. S. Mahabaleshwar, S. N. R. Nayakar, H. S. R. Patil, and B. Souayeh "Darcy–Brinkman model for Boussinesq–Stokes suspension tetra dusty nanofluid flow over stretching/shrinking sheet with suction/injection," *Numer. Heat Transf. Part Appl.* vol. 0, no. 0, pp. 1–23, 2024, doi: 10.1080/10407782.2024.2359047.

5 Axisymmetric Nanofluid Flow with Non-Uniform Magnetic Field

Dhanapal Prakash, Elumalai Ragupathi, Narsu Sivakumar, Murugan Muthtamilselvan, and Ikhyun Kim

5.1 INTRODUCTION

Nanotechnology has garnered significant attention from researchers due to its wide range of applications, including microelectronics, transportation, and medicine. Choi [1] introduced the concept of nanofluids, which are homogeneous mixtures of base liquids and nanoparticles ranging from 1–100 nm in size, to enhance thermal conductivity. The thermal conductivity of base liquids alone is often insufficient to meet the demands of modern technologies. This gap can be addressed by incorporating nanoparticles into base liquids, a solution that has been proven by various studies to enhance the convective heat transfer performance of these liquids. Among many, Masuda *et al.* [2] observed an enhancement of thermal conductivity in nanofluids. In 2006, Buongiorno [3] demonstrated that the two commonly used theoretical models—homogeneous flow and dispersion models were inadequate for accurately predicting the heat transfer behavior of nanofluids. To address the limitations of these models, Buongiorno developed a new approach that better accounted for the unusual convective heat transfer properties observed in nanofluids. Recognizing the inherent inhomogeneity of nanoliquids, he highlighted the significance of Brownian diffusion and thermophoresis as key slip mechanisms. Based on this, Buongiorno formulated a comprehensive transport model to better describe the convective heat transfer in nanoliquids. This model has been used by several researchers [4–10]. Yang *et al.* [11] modified Buongiorno's model in order to fully account for the effects of nanoparticle volume fraction. In recent years, several researchers reported the numerical and experimental results for single phase nanofluid flow over the different geometries, such as, cavity, cylinder and stretching sheets [12–17].

Besides, a non-uniform magnetic field refers to a magnetic field in which the strength and/or direction of the magnetic force varies with position. Unlike uniform magnetic fields, where the field lines are parallel and evenly spaced, non-uniform fields exhibit more complex patterns, with varying intensities and orientations. These fields are a fundamental aspect of many natural and engineered systems, influencing phenomena ranging from the behavior of charged particles to the functioning of various magnetic devices. The study of non-uniform magnetic fields plays a

DOI: 10.1201/9781003611820-6

crucial role in advancing fields like electromagnetism, materials science, and medical imaging. By exploring the intricate dynamics of these fields, researchers are able to better understand their impact on particle motion, fluid flow, and heat transfer, as well as to design more efficient technologies in diverse applications, such as magnetic confinement in fusion reactors, electromagnetic sensors, and magnetic resonance imaging (MRI) systems. The understanding of non-uniform magnetic fields opens new doors to innovations in both theoretical and applied physics. Kandaswamy *et al.* [18] have investigated magnetic boundary layer flow of nanofluid past a vertical stretching surface by using scaling transformations. The flow characteristics with the combination of nanofluid parameters with buoyancy force on the MHD nanoliquid through the stretching surface has been examined by Makinde *et al.* [19]. Das *et al.* [20] disclosed the effect of non-uniform heat generation on the MHD flow numerically and Mabood *et al.* [21] examined the influence of viscous dissipation on MHD flow of an electrically conducting water-based nanoliquid past a nonlinear stretching sheet. They found that the magnetic parameter enforces a decrease in the dimensionless velocity and an increase in the temperature. Mahanthesh *et al.* [22] developed the numerical solutions of three-dimensional flow of magneto-nanofluid over a nonlinear stretching surface. Further, Gupta *et al.* [23] applied the OHAM method to visualize the effect of chemical reaction past the MHD nonlinear sheet of incompressible nanoliquid and Ramya *et al.* [24] used the Keller box method to find the essence of velocity and thermal slip on the MHD nanoliquid nonlinear sheet. Finite-element analysis is made to study the variable viscosity on MHD nanoliquid flow with the Cattaneo-Christov model by Ali *et al.* [25]. Recently, Uddin *et al.* [26] employed the particle swarm optimization with the finite-difference method to develop dual solutions of MHD nanoliquid stagnation point flow. Very recently, Raghunath [27] explored the time-dependent MHD nanofluid flow over the vertical porous plate with the presence of the chemical reaction, radiation, and soret effects.

In recent years, regression analysis has been a powerful and widely used statistical technique that plays a pivotal role in understanding and modeling the relationships between variables. It serves as a fundamental tool in the fields of statistics, economics, finance, social sciences, and many other disciplines. At its core, regression analysis explores the connections between a dependent variable and one or more independent variables, aiming to uncover patterns, make predictions, and gain insights into how changes in the independent variables affect the dependent variable. Regression analysis provides a systematic framework for quantifying and describing the association between variables, allowing researchers, analysts, and decision-makers to make informed decisions, develop accurate predictions, and gain a deeper understanding of the underlying dynamics within their data. This introductory exploration will delve into the key concepts, methods, and applications of regression analysis, shedding light on its significance and versatility in addressing a wide range of real-world problems. Darbari *et al.* [28] adopted the response surface methodology to analyze the sensitivity of the nanoparticle diameter and solid volume fraction on the nanofluid flow over the channel. Moreover, Kumar *et al.* [29] explored the significance of micropolar nanofluid flow through the porous stretching sheet with the presence of the nonlinear thermal radiation. They discussed the importance of the skin friction coefficient and

Nusselt number rate via the quadratic multiple linear regression model. In recent years, statistical analysis for the nanofluid flow through the various geometries has been inspected by several researchers [30–32].

The previous analyses clearly indicate that the concentration of nanoparticles on the plate is actively influenced, yet it remains unclear how the volume fraction of nanoparticles is precisely distributed across the surface. To address this uncertainty, it is crucial to establish a new boundary condition for nanoparticle concentration, as initially proposed by Kuznetsov and Nield [33]. Furthermore, the existing literature is mostly confined to studies involving uniform magnetic fields, whereas in real-world scenarios, the majority of electrically conducting fluid flows occur under non-uniform or inhomogeneous magnetic fields [34]. This gap in research has prompted us to explore the effects of non-uniform external magnetic fields on the three-dimensional stagnation point flow and heat transfer within nanofluid-saturated porous media. Such flows are observed when a fluid jet impinges on a rigid body [35], during boundary layer development along material handling conveyors, and in applications such as the aerodynamic extrusion of plastic sheets and in petrochemical industries [36]. Additionally, the impact of non-uniform magnetic fields has been successfully applied in various fields, including magnetic drug targeting through tracheobronchial airways [37], nanofluid flow in tubes and square domains [38–40], and ferro-nanofluid flow through U-turn pipes in solar heat exchangers [41]. Mistrangelo and Bühler [42] experimentally examined the MHD fluid flow over the curved pipes under the influence of non-uniform magnetic field. Zitouni *et al.* [43] explored the entropy generation, flow and heat transport aspects of mixed convection ferro-fluid flow under the impact of non-uniform magnetic field. In addition, we assume that the rigid plate transfers heat through convection with an external nanofluid, and that the nanoparticle mass flux on the plate is zero. To simplify the transport equations, a transformation technique is applied, converting them into nonlinear ordinary differential equations (ODEs). These ODEs are then solved both analytically and numerically. The analytical solution is obtained using the homotopy analysis method (HAM), while the numerical solution is found using the Runge–Kutta–Fehlberg method combined with the shooting technique.

5.2 MATHEMATICAL ANALYSIS

A steady 3D viscous electrically conducting nanofluid through porous medium in the half-space S (see Figure 5.1), given by

$$S = \{(x,y,z) \in \mathbb{R}^3 : (x,z) \in \mathbb{R}^2, y > 0\}, \tag{5.1}$$

and the boundary of S, i.e., $y = 0$, is a rigid occupying

$$S^- = \{(x,y,z) \in \mathbb{R}^3 : (x,z) \in \mathbb{R}^2, y < 0\}, \tag{5.2}$$

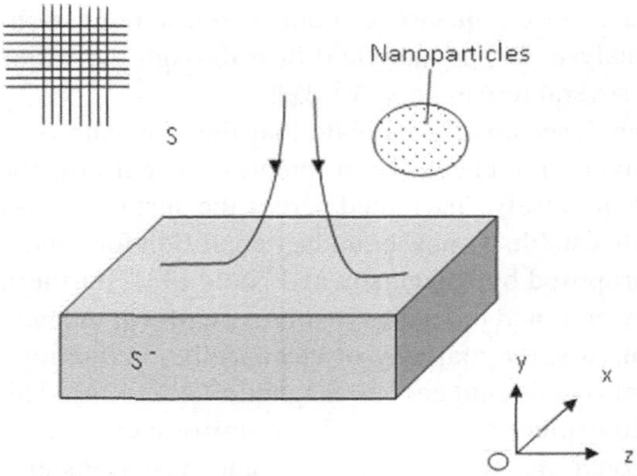

FIGURE 5.1 Physical configuration and geometrical co-ordinates.

where (e_x, e_y, e_z) as the coordinate axes in order to have the orthodox base of $\mathbb{R}^3$. With the consideration of negligible external electric field E_e, the external magnetic field is of the form [44]

$$H_e = H_\infty \left[x\, e_x - (c+1)\, y\, e_y + c\, z\, e_z\right],\ H_\infty = \text{ constant},\ (x,y,z) \in \mathbb{R}^3, \quad (5.3)$$

and the parameter c is a measure of the 3D motion.

It is well known that H_e concurs with the total magnetic field, which is parallel to the speed of the fluid in the half-space $y > 0$. With the assumption of no-slip on the boundary, the alignment can be disturbed [45]. Many authors have included only the electromagnetic field in the fluid, but it is proved that the electromagnetic field in the solid is formally the same as that of the fluid model over the solid [44]. Thus, it can be represented by

$$H = H_\infty \left[x\, h'(y)\, e_x - \left[ck(y) + h(y)\right]\, e_y + c\, z\, k'(y)\, e_z\right],\quad y \geq 0,$$

$$H_s = H_\infty \left[x\, h_s'(y)\, e_x - \left[ck_s(y) + h_s(y)\right]\, e_y + c\, z\, k_s'(y)\, e_z\right],\quad y \leq 0, \quad (5.4)$$

where h, k, h_s, and k_s are sufficiently regular unknown functions to be determined.

In most of the recent studies, the nanoliquid has been assumed as homogeneous which is not realistic and it may lead to inappropriate results being produced. In fact, there are two important reasons for inhomogeneity in a small dimension of the nanoparticles. They are Brownian motion and thermophoretic diffusivity [3] and it can be described by the Brownian diffusion coefficient D_B, and thermal diffusion coefficient D_T, respectively, as

$$D_B = \frac{k_B T}{3\pi \mu_{bf} d_p} \quad \text{and} \quad D_T = \beta \frac{\mu_{bf}}{\rho_{bf}} C,$$

where k_B is Boltzmann's constant, μ_{bf} is the dynamic viscosity of the base fluid, d_p is the nanoparticle diameter and $\beta = 0.26\left(k_{bf}/\left(2k_{bf}+k_p\right)\right)$ is a proportionality factor between the slip velocity due to thermophoresis and the temperature factor $\nabla T/T_\infty$ (McNab and Meisen [46], Duhr and Braun [47]). The corresponding nanoparticle fluxes due to Brownian diffusion and thermophoresis can be calculated as

$$\boldsymbol{J}_{p,B} = -\rho_p D_B \nabla C \quad and \quad \boldsymbol{J}_{p,T} = -\rho_p D_T \nabla T/T_\infty. \tag{5.5}$$

Therefore, the total nanoparticle flux can be given as

$$\boldsymbol{J}_p = \boldsymbol{J}_{p,B} + \boldsymbol{J}_{p,T} = -\rho_p \left(D_B \nabla C + D_T \frac{\nabla T}{T_\infty} \right). \tag{5.6}$$

Further, the energy flux relative to the nanofluid velocity can be expressed as

$$\boldsymbol{q} = -k\nabla T + h_p \boldsymbol{J}_p, \tag{5.7}$$

where h_p is the sepcific enthalpy of nanoparticles.

With the inclusion of Brownian motion and thermophoretic force, the governing equations may be written as

$$\nabla.\boldsymbol{V} = 0, \tag{5.8}$$

$$\boldsymbol{V}.\nabla\boldsymbol{V} = -\frac{1}{\rho}\nabla p + \nu\nabla^2\boldsymbol{V} - \frac{\nu}{K}\boldsymbol{V} + \frac{\mu_e}{\rho}\left(\nabla \times \boldsymbol{H}\right) \times \boldsymbol{H}, \tag{5.9}$$

$$\nabla \times H = \sigma_e\left(\boldsymbol{E} + \mu_e \boldsymbol{V} \times \boldsymbol{H}\right), \tag{5.10}$$

$$\nabla \times \boldsymbol{E} = 0, \quad \nabla.\boldsymbol{E} = 0, \quad \nabla.\boldsymbol{H} = 0 \quad in \;\; \mathrm{S}, \tag{5.11}$$

$$\boldsymbol{V}.\nabla T = -\frac{1}{(\rho c)_f}\left[\nabla.\boldsymbol{q} + h_p \nabla.\boldsymbol{J}_p\right] = \alpha\nabla^2 T + \tau\left[D_B \nabla C.\nabla T + \frac{D_T}{T_\infty}\nabla T.\nabla T\right], \tag{5.12}$$

$$\boldsymbol{V}.\nabla C = -\frac{1}{\rho_p}\nabla.\boldsymbol{J}_p = D_B \nabla^2 C + \frac{D_T}{T_\infty}\nabla^2 T, \tag{5.13}$$

where $\boldsymbol{V} = (u, v, w)$ is the velocity field along the directions (x, y, z), respectively, p is the pressure, T is the temperature field, C is the volumetric concentration field, $\boldsymbol{E}$ and $\boldsymbol{H}$ are electric and magnetic fields, ρ is the mass density, μ_e is the magnetic permeability, σ_e is the electrical conductivity, ν is the kinematic Newtonian viscosity co-efficient, K is the permeability of the porous medium, τ is the heat capacity ratio of the base fluid and nanoparticles. The nanofluid model, which is used in the present study, assumes a dilute suspension (less than 5% of volume fraction) of non-interactive nanoparticles in the base fluid.

The relevant flow, heat, and mass transfer fields at $y=0$ can be defined as

i. Strict adherence condition: $V=0$.
ii. Boundary conditions of the third kind: $-k\nabla T = h_T\left(T_f - T\right)$, i.e., heat flux from the plate is proportional to the difference between the temperature of the hot fluid below the plate surface and the plate temperature.
iii. Zero nanoparticle flux: $D_B \nabla C + \frac{D_T}{T_\infty}\nabla T = 0$.

The electromagnetic field in the solid $(\boldsymbol{H_s}, \boldsymbol{E_s})$ is considered and it is aligned to the velocity at infinity. In [45], it is proved that if $\boldsymbol{H_s}$ is non-uniform and its non-degenerate field lines belong to surfaces, which asymptote to the plane $y=0$ as $|x|, |z| \to \infty$, then $\boldsymbol{E_s}=0$ and

$$H_s = H_\infty\left[x\,h'(0)\,e_x - \left[ck'(0)+h'(0)\right]e_y + c\,z\,k'(0)\,e_z\right], \quad y \le 0, \qquad (5.14)$$

and $ck(0)+h(0)=0, \quad c\in(-1,+\infty), \quad c\neq 0$, due to the normal component of $\mu_e \boldsymbol{H}$ across the boundary $y=0$ is continuous. This implies $h(0)=0, k(0)=0$.

The three-dimensional stagnation-point flow, heat and mass transfer is determined by

$$\eta = \sqrt{\frac{a}{\nu}}y, \quad V = \left[a\,x f'(\eta)\,e_x - \sqrt{a\nu}\left(f(\eta)+cg(\eta)\right)e_y + cazg'(\eta)\,e_z\right],$$

$$\boldsymbol{H} = H_\infty\left[x\psi_1{}'f'(\eta)\,e_x - \sqrt{(\nu/a)}\left(\psi_1(\eta)+c\psi_2(\eta)\right)\boldsymbol{e_y} + cz\psi_2{}'f'(\eta)\,\boldsymbol{e_z}\right], \qquad (5.15)$$

$$T = T_\infty + \left(T_f - T_\infty\right)\theta(\eta), \quad C = C_\infty + C_\infty\phi(\eta).$$

Substituting the above transformations into the governing Eqs. (5.8)–(5.11), we get

$$a^2x\left[f''' + (f+cg)f'' - f^2 - \frac{\nu}{Ka}f - \frac{\mu_e H_\infty^2}{\rho a^2}\left(\psi_1 + c\psi_2\right)\psi_1''\right] = \frac{1}{\rho}\frac{\partial p}{\partial x}, \qquad (5.16)$$

$$a\sqrt{a\nu}\left[(f+cg)\left(f'+cg'\right) + \left(f''+cg''\right) + \frac{\nu}{Ka}(f+cg)\right] +$$

$$\frac{\mu_e H_\infty^2}{\rho}\sqrt{\frac{a}{\nu}}\left[x^2\psi_1{}'\psi_1'' + c^2z^2\psi_2{}'\psi_2''\right] = -\frac{1}{\rho}\sqrt{\frac{a}{\nu}}\frac{\partial p}{\partial \eta}, \qquad (5.17)$$

$$a^2cz\left[g''' + (f+cg)\,g'' - cg'^2 - \frac{\nu}{Ka}g' - \frac{\mu_e H_\infty^2}{\rho a^2}\left(\psi_1 + c\psi_2\right)\psi''_2\right] = \frac{1}{\rho}\frac{\partial p}{\partial z}, \quad (5.18)$$

$$\psi_1'' + \sigma_e\mu_e\nu\left[(f+cg)\,\psi_1' - (\psi_1 + c\psi_2)f'\right] = 0, \quad (5.19)$$

$$\psi_2'' + \sigma_e\mu_e\nu\left[(f+cg)\,\psi_2' - (\psi_1 + c\psi_2)g'\right] = 0, \quad (5.20)$$

It is known that, far from the wall, the pressure p has the same behavior as for an inviscid fluid. So the corresponding velocity and pressure fields for an inviscid fluid are given by

$$V = \left[a\,x\boldsymbol{e}_x - \sqrt{a\,\nu}\,(c+1)\,(\eta - L)\,\boldsymbol{e}_y + acz\boldsymbol{e}_z\right], \quad (5.21)$$

$$p = -\frac{1}{2}\rho\left[a^2x^2 + a\nu(c+1)^2(\eta - L)^2 + a^2c^2z^2\right] + p_0, \quad \eta \geq L \quad (5.22)$$

where p_0 and L are constants. Integrating Eq. (5.17) with respect to η and by using Eqs. (5.21)–(5.22), we find the pressure field as

$$p = -\rho\frac{a^2}{2}\left[x^2 + \frac{\nu}{a}(f+cg)^2 + c^2z^2\right] - \rho a\nu\left(f' + cg'\right) - \frac{\nu^2}{K}\int_0^\eta\left[f(s) + cg(s)\right]ds$$

$$-\mu_e\frac{H_\infty^2}{2}\left[x^2\left(\psi_1'^2 - 1\right)\right] - \mu_e\frac{H_\infty^2}{2}\left[c^2z^2\left(\psi_2'^2 - 1\right)\right] + p_0, \quad (5.23)$$

On substituting Eq. (5.23) into the Eqs. (5.16) and (5.18), we obtain the velocity and magnetic fields along the directions x, and z respectively as

$$f''' + (f+cg)f'' - f'^2 + 1 - \Gamma f' - \beta_m\left[\left(\psi_1 + c\psi_2\right)\psi_1'' - \psi_1'^2 + 1\right] = 0, \quad (5.24)$$

$$g''' + (f+cg)\,g'' - cg'^2 + c - \Gamma g' - \beta_m\left[\left(\psi_1 + c\psi_2\right)\psi_2'' - c\psi_2'^2 + c\right] = 0, \quad (5.25)$$

$$\psi_1'' + R_m\left[(f+cg)\,\psi_1' - (\psi_1 + c\psi_2)f'\right] = 0, \quad (5.26)$$

$$\psi_2'' + R_m\left[(f+cg)\,\psi_2' - (\psi_1 + c\psi_2)g'\right] = 0, \quad (5.27)$$

and the usual adherence condition on the boundary provides

$$f(0) = 0, \quad f'(0) = 0, \quad g(0) = 0, \quad g'(0) = 0, \quad \psi_1(0) = 0, \quad \psi_2(0) = 0,$$

$$f'(\infty) = 1, \quad g'(\infty) = 1, \quad \psi_1'(\infty) = 1, \quad \psi_2'(\infty) = 1. \tag{5.28}$$

However, in [45], it is proved that the total magnetic field [Eq. (5.3)] and the velocity are parallel at infinity only if the flow is axisymmetric $\left(\text{i.e., } c = 1, f = g, \psi_1 = \psi_2 \text{ in Eqs. } (5.24)-(5.28)\right)$. So, the velocity and magnetic fields for such axisymmetric flow can be described as

$$f''' + 2ff'' - f'^2 + 1 - \Gamma f' - \beta_m \left(2\psi_1 \psi_{1''} - \psi_{1'}{}^2 + 1\right) = 0, \tag{5.29}$$

$$\psi_1'' + 2R_m \left(f\psi_1' - \psi_1 f'\right) = 0, \tag{5.30}$$

and the boundary conditions become

$$f(0) = 0, \quad f(0) = 0, \quad \psi_1(0) = 0,$$

$$f(\infty) = 1, \quad \psi_1'(\infty)=1, \tag{5.31}$$

where $\beta_m = \frac{\mu_e H_\infty^2}{\rho a^2}$ is the Alfv e' n number, $R_m = \nu\sigma_e\mu_e$ is the magnetic Prandtl number and $\Gamma = \frac{\nu}{Ka}$ is the permeability parameter.

The heat and mass transfer analyses are made by substituting Eq. (5.15) into the governing Eqs. (5.12)–(5.13) and we obtain

$$\theta'' + 2\,\mathrm{Pr}\, f\theta' + \mathrm{Nb}\theta'\phi' + \mathrm{Nt}\theta'^2 = 0, \tag{5.32}$$

$$\phi'' + \frac{\mathrm{Nt}}{\mathrm{Nb}}\theta'' + 2\mathrm{Le}\,\mathrm{Pr}\, f\phi' = 0, \tag{5.33}$$

and the relevant boundary conditions become

$$\theta'(0) = -\mathrm{Bi}\left[1 - \theta(0)\right], \quad \mathrm{Nb}\,\phi'(0) + \mathrm{Nt}\theta'(0) = 0, \tag{5.34}$$

$$\theta(\infty) = 0, \quad \phi(\infty) = 0, \tag{5.35}$$

where $\mathrm{Pr} = \frac{\nu}{\alpha}$ is the Prandtl number, $\mathrm{Nb} = \frac{\tau D_B C_\infty}{\alpha}$ is the Brownian motion parameter, $\mathrm{Nt} = \frac{\tau D_T (T_f - T_\infty)}{\alpha T_\infty}$ is the thermophoresis parameter, $\mathrm{Le} = \frac{\alpha}{D_B}$ is the Lewis number, $\mathrm{Bi} = \frac{h}{k}\sqrt{\frac{\nu}{a}}$ is the Biot number.

The physical quantities of interest in this problem are the local skin friction τ_0, magnetic field H_0, and Nusselt number Nu_x, which are defined as

$$\tau_0 = \rho a\sqrt{a\nu} f''(0)\left(xe_x + ze_z\right),\ H_0 = H_\infty \psi_1'(0)\left(xe_x + ze_z\right),\ Nu_x = -\sqrt{\frac{a}{\nu}}x\theta'(0). \tag{5.36}$$

It is worth mentioning that the dimensionless mass flux is identically zero by applying physically more realistic boundary, i.e., the normal flux of nanoparticles is zero at the boundary [45].

5.3 NUMERICAL METHOD

Here, we solved the coupled system of equations with the boundaries (5.29)–(5.35) numerically via the R-K-F method with shooting process. First, the higher order differential equations are transformed to a set of first order equations with nine unknowns. They are listed as follows:

$$\begin{aligned}
&y_1' = y_2,\quad y_2' = y_3,\\
&y_3' = y_2^2 - 2y_1 y_3 - 1 + \Gamma y_2 - \beta_m\left[y_5^2 + 4R_m y_4\left(y_5 y_1 - y_4 y_2\right) - 1\right]\\
&y_4' = y_5,\quad y_5' = -2R_m\left(y_1 y_5 - y_4 y_2\right),\\
&y_6' = y_7,\quad y_7' = -2\mathrm{Pr}\, y_1 y_7 - \mathrm{Nb}\, y_7 y_9 - \mathrm{Nt}\, y_7^2,\\
&y_8' = y_9,\quad y_9' = \tfrac{\mathrm{Nt}}{\mathrm{Nb}}\left[2\mathrm{Pr}\, y_1 y_7 + \mathrm{Nb}\, y_7 y_9 + \mathrm{Nt}\, y_7^2\right] - 2\mathrm{Le}\ \mathrm{Pr}\, y_1 y_9\ ,
\end{aligned} \tag{5.37}$$

and the boundary conditions become

$$\begin{aligned}
&y_1 = 0,\quad y_2 = 0,\quad y_4 = 0,\quad y_7 = -\mathrm{Bi}\left(1 - y_3\right),\quad y_9 = \frac{\mathrm{Nt}}{\mathrm{Nb}}\left[\mathrm{Bi}\left(1 - y_3\right)\right]\quad at\ \ \eta = 0,\\
&y_2 = 1,\quad y_5 = 1,\quad y_6 = 0,\quad y_8 = 0\quad as\ \ \eta \to \infty,
\end{aligned} \tag{5.38}$$

where $y_1 = f, y_2 = f', y_3 = f'', y_4 = \psi_1, y_5 = \psi_1', y_6 = \theta, y_7 = \theta', y_8 = \phi, y_9 = \phi'$.

Next, to retrieve solutions of the system (5.37)–(5.38), we require nine initial conditions, but we have only five initial conditions. So the other four initial conditions $\left(y_3(0), y_5(0), y_6(0), y_8(0)\right)$ are found by processing the shooting method with the satisfaction of four end boundary conditions given in Eq. (5.38) and the secant method is used to adjust the guess values. Now the system with a proper initial guess is integrated by Runge–Kutta–Fehlberg method with a step size $\Delta\eta = 0.01$. The process is repeated until we get the converged results within a tolerance level of 10^{-5}. The numerical solution $(f, \psi_1, \theta, \phi)$ of the problem satisfies

$$\lim_{\eta\to\infty}\left[f(\eta) - \eta\right] = -\alpha,\quad \lim_{\eta\to\infty}\left[\psi_1(\eta) - \eta\right] = -\alpha,\quad \lim_{\eta\to\infty}\theta(\eta) = 0,\quad \lim_{\eta\to\infty}\phi(\eta) = 0,$$

TABLE 5.1
Comparison of Descriptive Quantities of the Motion under Certain Limiting Conditions (Nb=Nt=10^{-5},Pr=Le=Bi=1). The Values Reported by Borrelli *et al.* [44] are Given in Brackets

		$f'(0)$			$\psi_1'(0)$		
R_m	β_m	Borelli *et al.* [44]	bvp4c	HAM (8th order)	Borelli *et al.* [44]	bvp4c	HAM (8th order)
0.1	0.2	1.228277	1.2283	1.2285	0.755561	0.7556	0.7554
	0.5	1.060342	1.0608	1.0601	0.706264	0.7068	0.7069
1	0.2	1.181151	1.1812	1.1810	0.531153	0.5312	0.5310
	0.5	0.937041	0.9370	0.9371	0.471837	0.4718	0.4721
100	0.2	1.171052	1.1711	1.1709	0.161306	0.1613	0.1615
	0.5	0.922009	0.9220	0.9221	0.139184	0.1392	0.1396

where the parameter α represents the displacement thickness. The momentum, thermal, and nanoparticle concentration layer thickness can be defined as

i. η_f, the value of η such that $f'(\eta) = 0.99$.
ii. η_θ, the value of η such that $\theta(\eta) = 0.01$.
iii. η_ϕ, the value of η such that $\phi(\eta) = 0.01$.

In order to validate the obtained results, comparisons have been done with Borrelli *et al.* [25] for $f''(0)$ and $\psi_{1'}(0)$ in Table 5.1 and they are in better agreement.

5.4 HOMOTOPY ANALYSIS APPROACH

The homotopy analysis method (HAM) is a powerful mathematical technique that has gained significant attention in recent years for solving nonlinear differential equations. It offers a systematic and effective approach for finding approximate analytical solutions to a wide range of problems in various fields, including physics, engineering, and applied mathematics. The method was first introduced by Dr. Shijun Liao, a Chinese mathematician, in 1992. Inspired by homotopy theory and perturbation methods, HAM provides a unified framework for handling nonlinear equations by constructing a suitable homotopy equation. This equation contains an auxiliary parameter called the homotopy parameter, which allows us to continuously deform the problem from an initial guess (often a simpler equation) to the desired nonlinear equation. The key idea behind HAM is to express the solution of the original equation as a convergent series in powers of the homotopy parameter. By applying the homotopy analysis method, one can systematically compute successive approximations of the solution, improving the accuracy with each iteration. The flexibility of HAM allows researchers to control and adjust the convergence region and rate, making it adaptable to a wide range of nonlinear problems. One of the notable advantages of the homotopy analysis method is its ability to handle nonlinear problems without relying on small parameters or restrictive assumptions. Unlike traditional perturbation techniques, HAM does not require the presence of a small parameter in the equation. This makes it particularly useful for tackling highly nonlinear problems that are challenging to solve using other analytical or numerical methods.

By following the procedure of the HAM approach, the auxiliary linear operator and initial approximation are taken as

$$f_0(\eta) = 1-\eta-e^{-\eta}, \quad (\psi_1)_0(\eta) = 1-e^{-\eta}, \tag{5.39}$$

$$\theta_0(\xi) = \frac{\text{Bi}}{1+\text{Bi}}e^{-\eta}, \quad \phi_0(\xi) = \frac{-\text{Nt}}{\text{Nb}}\frac{\text{Bi}}{1+\text{Bi}}e^{-\eta},$$

and the auxiliary linear operator can be expressed as

$$\mathcal{L}_f f = f_{\eta\eta\eta} + f_{\eta\eta}, \qquad \mathcal{L}_{\psi_1}\psi_1 = (\psi_1)_{\eta\eta} + (\psi_1)_{\eta}, \tag{5.40}$$

$$\mathcal{L}_\theta \theta = \theta_{\eta\eta} - \theta, \qquad \mathcal{L}_\phi \phi = \phi_{\eta\eta} - \phi,$$

satisfying the following characteristics:

$$\mathcal{L}_f\left(\breve{C}_1 + \eta\breve{C}_2 + \breve{C}_3 e^{-\eta}\right) = 0,$$

$$\mathcal{L}_{\psi_1}\left(\breve{C}_4 + \breve{C}_5 e^{-\eta}\right) = 0,$$

$$\mathcal{L}_\theta\left(\breve{C}_6 e^{\eta} + \breve{C}_7 e^{-\eta}\right) = 0,$$

$$\mathcal{L}_\phi\left(\breve{C}_8 e^{\eta} + \breve{C}_9 e^{-\eta}\right) = 0, \tag{5.41}$$

where the boundary constraints are employed to compute the constant values $\breve{C}_i, i = 1,...9$.

5.4.1 Convergence Analysis

The region of convergence of series solutions can be modified using auxiliary parameters $\hbar_f, \hbar_{\psi_1}, \hbar_\theta$, and $\hbar_\phi$. For this reason, we have plotted $\hbar$ curves (see Figures 5.2–5.5) of $f''(0), \psi_1'(0), \theta'(0)$, and $\phi'(0)$. Acceptable ranges of auxiliary parameters $\hbar_f, \hbar_{\psi_1}, \hbar_\theta$, and $\hbar_\phi$ are $[-0.8, 0.7]$, $[-0.7, -0.8]$, $[-0.7, 0.6]$, and $[-0.8, 0.6]$, respectively.

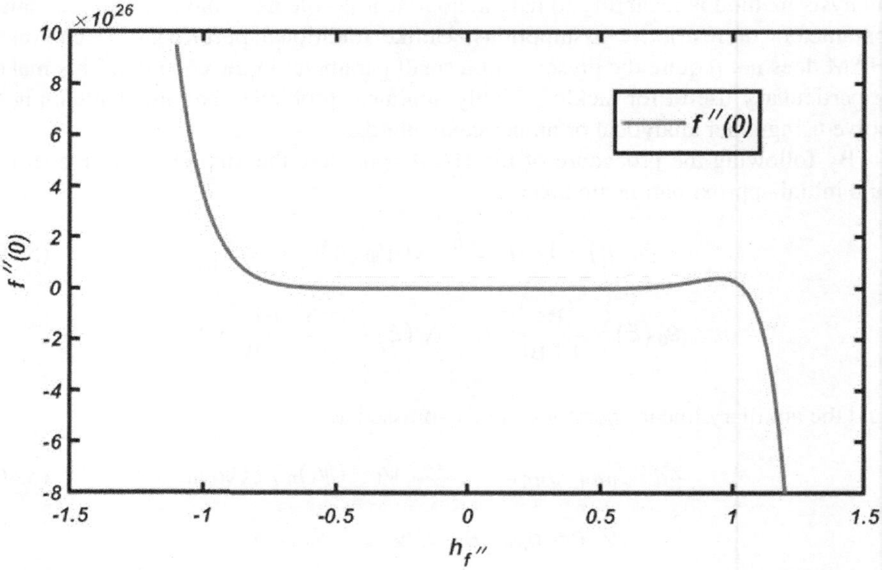

FIGURE 5.2 $\hbar$-curve for $f''(0)$.

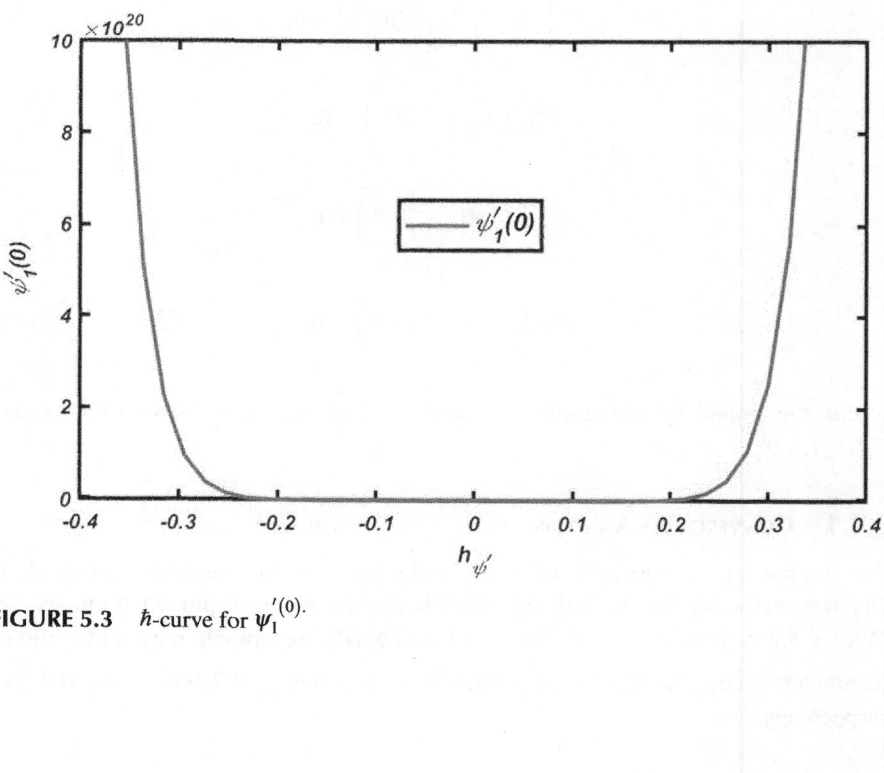

FIGURE 5.3 $\hbar$-curve for $\psi_1'(0)$.

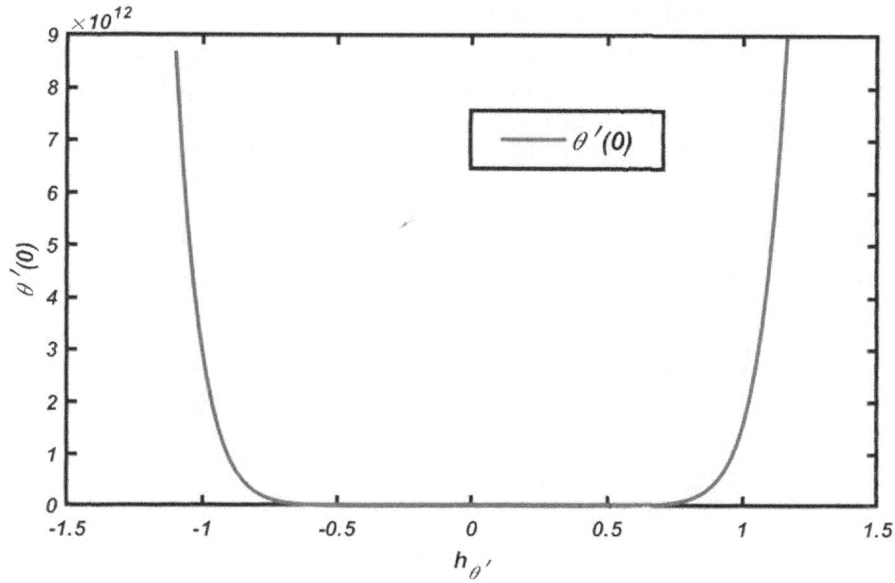

FIGURE 5.4 $\hbar$-curve for $\theta'(0)$.

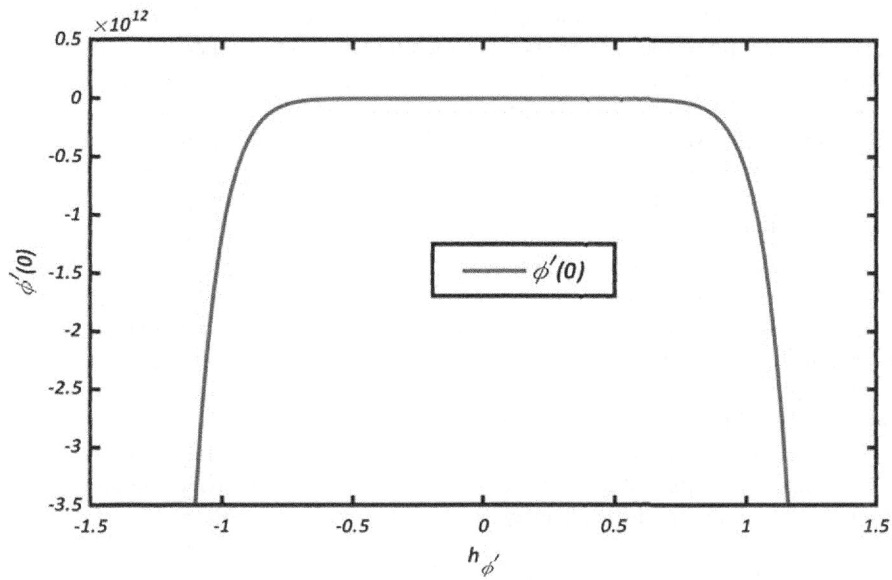

FIGURE 5.5 $\hbar$-curve for $\phi'(0)$.

5.5 RESULTS AND DISCUSSION

The numerical simulation of the transport equations with the boundary conditions (29)–(35) has been explored with the following range of key parameters: magnetic Prandtl number ($R_m = 10^{-2} - 10^2$), Alfvén number ($\beta_m = 0 - 1$), permeability parameter ($\Gamma = 10^{-2} - 10^{-1}$), Prandtl number ($\mathrm{Pr} = 1 - 7$), Lewis number ($1 - 10$), Brownian motion parameter ($\mathrm{Nb} = 0.1 - 0.5$), thermophoresis parameter ($\mathrm{Nt} = 0.1 - 0.5$), Biot number ($\mathrm{Bi} = 0.1 - 100$). The values of these parameters are chosen according to Glauert [48], Nield and Kuznetsov [5]. It is worth mentioning that for $\beta > 1$ the whole formulation of this problem breaks down, since the Alfvén speed is greater than the speed of the undisturbed flow. Further, for $\mathrm{Bi} \to \infty$, the convective boundary condition reduces to the constant surface temperature boundary condition analyzed by Nield and Kuznetsov [5]. The values of the pertinent parameters are assumed as $\Gamma = 0.01, R_m = 1, \beta_m = 0.2, \mathrm{Le} = 1, \mathrm{Nb} = \mathrm{Nt} = 0.1, \mathrm{Pr} = 1$ and $\mathrm{Bi} = 1$, unless otherwise specified.

5.5.1 Hydrodynamic Aspects

The effects of Alfvén number (β_m) and permeability parameter (Γ) on the dimensionless axisymmetric velocity and magnetic field are shown in Figure 5.6. It can be seen that the Alfvén number causes the velocity to be enhanced as well as magnetic fields, which forces the boundary layer thickness to grow (see Table 5.2). Similar behavior has been observed by Borrelli *et al.* [44, 49] in their study of boundary layer flow of Newtonian and micropolar fluids. Physically, the Alfvén number is the square of the ratio of the Alfvén speed to the undisturbed flow speed. In fact, Alfvén speed gives more restoring force to provide an effective stress on the magnetic field lines and velocity field. Thus, the parameter β_m made the layer and displacement thickness to increase (see Table 5.2). The Darcian body force has reduced in magnitude by increasing the permeability of the medium and hence, progressively less drag is experienced by the flow and flow retardation is thereby decreased. Thus, with the fixed value of β_m, the permeability parameter Γ enforces the velocity to be enhanced, as well as magnetic fields and subsequently its thickness increases.

Figure 5.7 illustrates the influence of magnetic Prandtl number R_m and permeability parameter Γ on the velocity and magnetic field. It is found that an increase in the magnetic Prandtl number R_m leads to a decrease in the momentum and displacement thickness of the boundary layer (see Table 5.2). Like thermal Prandtl number, magnetic Prandtl number is the ratio of the viscous to the magnetic diffusivity and it is proportional to the conductivity of the nanofluid. A rise in R_m decreases the magnetic field strength at the plate due to the dominance of viscous diffusivity. Meanwhile, the magnetic field lines are diffused away to a uniform field, i.e., in the region of constant flux [48]. Further, the permeability parameter leads the velocity boundary layer to become more thinner in the viscous dominated zone.

The variation of magnetic Prandtl number R_m, Alfvén number β_m and permeability parameter Γ on the dimensionless wall shear and total magnetic field are shown in Figure 5.8 and Table 5.2. The quantity $f''(0)$ and $\psi_{1'}(0)$ measure the friction and

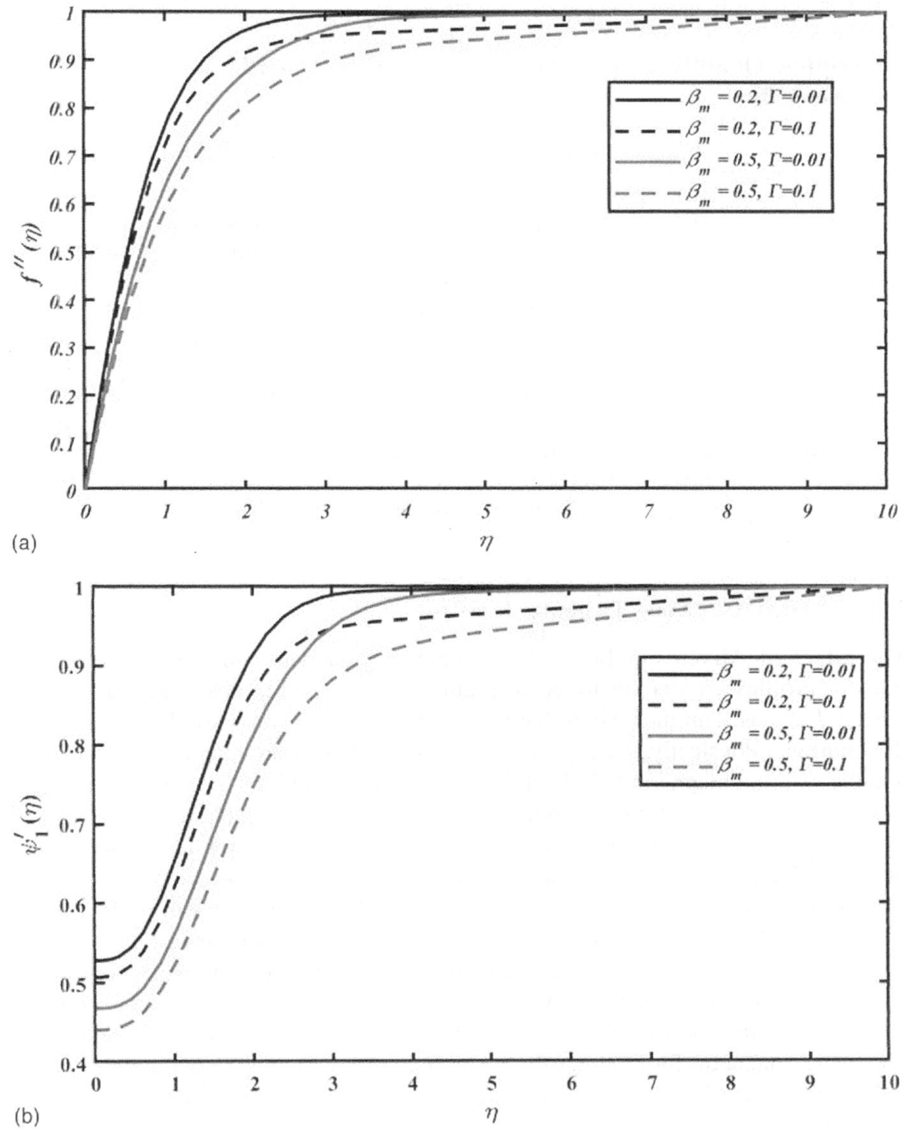

FIGURE 5.6 Dimensionless (a) velocity and (b) magnetic profile for different values of Alfvén number (β_m) and permeability parameter (Γ).

magnetic field experienced by the nanofluid at the boundary of the obstacle as it flows over the rigid flat plate. It is seen that the wall shear stress and magnetic field decrease with magnetic Prandtl number and Alfvén number due to the reduction in the slope of the velocity and magnetic profiles. A similar conclusion on R_m and β_m was derived by Dorrepaal and Moosavizadeh [45].

TABLE 5.2
Descriptive Quantities of the Motion when Nb=0.1,Nt=0.1, Pr=1,Le=1,Bi=1

R_m	β_m	Γ	$f'(0)$	$\psi_1'^{(0)}$	α	η_f
1	0.2	0.01	1.181151	0.531153	0.668506	2.505267
		0.05	1.152547	0.521572	0.735113	3.665221
		0.1	1.124841	0.512193	0.801286	4.324465
	0.5	0.01	0.937041	0.471837	0.930553	3.581925
		0.05	0.904658	0.460161	1.024800	4.339320
		0.1	0.870414	0.446094	1.173094	5.517491
10	0.2	0.01	1.168055	0.311822	0.654812	2.342751
		0.05	1.135539	0.304706	0.724123	3.456122
		0.1	1.104407	0.297801	0.795231	4.134654
	0.5	0.01	0.912187	0.270986	0.821413	2.912458
		0.05	0.873356	0.261677	0.995814	3.739014
		0.1	0.836998	0.252784	1.023409	4.952481

5.5.2 Heat and Mass Transfer Aspects

The effect of Alfvén number and permeability parameter on the heat and mass transfer profiles are demonstrated in Figures 5.9–5.10. It is observed that the parameter β_m causes an increase in the temperature and the nanoparticle concentration distributions. Physically, β_m represents the ratio between the magnetic permeability and the external magnetic field [49]. For $\beta < 1$, the strength of the external magnetic field can be greater and it gives more magnetic force perpendicular to the rigid plate. As a result of magnetic field, the fluid temperature and the nanoparticle concentration observe the additional heating due to the resistance of fluid flow. Moreover with the fixed value of β_m, a rise in the permeability leads to an improvement in the thermal and nanoparticle layer thickness due to decrement in the flow retardation.

Figures 5.11–5.12 demonstrate the influence of magnetic Prandtl number R_m and permeability parameter Γ on the dimensionless temperature and concentration fields. Evidently, increasing $R_m(> 1)$ leads to a rise in the viscous forces, which dominates magnetic force along the plate and so the magnetic conductivity on the plate is reduced. Thus, it leads to a reduction in the temperature and thermal layer thickness as well. However, the reduction in the temperature profile is delayed by the permeability. The reverse behavior has been observed in the concentration profile by increasing the magnetic Prandtl number and thereby increases the concentration layer thickness.

The effect of Alfvén number and the thermophoresis parameter on the dimensionless temperature are shown in Figure 5.13 for the different values of Prandtl number. It can be seen that the local temperature in the thermal boundary layer rises as the thermophoresis parameter increases due to an increment in the thermophoretic force, which is created by the thermal gradient and it creates a flow speed away from the rigid plate. By continuing

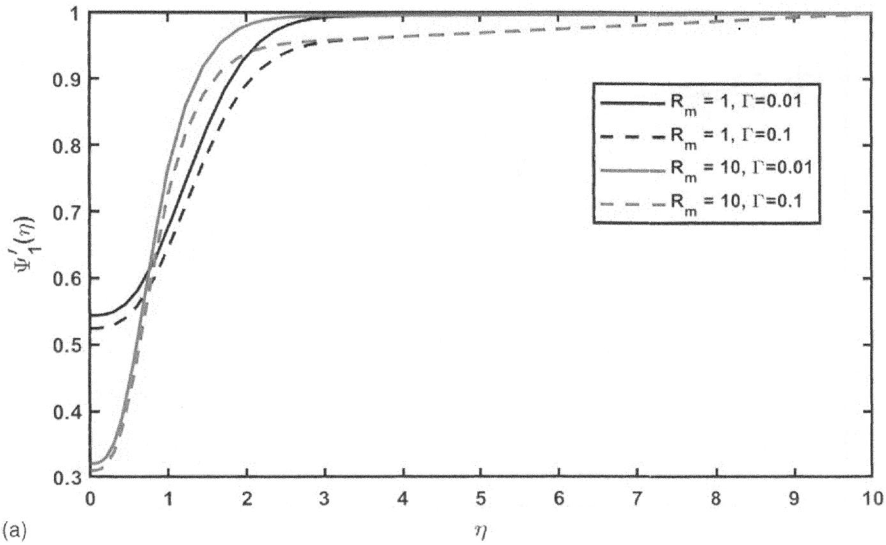

(a)

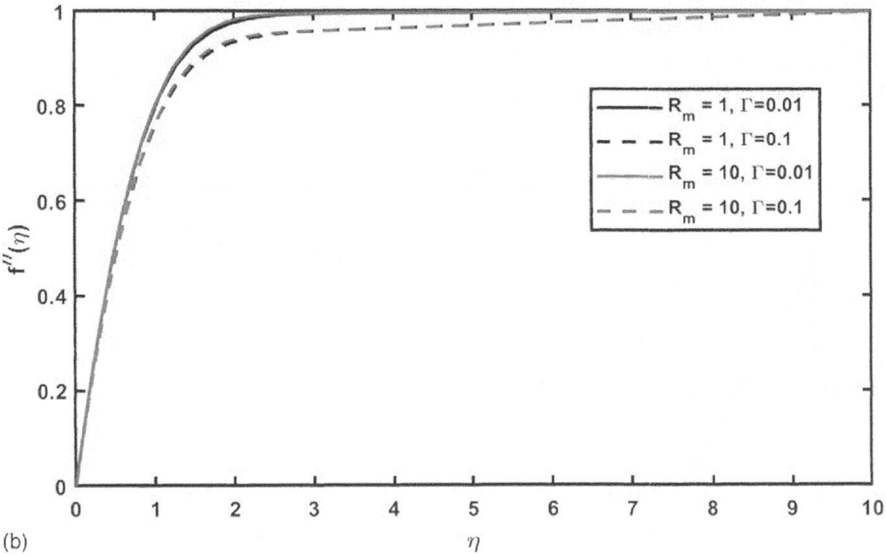

(b)

FIGURE 5.7 Dimensionless (a) velocity and (b) magnetic profile for different values of magnetic Prandtl number (R_m) and permeability parameter (Γ).

the process, more heated fluid flows away from the plate and the temperature within the layer increases. However, this effect is more pronounced by applying external magnetic force on the rigid plate, i.e., by increasing the Alfvén number. Physically, Pr relates to the ratio between the thermal and momentum diffusivities. With Pr < 1, the heat will diffuse more quickly than momentum. Generally, smaller Prandtl numbers are more conductive and low viscosity such as air and other gases. An increase in the Prandtl number forces

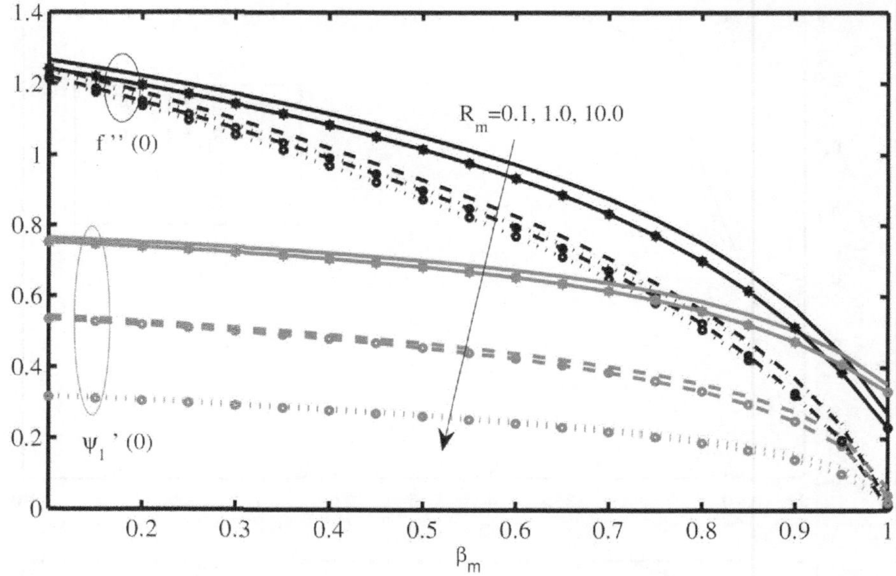

FIGURE 5.8 Plot of $f'(0)$ and $\psi_{1'}(0)$ for different values of magnetic Prandtl number (R_m), Alfvén number (β_m) and permeability parameter (Γ).

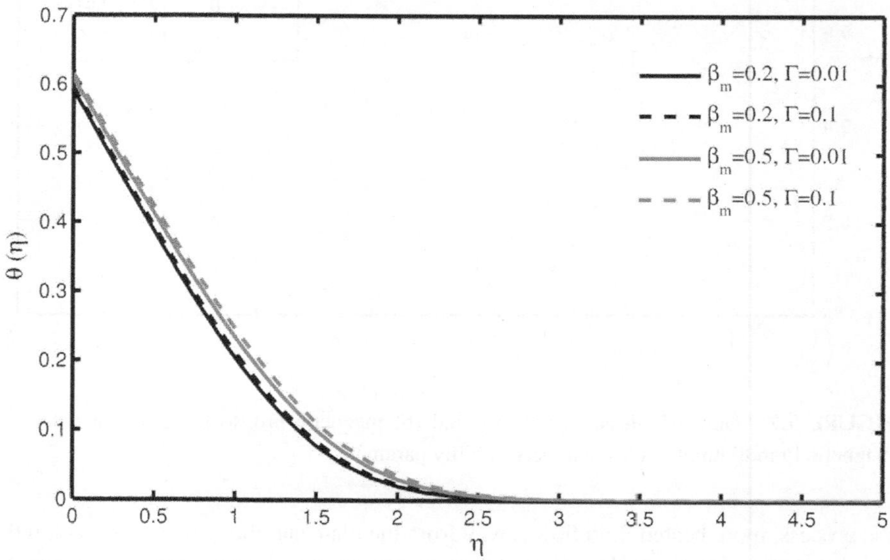

FIGURE 5.9 Dimensionless temperature profile for different values of Alfvén number (β_m) and permeability parameter (Γ).

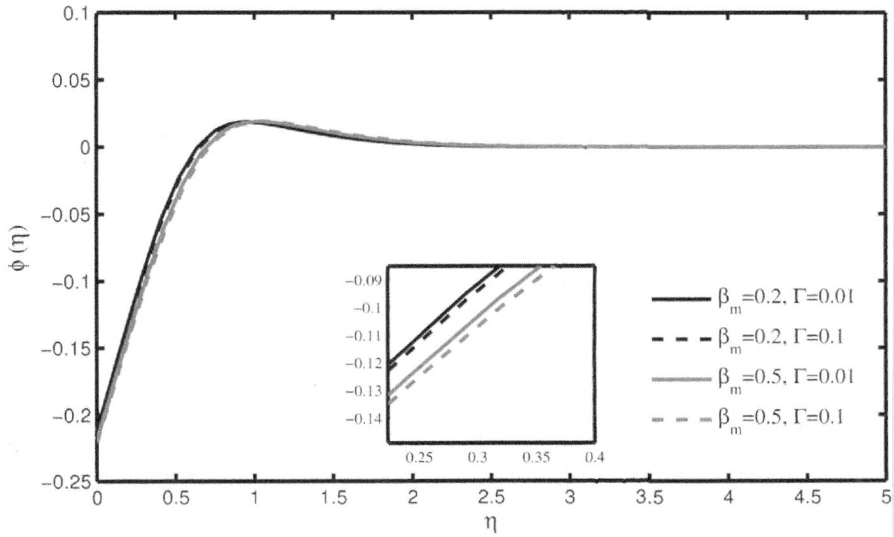

FIGURE 5.10 Dimensionless nanoparticle volumetric profile for different values of Alfvén number (β_m) and permeability parameter (Γ).

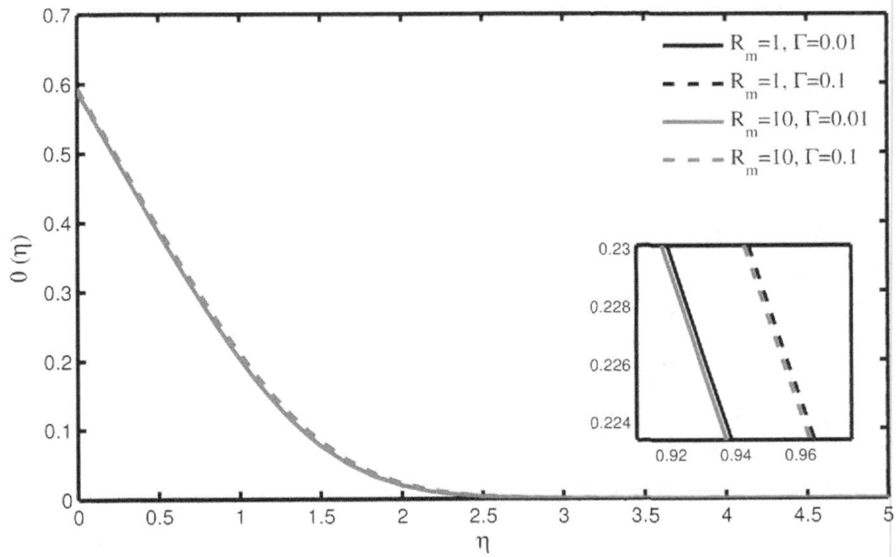

FIGURE 5.11 Dimensionless temperature profile for different values of magnetic Prandtl number (R_m) and permeability parameter (Γ).

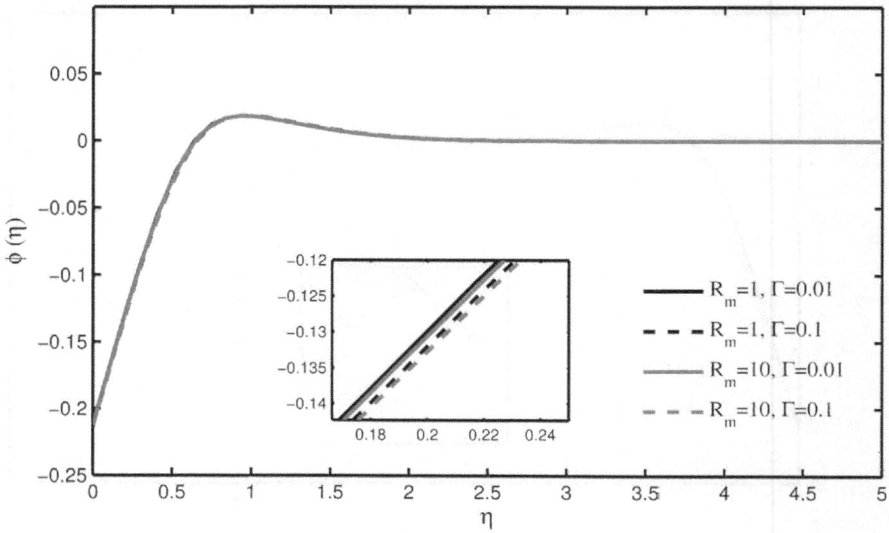

FIGURE 5.12 Dimensionless nanoparticle volumetric profile for different values of magnetic Prandtl number (R_m) and permeability parameter (Γ).

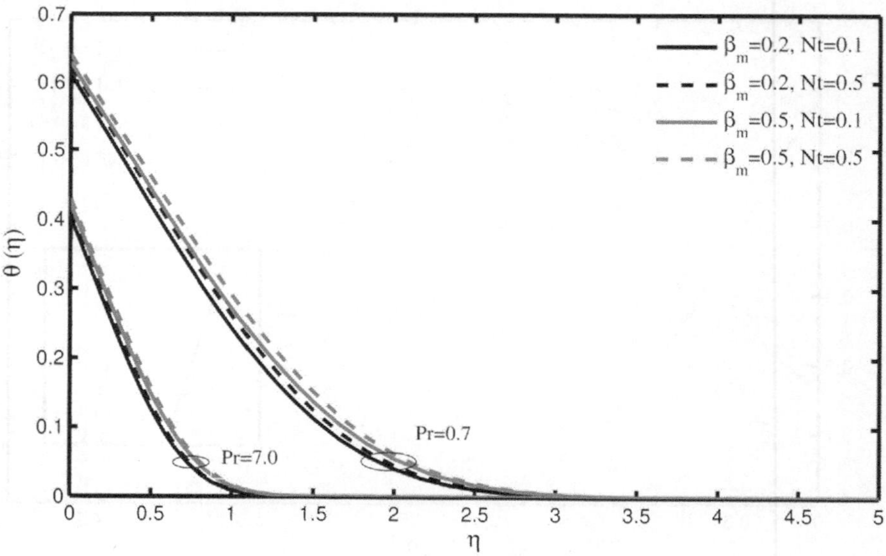

FIGURE 5.13 Dimensionless temperature profile for different values of Alfvén number (β_m), thermophoresis parameter (Nt) and Prandtl number (Pr).

an improvement of the fluid viscosity but reducing the flow velocity. Hence, it decreases the temperature as well. This is consistent with the fact that the thermal layer thickness decreases with an increase in the Prandtl number.

In accordance with Figure 5.14, the nanoparticle volume fraction boundary layer increases with an increase in the thermophoresis parameter or Alfvén number. For the hot rigid plate, thermophoresis tends to blow the nanoparticle volume fraction boundary layer away from the plate and such a hot plate repels the sub-micron sized particles, thereby forming a relatively particle-free layer near the plate. Hence the boundary layer thickness increases with an increase in the thermophoresis parameter. However, the effect of Nt on the volumetric profile is more pronounced for the lower value of Prandtl number.

The effect of Biot and Lewis numbers on the temperature profile are shown in Figure 5.15. In fact, the prescribed wall temperature can be attained with the large Biot number. An increase in the values of the Biot number leads to an increase in the heat transfer coefficient (see Figure 5.17) and consequently increases the temperature. From Table 5.3, it is observed that the enhancement in the Biot number is to increase both thermal and concentration layer thickness. On the other hand, the Lewis number defines the rate between the thermal and the species diffusions in the boundary layer regime. Thus, an increase in Le increases the thermal boundary layer thickness. However, this increment is more significant in the case of constant wall temperature. The influence of the Lewis number and Brownian diffusion parameter on the nanoparticle volumetric profile are elucidated in Figure 5.16. As the Lewis number or Brownian diffusion parameter increases, there is a significant decrease in the concentration of the nanoparticles in the concentration boundary layer.

Figure 5.17 is plotted to study the variation of thermophoresis parameter, Biot number and magnetic parameters on the dimensionless heat flow from the rigid

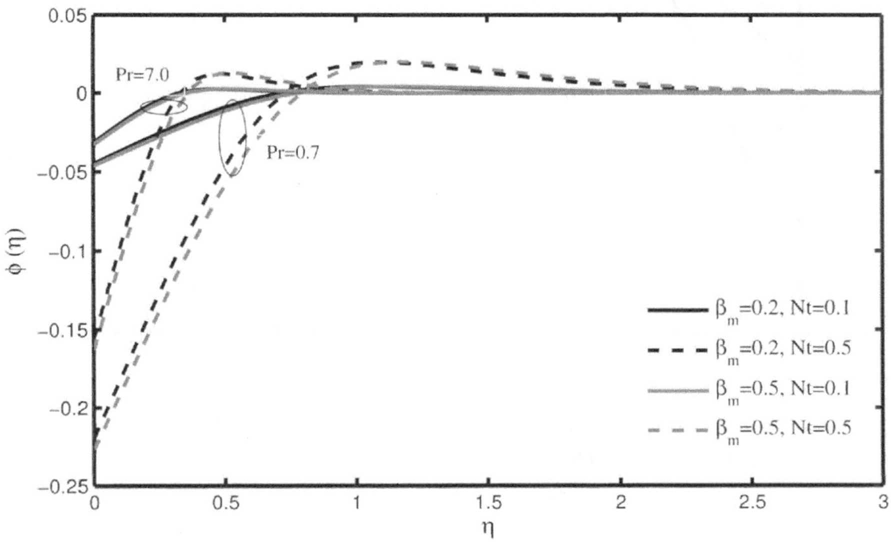

FIGURE 5.14 Dimensionless nanoparticle volumetric profile for different values of Alfvén number (β_m), thermophoresis parameter (Nt) and Prandtl number (Pr).

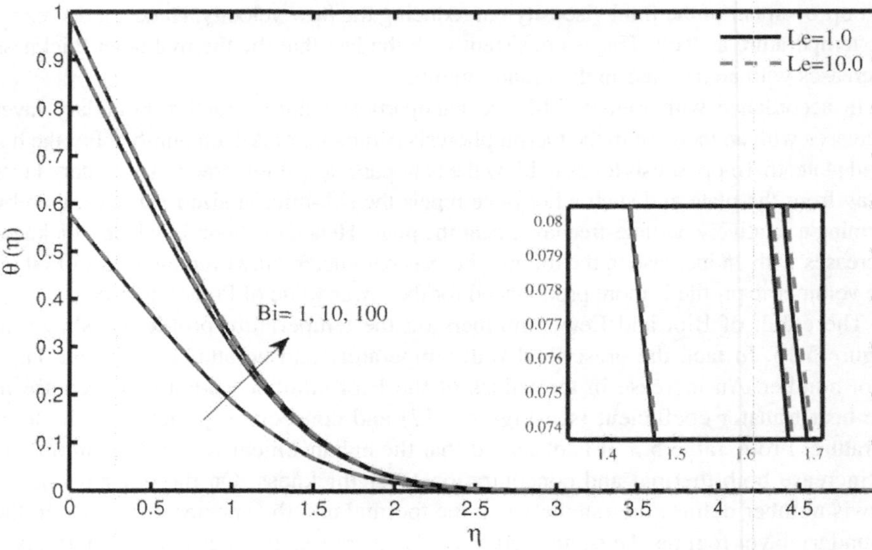

FIGURE 5.15 Dimensionless temperature profile for different values of Lewis number (Le) and Biot number (Bi).

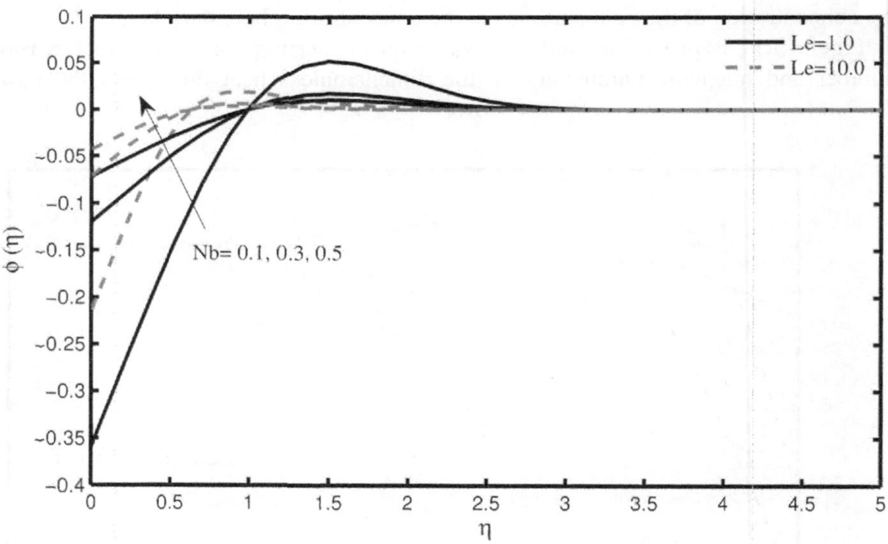

FIGURE 5.16 Dimensionless nanoparticle volumetric profile for different values of Lewis number (Le) and Brownian parameter (Nb).

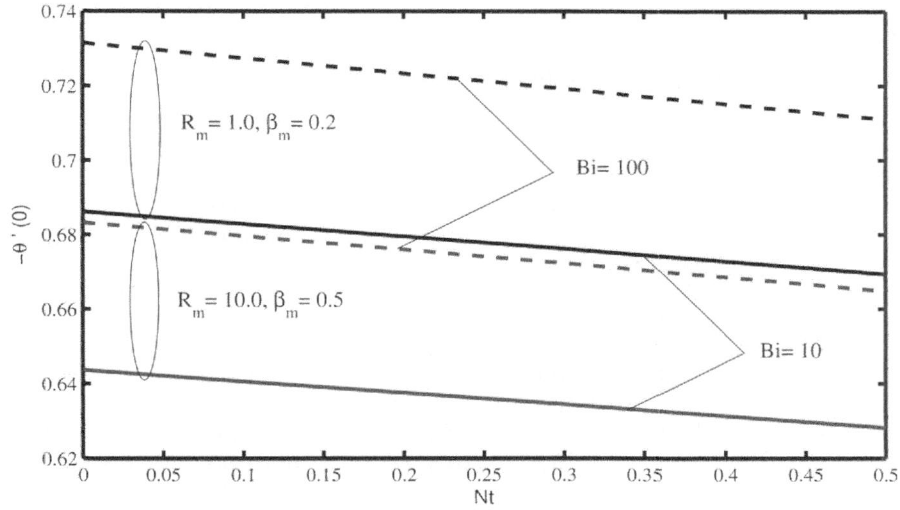

FIGURE 5.17 Local temperature gradient for different values of thermophoresis parameter (Nt) and Biot number (Bi).

TABLE 5.3
Thermal and Nanoparticle Concentration Boundary Layer Thickness with Variation of Bi when R_m=1, β_m=0.2, Γ=0.01, Nb=0.1, Nt=0.1, Pr=1, Le=1

Bi	0.01	0.1	1	10	∞
η_θ	0.3448	1.6352	2.1595	2.2933	2.3121
η_ϕ	1.8051	2.7084	3.1062	3.2155	3.2245

plate into the nanofluid. It is observed that the wall temperature is decreased with an increase in the thermophoresis parameter. It is worth mentioning that the same behavior has been observed in the case of isothermal [5] and constant flux [48] boundary conditions. In addition, the stronger convective heating ($\mathrm{Bi} \to \infty$) on the rigid plate gives higher heat dissipation to the nanofluid, which offers the higher heat transfer from the plate to the nanofluid. On the other hand, the magnetic field parameters decrease the heat transfer rate, which is because the energy injection velocity is smaller than the Alfvén speed. From Table 5.4, it is interesting to notice that the heat transfer rate is independent of Brownian motion parameter Nb when zero nanoparticle flux is imposed on the boundary. A similar conclusion was obtained by Nield and Kuznetsov [4] in the case of constant wall temperature.

5.5.3 Regression Analysis

The linear multiple regression estimate of the heat transfer rate is found here to develop algebraic correlations and it can be written as

TABLE 5.4
Numerical Values of Wall Temperature Gradient ($-\theta'(0)$) when Γ=0.01, Pr=1.0

R_m	β_m	Le	Nb	Nt	$-\theta'(0)$ Bi = 0.1	Bi = 1	Bi = 10
1	0.2	1	0.1	0.1	0.088043	0.423422	0.682856
				0.5	0.088014	0.420198	0.669321
			0.5	0.1	0.088043	0.423422	0.682856
				0.5	0.088014	0.420198	0.669321
		10	0.1	0.1	0.088026	0.421524	0.674906
				0.5	0.087928	0.410349	0.629922
			0.5	0.1	0.088026	0.421524	0.674906
				0.5	0.087928	0.410349	0.629922
	0.5	1	0.1	0.1	0.087259	0.405830	0.638214
				0.5	0.087227	0.402590	0.625605
			0.5	0.1	0.087259	0.405830	0.638214
				0.5	0.087227	0.402591	0.625605
		10	0.1	0.1	0.087240	0.403904	0.630736
				0.5	0.087131	0.392609	0.588567
			0.5	0.1	0.087240	0.403904	0.630737
				0.5	0.087131	0.392609	0.588567
2	0.2	1	0.1	0.1	0.088035	0.423380	0.682754
				0.5	0.088007	0.420174	0.669296
			0.5	0.1	0.088035	0.423380	0.682754
				0.5	0.088007	0.420174	0.669296
		10	0.1	0.1	0.088018	0.421486	0.674822
				0.5	0.087921	0.410345	0.629975
			0.5	0.1	0.088018	0.421486	0.674822
				0.5	0.087921	0.410345	0.629975
	0.5	1	0.1	0.1	0.087239	0.405986	0.638617
				0.5	0.087208	0.402799	0.626203
			0.5	0.1	0.087239	0.405986	0.638616
				0.5	0.087208	0.402799	0.626203
		10	0.1	0.1	0.087220	0.404071	0.631180
				0.5	0.087111	0.392880	0.589344
			0.5	0.1	0.087220	0.404072	0.631180
				0.5	0.087111	0.392880	0.589343

$$\mathrm{Nu}_{est} = \mathrm{Nur} + C_b\mathrm{Nb} + C_t\mathrm{Nt} \tag{5.42}$$

where Nur is the value found for the ordinary fluid. The quantity ε is the maximum relative error and it is calculated by

$$\varepsilon = \left|\left(\mathrm{Nu}_{est} - \mathrm{Nur}\right)/\mathrm{Nur}\right| \tag{5.43}$$

TABLE 5.5
Summary of the Linear Regression Estimate Nu_{est}=Nur + C_b Nb + C_t Nt, and ε is the Maximum Relative Error Defined by ϵ= | (Nu_{est}-Nur)/Nur |, for Nb, Nt each in [0.1 0.2 0.3 0.4 0.5]

R_m	β_m	Le	Bi	Nur	c_b	c_t	ε
1.0	0.2	1.0	1.0	0.4220	0.0000	-0.0081	0.0016
			10.0	0.6805	0.0000	-0.0337	0.0027
		10.0	1.0	0.4221	0.0000	-0.0280	0.0106
			10.0	0.6803	0.0000	-0.1118	0.0043
	0.5	1.0	1.0	0.4034	0.0000	-0.0081	0.0016
			10.0	0.6333	0.0000	-0.0313	0.0026
		10.0	1.0	0.4035	0.0000	-0.0284	0.0107
			10.0	0.6332	0.0000	-0.1044	0.0047
10.0	0.2	1.0	1.0	0.4222	0.0000	-0.0080	0.0015
			10.0	0.6809	0.0000	-0.0334	0.0027
		10.0	1.0	0.4223	0.0000	-0.0279	0.0105
			10.0	0.6807	0.0000	-0.1114	0.0042
	0.5	1.0	1.0	0.4042	0.0000	-0.0080	0.0016
			10.0	0.6353	0.0000	-0.0307	0.0026
		10.0	1.0	0.4043	0.0000	-0.0280	0.0106
			10.0	0.6352	0.0000	-0.1036	0.0042

For the fixed values of physical parameters $R_m = 1, \beta_m = 0.2, \mathrm{Le} = 1, \mathrm{Bi} = 1$, the regression estimate of the reduced Nusselt number is obtained in the following form:

$$\mathrm{Nu}_{est} = 0.4220 + 0.0000\ \mathrm{Nb} - 0.0081\ \mathrm{Nt} \tag{5.44}$$

with the maximum error about 1%. The negative value of the co-efficient shows that the reduced Nusselt number has a decrement with Nt. Also, it is evident that the Nusselt number is fully independent of Nb. The same analysis is repeated for different values of physical parameters and is presented in Table 5.5. We found that for any kind of combined parameters, the maximum error is about 1%. The standard error in the table elucidates that the linear regression model of Eq. (5.39) is an accurate representation of the present numerical data.

5.6 CONCLUSION

A deformity in the non-uniform magnetic field on the axisymmetric boundary layer flow of nanofluid saturated porous media over a rigid body has been studied in detail. The alignment of magnetic and velocity fields are preserved at infinity. The graphical and tabular representation of the characteristics of the flow from the plate to the fluid are given. The important outcomes of the present context are as follows:

1. Unlike uniform magnetic field, non-uniform magnetic field is represented by two different parameters R_m and β_m and the flow behavior have the decreasing and increasing effects on the performance of these magnetic parameters, respectively.
2. Enhancement in the permeability parameter leads to an increase in the momentum, thermal and concentration layers. However, the significant increment is observed in the momentum layer.
3. The higher heat transfer rate from the plate to the nanofluid has been observed in the case of stronger convective heating on the rigid plate.
4. Compared to thermophoresis parameter, the Brownian motion parameter has the negligible effect on the heat transfer rate.
5. The magnetic parameters are manipulated and the suspended nanoparticles' concentration and heat transfer characteristics in the fluid are rearranged.

Future studies on nanofluid flow over a vertical plate with a non-uniform magnetic field can explore several extensions to enhance the understanding of this complex phenomenon. One promising direction is the incorporation of time-dependent effects to analyze transient behavior, which is crucial for practical engineering applications. Additionally, investigating the influence of variable thermophysical properties of nanofluids, such as temperature-dependent viscosity and thermal conductivity, can improve model accuracy. Further, the inclusion of non-Newtonian nanofluids and their interaction with the magnetic field can provide insights into advanced industrial applications, such as cooling systems and biomedical technologies. Moreover, employing machine learning and optimization techniques could aid in predicting optimal conditions for heat transfer enhancement. Experimental validation of numerical results would also be beneficial to ensure real-world applicability.

REFERENCES

1. S. Choi, Enhancing thermal conductivity of fluids with nanoparticle, in: D. A. Siginer and H. P. Wang (Eds.), *Developments and applications of non-Newtonian flows*, ASME MD vol. 231 and FED, 66 (1995), 99–105.
2. H. Masuda, A. Ebata, K. Teramae, N. Hishinuma, Alteration of thermal conductivity and viscosity of liquid by dispersing ultra-fine particles, *Netsu Bussei*, 7 (1993), 227–233.
3. J. Buongiorno, Convective transport in nanofluids, *ASME J. Heat Transf.*, 128 (2006), 240–250.
4. D. A. Nield, A. V. Kuznetsov, The Cheng-Minkowycz problem for natural convective boundary-layer flow in a porous medium saturated by a nanofluid, *Int. J. Heat Mass Transf.*, 52 (2009), 5792–5795.
5. A. V. Kuznetsov, D. A. Nield, Natural convective boundary-layer flow of a nanofluid past a vertical plate, *Int. J. Therm. Sci.*, 49 (2010), 243–247.
6. W. A. Khan, I. Pop, Boundary-layer flow of a nanofluid past a stretching sheet, *Int. J. Heat Mass Transf.*, 53 (2010), 2477–2483.
7. A. Aziz, W. A. Khan, Natural convective boundary layer flow of a nanofluid past a convectively heated vertical plate, *Int. J. Therm. Sci.*, 52 (2012), 83–90.

8. W. Ibrahim, B. Shankar, MHD boundary layer flow and heat transfer of a nanofluid past a permeable stretching sheet with velocity, thermal and solutal slip boundary conditions, *Comput. Fluids*, 75 (2013), 1–10.
9. M. Muthtamilselvan, D. Prakash, Deog-Hee Doh, Effect of thermal non-equilibrium on transient hydromagnetic flow over a moving surface in a nanofluid saturated porous media, *J. Mech. Sci. Tech.*, 28 (2014), 3709–3718.
10. M. Muthtamilselvan, D. Prakash, Unsteady hydromagnetic slip flow and heat transfer of nanofluid over a moving surface with prescribed heat and mass fluxes, *Proc. IMechE Part C: J. Mech. Eng. Sci.*, 229 (2014), 703–715.
11. C. Yang, W. Li, Y. Sano, M. Mochizuki, A. Nakayama, On the anomalous convective heat transfer enhancement in nanofluids: A theoretical answer to the nanofluids controversy, *J. Heat Transf.*, 135 (2013), 54504–54504.
12. M. A. Teamah, M. M. K. Dawood, Ali Shehata, Numerical and experimental investigation of flow structure and behavior of nanofluids flow impingement on horizontal flat plate, *Exp. Thermal Fluid Sci.*, 74 (2016), 235–246.
13. F. B. Tehrani, A. Khabazipur, S. I. Vasefi, Flow and heat transfer analysis of TiO_2/water nanofluid in a ribbed flat-plate solar collector, *Renewable Energy*, 122 (2018), 406–418.
14. D. Pal, G. Mandal, Magnetohydrodynamic non-linear thermal radiative heat transfer of nanofluids over a flat plate in a porous medium in existence of variable thermal conductivity and chemical reaction, *Int. J. Ambient Energy*, 42(10) (2019), 1167–1177.
15. R. Khademi, A. Razminia, V. I. Shiryaev, Conjugate-mixed convection of nanofluid flow over an inclined flat plate in porous media, *Appl. Math. Comput.*, 366 (2020), 124761.
16. I. Waini, A. Ishak, I. Pop, Symmetrical solutions of hybrid nanofluid stagnation-point flow in a porous medium, *Int. Commun. Heat Mass Transf.*, 130 (2022), 105804.
17. T. R. Desisa, Experimental and numerical investigation of heat transfer characteristics in solar flat plate collector using nanofluids, *Int. J. Thermofluids*, 18 (2023), 100325.
18. R. Kandasamy, P. Loganathan, P. Puvi Arasu, Scaling group transformation for MHD boundary-layer flow of a nanofluid past a vertical stretching surface in the presence of suction/injection, *Nuclear Eng. Design*, 241 (2011), 2053–2059.
19. O. D. Makinde, W. A. Khan, Z. H. Khan, Buoyancy effects on MHD stagnation point flow and heat transfer of a nanofluid past a convectively heated stretching/shrinking sheet, *Int. J. Heat Mass Transf.*, 62 (2013), 526–533.
20. S.Das, R. N. Jana, O. D. Makinde, MHD boundary layer slip flow and heat transfer of nanofluid past a vertical stretching sheet with non-uniform heat generation/absorption, *Int. J. Nanoscience*, 13(3) (2014), 1450019.
21. F. Mabood, W. A. Khan, A. I. M. Ismail, MHD boundary layer flow and heat transfer of nanofluids over a nonlinear stretching sheet: A numerical study, *J. Magnetism Magnetic Materials*, 374 (2015), 569–576.
22. B. Mahanthesh, B. J. Gireesha, R. S. Reddy Gorla, F. M. Abbasi, S. A. Shehzad, Numerical solutions for magnetohydrodynamic flow of nanofluid over a bidirectional non-linear stretching surface with prescribed surface heat flux boundary, *J. Magnetism Magnetic Mater.*, 417 (2016), 189–196.
23. S. Gupta, D. Kumar, J. Singh, MHD mixed convective stagnation point flow and heat transfer of an incompressible nanofluid over an inclined stretching sheet with chemical reaction and radiation, *Int. J. Heat Mass Transf.*, 118 (2018), 378–387.
24. D. Ramya, R. Srinivasa Raju, J. Anand Rao, A. J. Chamkha, Effects of velocity and thermal wall slip on magnetohydrodynamics (MHD) boundary layer viscous flow and heat transfer of a nanofluid over a non-linearly-stretching sheet: A numerical study, *Propuls. Power Res.*, 7 (2018), 182–195.

25. L. Ali, X. Liu, B. Ali, Finite element analysis of variable viscosity impact on MHD flow and heat transfer of nanofluid using the Cattaneo-Christov model, *Coatings*, 10(4) (2020), 395.
26. Z. Uddin, K. Sai Vishwaka, S. Harmand, Numerical duality of MHD stagnation point flow and heat transfer of nanofluid past a shrinking/stretching sheet: Metaheuristic approach, *Chinese J. Phys.*, 73 (2021), 442–461.
27. K. Raghunath, Study of heat and mass transfer of an unsteady magnetohydrodynamic (MHD) nanofluid flow past a vertical porous plate in the presence of chemical reaction, radiation and soret effects, *J. Nanofluids*, 12 (2023), 767–776.
28. B. Darbari, S. Rashidi, J. A. Esfahani, Sensitivity analysis of entropy generation in nanofluid flow inside a channel by response surface methodology, *Entropy*, 18 (2016), 52.
29. B. Kumar, G. S. Seth, R. Nandkeolyar, Regression model and successive linearization approach to analyse stagnation point micropolar nanofluid flow over a stretching sheet in a porous medium with nonlinear thermal radiation, *Phys. Scr.*, 94 (2019), 115211.
30. R. Sharma, S. M. Hussain, C. S. K. Raju, G. S. Seth, A. J. Chamkha, Study of graphene Maxwell nanofluid flow past a linearly stretched sheet: A numerical and Statistical approach, *Chinese J. Phys.*, 68 (2020), 671–683.
31. A. M. Darbari, M. A. Alavi, S. R. Saleh, V. Nejati, Sensitivity analysis of nanofluid flow over different flat tubes confined between two parallel plates using Taguchi method and statistical analysis of variance, *Int. J. Thermal Sci.*, 173 (2022), 107428.
32. G. Revathi, S. Avadapu, C. S. K. Raju, M. Jayachandra Babu, A. M. Zidan, M. K. Alaoui, N. A. Shah, J. D. Chung, Dynamics of lorentz force and cross-diffusion effects on ethylene glycol based hybrid nanofluid flow amidst two parallel plates with variable electrical conductivity: A multiple linear regression analysis, *Case Stud. Thermal Eng.*, 41 (2023), 102603.
33. A. V. Kuznetsov, D. A. Nield, The Cheng-Minkowycz problem for natural convective boundary layer flow in a porous medium saturated by a nanofluid: A revised model, *Int. J. Heat Mass Transf.*, 65 (2013), 682–685.
34. J. M. Jalil, K. A. Al-Tae'y, The effect of nonuniorm magnetic field on natural convection in an enclosure, *Numer. Heat Transf., Part A*, 51 (2007), 899–917.
35. A. Borrelli, G. Giantesio, M. C. Patria, Three-dimensional MHD stagnation point flow of Newtonian and a micropolar fluid, *Int. J. Pure Appl. Math.*, 73 (2011), 165–188.
36. N. Bachok, A. Ishak, R. Nazar, I. Pop, Flow and heat transfer at a general three-dimensional stagnation point in a nanofluid, *Physica B*, 405 (2010), 4914–4918.
37. O. Pourmhrana, M. Rahimi-Gorjia, M. Gorji-Bandpya, T. B. Gorji, Simulation of magnetic drug targeting through tracheobronchial airway in presence of an external non-uniform magnetic field using Lagrangian magnetic particle tracking, *J. Magnet. Magnet. Mater.*, 393 (2015), 380–393.
38. S. Hariri, M. Mokhtari, M. B. Gerdroodbary, K. Fallah, Numerical investigation of the heat transfer of a ferrofluid inside a tube in the presence of a non-uniform magnetic field, *Eur. Phys. J. Plus*, 132 (2017), 65.
39. P. S. B. Szabo, W. G. Fruh, The transition from natural convection to thermomagnetic convection of a magnetic fluid in a non-uniform magnetic field, *J. Magnet. Magnet. Mater.*, 447 (2018), 116–123.
40. S. A. M. Mehryan, A. Tahmasebi, M. Izadi, M. Ghalambaz, Melting behavior of phase change materials in the presence of a non-uniform magnetic-field due to two variable magnetic sources, *Int. J. Heat Mass Transf.*, 149 (2020), 119184.
41. S. Li, L. Mao, A. Alizadeh, X. Zhang, S. V. Mousavi, The application of non-uniform magnetic field for thermal enhancement of the nanofluid flow inside the U-turn pipe at solar collectors, *Sci. Rep.*, 13 (2023), 8471.

42. C. Mistrangelo, L. Bühler, MHD flow in curved pipes under a nonuniform magnetic field, *IEEE Trans. Plasma Sci.*, 52(9) (2024), 3455–3461.
43. K. Zitouni, M. Boumaaza, L. Aidaoui, Y. Lasbet, H. E. Abdellatif, S. A. Khan, Y. Chetbani, A. Belaadi, I. M. H. Alshaikh, D. Ghernaout, Numerical assessment of heat transfer and entropy generation of a mixed convection ferrofluid flow under the effect of a non-uniform magnetic field, *Case Stud. Thermal Eng.*, 66 (2025), 105788.
44. A. Borrelli, G. Giantesio, M. C. Patria, Effect of a non-uniform external magnetic field on the 3D stagnation-point flow, *Eur. J. Mech. B/Fluids*, 48 (2014), 210–217.
45. J. M. Dorrepaal, S. Moosavizadeh, Steady incompressible magnetohydrodynamic flow near a point of reattachment, *Phys. Fluids*, 10 (1998), 1512–1518.
46. G. S. McNab, A. Meisen, Thermophoresis in liquids, *J. Colloid Interface Sci.*, 44 (1973), 339–346.
47. S. Duhr, D. Braun, Thermophoretic depletion follows Boltzmann distribution, *Phys. Rev. Lett.*, 96 (2006), 168301.
48. M. B. Glauert, On magnetohydrodynamic flows with aligned magnetic fields, *Fluid Mech.*, 19 (1963), 49–59.
49. A. Borrelli, G. Giantesio, M. C. Patria, An exact solution for the 3D MHD stagnation-point flow of a micropolar fluid, *Commun. Nonlinear Sci. Numer. Simulat.*, 20 (2015), 121–135.

Part II

Nanofluids and Heat Transfer Applications

6 Nanofluid Flows in Heat Exchangers

Abdul Jabbar Ansari and Akshay Kumar Maurya

6.1 INTRODUCTION

A common design for industrial applications and commercial applications is the DPHE, which has a simple flow configuration where hot and cold liquids can flow in parallel or counter to each other [1]. It finds utility in various sectors, such as hygienic and pharmaceutical applications for heating and cooling purposes. Moreover, its simple construction, cost-effectiveness, and ease of cleaning have made it a favored choice in renewable power systems, including solar power, incineration, air conditioning, and latent heat storage [2]. A notable benefit of the DPHE is its capability to handle particle-laden products without clogging. Industries such as chemicals, oil, food, and gas utilize DPHEs for tasks like pasteurization, preheating, fermenter heating, and wastewater heating [3]. Understanding heat exchangers, characteristics of pressure drop, and convection heat transfer functions become crucial to bring this technology to the market. Consequently, numerous methods are continuously explored to improve heat exchanger functionality, including design enhancements, novel system configurations, technological advancements, modifications, integration with other systems, utilization of different working mediums, and the incorporation of heat recovery systems [4]. While it is crucial to understand that enhancing heat transfer in heat exchangers often leads to a rise in pressure drop, necessitating more pumping power, it is equally important to balance the advantages and costs of improved heat transfer and increased pressure drop [5, 6]. Ding *et al.* conducted numerical simulations on TiO_2-water NFs within a DPHE, taking into account different flow rates and mass fractions of TiO_2. The research discovered that all proportions of TiO_2-water NFs demonstrated superior heat transfer capabilities compared to deionized water. However, these fluids also increased the flow resistance within the corrugated tube [7]. In another study by Jassim *et al.*, TiO_2 (titanium dioxide) and Al_2O_3 (alumina) NFs were experimentally evaluated to enhance heat exchange functionality while maintaining conventional fluid characteristics. This evaluation resulted in functional improvements ranging from 13% to 23% at a 3% concentration of NFs [8]. Over the past decade, numerous researchers have explored leveraging NFs as heat transfer fluids to improve the

DOI: 10.1201/9781003611820-8

functionality of DPHEs. A CuO-water NF was used as a fluid of heat transfer to improve DPHEs in an experimental study by Kavita *et al.* [9]. Mansoury *et al.* utilized NPs with a size of 20 nm and studied turbulent flow regimes. The coefficient of heat transfer of DPHE improved significantly by 60% according to the results, whereas the heat exchanger's plate showed an 11% enhancement [10]. According to Akbar *et al.*, the hybrid NF showed good thermophysical properties that could be used in real and industrial settings, resulting in better heat transfer functionality [11]. Using Al_2O_3 and Cu NFs, Jassim *et al.* tested the functionality and heat loss of the DPHEs experimentally. Nusselt number exhibited a direct proportionality to the Reynolds number under all conditions, indicating enhanced efficiency of the exchanger through the using of NPs in the base liquid [12]. A hybrid MgO/MWCNT-water NF was used by Henein *et al.* as the working fluid to improve the thermodynamic functionality of the transparent heat-pipe solar collector. The outcomes demonstrated increased energy efficiency and exergy as the weight ratio and volumetric flow rate of MWCNT NPs were increased. Specifically, the hybrid NF composed of 50% MgO/MWCNT showed energy efficiency and collector exergy improvements of 55.83% and 77%, respectively [13]. Vinceley *et al.* conducted a practical on the functionality of the flat-plate solar collector employing the concentric DPHE connected to a forced-moving graphene-water oxide NF [14]. Kore *et al.* conducted a numerical and computational study on enhancing heat transfer rates in heat exchangers. They investigated the use of obstructing geometries, including circular ring, circular ring with clearance, and hexagonal ring tabulators, with various pitch diameter ratios (PDRs) [15]. Kore *et al.* investigated how a single dimple combined with an arch-type turbulator affects heat transfer and fluid friction [16]. While previous research has predominantly focused on enhancing the coefficient of heat transfer, the significance of minimizing or preventing pressure drops has been overlooked. Furthermore, studies on the NFs used in heat exchangers have mainly concentrated on factors influencing heat transfer, with limited guidance provided on parameters affecting the pressure drop. The papers featured in this article have been selected based on specific criteria outlined in Table 6.1, while their year-by-year distribution is illustrated in Figure 6.1.

TABLE 6.1
Criterion for Selecting and Rejecting the Papers

S.N.	Selection Conditions	Rejection Conditions
1	Scholarly articles that underwent peer review from 1986 to 2024.	Articles that do not publish from 1986 to 2024.
2	Content that is relevant to enhancing the functionality of the DPHEs.	Content related to other disciplines not related to the industry. Enhancing the functionality of the DPHEs.
3	Written in the English language.	Articles that were not in English.
4	Articles exploring the impact of NFs on the functionality of DPHEs.	Material that is unrelated to how NFs influence the functionality of the DPHEs.

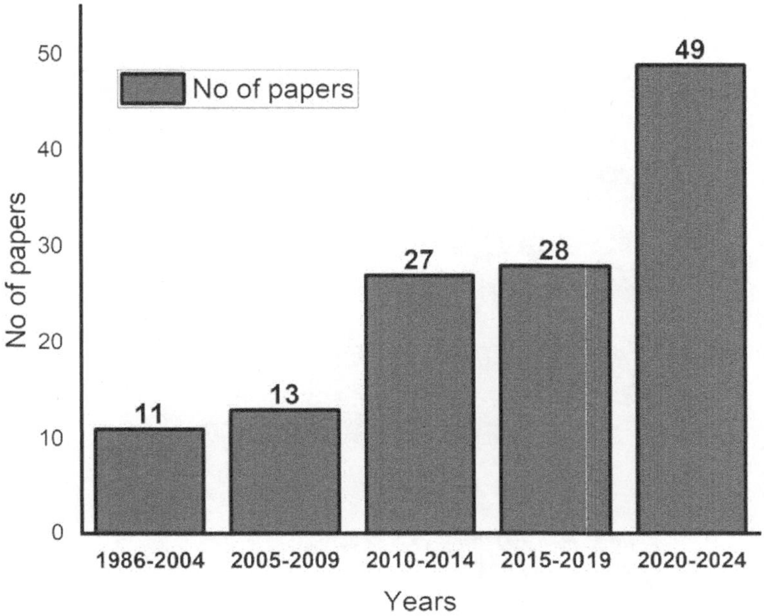

FIGURE 6.1 Year-wise distribution of papers.

6.2 HEAT EXCHANGER PERFORMANCE

Various approaches have been explored to enhance the functionality of heat exchangers. Table 6.2 provides a comprehensive overview of the active and passive methods employed [2]. Typically, selecting a fluid with higher viscosity for the annular region (the larger passage field) is preferable as it leads to lower pressure drop [5]. Figure 6.2 illustrates an introduction to the configuration of the DPHE. Two types of flow systems can be implemented: In parallel flow, both hot and cold fluids enter from the same side and move in the same direction (Figure 6.3a), and on the other hand, in counter flow, the fluids enter from different sides and move in opposite directions (Figure 6.3b) [5, 6].

There are numerous benefits associated with a heat exchanger that reduces pressure drop, including the elimination of the need for a large pump, cost savings, and reduced power consumption, thereby lowering operational expenses. Furthermore, lower velocities resulting from reduced pressure drops contribute to the prevention of wear, pipe shaking, noise, and further pressure drops [5].

Various components can impact the heat exchanger functionality, including fluid characteristics, design, and flow patterns (Figure 6.4). In contrast to active techniques, passive techniques offer a more gentle approach. Simple methods have been implemented to boost the heat transfer coefficient, as demonstrated in Table 6.2. Among these techniques, spiral wire and semi-porous wire showed higher pressure drop rates. However, the increase in pressure drop can be offset by an improvement in heat transfer efficiency. Passive methods outperform active methods due to their

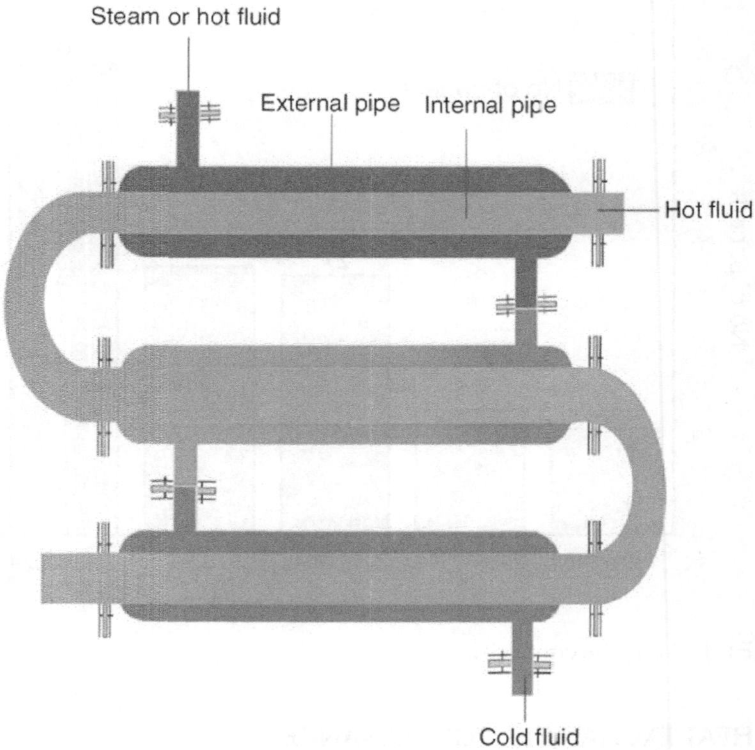

FIGURE 6.2 A simple DPHE [17].

resourcefulness, cost-effectiveness, and ability to achieve satisfactory pressure drop rates [19]. NPs represent a promising avenue within the realm of passive technologies for enhancing the functionality of heat exchangers.

The base liquid and specific factors of the NPs, like concentration, geometry (shape), and size, can influence thermophysical properties of NFs, as noted by Gupta *et al.* [20]. The utilization of NFs has emerged as an increasingly popular technique in recent times. Several NFs have been tested for improved functionality of a DPHE [21]. The system's pressure drop and heat transfer coefficient are determined by the unique diversities in the thermophysical characteristics of NFs, which include specific heat, density, and coefficient of expansion, viscosity, and thermal conductivity among others [22].

6.3 NANOFLUID IN DPHES

Targui and Kahalerras studied numerically the effect of the simultaneous use of porous baffles and different NFs in a DPHE. It was found that the particle volume fraction played a more significant role in heat transfer in comparison to the nature of NF. They also stated that the highest thermal effectiveness was for a time when the NFs

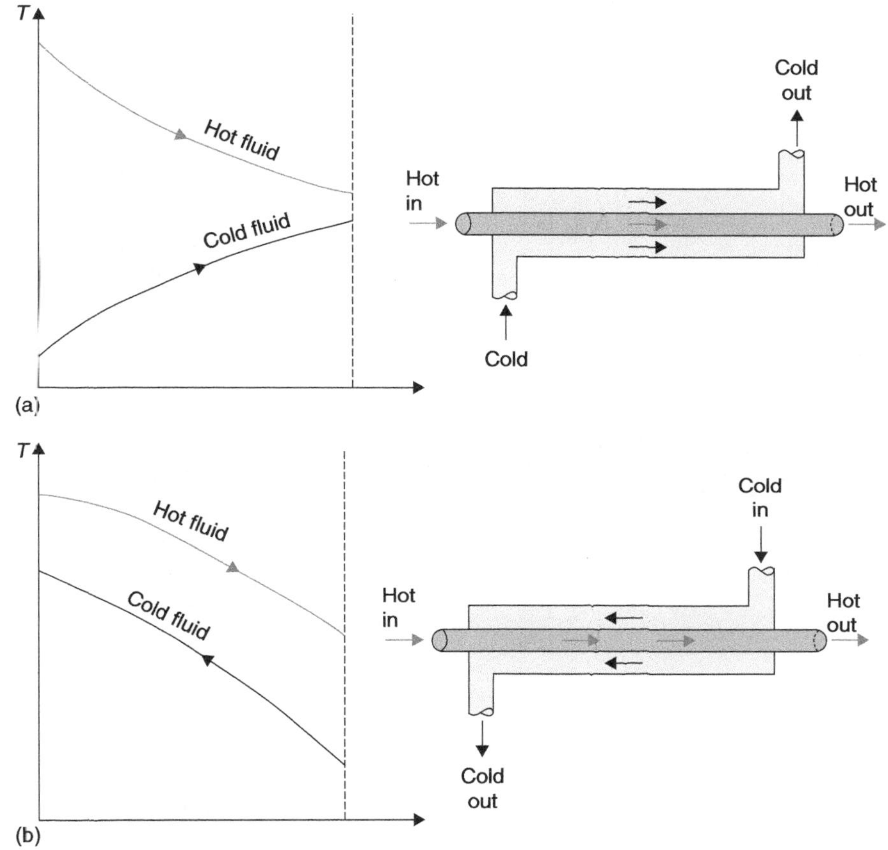

FIGURE 6.3 (a) Parallel flow in a DPHE; (b) Counter-flow in a DPHE [18].

flowed in the annulus of the DPHE [23]. Prasad *et al.* conducted a study on the effect of trapezoidal strips wrapped in DPHE. The outcomes demonstrated that the rate of heat transfer increase because of intermediate flows was greater in the annulus than in the plenum [24]. Saeedan *et al.* examined the computational dynamics and modeling of helix deflector DPHEs and found that an enhancement in the Nusselt number was observed when the volumetric concentration was enhanced. However, when the CNT/water concentration was increased, a reduction in the Nusselt number was observed [25]. Reddy and Rao conducted a study on the flow properties of DPHE fitted with pre-wired coils. The study revealed that the concentration of NPs in an NF at 0.02% resulted in a 10.73% improvement in the coefficient of heat transfer and an 8.73% comparison to base fluid, there was an improvement in friction [26]. Mohammed *et al.* utilized NFs such as CuO, Al_2O_3, ZnO, and SiO_2 with concentrations ranging from 1% to 4% and diameters ranging from 20 to 50 nm. The study showed that the rate of heat transfer enhanced when greater concentrations and smaller diameters of NFs were used [27]. Maddah *et al.* explored the influence of Al_2O_3 NF with enhanced

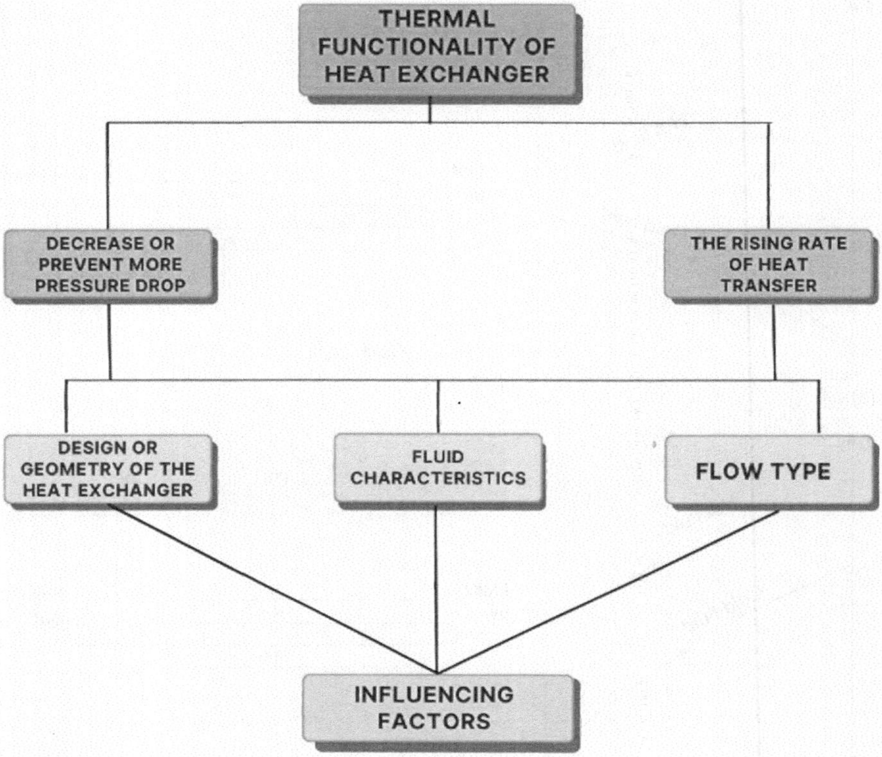

FIGURE 6.4 Scheme showing factors that affected heat exchanger functionality.

strips wrapped in DPHE. The researchers varied twisted risers' stroke length and rotation speed based on the GPR. The study revealed that using twisted band NFs with GPR 1 resulted in a reduction in both heat transfer and pressure drop [28]. Demir *et al.* determined the properties of heat transfer of Al_2O_3 and TiO_2 NFs in a DPHE where a constant wall temperature boundary condition. Its outcomes indicated that an increase in the concentration of NFs leads to an enhancement in both pressure drop and the rate of heat transfer. Furthermore, even NFs with low concentrations were found to increment the rate of heat transfer [29]. Shakiba and Vahedi conducted a study to examine the impact of magnetic areas on the hydrodynamic and thermal properties of DPHEs [30]. Wu *et al.* obtained that the NFs influenced critical Reynolds numbers. Furthermore, the study also revealed that the NFs' thermophysical properties had a greater impact on the heat transfer rate than their lateral effects, such as thermophoresis, diffusion, and Brownian motion. The study concluded that these NFs could be considered homogeneous phases [31]. In a DPHE, Bahiraei and Hangi conducted an experiment to test the MnAZn NF in the existence of a magnetic field. In the experiment, the NF functioned as a coolant within an air chamber, while the annulus was filled with water. However, the impact of the magnetic field was found to

TABLE 6.2
Examples of Methods used to Enhance the DPHE

S.N.	Enhancement Perspective	Passive Method (Observations (by Percentage, %))
1.	Metal foam	Nusselt number increased by 57.21% [119]
2.	NFs (MgO-ethylene glycol)	Increase in heat transfer coefficient by 27% at wt% = 0.3 and pressure drop by 35% at wt% = 0.3 [120]
3.	Helical wires	Increases Nusselt number by up to 2.64 times. Enhances the coefficient of friction by about 2.74 times [121]
4.	Porous media	44% increase in heat transfer at higher pressure drops [122]
5.	Using fins	Increase the heat transfer rate by about 90–98%. Excessive pressure drop in finned tube [123]
6.	Twisted tape	A significant increase (15%) in heat transfer rate. The friction coefficient has been enhanced by 10% [124]
S.N.	**Improvement Approach**	**Active Method (Observations (by Percentage, %))**
1.	Using ultrasonic vibration	Heat transfer is increased by almost 60% [125]
2.	Using surface vibrations	Increase the heat transfer coefficient by 9% [126]
3.	Using air bubble injection	The overall heat transfer coefficient can be improved ranging from 10.30% to 149.50% [127]
4.	Using magnetic field	320% increased heat transfer and slight pressure drop [128]

be short-lived at larger Reynolds numbers [32]. These NFs, which are suspensions of small, nano-sized solid particles in a base liquid, can be utilized as heat transfer fluids due to their generally higher thermal efficiencies compared to conventional fluids. They have been extensively researched as alternative working fluids. Common base fluids include water, motor oil, paraffin oil, ethylene glycol, and others [33]. Alklaibi *et al.* experimentally assessed Fe_3O_4–SiO_2/water hybrid NFs in plate heat exchangers, evaluating overall heat transfer coefficient (U), thermal performance factor (TPF), and friction factor (Ff). They found that using 1 volume percent hybrid NF enhanced U by 21.30%, Nu by 25.85%, Ff by 37.59%, and TPF by 18.1% compared to the base fluid [34].

6.4 INFLUENCE OF THE MORPHOLOGICAL CONCENTRATION OF NANOPARTICLES ON THE CHARACTERISTICS OF NANOLIQUIDS

6.4.1 Nanofluids in Heat Exchangers

At various Reynolds numbers, the impact of NFs on the pressure drop and the rate of heat transfer was conducted by Wang *et al.* Pressure drop, a fundamental physical property, was found to be affected by NFs and is influenced by factors such as

Brownian motion, diffusion, and thermophoresis exhibited by the particles. The conclusion drawn was that NF should be considered as inhomogeneous phases for future experiments [35]. The impact of heat transfer on a compact DPHE was achieved by Aghayari *et al.* Different concentrations of NPs were tested, and variations in heat transfer rates were observed. The study demonstrated efficient heat transfer with increasing Reynolds number, resulting in 12% heat rectification for the entire system [36]. Esfe *et al.* investigated the impacts of a very small percentage (0.005–0.02) of MgO NPs in water within the Reynolds number range of 2500–26000 [37]. Asirvatham *et al.* researched Ag NPs and discovered that the improvement of heat transfer gradually increased to 28.7% and 69.3% at volume concentrations of 0.3% and 0.9%, respectively [38]. Sarafraz *et al.* conducted an experiment in which they mixed organic green tea leaf NPs and silver nitrate with a base liquid such as water or ethylene glycol. The study found that a unit volume concentration resulted in 67% heat transfer rectification [39]. Sajadi *et al.* researched the use of TiO_2 NPs and found that a NP volume fraction of 0.25% resulted in a 22% enhancement of heat transfer [40]. Zamzamian *et al.* researched using CuO and Al_2O_3 NPs in ethylene glycol. According to the study, the rate of heat transfer enhanced by 26% for Al_2O_3 NFs and 37% for CuO NFs [41]. Dharmakkan *et al.* explored the impact of hybrid NFs on microplate heat exchangers, finding that TiO_2-ZnO/ethylene glycol NF significantly enhanced thermal conductivity and heat transfer, achieving a Nusselt number of 35.8 at 50°C and 4% volume fraction [42].

6.4.2 Impact of Nanoparticles on Nanofluid Properties

Average NP sizes of TiO_2/water, Ag/water, and ZnO/water NFs were found to range from 14 to 20 nm, according to a study on their thermophysical properties, which were determined in practical experiments conducted by Hozien *et al.* A study prepared three different NFs, each with a NP volume concentration of 0.25%. A study found that all three NFs exhibited increased viscosity, density, and thermal conductivity. The synthesized NPs were characterized using techniques such as TEM, SEM, XRD, and the Zetasizer test [43]. Pordanjani *et al.* found to change in the thermophysical properties of NFs, including their thermal conductivity, density, specific heat, and viscosity by the addition of NPs to the base liquid [44]. The study found that the transfer of energy within a liquid can be improved by the presence of a NP suspension, resulting in enhanced heat transfer properties and thermal conductivity [45]. According to other practical studies, the thermophysical properties of NFs are influenced by various factors, including the concentration, shape, size, and inherent NP properties [46]. Alqarni *et al.* developed an efficient heat exchanger using Al_2O_3-Cu-water hybrid NF. Using FVM (finite volume method) and SIMPLEC algorithm, they optimized the performance with ribbed tubes, to simulate the heat transfer [47].

6.4.2.1 Impact of Volume Concentration of Nanoparticles on Nanofluids Properties

Osman *et al.* conducted a study on the viscosity of the Al_2O_3-water NF was utilized to enhance as the volume concentration of NF was initiated and enhanced [48].

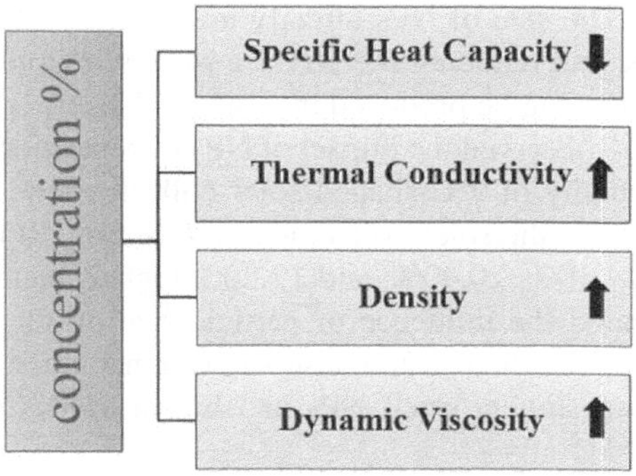

FIGURE 6.5 Influences of an increase in NP concentration in a base fluid on NF properties.

According to a study, the flow state (turbulent or laminar) was determined by the Reynolds number, which was influenced by the density of the NF [49]. Similarly, a relationship between NP concentration and NFs properties is illustrated in Figure 6.5. Rabienataj *et al.* examined practical tests using the Viscolite 2700 viscometer and showed that as the concentration of NPs in the base liquid increases, viscosity of NF also increases [50]. Saleh *et al.* obtained multi-walled carbon nanotubes/aqueous NFs (MWCNTs/aqueous NFs) and found that thermal conductivity, density, and viscosity of NF enhanced by 15.27%, 1%, and 9.15%, respectively, when compared to a concentration of 0.3% particles with water as the base liquid [51]. In a recent practical study on NFs (Al_2O_3-based liquid and CuO-based liquid), Asokan *et al.* found that as the concentration of NPs in the base liquid increased, so did thermal conductivity, viscosity, and the density of NF [52]. A study found that mixture of 60% ethylene glycol and 40% water was used as the base liquid [53]. Bhattad *et al.* studied Al_2O_3-TiO_2 hybrid nanofluids in plate heat exchangers, examining temperature (283–298 K) and particle volume ratio (0:5 to 5:0). They found that higher TiO_2 ratios decreased heat transfer performance, while higher inlet temperatures reduced pressure drop. The Al_2O_3(5:0) NF showed the best improvement in heat transfer metrics, whereas TiO_2 (0:5) increased pressure drop and Prandtl number [54]

Sundar and Mouli experimentally studied Fe_3O_4–SiO_2/water NFs in plate heat exchangers under laminar flow. They varied particle loadings (0.2%–1%) and flow rates (0.05–0.1166 kg s−1). Using a scaled conjugate gradient method, they found that increasing Re and NP volume fraction improved heat exchanger effectiveness and the number of transfer units. At 1 volume percentage hybrid NF and Re of 571.89, efficiency increased by 13.13% and transfer units by 24.51% compared to water [55].

6.4.2.2 Impact of Nanoparticle Size on Nanofluid Properties

The thermophysical characteristics of NFs are influenced by the morphology and concentration of NPs [56]. Similarly, a study examined the impact of NP size on the viscosity, density, and thermal conductivity of ionic liquids (ILs) as a NF function. Stepwise improvements in thermal conductivity were observed, with average increases of 9.73%, 6.53%, 6.41%, and 7.60% for 10 nm, 30 nm, 60 nm, and 90 nm

NPs, gradually [57]. The size of NPs directly affects thermal conductivity, density, and viscosity of NFs, and reducing the NP size results in improved thermophysical properties [58, 59]. NPs can be produced in a range of sizes, typically between 5 nm and 100 nm. Kim *et al.* observed the impact of NP concentration and size in an Al_2O_3-water NF on functionality of a U-shaped solar collector. The concentration of NF, and optimal efficiency of the solar collector using 20 nm, 50 nm, and 100 nm NFs was observed to be 24.10%, 20.40%, and 17.80% higher than that with water alone [60]. Zhang investigated the influence of particle size on the functionality of heat transfer of SiO_2-water NF. The research also showed that when compared to the base liquid, NFs containing smaller-sized NPs had higher values of dynamic viscosity [Figure 6.6(a), (b)] [61].

6.4.2.3 Impact of Nanoparticle Shape on Nanofluid Properties

Cui *et al.* investigated that TiO_2/water NFs containing club-shaped and sheet-shaped NPs exhibited higher relative thermal conductivity (RTC) compared to other shapes [62]. It was observed that plate-shaped and cylindrical NPs at specific particle concentrations led to increased viscosity [63]. Zahmatkesh *et al.* introduced formulas and observed the effect of varying aspect ratios on thermal conductivity and viscosity of NFs [64]. A study by Elias *et al.* examined how the shape of NPs in NFs affected the efficiency of heat exchangers. Specifically, cylindrical NPs have shown improved effectiveness in heat transfer, as indicated by their coefficient of heat transfer, overall heat transfer coefficient, and thermal conductivity (refer to Figure 6.7) [65].

The viscosity and thermal conductivity of the final nanomaterials are directly influenced by the shapes of the NPs commonly used in NF preparation [66]. Shahsavar *et al.* investigated hydraulic and thermal properties of a boehmite-alumina NF in mini-channel DPHE, taking into account the effect of various shapes of NPs. The study revealed that the shape of NPs has an impact on the thermal conductivity of NF. Specifically, plate-shaped NPs exhibited the lowest thermal conductivity, while cylindrical-shaped NPs showed the highest thermal conductivity [67]. Bouselsal *et al.* enhanced small tube/shell heat exchangers by changing tube shapes and using Al_2O_3-MWCNT/water nanofluids. Diamond-shaped tubes showed the best heat transfer, with up to 103.77% improvement at 2% nanoparticle concentration and minimal entropy [68]. Ikhtiar *et al.* numerically studied the lamella heat exchanger for engine intake air cooling. They found that smaller internal holes in the lamella increased heat transmission, temperature drop, number of transfer units, and pressure drop, with all parameters except LMTD being inversely related to internal openness and directly proportional to the aspect ratio [69]. Kücük experimentally studied mini-channel STHEs (shell and tube heat exchange), finding that their overall heat transfer coefficient (U) was 1.1–6.6 times higher than traditional heat exchangers. The mini-channels achieved a U of 6700 W/m²K and caused a pressure drop over 14,000 psi higher than macro-tube heat exchangers [70]. In a triple-tube heat exchanger (TTHX), Gaur *et al.* investigated MXene-based hybrid NFs with dissimilar-shaped NPs (CuO spherical, MWCNT cylindrical, Al2O3 blade, graphene platelet, and Fe_3O_4 brick) at a 1% volume ratio. The study found that MXene-spherical NFs improved efficiency by 5.16% compared to water. MXene-cylindrical NFs showed an 11.23% increase in

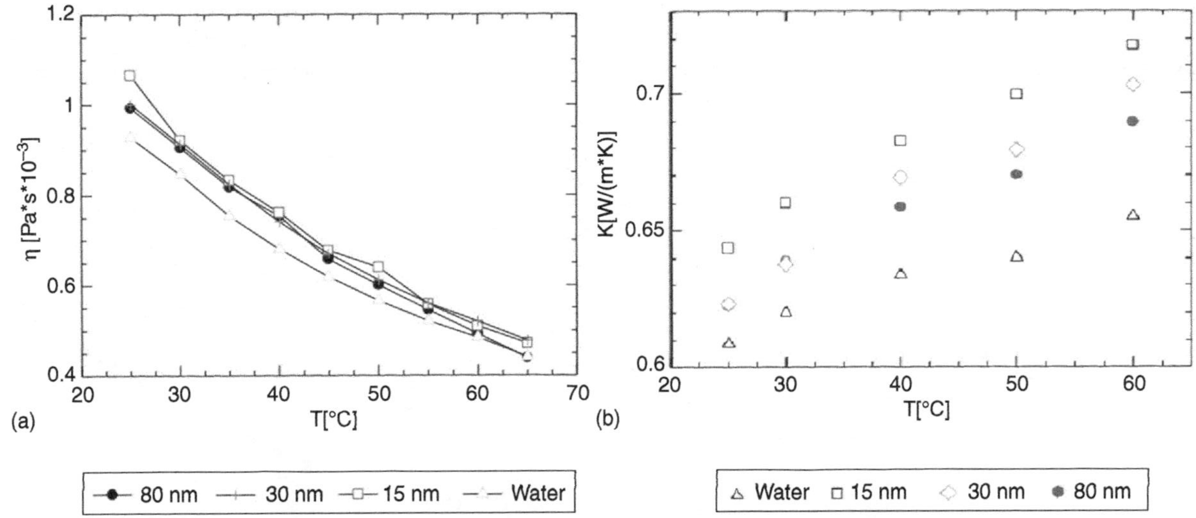

FIGURE 6.6 (a) Viscosity differentiation of SiO_2-water NFs as a function of temperature and particle size; (b) thermal conductivity differentiation of SiO_2-water NFs as a function of temperature and particle size [61].

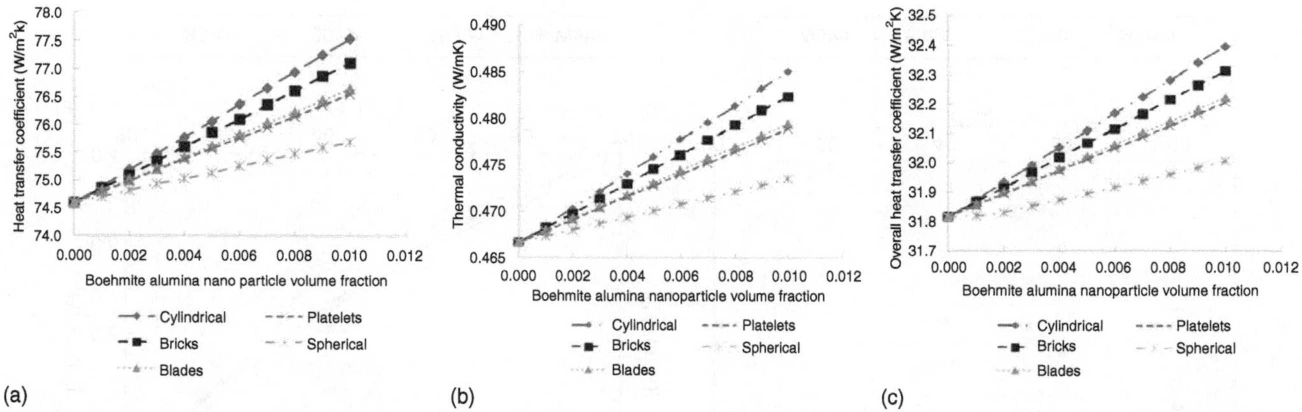

FIGURE 6.7 Influence of different particle shapes on (a) the heat transfer coefficient of NFs, (b) thermal conductivity of NFs, and (c) overall heat transfer coefficient of NFs [65].

heat transfer and a 12.69% higher performance index at the lowest mass flow rate. MXene-cylindrical nanofluids also generated 24.11% less entropy than DI water, achieving a maximum thermal performance factor of 1.65 [71].

6.5 HEAT TRANSFER ENLARGEMENT METHODS

There are typically three primary techniques for improving heat transfer.

6.5.1 Active Method

Zhang *et al.* enhance heat transfer by using rotor variants. Three types of geometries, namely discrete impeller, helical impeller, and ladder impeller, were employed. A percentage enhancement in the Nusselt number from about 71.5 to 123.1 compared to plain tubes was observed, and the corresponding elongation of the coefficient of friction ranged from 37.4 to 74.8% [72]. El-Maghlanys *et al.* analyzed a DPHE that had a rotating internal tube. It has frequently been mentioned that the rate of the heat transfer, efficiency, and number of transfer units (NTUs) enhance with increasing speed [73]. Rai *et al.* enhanced the thermal performance of plain tube heat exchangers using twisted turbulator inserts (TTI) and perforated twisted turbulator inserts (PTTIs) with water-based ternary hybrid nanofluid (THNF) containing CuO, Al_2O_3, and TiO_2. THNF at 0.06%, 0.09%, and 0.12% volume concentrations improved thermal and hydraulic efficiency. The combination of 0.12% THNF with TTI and PTTI significantly boosted heat transmission, exergy change, friction factor, and reversibility [74]. Examples of this approach include using magnetic fields to disrupt the flow, introducing surface or flow vibrations, using reciprocating pistons, and applying electromagnetic fields.

6.5.2 Passive Method

The following sections discuss in detail the research related to each of these methods.

6.5.2.1 Wired Coils

Naphon investigated the impact of pre-wired coil cartridges on the heat transfer and pressure drop in concentric DPHE that used water as the working medium. It found that the utilization of these wired coils has a considerable impact on the heat transfer rate in a system with laminar flow [75]. Akpinar researched the dimensionless exergy losses, heat transfer, and pressure drop in concentric DPHEs that had wire coils put in the internal pipe. Cold water and hot air served as the working fluids, flowing through an annulus and an air chamber, respectively. The study revealed that Nusselt number and friction coefficient improved, leading to an exergy loss that was 2.64–2.74 times greater than a smooth pipe [76]. Essentially, the impact of this decreases as the Reynolds number gets larger. It is important to note that there have also been other studies that used coil inserts in DPHEs that are comparable [77, 78]. Anitha *et al.* analyzed Al_2O_3-CuO/water hybrid NF in a SHCHE (shell and helically coiled heat exchanger), varying the HYNF (hybrid nanofluid)'s Reynolds number from 6000 to 15,000. With a fixed oil Reynolds number of 900, 0.1 vol% HYNF showed the best thermal performance, increasing the Nusselt number by 20% and exergetic efficiency by 48% compared to the base fluid and other NFs [79]. Marzouk *et al.* experimentally and numerically investigated unique tube designs for helical tube heat exchangers. Their findings showed that novel configurations significantly improved heat transfer coefficients compared to uniform tube distributions. Higher Reynolds numbers led to decreased pressure, with only marginal changes in pressure drop and pumping power, but significantly enhanced heat transfer performance [80]. Painuly *et al.* evaluated heat transfer and flow characteristics of SiC-MWCNT and Al_2O_3–MWCNT hybrid NFs in HCTs (helically corrugated tubes) under laminar flow. They studied various factors, including volume fraction, pseudo plastic behavior, fluid selection, and tube geometry. HCT-3 showed the highest Nu increase (61.3% from smooth tube). At volume fraction = 0.1%, NFs changed from Newtonian to shear thinning. Both NFs had TPF (thermal performance factor) values >1 [81].

6.5.2.2 Twisted Tapes Inserted in DPHEs

Yadav investigated the impact of half-length braided ribbons on both heat transfer and pressure drop within U-tube DPHE. It is important to mention that the impact of fabric strip inserts within DPHEs has not yet been thoroughly studied [82]. Twisted tape cladding is commonly acknowledged as a highly effective method for improving heat transfer and is widely used due to its affordability, ease of maintenance, simplicity, and uncomplicated installation [83]. Naphon investigated heat transfer and pressure drop in a horizontal DPHE, both with and without the presence of braided bands, confirming its practical feasibility. The twisted bands used were made of 1 mm thick aluminum. The findings showed that twisted ribbon jacket had a considerable effect on both heat transfer and pressure drop within heat exchanger [84]. Typically, the twisted mesh acts similarly to a continuous vortex generator, generating turbulence in the flow. This promotes better fluid mixing, ultimately leading to higher heat transfer rates. However, previous studies have indicated that twisted tapes exhibit optimal performance in laminar flow systems [85]. Balaga *et al.* studied the impact of nanoparticles and different inserts on heat transfer and pressure drop in f-MWCNT-Fe_2O_3/water hybrid NF. Increasing NP concentration from 0.01% to 0.03% improved heat transfer by 21.66% to 31.66% and pressure drop by 1.14 times. Inserts with a 3:1 twist and coil ratio further enhanced heat transfer by 32.31% to 94.03%, with only a 1.09 to 1.13 times increase in friction factor [86]. Al-Obaidi and Alhamid analyzed the thermal-hydraulic performance of circular pipe heat exchangers with various twisted tape inserts. These inserts increased flow resistance and pressure differential, causing more vortex motion. Their findings showed that increasing the number of twisted tape inserts from 1 to 5 resulted in temperature differences of 38.1%, 46.11%, and 50.52% compared to a smooth pipe [87]. Kumar and Kumar examined the performance of a single-pass heat exchanger using a hybrid metal matrix NF (70% Al2O3, 30% CuO) and twisted tapes (ratios y/W 4 and 3). Using the RNG k–ε model and Ansys 2020 R2, they found a significant Nu increase of 370% for twist ratio 3. While the friction factor rose by 294% with twisted tapes, volume concentration had no effect. The thermal performance factor (TPF) was 3.52 times higher than a normal tube heat exchanger for twist ratio 3 [88].

6.5.2.3 Other Kinds of Tabulators

Akpinar and Bicer investigated research on diameter, quantity of holes, and arrangement of the turbulators. Two patterns were considered: linear and zigzag. It was observed that Nusselt number showed the greatest increase, being 130% higher compared to a smooth tube [89]. Numerical studies have also examined the utilization of fins as turbulators within DPHEs to improve heat transfer. AlKayiem and El-Rahman conducted studies specifically focused on the coastal region [90]. Zhang *et al.* improved the heat transfer shell side of a DPHE by utilizing vortex generators and spiral fins. In this instance, the working medium consisted of cold air circulating through the annulus and steam flowing through the inner tube. Rectangular fin pair (RWP), delta wing (DW), rectangular wing (RW), and delta fin pair (DWP) were the four various types of geometries of vortex generators that were used [91]. Yildiz *et al.* achieved an empirical exploration into heat transfer and flow of pressurized fluids within DPHE. Rotation was induced by loosely mounted propellers in the flow. The observation was made that the rate of heat transfer was 250% greater in comparison to a plain tube, while pressure drop in this system was approximately 500–1000% higher than that of a plain tube. These figures depended on the number of helices and the Reynolds number were fixed at 5 and 10, respectively [92]. Alkam and Al-Nimr conducted research into the impact of porous substrates within the substrates placed on both sides, resulting in increased heat transfer rates in both counterflow and parallel configurations. The improvement was particularly significant when the thermal conductivity coefficients were higher [93]. It should be noted that other instances of simulations that incorporate porous substrates have been demonstrated [94]. Eiamsa-ard *et al.* investigated the impact of louvered strips and both front and rear configurations were available for the louvered strips. When using the rear configuration, there was a 263% enhancement in the mean Nusselt number and a 233% increase in the enhanced coefficient of friction, compared to a plain tube. In the front configuration of the louvered strips, there were increases of 284% and 413% in mean Nusselt number and coefficient of friction, respectively. Even for front configuration, the criterion for evaluating functionality improved by approximately 9–24% [95]. Durmus calculated the exergy and heat transfer losses on concentric coil DPHE that had a snail inlet. By generating a turbulent flow, the snail inlet, which was situated in the inlet area of the bladder, increased heat transfer. When the results were combined, it was found that the Nusselt number increased by 85% to 200% on both sides of the counterflow heat exchanger, with the reported values primarily dependent on rotation angles. An observation was made that the counterflow and parallel flow pressure drops were about 110% higher than those in a smooth tube. According to the criterion of exergy losses, counterflow was identified as the most desirable case [96]. A study was carried out by Raj *et al.* to examine how double expansion tubes impact the flow properties when ethylene glycol and water are used as working fluids. Under constant pumping power, the largest enhancement in internal heat transfer rate was noted for ethylene glycol at 6% in the laminar regime and 18% in the turbulent water regime.

The double-width tubes were found to be more efficient in turbulent flow regimes [97]. Utilizing rough surfaces is another method to enhance the functionality of a DPHE [98–100]. Baffle plates are frequently used as a heat augmentation method in tube and shell heat exchangers, but have not been extensively studied. Based on the author database, the only numerical evaluation studies were conducted by Targui *et al.* [101, 102]. Abidi *et al.* investigated turbulent Fe_3O_4-MWCNT/water hybrid NF in an STHE (shell-and-tube heat exchanger), using baffles and varying Reynolds number (5000–12,500) and volume fraction (0–4%). Results showed the highest friction coefficient with a 180° baffle, 4% volume fraction, and Reynolds number 5000 [103]. Gaur *et al.* examined various water-based hybrid NFs in a triple concentric tube heat exchanger (TCTHX). Shapes studied included MWCNT-cylindrical, Al_2O_3-blade, graphene-platelet, and Fe_3O_4-brick. The brick-blade hybrid showed the best performance in terms of performance index (η), thermal performance factor (TPF), and cost benefit ratio (CBR). At 0.1 kg s^{-1}, the brick-platelet hybrid was 7.2% less efficient than water. The platelet-cylindrical hybrid had a 12.71% lower heat transfer rate and a 5.09% lower performance index than DI water. The blade-platelet hybrid produced 19.49% less entropy than pure water [104]

6.5.2.4 Extended Surfaces (Fins)

Kumar *et al.* investigated, both numerically and experimentally, longitudinal fins of a DPHE, where fins had three various geometries: rectangular, parabolic, and triangular. It has been found that the longitudinal parabolic fins caused lower pressure drops due to their heat transfer characteristics, while the efficiency of the rectangular longitudinal ribs was superior to the others [105]. An optimal longitudinal rib pattern resulting in increased coupled heat transfer was achieved by Iqbal *et al.* The geometries of the fins were parabolic, trapezoidal, and rectangular. Optimization depending upon the identical diameter resulted in a 289% enhancement in the coefficient of heat transfer, whereas the hydraulic diameter showed a 70% improvement [106]. Note that equivalent optimization studies on fin DPHEs also published by Iqbal *et al.* were found [107, 108]. Barga and Saboya conducted under turbulent flow conditions using air as the working fluid in annulus, while water was used in the inner tube. Nusselt number ratio for ribbed ring to connecting tube number found to be less than one and to decrease as Reynolds number increased. In conclusion the use of these fins had a negative impact on heat transfer in the annulus [109]. Kahalerras and Targui achieved enriching heat exchange in DPHE by employing porous fins on the bladder's exterior surface. It was noted that the pertinent findings mentioned could be easily verified if the fluids moving through the two tubes are the same and have the same mass flow rate [110]. Sahiti *et al.* conducted studies on minimizing entropy generation in fin DPHE, considering various pin lengths and flow lengths as a function of the Reynolds number. They indicated that shorter pin heights in more channels took precedence over smaller heat exchanger traces with higher pin heights [111]. A study was conducted by Taborek to determine the optimal operating conditions for rectangular and smooth-finned double-tube and multi-tube heat exchanger. Numerical simulations were used to obtain further research results on the modified DPHE, which included three different fin geometries: rectangular, parabolic, and triangular. The use of parabolic fins was found to result in a higher criterion for evaluating the functionality of heat exchangers [112]. Syed *et al.* performed laminar convection in DPHE with interchangeable fin tip thickness. DG-FEM was used in the study. The DPHE's efficiency was maintained while keeping the Nusselt number, pressure drop, and coefficient *J*. An enhancement of 89% in the post-augmentation *J*-factor and 178% in the Nusselt number was observed with a rectangular cross-section. For the triangular section, the improved *J*-factor and Nusselt number were 19% and 9.5%, respectively [113]. Nazari *et al.* numerically analyzed the effects of MWCNT–Fe_3O_4/water hybrid NF (0.3% volume concentration) and winglet vortex generators (VG) in a fin-plate heat exchanger using SST (shear stress transport) turbulence models. The study found that NFs and VGs enhance heat transfer by increasing thermal conductivity and reducing boundary layers. The highest heat transfer improvement was 5.2% with NF alone, 69.2% with VG, and 74.6% with both [114]. It should be noted that other numerical studies on finned DPHEs have also been conducted [115, 116].

6.5.3 Compound Method

Omkar *et al.* investigated spiral vanes on the outer surface of a rotating internal tube, with water and glycerine gradually circulated through the annulus and inner tube, respectively. An increase of 64% in the Nusselt number was observed at 100 rpm of the internal tube compared to a constant internal tube [117]. This study employed a technique that combined passive and active methods to enhance heat transfer. The simultaneous utilization of wire coils and liquid vibrations serves as an excellent example of this method, which has already been extensively researched for heat exchangers [118].

6.6 CONCLUSIONS

This review analyzes various factors, including NP and NF properties, and their impact on the DPHE's functionality.

- Some research has shown that heat improvement techniques, such as the use of spiral wires, can be effective in DPHEs. Active increment technology is not commonly used in DPHEs. Further research is needed to combine NFs and passive heat enhancement techniques to address various issues.
- The concentration of NPs in NFs affects their thermophysical properties, leading to increased pressure drop and enhanced heat transfer. Smaller NPs result in higher viscosity and improved thermal conductivity. Platelet and cylindrical-shaped NPs enhance heat transfer and thermal conductivity, while spherical NPs exhibit a lower pressure drop. A study using a mathematical model and a DPHE with TiO_2-water NFs found that 0.8% volumetric concentration of 20 nm spherical NPs increased viscosity, thermal conductivity, and density by 2.03%, 1.98%, and 2.61%, respectively.

- Many studies have focused on achieving lower friction coefficients and a greater rate of heat transfer through use of passive heat transfer enhancement techniques. These techniques have been shown to reduce pressure drop by up to 1000% and increase heat transfer rate by approximately 400% compared to smooth tubes under certain conditions. Changing the geometry in DPHE is one way to improve performance, and further research is needed in this area.

ACKNOWLEDGMENT

The authors would like to acknowledge the Department of Mechanical Engineering at BIET Lucknow, and Department of Mechanical Engineering, Visvesvaraya National Institute of Technology, Nagpur for providing their support.

DATA AVAILABILITY

Data will be made available on request.

APPENDIX I

S.N	Abbreviation	Description
1	DPHEs	Double pipe heat exchangers
2	DG-FEM	Discontinuous Galerkin finite element method
3	GPR	Growth progression rate
4	MWCNT	Multi-walled carbon nanotube
5	NPs	Nanoparticles
6	NFs	Nanofluids

REFERENCES

[1] Rehman, HA, Wang B, Fatima S, Gan Z. Progress in measuring techniques and thermal radiative properties of metals at cryogenic temperatures: a review. *Cryogenics*. 2024;143: 103950. https://doi.org/10.1016/j.cryogenics.2024.103950

[2] Souli M, Fouaidy M, Hammoudi N. Experimental test of a supercritical helium heat exchanger dedicated to EUROTRANS 150 kW CW power coupler. *Cryogenics*. 2010;50(5):295–303. https://doi.org/10.1016/j.cryogenics.2010.01.006

[3] Omidi M, Farhadi M, Jafari M. A comprehensive review on double pipe heat exchangers. *Applied Thermal Engineering*. 2017 Jan 5;110:1075–90.

[4] Alshukri MJ, Hussein AK, Eidan AA, Alsabery AI. A review on applications and techniques of improving the performance of heat pipe-solar collector systems. *Solar Energy*. 2022 Apr 1;236:417–33.

[5] Louis SP, Ushak S, Milian Y, Nemś M, Nemś A. Application of nanofluids in improving the performance of double-pipe heat exchangers—a critical review. *Materials*. 2022 Oct 3;15(19):6879.

[6] Yasukawa, Y, Matsumoto N, Ueda Y. Experimental study on the effect of aftercooler configuration on the performance of pulse tube cryocoolers. *Cryogenics*. 2022; 121: 103408. https://doi.org/10.1016/j.cryogenics.2021.103408

[7] Ding Z, Qi C, Luo T, Wang Y, Tu J, Wang C. Numerical simulation of nanofluids forced convection in a corrugated double-pipe heat exchanger. *The Canadian Journal of Chemical Engineering*. 2022 Aug;100(8):1954–64.

[8] Jassim EI, Ahmed F. Assessment of nanofluid on the performance and energy-environment interaction of Plate-Type-Heat exchanger. *Thermal Science and Engineering Progress*. 2021 Oct 1;25:100988.

[9] Kavitha R, Abd Algani YM, Kulkarni K, Gupta MK. Heat transfer enhancement in a double pipe heat exchanger with copper oxide nanofluid: an experimental study. *Materials Today: Proceedings*. 2022 Jan 1;56:3446–9.

[10] Mansoury D, Doshmanziari FI, Kiani A, Chamkha AJ, Sharifpur M. Heat transfer and flow characteristics of Al2O3/water nanofluid in various heat exchangers: experiments on counter flow. *Heat Transfer Engineering*. 2019 Jun 11;41:220–34.

[11] Akbar HM, Mohammed AS, Ezzat SB. Hybrid nanofluid to improve heat transfer and pressure drop through horizontal tube. *Materials Today: Proceedings*. 2021 Jan 1;42:1885–8.

[12] Jassim EI, Ahmed F. Experimental assessment of Al 2 O 3 and Cu nanofluids on the performance and heat leak of double pipe heat exchanger. *Heat and Mass Transfer*. 2020 Jun;56:1845–58.

[13] Henein SM, Abdel-Rehim AA. The performance response of a heat pipe evacuated tube solar collector using MgO/MWCNT hybrid nanofluid as a working fluid. *Case Studies in Thermal Engineering*. 2022 May 1;33:101957.

[14] Vincely DA, Natarajan EJ. Experimental investigation of the solar FPC performance using graphene oxide nanofluid under forced circulation. *Energy Conversion and Management*. 2016 Jun 1;117:1.

[15] Kore SS, Chaudhary MK, Patil AS, Kolate VD. Computational study and enhancement of heat transfer rate by using inserts introduced in a heat exchanger. *Journal of Advanced Research in Fluid Mechanics and Thermal Sciences*. 2023;103(1):16–29.

[16] Kore SS, Chaudhary MK, Bhambare PS, Kaithari DK. The heat transfer and fluid flow investigations of single dimple with straight and curved arch turbulator within in a duct. *Journal of Advanced Research in Fluid Mechanics and Thermal Sciences*. 2024;115(1):206–16.

[17] Okonkwo EC, Wole-Osho I, Almanassra IW, Abdullatif YM, Al-Ansari T. An updated review of nanofluids in various heat transfer devices. *Journal of Thermal Analysis and Calorimetry*. 2021 Sep;145:2817–72.

[18] Ezgi C. Basic design methods of heat exchanger. *Heat Exchangers–Design, Experiment and Simulation*. 2017 Apr 27;9:9–35.

[19] Babu R, Kumar P, Roy S, Ganesan R. A comprehensive review on compound heat transfer enhancement using passive techniques in a heat exchanger. *Materials Today: Proceedings*. 2022 Jan 1;54:428–36.

[20] Gupta M, Singh V, Kumar R, Said Z. A review on thermophysical properties of nanofluids and heat transfer applications. *Renewable and Sustainable Energy Reviews*. 2017 Jul 1;74:638–70.

[21] Okonkwo EC, Wole-Osho I, Almanassra IW, Abdullatif YM, Al-Ansari T. An updated review of nanofluids in various heat transfer devices. *Journal of Thermal Analysis and Calorimetry*. 2021 Sep;145:2817–72.

[22] Ambreen T, Kim MH. Heat transfer and pressure drop correlations of nanofluids: a state of art review. *Renewable and Sustainable Energy Reviews*. 2018 Aug 1;91:564–83.

[23] Targui N, Kahalerras H. Analysis of a double pipe heat exchanger performance by use of porous baffles and nanofluids. *International Journal of Mechanical, Aerospace, Industrial, Mechatronic and Manufacturing Engineerig*. 2014 Aug 1;8(9):1581–6.

[24] Prasad PD, Gupta AV, Deepak K. Investigation of trapezoidal-cut twisted tape insert in a double pipe u-tube heat exchanger using Al2O3/water nanofluid. *Procedia Materials Science*. 2015 Jan 1;10:50–63.

[25] Templeton JD, Hassani F, Ghoreishi-Madiseh SA. Study of effective solar energy storage using a double pipe geothermal heat exchanger. *Renewable Energy*. 2016 Feb 1;86:173–81.

[26] Reddy MC, Rao VV. Experimental investigation of heat transfer coefficient and friction factor of ethylene glycol water based TiO2 nanofluid in double pipe heat exchanger with and without helical coil inserts. *International Communications in Heat and Mass Transfer*. 2014 Jan 1;50:68–76.

[27] Bahiraei M, Hangi M. Investigating the efficacy of magnetic nanofluid as a coolant in double-pipe heat exchanger in the presence of magnetic field. *Energy Conversion and Management*. 2013 Dec 1;76:1125–33.

[28] Maddah H, Alizadeh M, Ghasemi N, Alwi SR. Experimental study of Al2O3/water nanofluid turbulent heat transfer enhancement in the horizontal double pipes fitted with modified twisted tapes. *International Journal of Heat and Mass Transfer*. 2014 Nov 1;78:1042–54.

[29] Omidi M, Farhadi M, Jafari M. A comprehensive review on double pipe heat exchangers. *Applied Thermal Engineering*. 2017 Jan 5;110:1075–90.

[30] Shakiba A, Vahedi K. Numerical analysis of magnetic field effects on hydro-thermal behavior of a magnetic nanofluid in a double pipe heat exchanger. *Journal of Magnetism and Magnetic Materials*. 2016 Mar 15;402:131–42.

[31] Wu Z, Wang L, Sundén B. Pressure drop and convective heat transfer of water and nanofluids in a double-pipe helical heat exchanger. *Applied Thermal Engineering*. 2013 Oct 2;60(1–2):266–74.

[32] Bahiraei M, Hangi M. Investigating the efficacy of magnetic nanofluid as a coolant in double-pipe heat exchanger in the presence of magnetic field. *Energy Conversion and Management*. 2013 Dec 1;76:1125–33.

[33] Porgar S, Vafajoo L, Nikkam N, Vakili-Nezhaad G. A comprehensive investigation in determination of nanofluids thermophysical properties. *Journal of the Indian Chemical Society*. 2021 Mar 1;98(3):100037.

[34] Alklaibi AM, Mouli KVC, Sundar LS. Heat transfer, and friction factor of Fe3O4–SiO2/water hybrid nanofluids in a plate heat exchanger: experimental and ANN predictions. *International Journal of Thermal Sciences*. 2024;195: 108608.

[35] Wu Z, Wang L, Sundén B. Pressure drop and convective heat transfer of water and nanofluids in a double-pipe helical heat exchanger. *Applied Thermal Engineering*. 2013 Oct 2;60(1–2):266–74.

[36] Aghayari R, Maddah H, Ashori F, Hakiminejad A, Aghili M. Effect of nanoparticles on heat transfer in mini double-pipe heat exchangers in a turbulent flow. *Heat and Mass Transfer*. 2015 Mar;51:301–6.

[37] Hemmat EM, Saedodin S. Turbulent forced convection heat transfer and thermophysical properties of Mgo–water nanofluid with consideration of different nanoparticles

diameter, an empirical study. *Journal of Thermal Analysis and Calorimetry*. 2015 Feb;119:1205–13.

[38] Asirvatham LG, Raja B, Lal DM, Wongwises S. Convective heat transfer of nanofluids with correlations. *Particuology*. 2011 Dec 1;9(6):626–31.

[39] Sarafraz MM, Hormozi FJ. Intensification of forced convection heat transfer using biological nanofluid in a double-pipe heat exchanger. *Experimental Thermal and Fluid Science*. 2015 Sep 1;66:279–89.

[40] Sajadi AR, Kazemi MH. Investigation of turbulent convective heat transfer and pressure drop of TiO2/water nanofluid in circular tube. *International Communications in Heat and Mass Transfer*. 2011 Dec 1;38(10):1474–8.

[41] Zamzamian A, Oskouie SN, Doosthoseini A, Joneidi A, Pazouki M. Experimental investigation of forced convective heat transfer coefficient in nanofluids of Al2O3/EG and CuO/EG in a double pipe and plate heat exchangers under turbulent flow. *Experimental Thermal and Fluid Science*. 2011 Apr 1;35(3):495–502.

[42] Dharmakkan N, Srinivasan PM, Muthusamy S, Jomde A, Shamkuwar S, Sonawane C, Sharma K, Alrubaie AJ, El Shafay AS, Panchal H. A case study on analyzing the performance of microplate heat exchanger using nanofluids at different flow rates and temperatures. *Case Studies in Thermal Engineering*. 2023;44:102805.

[43] Hozien O, El-Maghlany WM, Sorour MM, Mohamed YS. Experimental study on thermophysical properties of TiO2, ZnO and Ag water base nanofluids. *Journal of Molecular Liquids*. 2021 Jul 15;334:116128.

[44] Pordanjani AH, Aghakhani S, Afrand M, Mahmoudi B, Mahian O, Wongwises S. An updated review on application of nanofluids in heat exchangers for saving energy. *Energy Conversion and Management*. 2019 Oct 15;198:111886.

[45] Okonkwo EC, Wole-Osho I, Almanassra IW, Abdullatif YM, Al-Ansari T. An updated review of nanofluids in various heat transfer devices. *Journal of Thermal Analysis and Calorimetry*. 2021 Sep;145:2817–72.

[46] Porgar S, Vafajoo L, Nikkam N, Vakili-Nezhaad G. A comprehensive investigation in determination of nanofluids thermophysical properties. *Journal of the Indian Chemical Society*. 2021 Mar 1;98(3):100037.

[47] Alqarni MM, Ibrahim M, Assiri TA, Saeed T, Abd Allah AM, Ali V. Two-phase simulation of a shell and tube heat exchanger filled with hybrid nanofluid. *Engineering Analysis with Boundary Elements*. 2023;146:80–88.

[48] Osman S, Sharifpur M, Meyer JP. Experimental investigation of convection heat transfer in the transition flow regime of aluminium oxide-water nanofluids in a rectangular channel. *International Journal of Heat and Mass Transfer*. 2019 Apr 1;133: 895–902.

[49] Bakthavatchalam B, Habib K, Saidur R, Saha BB, Irshad K. Comprehensive study on nanofluid and ionanofluid for heat transfer enhancement: a review on current and future perspective. *Journal of Molecular Liquids*. 2020 May 1;305:112787.

[50] Darzi AR, Farhadi M, Sedighi K. Heat transfer and flow characteristics of Al2O3–water nanofluid in a double tube heat exchanger. *International Communications in Heat and Mass Transfer*. 2013 Oct 1;47:105–12.

[51] Saleh B, Sundar LS. Thermal performance, embodied energy and environmental CO2 emissions analyses for double pipe U-bend heat exchanger working with MWCNT/water nanofluid. *International Journal of Thermal Sciences*. 2021 Nov 1;169:107094.

[52] Asokan N, Gunnasegaran P, Wanatasanappan VV. Experimental investigation on the thermal performance of compact heat exchanger and the rheological properties of low concentration mono and hybrid nanofluids containing Al2O3 and CuO nanoparticles. *Thermal Science and Engineering Progress*. 2020 Dec 1;20:100727.

[53] Ciornei FC, Alaci S, Amarandei D, Irimescu L, Romanu IC, Acsinte LI. Method and device for measurement of dynamic viscosity. In Minodora RIPA (Ed.) *IOP Conference Series: Materials Science and Engineering 2017 Feb* 1 (Vol. 174, No. 1, p. 012041). IOP Publishing.

[54] Bhattad A, Rao BN, Atgur V, Veza I, Zamri MFMA, Fattah IMR. Thermal performance evaluation of plate-type heat exchanger with alumina–titania hybrid suspensions. *Fluids*. 2023;8(4):120.

[55] Sundar LS, Mouli KVC. Effectiveness and number of transfer units of plate heat exchanger with Fe3O4–SiO2/Water hybrid nanofluids: experimental and artificial neural network predictions. *Case Studies in Thermal Engineering*. 2024;53:103949.

[56] Okonkwo EC, Wole-Osho I, Almanassra IW, Abdullatif YM, Al-Ansari T. An updated review of nanofluids in various heat transfer devices. *Journal of Thermal Analysis and Calorimetry*. 2021 Sep;145:2817–72.

[57] Main KL, Eberl BK, McDaniel D, Tikadar A, Paul TC, Khan JA. Nanoparticles size effect on thermophysical properties of ionic liquids based nanofluids. *Journal of Molecular Liquids*. 2021 Dec 1;343:117609.

[58] Gupta M, Singh V, Kumar R, Said Z. A review on thermophysical properties of nanofluids and heat transfer applications. *Renewable and Sustainable Energy Reviews*. 2017 Jul 1;74:638–70.

[59] Zhang L, Zhang A, Jing Y, Qu P, Wu Z. Effect of particle size on the heat transfer performance of SiO2–water nanofluids. *The Journal of Physical Chemistry C*. 2021 Jun 10;125(24):13590–600.

[60] Kim H, Kim J, Cho H. Experimental study on performance improvement of U-tube solar collector depending on nanoparticle size and concentration of Al2O3 nanofluid. *Energy*. 2017 Jan 1;118:1304–12.

[61] Zhang L, Zhang A, Jing Y, Qu P, Wu Z. Effect of particle size on the heat transfer performance of SiO2–water nanofluids. *The Journal of Physical Chemistry C*. 2021 Jun 10;125(24):13590–600.

[62] Cui W, Cao Z, Li X, Lu L, Ma T, Wang Q. Experimental investigation and artificial intelligent estimation of thermal conductivity of nanofluids with different nanoparticles shapes. *Powder Technology*. 2022 Jan 1;398:117078.

[63] Saranya S, Al-Mdallal QM. Computational study on nanoparticle shape effects of Al2O3-silicon oil nanofluid flow over a radially stretching rotating disk. *Case Studies in Thermal Engineering*. 2021 Jun 1;25:100943.

[64] Zahmatkesh I, Sheremet M, Yang L, Heris SZ, Sharifpur M, Meyer JP, Ghalambaz M, Wongwises S, Jing D, Mahian O. Effect of nanoparticle shape on the performance of thermal systems utilizing nanofluids: a critical review. *Journal of Molecular Liquids*. 2021 Jan 1;321:114430.

[65] Elias MM, Miqdad M, Mahbubul IM, Saidur R, Kamalisarvestani M, Sohel MR, Hepbasli A, Rahim NA, Amalina MA. Effect of nanoparticle shape on the heat transfer and thermodynamic performance of a shell and tube heat exchanger. *International Communications in Heat and Mass Transfer*. 2013 May 1;44:93–9.

[66] Gupta M, Singh V, Kumar R, Said Z. A review on thermophysical properties of nanofluids and heat transfer applications. *Renewable and Sustainable Energy Reviews*. 2017 Jul 1;74:638–70.

[67] Shahsavar A, Rahimi Z, Salehipour H. Nanoparticle shape effects on thermal-hydraulic performance of boehmite alumina nanofluid in a horizontal double-pipe minichannel heat exchanger. *Heat and Mass Transfer*. 2019 Jun 1;55:1741–51.

[68] Bouselsal M, Mebarek-Oudina F, Biswas N, Ismail AAI. Heat transfer enhancement using Al2O3-MWCNT hybrid-nanofluid inside a tube/shell heat exchanger with different tube shapes. *Micromachines*. 2023;14(5):1072.

[69] Ikhtiar U, Hairuddin AAB, Asarry AB, Rezali KABM, Ali HM. Numerical investigation of lamella heat exchanger for engine intake charge air cooling utilizing refrigerant as coolant medium. *Alexandria Engineering Journal.* 2023;65:661–73.

[70] Kücük H. The effect of minichannels on the overall heat transfer coefficient and pressure drop of a shell and tube heat exchanger: experimental performance comparison. *International Journal of Thermal Sciences.* 2023;188:108217.

[71] Gaur SK, Sahoo RR, Sarkar J. Thermo-hydraulic performance investigation of triple-tube heat exchanger with MXene-based various shape hybrid nanofluids. *Journal of Thermal Analysis and Calorimetry.* 2024;149(7):3071–84.

[72] Zhang Z, Ding Y, Guan C, Yan H, Yang W. Heat transfer enhancement in double-pipe heat exchanger by means of rotor-assembled strands. *Chemical Engineering and Processing: Process Intensification.* 2012 Oct 1;60:26–33.

[73] El-Maghlany W, Eid E, Teamah M, Shahrour I. Experimental study for double pipe heat exchanger with rotating inner pipe. *International Journal of Advanced Scientific and Technical Research.* 2012;4(2):507–24.

[74] Rai R, Kumar V, Sahoo RR. Energy, exergy-emission performance investigation of heat exchanger with Turbulators inserts and ternary hybrid nanofluid. *Journal of Thermal Analysis and Calorimetry.* 2024;149(9):3999–4017.

[75] Naphon P. Effect of coil-wire insert on heat transfer enhancement and pressure drop of the horizontal concentric tubes. *International Communications in Heat and Mass Transfer.* 2006 Jul 1;33(6):753–63.

[76] Akpinar EK. Evaluation of heat transfer and exergy loss in a concentric double pipe exchanger equipped with helical wires. *Energy Conversion and Management.* 2006 Nov 1;47(18–19):3473–86.

[77] Choudhari SS, Taji S. Experimental studies on effect of coil wire insert on heat transfer enhancement and friction factor of double pipe heat exchanger. *International Journal of Computer Engineering Research.* 2013;3:32–9.

[78] Zohir AE, Habib MA, Nemitallah MA. Heat transfer characteristics in a double-pipe heat exchanger equipped with coiled circular wires. *Experimental Heat Transfer.* 2015 Nov 2;28(6):531–45.

[79] Anitha S, Shasthick M, Senthilkumar B, Chandramohan P, Pichumani M. How the estimation of entropy generation and exergy loss of hybrid nanofluids governs the thermal performance of heat exchanger. *International Journal of Fluid Mechanics Research.* 2023;50(1):23–39.

[80] Marzouk SA, Abou Al-Sood MM, El-Said EM, El-Fakharany MK, Younes MM. Study of heat transfer and pressure drop for novel configurations of helical tube heat exchanger: a numerical and experimental approach. *Journal of Thermal Analysis and Calorimetry.* 2023;148(13):6267–82.

[81] Painuly A, Mishra NK, Zainith P. Thermo-hydraulic performance evaluation and comparison of SiC-MWCNT and Al2O3-MWCNT Non-Newtonian hybrid nanofluids using a heat exchanger equipped with helically corrugated tubes: an experimental study. *Journal of Thermal Analysis and Calorimetry.* 2024;149(9):3965–80.

[82] Yadav AS. Effect of half length twisted-tape turbulators on heat transfer and pressure drop characteristics inside a double pipe u-bend heat exchanger. *JJMIE.* 2009 Mar;3(1):17–22.

[83] Liu S, Sakr M. A comprehensive review on passive heat transfer enhancements in pipe exchangers. *Renewable and Sustainable Energy Reviews.* 2013 Mar 1;19:64–81.

[84] Naphon P. Heat transfer and pressure drop in the horizontal double pipes with and without twisted tape insert. *International Communications in Heat and Mass Transfer.* 2006 Feb 1;33(2):166–75.

[85] Bharadwaj P, Khondge AD, Date AW. Heat transfer and pressure drop in a spirally grooved tube with twisted tape insert. *International Journal of Heat and Mass Transfer* 2009 Mar 1;52(7–8):1938–44.

[86] Balaga R, Koona R, Tunuguntla S. Heat transfer enhancement of the f-MWCNT-Fe2O3/water hybrid nanofluid with the combined effect of wire coil with twisted tape and perforated twisted tape. *International Journal of Thermal Sciences*. 2023;184:108023.

[87] Al-Obaidi AR, Alhamid J. The effect of different twisted tape inserts configurations on fluid flow characteristics, pressure drop, thermo-hydraulic performance and heat transfer enhancement in the 3D circular tube. *International Journal of Ambient Energy*. 2023;44(1):57–72.

[88] Kumar R, Kumar P. Effect of twist ratio on the thermal-hydraulic behavior of an aluminum oxide/cupric oxide nanofluid heat exchanger. *Journal of Thermal Analysis and Calorimetry*. 2024;149(9):4103–17.

[89] Akpinar EK, Bicer Y. Investigation of heat transfer and exergy loss in a concentric double pipe exchanger equipped with swirl generators. *International Journal of Thermal Sciences*. 2005 Jun 1;44(6):598–607.

[90] Al-Kayiem HH, El-Rahman M. Ribbed double pipe heat exchanger: analytical analysis. *Journal of Engineering Science and Technology*. 2011 Feb 20;6(1):39–49.

[91] Zhang L, Guo H, Wu J, Du W. Compound heat transfer enhancement for shell side of double-pipe heat exchanger by helical fins and vortex generators. *Heat and Mass Transfer*. 2012 Jul;48:1113–24.

[92] Yildiz C, BíçerY, Pehlívan D. Influence of fluid rotation on the heat transfer and pressure drop in double-pipe heat exchangers. *Applied Energy*. 1996 May 1;54(1):49–56.

[93] Alkam MK, Al-Nimr MA. Improving the performance of double-pipe heat exchangers by using porous substrates. *International Journal of Heat and Mass Transfer*. 1999 Oct 1;42(19):3609–18.

[94] Allouache N, Chikh S. Second law analysis in a partly porous double pipe heat exchanger. *Journal of Applied Mechanics*. 2006;73:60–5.

[95] Targui N, Kahalerras H. Analysis of fluid flow and heat transfer in a double pipe heat exchanger with porous structures. *Energy Conversion and Management*. 2008 Nov 1;49(11):3217–29.

[96] Eiamsa-ard S, Pethkool S, Thianpong C, Promvonge P. Turbulent flow heat transfer and pressure loss in a double pipe heat exchanger with louvered strip inserts. *International Communications in Heat and Mass Transfer*. 2008 Feb 1;35(2):120–9.

[97] Durmuş A. Heat transfer and energy loss in a concentric heat exchanger with snail entrance. *International Communications in Heat and Mass Transfer*. 2002 Apr 1;29(3):303–12.

[98] Raj R, Lakshman NS, Mukkamala Y. Single phase flow heat transfer and pressure drop measurements in doubly enhanced tubes. *International Journal of Thermal Sciences*. 2015 Feb 1;88:215–27.

[99] Hsieh SS, Liauh CT, Ku AC. Thermal analysis of the performances of helical-type, roughened, double-pipe heat exchangers. *Applied Energy*. 1987 Jan 1;26(1):67–73.

[100] Hsieh SS, Liauh CT, Ku AC. Heat transfer coefficients of double pipe heat exchanger with helical type roughened surface. *Heat Recovery Systems and CHP*. 1987 Jan 1;7(2):119–27.

[101] Targui N, Kahalerras H. Analysis of a double pipe heat exchanger performance by use of porous baffles and nanofluids. *International Journal of Mechanical, Aerospace, Industrial, Mechatronic and Manufacturing Engineerig*. 2014 Aug 1;8(9):1581–6.

[102] Targui N, Kahalerras H. Analysis of a double pipe heat exchanger performance by use of porous baffles and pulsating flow. *Energy Conversion and Management*. 2013 Dec 1;76:43–54.

[103] Abidi A, Sajadi SM. Numerical assessment of hydraulic behavior and thermal efficiency of multiphase hybrid nanofluid in a shell-and-tube heat exchanger with inclined baffles. *Engineering Analysis with Boundary Elements*. 2023;156:114–25.

[104] Gaur SK, Sahoo RR, Sarkar J. Numerical investigation on assessing the influence of diverse-shaped hybrid nanofluids on thermal performance of triple tube heat exchanger. *Powder Technology*. 2024;439:119690.

[105] Kumar S, Karanth KV, Murthy K. Numerical study of heat transfer in a finned double pipe heat exchanger. *World Journal of Modelling and Simulation*. 2015 Jan 1;11(1):43–54.

[106] Iqbal Z, Syed KS, Ishaq M. Fin design for conjugate heat transfer optimization in double pipe. *International Journal of Thermal Sciences*. 2015 Aug 1;94:242–58.

[107] Iqbal Z, Syed KS, Ishaq M. Optimal fin shape in finned double pipe with fully developed laminar flow. *Applied Thermal Engineering*. 2013 Mar 1;51(1-2):1202–23.

[108] Iqbal Z, Syed KS, Ishaq M. Optimal convective heat transfer in double pipe with parabolic fins. *International Journal of Heat and Mass Transfer*. 2011 Dec 1;54(25–26):5415–26.

[109] Braga CV, Saboya FE. Turbulent heat transfer, pressure drop and fin efficiency in annular regions with continuous longitudinal rectangular fins. *Experimental Thermal and Fluid Science*. 1999 Oct 1;20(2):55–65.

[110] Kahalerras H, Targui N. Numerical analysis of heat transfer enhancement in a double pipe heat exchanger with porous fins. *International Journal of Numerical Methods for Heat & Fluid Flow*. 2008 Jun 27;18:593–617.

[111] Sahiti N, Krasniqi F, Fejzullahu X, Bunjaku J, Muriqi A. Entropy generation minimization of a double-pipe pin fin heat exchanger. *Applied Thermal Engineering*. 2008 Dec 1;28(17–18):2337–44.

[112] Taborek J. Double-pipe and multitube heat exchangers with plain and longitudinal finned tubes. *Heat Transfer Engineering*. 1997 Jan 1;18(2):34–45.

[113] Syed KS, Ishaq M, Iqbal Z, Hassan A. Numerical study of an innovative design of a finned double-pipe heat exchanger with variable fin-tip thickness. *Energy Conversion and Management*. 2015 Jul 1;98:69–80.

[114] Alhuyi Nazari M, Mukhtar A, Mehrabi A, Ahmadi MH, Sharifpur M, Luong, TNL.. Effect of using hybrid nanofluid and vortex generator on thermal performance of plate–fin heat exchanger: numerical investigation. *Journal of Thermal Analysis and Calorimetry*. 2024;149(9):4227–37.

[115] Syed KS, Ishaq M, Bakhsh M. Laminar convection in the annulus of a double-pipe with triangular fins. *Computers & Fluids*. 2011 May 1;44(1):43–55.

[116] Iqbal M, Syed KS. Thermally developing flow in finned double-pipe heat exchanger. *International Journal for Numerical Methods in Fluids*. 2011 Apr 10;65(10): 1145–59.

[117] Omkar MS, Pravin AM, Sajid AG, Pradeep AP. Experimental investigation of double-pipe heat exchanger with helical fins on the inner rotating tube. *International Journal of Research Engineering and Technology*. 2014;3(7):98–102.

[118] Duangthongsuk W, Wongwises S. An experimental investigation of the heat transfer and pressure drop characteristics of a circular tube fitted with rotating turbine-type swirl generators. *Experimental Thermal and Fluid Science*. 2013 Feb 1;45:8–15.

[119] Shirvan KM, Ellahi R, Mirzakhanlari S, Mamourian M. Enhancement of heat transfer and heat exchanger effectiveness in a double pipe heat exchanger filled with porous media: numerical simulation and sensitivity analysis of turbulent fluid flow. *Applied Thermal Engineering*. 2016 Oct 25;109:761–74.

[120] Arya H, Sarafraz MM, Pourmehran O, Arjomandi M. Heat transfer and pressure drop characteristics of MgO nanofluid in a double pipe heat exchanger. *Heat and Mass Transfer*. 2019 Jun 1;55:1769–81.

[121] Akpinar EK. Evaluation of heat transfer and exergy loss in a concentric double pipe exchanger equipped with helical wires. *Energy Conversion and Management*. 2006 Nov 1;47(18–19):3473–86.

[122] Akbarzadeh M, Rashidi S, Keshmiri A, Shokri N. The optimum position of porous insert for a double-pipe heat exchanger based on entropy generation and thermal analysis. *Journal of Thermal Analysis and Calorimetry*. 2020 Jan;139:411–26.

[123] Baba MS, Raju AS, Rao MB. Heat transfer enhancement and pressure drop of Fe3O4-water nanofluid in a double tube counter flow heat exchanger with internal longitudinal fins. *Case Studies in Thermal Engineering*. 2018 Sep 1;12:600–7.

[124] Naphon P. Heat transfer and pressure drop in the horizontal double pipes with and without twisted tape insert. *International Communications in Heat and Mass Transfer*. 2006 Feb 1;33(2):166–75.

[125] Setareh M, Saffar-Avval M, Abdullah A. Experimental and numerical study on heat transfer enhancement using ultrasonic vibration in a double-pipe heat exchanger. *Applied Thermal Engineering*. 2019 Aug 1;159:113867.

[126] Hosseinian A, Meghdadi Isfahani AH. Experimental study of heat transfer enhancement due to the surface vibrations in a flexible double pipe heat exchanger. *Heat and Mass Transfer*. 2018 Apr;54(4):1113–20.

[127] Heyhat MM, Abdi A, Jafarzad A. Performance evaluation and exergy analysis of a double pipe heat exchanger under air bubble injection. *Applied Thermal Engineering*. 2018 Oct 1;143:582–93.

[128] Bezaatpour M, Goharkhah M. Convective heat transfer enhancement in a double pipe mini heat exchanger by magnetic field induced swirling flow. *Applied Thermal Engineering*. 2020 Feb 25;167:114801.

7 Double-Diffusive Effects in Williamson Nanofluid Flow

Raju Snehashree, Mallikarjun Basavaraju Patil, and Basma Souayeh

7.1 INTRODUCTION

One of the core concepts in fluid dynamics is fluid flow through a channel, which deals with movement of fluid inside a shape. Vertical channels play a crucial role in fluid mechanics, influencing fluid flow dynamics, heat and mass transfer. They are essential in heat exchangers, cooling towers, chemical reactors, natural convection studies, solar collectors, fluid transport and distribution, desalination plants, and filtration systems. They facilitate efficient heat transfer between fluids with different temperatures, optimize mixing and reaction rates, and aid in the capture and transfer of solar radiation. Vertical channels also facilitate the movement of fluids over varying elevations and distances in civil and environmental engineering. They also play a role in desalination processes, such as multistage flash distillation and reverse osmosis, and in filtration systems, directing fluid flow through filter media. Key fluid mechanics concepts in vertical channels include laminar and turbulent flow, boundary layer development, buoyancy-driven flows, and pressure drop. Understanding these principles is essential for designing efficient and effective systems in various industries. In order to enhance comprehension of fluid flow and heat transfer in the fluid dynamics systems, several studies focus on the natural convection of viscous fluids between two vertical flat plates, among them Aung and Worku [1], Cheng *et al.* [2], Chamkha [3], Barletta and Zanchini [4], Nield *et al.* [5] Grosan and Pop [6], Kuznetsov and Nield [7], Umavathi *et al.* [8], Das *et al.* [9], Mahabaleshwar *et al.* [10].

The growing demand for technologies involving nanoparticle-sized solid particles in various industries, including engines, heat exchangers, pharmaceutical processes, hybrid microelectronics, and solar collectors, necessitates improved thermal performance. The properties of fluid flow dynamics and heat transfer characteristics of convectional base fluids are significantly altered when nanoparticles are mixed with any conventional fluid, according to previous research. The term "nanofluid" was provided by Choi [11]. Brownian motion and thermophoresis are the two main processes that Buongiorno [12] identified as being responsible for the anomalous features of nanofluids in his model he developed to explain the thermal conductivity enhancement of such fluids. Pseudoplastic fluids are non-Newtonian fluids that arise in a variety of settings, including the creation of emulsions, the ejection of polymer sheets, and several industrial applications. Since it is difficult to study

DOI: 10.1201/9781003611820-9

the characteristics of non-Newtonian fluids, a number of models are put up to eliminate them. In 1929, Williamson [13] first submitted a model to investigate the flow of pseudoplastic materials. Lyubimov and Perminov [14] analyzed Williamson fluid motion over an inclined surface and shows that, even in mild gravitational fields, tangential vibrations cause a significant fluid flow with a mean flow velocity that is non-monotonic and varies with vibration frequency. Muthuraj *et al.* [15] examine unsteady MHD flow of a Williamson nanofluid in a vertical channel with thermal radiation in the presence of chemical reaction and oscillating wall temperature. Chamkha *et al.* [16] studied natural convection from an inclined plate within a porous media with varying porosity impacted by solar radiation thus the effects of varied porosity on fluid flow and heat transfer properties. Sharma *et al.* [17] examine that the unsteady magnetohydrodynamic Williamson flow in a vertical channel including porous material and heat radiation is important in many fields of science and engineering. Numerous studies on the flow of Williamson nanofluids in various geometries were recently published in the literature [18, 19, 20, 21, 22].

Thermal radiation is the process by which heat is transferred through fluid flow systems using electromagnetic waves. Temperature distribution, flow behavior, and heat transfer rates are all greatly impacted. Applications for it are many and include heat exchangers, combustion engines, reactors, weather prediction, energy transfer in climate studies, dispersion of pollutants, thermal loads in spacecraft and reentry vehicles, and bio-medicinal devices. By considering thermal radiation, engineers and scientists may maximize energy usage in a variety of applications, enhance system safety, and improve system performance. The effect of magnetohydrodynamics boundary layer flow of Williamson nanofluid on thermal radiation along a stretched surface with the medium of porosity considering thermal slippage and velocity. Gireesha *et al.* [23] examines the impact of thermal radiation, Joule heating, and heat production on magnetically driven Carreau fluid flow across a permeable sensor surface. The intricate relationships between magnetic fields, thermal effects, and the non-Newtonian characteristics of the Carreau fluid are observed. Dharmendar *et al.* [24] examine the phenomena of thermophoresis and Brownian motion-induced mass and heat transfer. The heat source or sink, non-linear thermal radiation has a significant impact on heat transfer rates and fluid dynamics. Khan *et al.* [25] examines the effects on velocity and temperature profiles highlights the intricate interactions between radiation, heat generation/absorption, and the non-Newtonian nature of Williamson fluid. Comparatively Hamid *et al.* [26] investigates the unsteady MHD flow of Williamson nanofluid in a permeable channel, revealing that skin friction is reduced by the Hartmann number and ratio parameter and increased by buoyancy. Umar *et al.* [27] investigates the effects of activation energy, microorganisms, bioconvection, and WU's slip on Williamson nanofluid using a vertical thin cylinder, which shows that it combines metallic nanoparticles with water. The Rosseland diffusion technique was widely employed in a number of studies concerning thermal radiation for estimating heat fluxes caused by radiation [28, 29, 30, 31].

The Soret and Dufour effects are interdependent transport phenomena in multicomponent systems, particularly in chemical engineering, petroleum engineering, and atmospheric sciences. The Soret effect, also known as thermal diffusion, involves the movement of particles or molecules in a fluid from regions of higher to lower temperature, leading to a concentration gradient. This phenomenon is relevant in fields such as petroleum engineering, chemical processing, and biological systems. The Dufour effect, also known as the diffusion-thermo effect, involves a concentration gradient causing a heat flux. This means that variations in concentration within a mixture can lead to heat transfer, analogous to how a temperature gradient can cause mass transfer in the Soret effect. Kafoussias and Williams [32] explores the impact of thermal-diffusion and diffusion-thermo on mixed free forced convective and mass transfer steady laminar boundary-layer flow along a vertical semi-infinite isothermal flat plate. Makinde *et al* [33] study an electrically conducting fluid flows steadily and incompressible through the laminar boundary layer past a moving vertical plate, which has applications in magnetic materials processing. Additionally Ameer and Veera Krishna [34] explores reactions kinetics, reactor models, and temperature, mass, and momentum transfer processes in chemical reactions. Soret and Dufour consequences, computational solutions, and numerical quantities, which are useful in chemical engineering. Sreedevi *et al.* [35] examine the effects of convective heat and mass transfer flow of an electrically conducting fluid on a permeable vertically stretched sheet, considering factors such as magnetic field, Joule heating, thermal radiation absorption, and buoyancy forces. Recent researches [36, 37, 38] on these phenomena in the context of heat transfer has been published.

By considering the above studies one can understand that these coupled effects are crucial for accurately modeling and predicting the behavior of multicomponent systems under non-isothermal conditions, designing efficient industrial processes, optimizing energy systems, and addressing environmental challenges. We can develop more effective procedures and acquire greater awareness of natural phenomena by taking these impacts into account, in this article, the Williamson nanofluid flow through a vertical channel susceptible to effects of Soret and Dufour with thermal radiation is analyzed systematically. This is achieved by formulating the nanofluid flow using Buongiorno's model and including thermophoresis and Brownian motion parameter. Influence of the different physical characteristics on the fluid flow.

7.2 MATHEMATICAL FORMULATION OF THE PROBLEM

Consider the fully developed, continuous fluid flow of an steady, incompressible Williamson nanofluid in the area enclosed by two vertical, infinite parallel flat plates spaced L apart (as shown in Figure 7.1). The y-axis is perpendicular to the gravitational field, and we select the x-axis to be parallel to the flow and opposed to it. Soret, Dufour, and radiation effects are determined by applying Buongiorno's model to the flow of nanofluids. The fluid's thermal conductivity is assumed to be variable and to change exponentially with temperature.

The governing mathematical equations for the conservation of mass, momentum, energy, and nanoparticle concentration according to the Boussenique approximation [39].

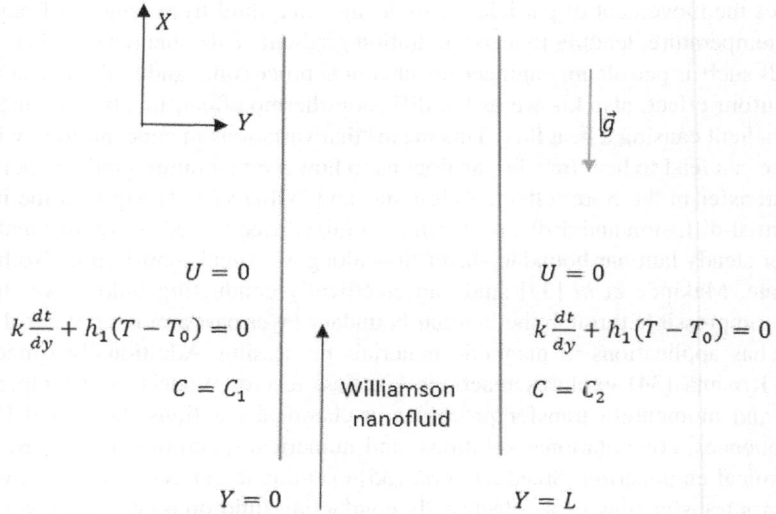

FIGURE 7.1 Physical configuration.

$$\frac{-dp}{dx}+\mu\frac{d^2u}{dy^2}+\frac{\Gamma}{\sqrt{2}}\frac{d}{dy}\left[\left(\frac{du}{dy}\right)^2\right]+\left[(\rho_{f_0}-\rho_f)(C-C_0)+(1-C_0)\rho_f\beta(T-T_0)\right]g=0 \tag{7.1}$$

$$k\frac{d^2T}{dy^2}+\left[1+\frac{\Gamma}{\sqrt{2}}\frac{du}{dy}\right]\left(\frac{du}{dy}\right)^2+(\rho C)_p\left[D_B\frac{dC}{dy}\frac{dT}{dy}+\frac{D_T}{T_0}\left(\frac{dT}{dy}\right)^2\right]$$
$$+\frac{DK_T}{C_sC_p}\frac{d^2C}{dy^2}-\frac{dq_R}{dy}=0 \tag{7.2}$$

$$D_B\frac{d^2C}{dy^2}+\left(\frac{D_T}{T_0}+\frac{DK_T}{T_0}\right)\frac{d^2T}{dy^2}=0 \tag{7.3}$$

With relevant boundary constraints are specified through,

$$\text{At the wall}\quad y=0:\quad u=0,\quad C=C_1,\quad k\frac{dT}{dy}+h_1(T-T_0)=0$$

$$\text{At the wall}\quad y=L:\quad u=0,\quad C=C_2,\quad k\frac{dT}{dy}+h_1(T-T_0)=0 \tag{7.4}$$

where u is the axial velocity of the fluid, μ is the viscosity, ρ is the density, T is the temperature of the fluid, C is the concentration of nanoparticle, $u_0 = \frac{-AL^2}{48\mu}$ is the reference velocity, $T_0 = \frac{T_1+T_2}{2}$ is the reference temperature, $C_0 = \frac{C_1+C_2}{2}$ is the reference nanoparticle concentration, Γ is the time constant, k is the thermal conductivity, D_B is the Brownian parameter, D_T is the thermophoresis coeffecient. The radiative heat flux is calculated using the Rosseland diffusion can be expressed as.

$$q_R = -\frac{4\sigma}{3\zeta}\left(\frac{dT^4}{dy}\right) \tag{7.5}$$

By using Taylor series expansion T^4 can be expressed below and where σ denotes the Stefan–Boltzmann constant and ζ is the mean absorption Rosseland coefficient.

$$T_0^2 + 4T_0^3\,(T - T_0) + 6T_0^2 (T - T_0)^2 + \ldots\ldots \approx T^4 \tag{7.6}$$

Considering that there exists a very small thermal gradient within the flow. It is evident that linear temperature T^4 is linear in temperature, we get

$$T^4 \approx 4T_0^3 T - 3T^4 \tag{7.7}$$

A thermal field is assumed to cause an exponential change in the fluid's thermal conductivity [40].

$$K = K_0 e^{-b(T-T_0)} = K_0\left(1 + b\left(T_0 - T\right)\right) \tag{7.8}$$

We define the following dimensionless parameters to change the above governing equations into their dimensionless form:

$$U = \frac{u}{u_0},\quad Y = \frac{y}{L},\quad \theta = \frac{T - T_0}{T_2 - T_0},\quad \phi = \frac{C - C_0}{C_2 - C_0},\quad Bi_1 = \frac{h_1 L}{k},\quad Bi_2 = \frac{h_2 L}{k},$$

$$\mathrm{Du} = \frac{DK_T\left(C_2 - C_0\right)}{C_s C_p k_0\left(T_2 - T_0\right)},\quad \mathrm{Sr} = \frac{DK_T\left(T_2 - T_0\right)}{\nu T_0\left(C_2 - C_0\right)},\quad \mathrm{Br} = \frac{\mu u_0^2}{k\left(T_2 - T_0\right)},$$

$$R_p = \frac{16\sigma {T_0}^3}{3\zeta k_0},\quad b_k = b\left(T - T_0\right),\quad \mathrm{We} = \frac{\sqrt{2}u_0\Gamma}{\mu L},\quad \mathrm{Pr} = \frac{\nu}{\alpha},\quad \mathrm{Le} = \frac{\alpha}{D_B}. \tag{7.9}$$

Using these parameters (7.9), equations (7.1) to (7.3) reduces as in the non-dimensional form

$$\frac{d^2U}{dY^2} + \mathrm{We}\frac{dU}{dY}\frac{d^2U}{dY^2} + \frac{Gr}{Re}\theta - \mathrm{Nr}\phi + 48 = 0 \tag{7.10}$$

$$\left(1-b_k\theta+R_p\right)\frac{d^2\theta}{dY^2}+(N_t-b_k)\left(\frac{d\theta}{dY}\right)^2+\mathrm{Nb}\frac{d\theta}{dY}\frac{d\phi}{dY}$$
$$+\mathrm{Br}\left[1+\frac{We}{2}\frac{dU}{dY}\right]\left(\frac{dU}{dY}\right)^2+\mathrm{Du}\frac{d^2\phi}{dY^2}=0 \tag{7.11}$$

$$\frac{d^2\phi}{dY^2}+\left(\frac{\mathrm{Nt}}{\mathrm{Nb}}+\mathrm{PrLeSr}\right)\frac{d^2\theta}{dY^2}=0 \tag{7.12}$$

The corresponding boundary constraints takes the non-dimensional form as

$$\text{At,}\quad Y=0:\quad U(0)=0,\quad \left(1-b_k\theta(0)\right)\frac{d\theta(0)}{dY}+\mathrm{Bi}_1\theta(0)=0,\quad \phi(0)=-1$$

$$\text{At,}\quad Y=1:\quad U(1)=0,\quad \left(1-b_k\theta(1)\right)\frac{d\theta(1)}{dY}+\mathrm{Bi}_2\theta(1)=0,\quad \phi(1)=1 \tag{7.13}$$

where Gr denotes the Grashof number given by $\mathrm{Gr}=\frac{(1-C_0)g\beta(T_2-T_0)L^3}{\nu^2}$, Re denotes the Reynold's number gievn by $\mathrm{Re}=\frac{u_0L}{\nu}$, Nr denotes the buoyancy ratio parameter given by $\mathrm{Nr}=\frac{g\left(\rho_f-\rho_{f_0}\right)(C_2-C_0)L^2}{\mu u_0}$, Nb denotes the Brownian motion parameter given by $\mathrm{Nb}=\frac{D_B(C_2-C_0)}{k/(\rho c)_p}$ and Nt denotes the thermophoresis parameter given by $\mathrm{Nt}=\frac{D_T(T_2-T_0)}{T_0k/(\rho c)_p}$

7.3 NUMERICAL APPROACH

In this section, the numerical algorithm nonlinear shooting method has been used to solve the system of nonlinear non-dimensional converting ordinary differential equations (ODEs) ((10)–(12)) over the accompanying boundary conditions (13) with the MATLAB® mathematical methodology bvp4c for the solving of boundary value problems using conventional differential equations. In order to solve a broad range of boundary value problems for ordinary differential equations, Kierzenka *et al.* [41] described how to simplify the MATLAB® problem-solving environment (PSE) [42–44].

A mathematical approach is utilized to determine a reduced set of ODEs based on the shooting analysis. A shooting technique requires the ability to integrate differential equations as an initial value problem with estimates for uncertain beginning values. This approach has the advantage of being highly consistent and takes less time. Higher-order ODEs are converted into first-order ODEs in order to use this approach.

Let $y(1) = U$, $y(2) = U'$, $y(3) = \theta$, $y(4) = \theta'$, $y(5) = \phi$, $y(6) = \phi'$.

The first order ordinary differential equations are

$$y'(1) = y(2),$$

$$y'(2) = -\mathrm{We}\,y(2)\,y(2)^2 - \frac{\mathrm{Gr}}{\mathrm{Re}}y(3) + \mathrm{Nr}y(5) - 48,$$

$$y'(3) = y(4),$$

$$y'(4) = \frac{1}{\left(1 - b_k\theta + R_p - \frac{N_t}{N_b} - \mathrm{PrLeSr}\right)} + (b_k - N_t)y(4)^2 - N_b y(4)y(6)$$

$$+B_r\left(1 + \frac{\mathrm{We}}{2}y(2)\right)y(2)^2 - \mathrm{Du}y(6)^2,$$

$$y'(5) = y(6),$$

$$y'(6) = \frac{-\left(\frac{N_t}{N_b} + P_r L_e S_r\right)}{1 - b_k\theta + R_p - \frac{N_t}{N_b} - PrLeSr} + (b_k - N_t)\,y(4)^2 - N_b y(4)y(6)$$

$$+B_r\left(1 + \frac{We}{2}y(2)\right)y(2)^2 - Du\,y(6)^2$$

with the corresponding boundary constraints

$$y0(1); \quad y1(1); \quad (1 - b_k y0(4)) + Bi_1 y0(3); \quad (1 - b_k y1(4))\,\mathrm{Bi}_1 y1(3);$$

$$y0(5) + 1; \quad y1(5) - 1.$$

See Figure 7.2.

7.4 RESULTS AND DISCUSSION

The focus of the present study is on analyzing the effects of Soret and Dufour in the Williamson nanofluid flow with thermal radiation through a vertical channel. The obtained numerical solutions for all flow characteristics by changing various physical attributes such as thermophoresis parameter Nt, Brownian parameter Nb, Buoyancy parameter Nr, Williamson parameter Γ, Radiation parameter R_p, Soret parameter Sr, Dufour parameter Du, Lewis number Le, and Prandtl number Pr covers significant

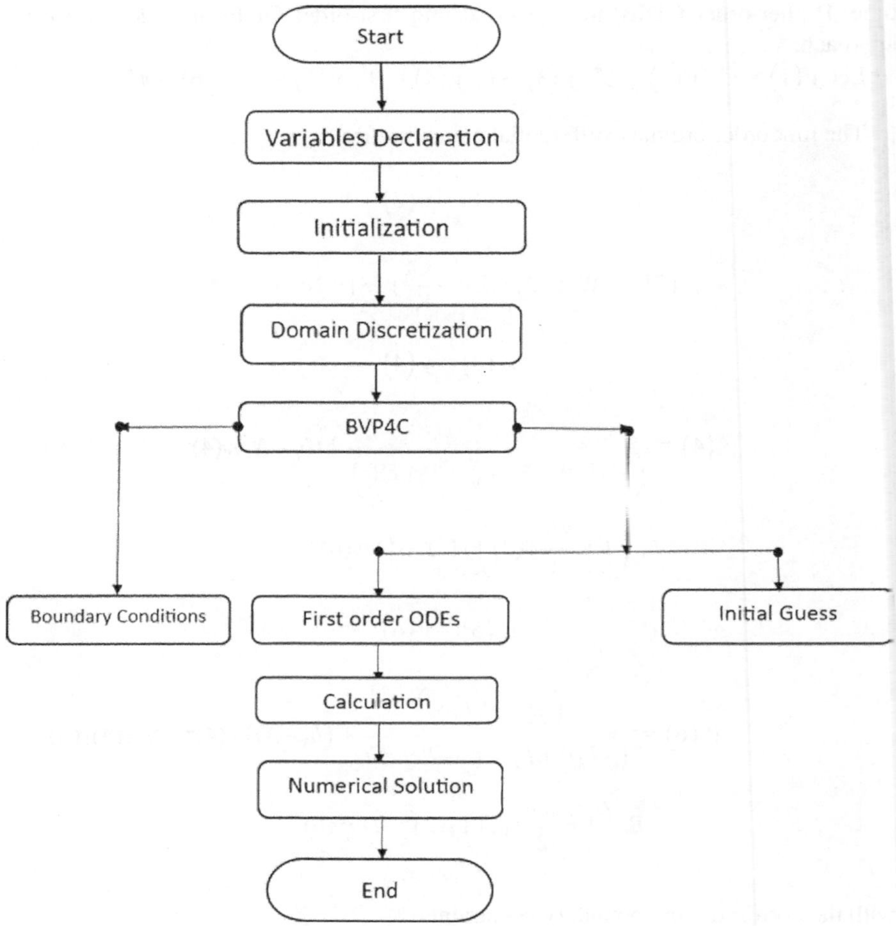

FIGURE 7.2 Systematic representation of the problem.

factors. On velocity, temperature and concentration fields with fixed values of $\frac{\mathrm{Gr}}{\mathrm{Re}} = 10$, $\mathrm{Nt} = 0.5$, $\mathrm{Nb} = 0.5$, $\mathrm{Nr} = 100$, $b_k = 0.2$, $\mathrm{Br} = 0.01$, $R_p = 1$, $\mathrm{Du} = 0.1$, $\mathrm{Sr} = 1$, $\mathrm{We} = 0.01$, $\mathrm{Bi1} = 100$, $\mathrm{Bi2} = 100$, $\mathrm{Le} = 1$ and $\mathrm{Pr} = 0.71$ are discussed using graphs.

7.4.1 Effects of Physical Parameters Nr, Nb, Nt, Gr, Br, Du, Sr, Le and We on Velocity Distribution

Figure 7.3 represents the velocity distribution for the fixed values of the physical parameters such as Nr, Nb, Nt, Gr, Br, Du, Sr, Le and We. From Figure 9.3(a) to 7.3(i) has been plotted in order to see the effects of physical attributes on the dimensionless velocity distribution. Figure 7.3(a) shows the influence with varying values of buoyancy parameter Nr, with the other values held constant, as shown in the graph.

It reveals that for the maximum velocity peak changes toward the left wall, the buoyancy parameter represents the relative influence of the buoyancy force of temperature to the viscous force, which boosts the velocity of the flow. Moreover, the flow reversal nature at larger buoyancy parameters is evident. Figure 7.3(b) shows the impact of Brownian parameter Nb increases as the fluid velocity grows because convective transport, caused by Brownian motion that predominates over diffusive transport. This indicates that convection is a more important mode of particle transport at higher velocities, while Brownian diffusion has more of an effect on particle distribution and fluid flow dynamics. Figure 7.3(c) describes the impact of thermophoresis parameter Nt on velocity. It indicates that with increasing values of Nt, because of the faster-moving particles, there was more resistance or drag. The fluid flow slows considerably with increasing Nt. The fluid velocity reduces as a result of a part of the flows pushing energy being used to overcome the particles' thermophoretic resistance. Figure 7.3(d) shows the velocity curve appears for various values of Gr. It was shown that as Gr grows it reduces. A Lorentz force, which is defined as the force that tends to slow the flow around the surface at the expense of an increased velocity. Figure 7.3(e) shows the velocity profile reduces as the Brinkman number is enhanced. Due to higher values improving the Brownian motion, where Nb results in more thermal diffusion compared to velocity. Figure 7.3(f) shows that a drop in the velocity profile results from an increase in the Dufour parameter Du with unchanged free convection parameters. It is also observed that the velocity (u) profile extends to a specific point on the channel width and subsequently declines with an increase in free convection parameters with constant Dufour parameter. It is obvious from the velocity line graph in Figure 7.3(g) that the relationship between velocity and Soret number is one of direct proportion. Moreover, it turns out that the velocity gradient increases closer to the heated plate. Whereas in Figure 7.3(h) represents the Lewis number, which shows that as the Lewis number increases velocity declines the flow rate. Figure 7.3(i) shows that the Williamson parameter affects velocity distribution, shrinking the velocity field. An increase in the parameter value leads to increased stress relaxation time, reducing fluid flow and lowering the velocity profile.

7.4.2 Effects of Physical Parameters Nb, Nt, R_p, Du, Sr, Le, and We on Temperature Distributions

Figure 7.4(a) shows that the impact of Brownian motion Nb on the temperature profile. When the magnitude of the Brownian motion is increased, the temperature pattern is enhanced. Physically, as a result of Brownian motion evolution occurs, the boundary layer warms up, which makes it easier for nanoparticles to move from the expansion sheet to the liquid region with less motion. As a result, the absorption of nanoparticles reduces. Figure 7.4(b) shows an increase in the thermophoresis parameter Nt typically leads to a decrease in temperature due to enhanced particle-driven heat transport away from hot regions. Figure 7.4(c) shows that as the thermal radiation parameter R_p increases, the temperature decreases due to enhanced radiative heat transfer, which allows more heat to escape and reduces the thermal energy. This cooling effect is more pronounced in confined channel flows. Figure 7.4(d) shows the velocity plots

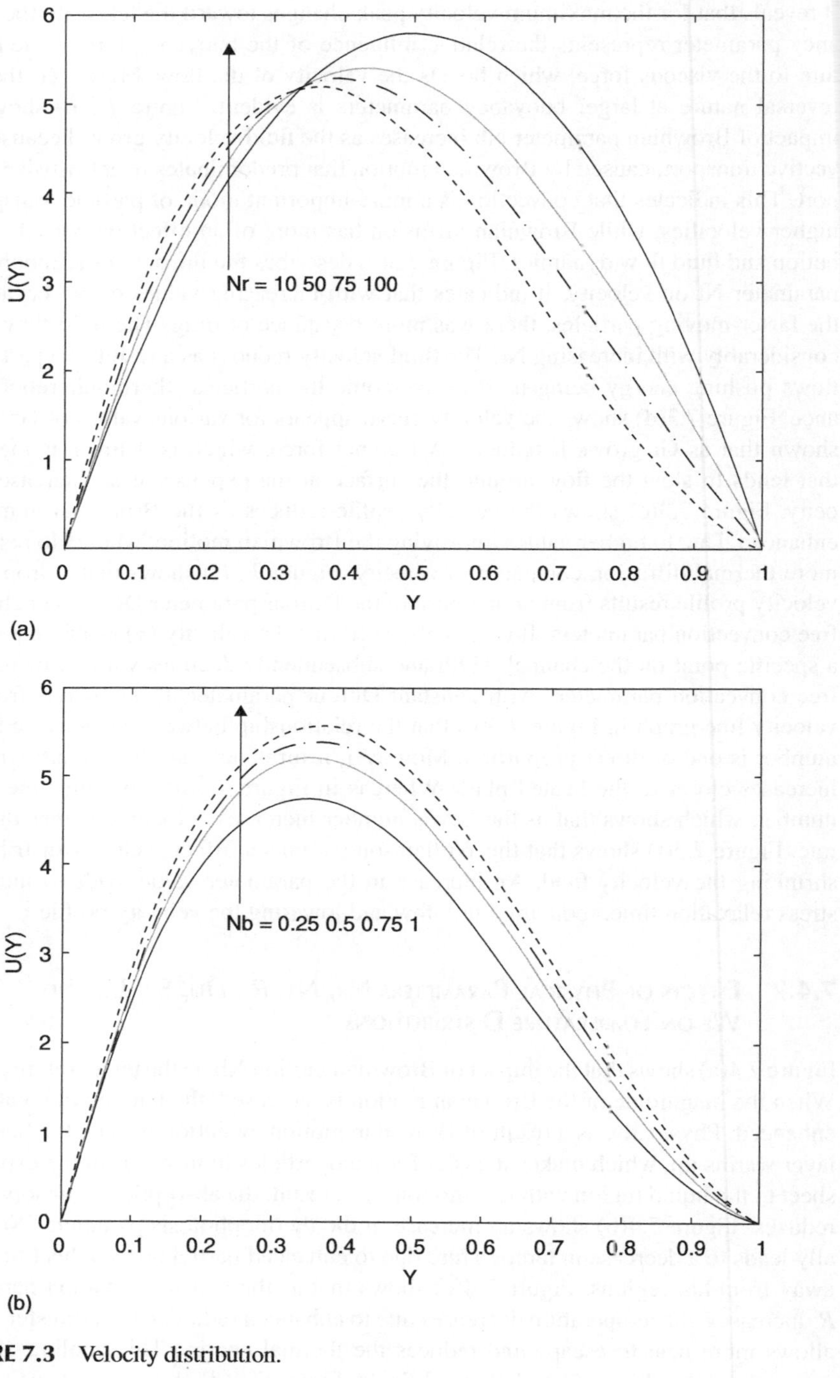

FIGURE 7.3 Velocity distribution.

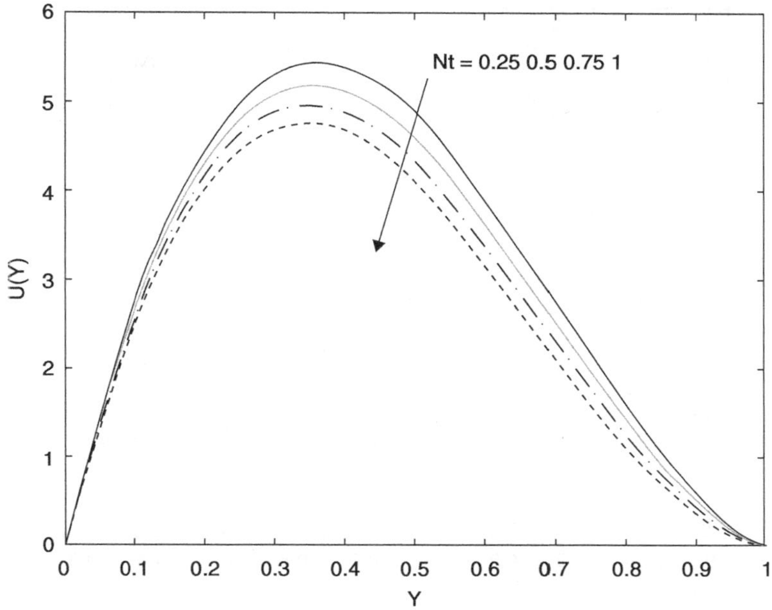

(c)

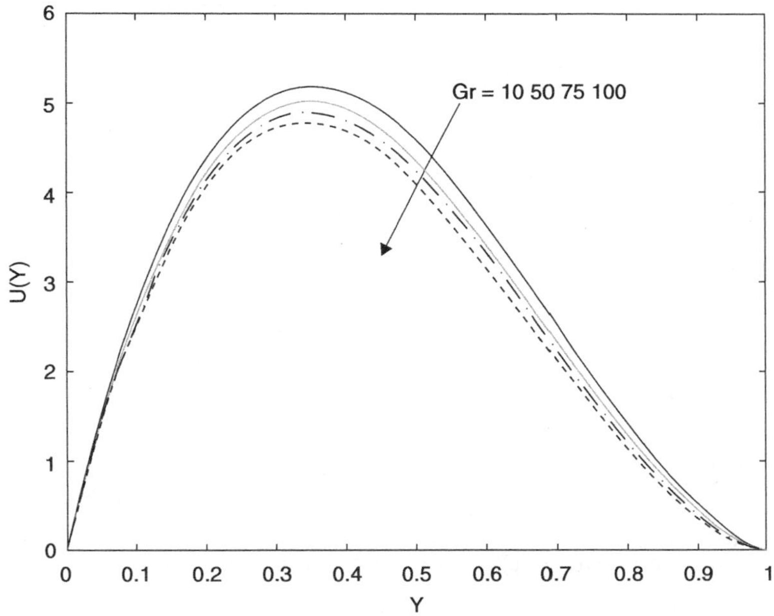

(d)

FIGURE 7.3 (Continued)

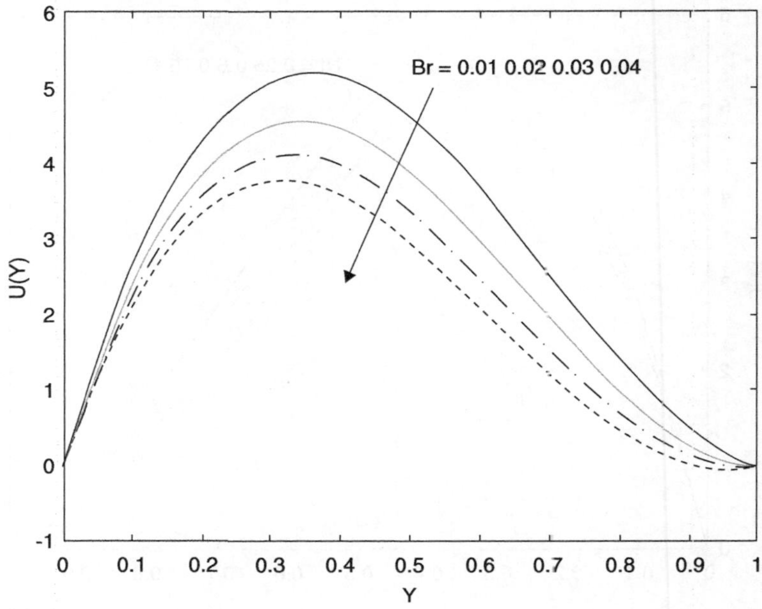

(e)

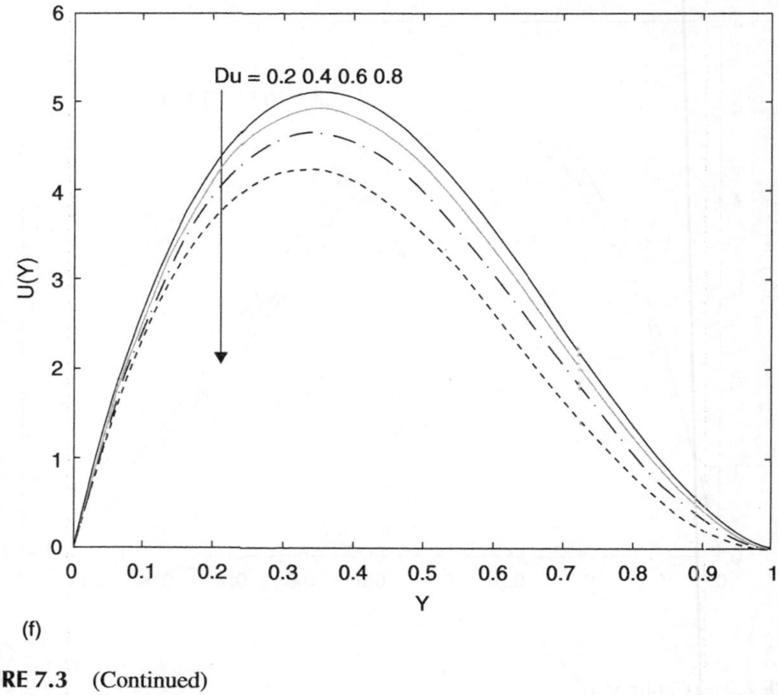

(f)

FIGURE 7.3 (Continued)

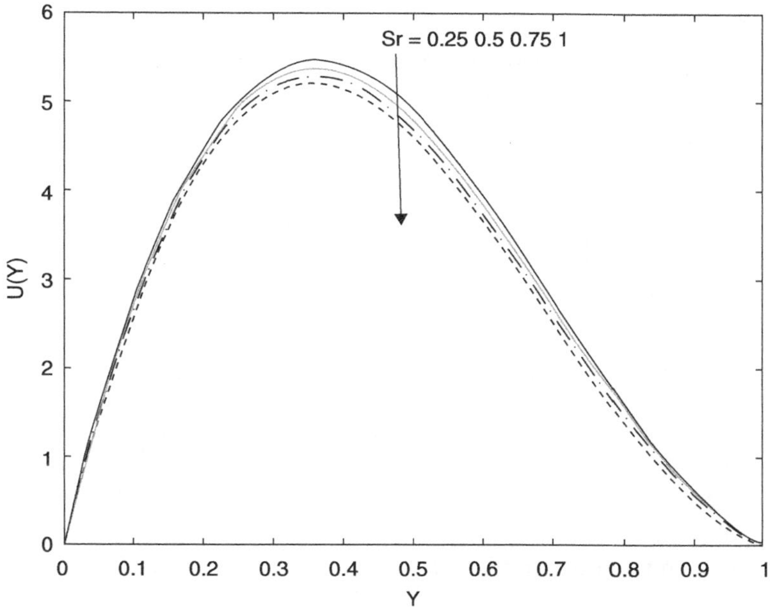

(g)

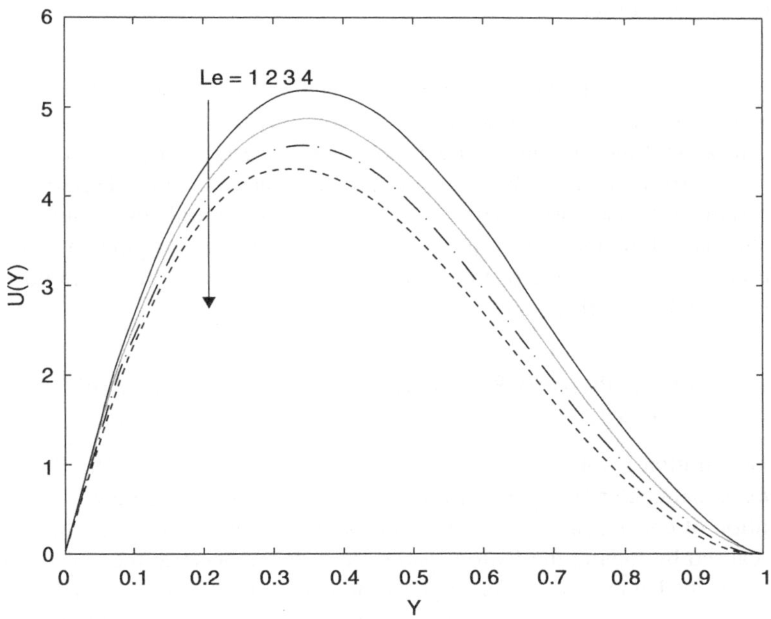

(h)

FIGURE 7.3 (Continued)

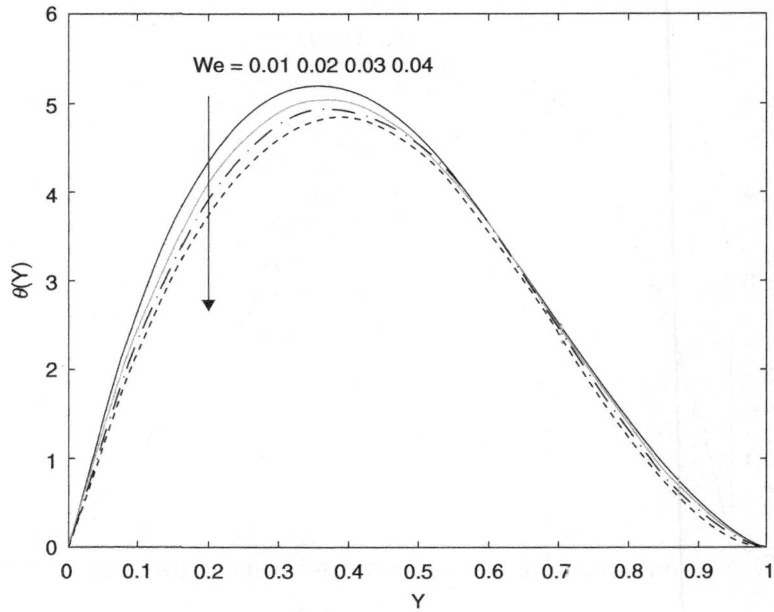

(i)

FIGURE 7.3 (Continued)

for different values of the Dufour parameter. The impact of concentration differences to thermal energy flux in the flow field is measured by the Dufour number. The illustration shows that the thermal field gets stronger as the Dufour parameter increases. Figure 7.4(e) shows that the fluid temperature drops across the fluid region when the Soret parameter Sr increases. Whereas in Figure 7.4(f),natural convection is particularly explained, thus the Lewis number decreases temperature gradients and buoyancy forces, but its impact depends on fluid system conditions and the importance of the thermal and mass diffusion.

7.4.3 Effects of Physical Parameters Du, Sr, and We on Concentration Distribution

The concentration graphs of nanoparticles with influencing Dufour Du, Soret Sr, and Williamson parameter We are shown in Figures 7.5(a) to (d). Figure 7.5(a) shows the distribution of concentrations with Dufour parameter Du. An increase in concentration is caused by raising the Dufour parameter. This is because the walls' thickness improves as the Du increases. The impact of temperature differences that have noticeable effects in mass diffusion is reflected in the Soret number in Figure 7.5(b). It shows that the concentration of nanoparticles increases as the Soret parameter increases. Furthermore, the ratio of the concentration difference, diffusion is the Soret parameter. The Williamson parameter We in Figure 7.5(c) shows that as concentration

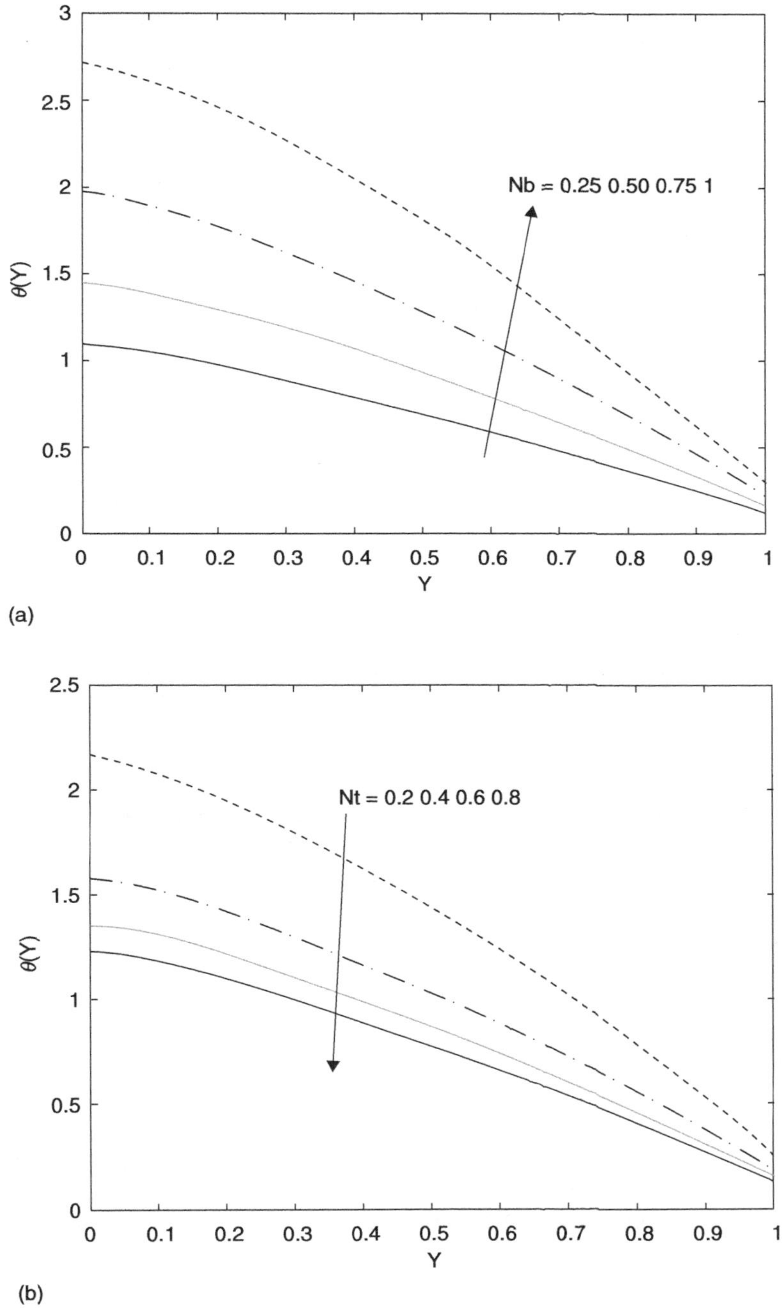

FIGURE 7.4 Temperature distribution.

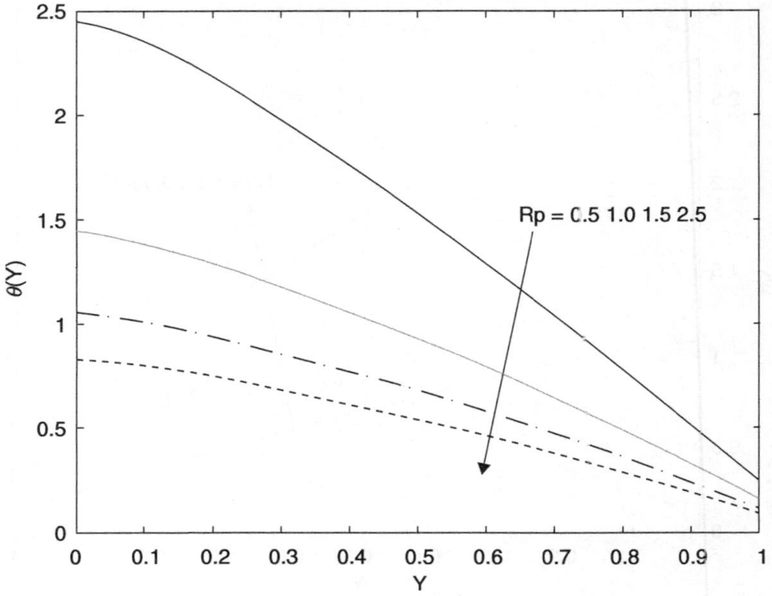

(c)

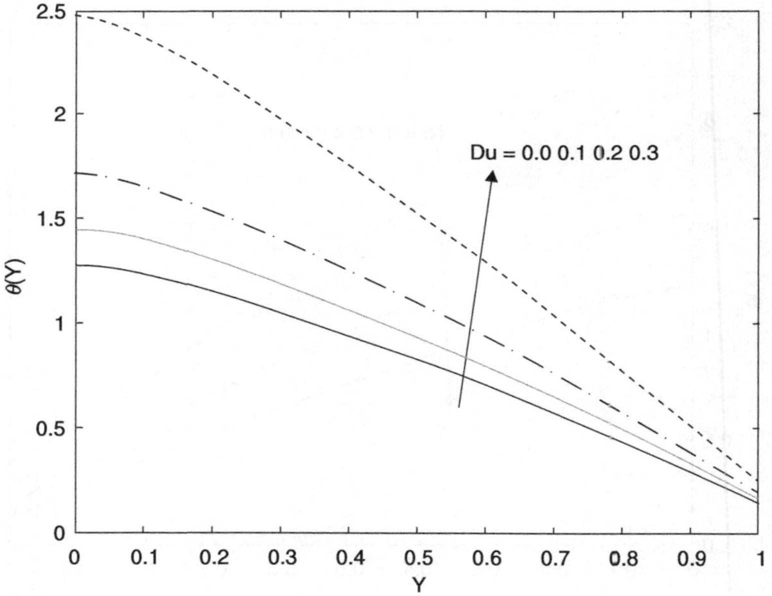

(d)

FIGURE 7.4 (Continued)

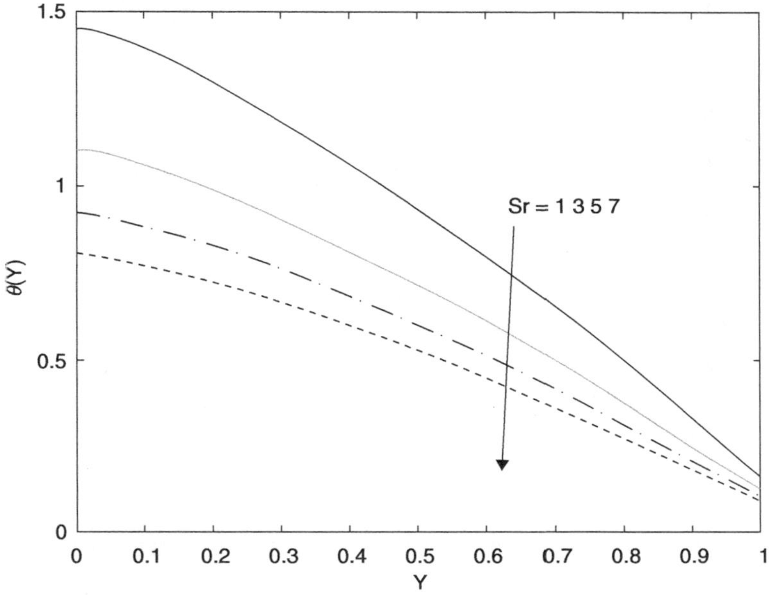

(e)

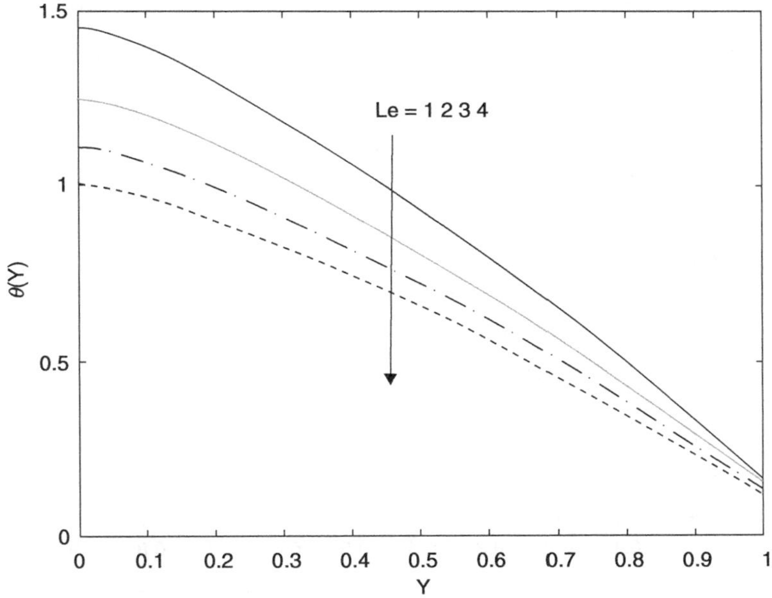

(f)

FIGURE 7.4 (Continued)

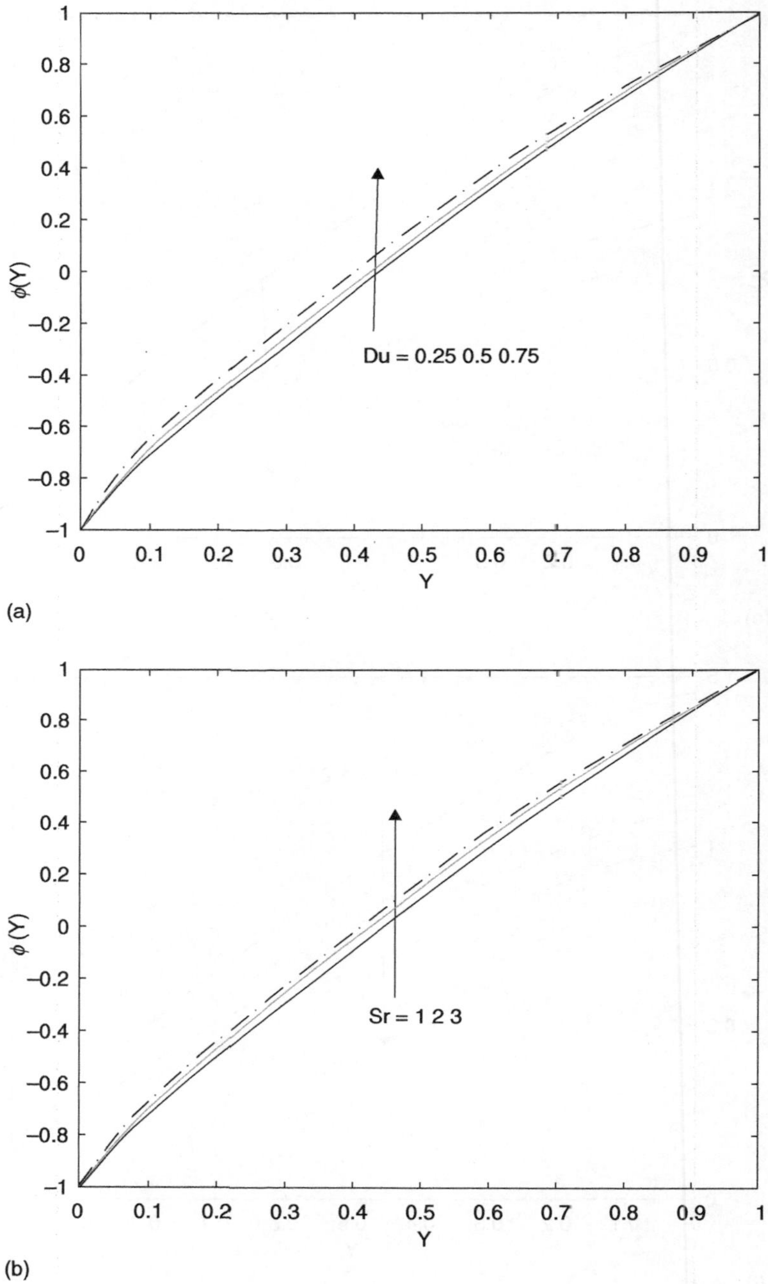

FIGURE 7.5 Concentration distribution.

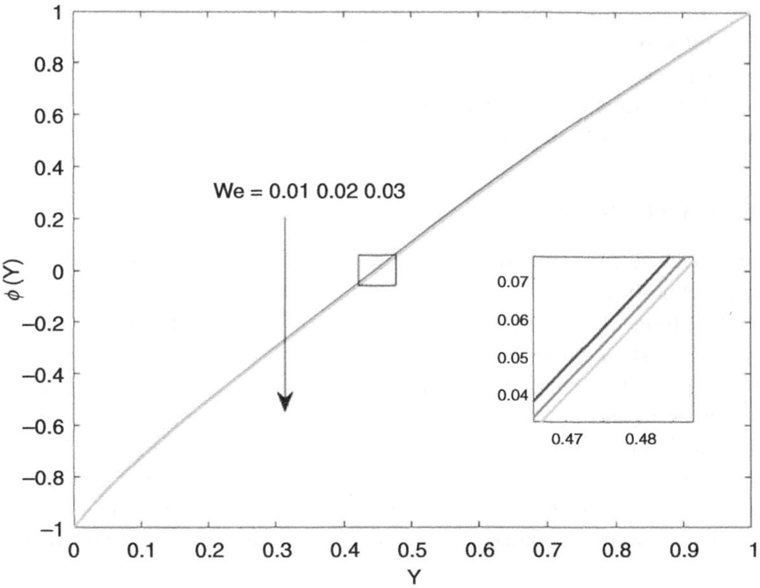

(c)

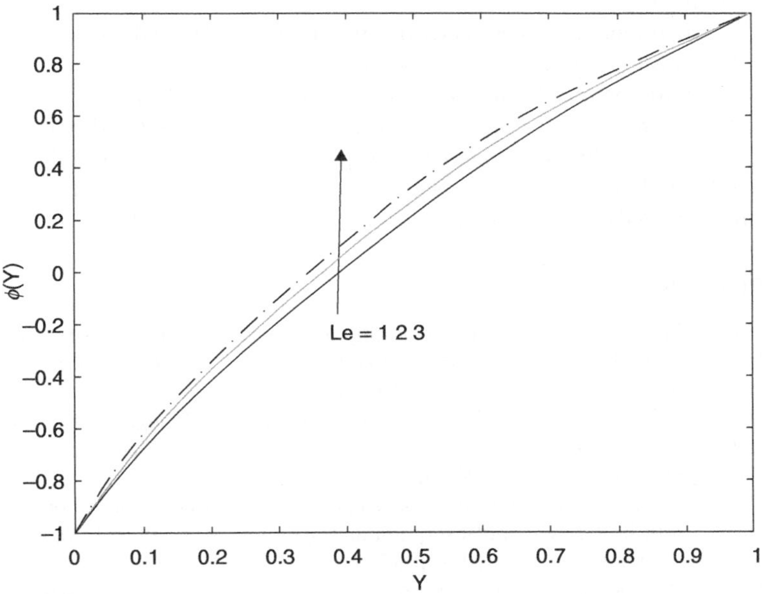

(d)

FIGURE 7.5 (Continued)

increases, the Williamson parameter enhances its mass diffusion rate within the fluid flow. The influence of Le on the energy profile $\varphi(Y)$ is analyzed in Figure 7.5(d) with increasing Lewis number Le values, the concentration distribution profile $\varphi(Y)$ is enhanced. The Lewis number represents the relationship between viscosity and the Brownian motion coefficient with the energy profile.

7.5 CONCLUSION

In this research framework, we investigate the consequences of an incompressible Williamson nanoliquid flow, taking into account the significance of heat radiation, Soret, and Dufour effects along a vertical channel. The importance of thermophoresis effects and Brownian motion are also examined. Williamson fluid is significant in dynamics, electrochemistry, and biomedical and automotive solutions. The important applications of nanofluid research in the fields of biotechnology, nanotechnology, cell biology, petroleum, metallurgy, renewable energy, and engineering difficulties have drawn the attention of scientists and researchers in recent years. The most current results are precise and clear, which increases their applicability in manufacturing-related fields. From the obtained results, the major findings of the study are as follows:

- Greater values of buoyancy ratio increase the flow and heat transfer.
- Brownian motion parameter enhances both velocity field and the temperature field.
- The thermophoresis parameter declines both velocity and thermal fields.
- Williamson number declines the velocity and thermal field, whereas in the concentration profiles it grows the mass diffusion slightly.
- Dufour and Soret parameters enhance the thermal field and mass diffusion rate whereas the Soret parameter drops the thermal field to a slight extent.
- If the Willaimson parameter and thermal radiation are neglected, then the case reduces to the problem considered by Patil Mallikarjun [45].

REFERENCES

[1] Aung, Win, and G. Worku. "Theory of fully developed, combined convection including flow reversal." *International Journal of Heat and Mass Transfer* 29.10 (1986): 1443–1458. https://doi.org/10.1016/0017-9310(86)90214-7.

[2] Cheng, Chin-Hsiang, Hong-Sen Kou, and Wen-Hsiung Huang. "Flow reversal and heat transfer of fully developed mixed convection in vertical channels." *Journal of Thermophysics and Heat Transfer* 4.3 (1990): 375–383.

[3] Chamkha, Ali J. "Non-Darcy fully developed mixed convection in a porous medium channel with heat generation/absorption and hydromagnetic effects." *Numerical Heat Transfer, Part A Applications* 32.6 (1997): 653–675.

[4] Barletta, A., and E. Zanchini. "On the choice of the reference temperature for fully-developed mixed convection in a vertical channel." *International Journal of Heat and Mass Transfer* 42.16 (1999): 3169–3181.

[5] Nield, D. A., A. V. Kuznetsov, and Ming Xiong. "Thermally developing forced convection in a porous medium: parallel plate channel with walls at uniform temperature, with

axial conduction and viscous dissipation effects." *International Journal of Heat and Mass Transfer* 46.4 (2003): 643–651.
[6] Grosan, T., and I. Pop. "Thermal radiation effect on fully developed mixed convection flow in a vertical channel." *Technische Mechanik-European Journal of Engineering Mechanics* 27.1 (2007): 37–47.
[7] Kuznetsov, A. V., and D. A. Nield. "Natural convective boundary-layer flow of a nanofluid past a vertical plate." *International Journal of Thermal Sciences* 49.2 (2010): 243–247.
[8] Umavathi, Jawali C., J. Prathap Kumar, Ali J. Chamkha, and Ioan Pop. "Mixed convection in a vertical porous channel." *Transport in Porous Media* 61 (2005): 315–335.
[9] Das, Sanatan, Mrinal Jana, and Rabindra Nath Jana. "Radiation effect on natural convection near a vertical plate embedded in porous medium with ramped wall temperature." *Open Journal of Fluid Dynamics* 1.01 (2011): 1
[10] Mahabaleshwar, U. S., M. B. Rekha, P. N. Vinay Kumar, F. Selimefendigil, P. H. Sakanaka, G. Lorenzini, and S. N. Ravichandra Nayakar. "Mass transfer characteristics of MHD Casson fluid flow past stretching/shrinking sheet." *Journal of Engineering Thermophysics* 29 (2020): 285–302.
[11] Choi, S. U. S., and Jeffrey A. Eastman. Enhancing thermal conductivity of fluids with nanoparticles. *No. ANL/MSD/CP-84938; CONF-951135-29*. Argonne, IL, USA: Argonne National Lab.(ANL), 1995.
[12] Buongiorno, Jacopo. "Convective transport in nanofluids." *Journal of Heat Transfer*, 128.3 (2006): 240–250. https://doi.org/10.1115/1.2150834
[13] Williamson, R. Vo. "The flow of pseudoplastic materials." *Industrial and Engineering Chemistry* 21.11 (1929): 1108–1111.
[14] Lyubimov, D. V., and A. V. Perminov. "Motion of a thin oblique layer of a pseudoplastic fluid." *Journal of Engineering Physics and Thermophysics* 75.4 (2002): 920–924.
[15] Muthuraj, R., D. Immaculate, Anant Shukla, and S. Srinivas. "MHD unsteady flow of a williamson nanofluid in a vertical porous space with oscillating wall temperature." *Frontiers in Heat and Mass Transfer (FHMT)* 7.1 (2016): 1–9. https://doi.org/10.5098/hmt.7.5
[16] Chamkha, Ali J., Camille Issa, and Khalil Khanafer. "Natural convection from an inclined plate embedded in a variable porosity porous medium due to solar radiation." *International Journal of Thermal Sciences* 41.1 (2002): 73–81.
[17] Sharma, Kalu Ram, and Shalini Jain. "An unsteady MHD Williamson fluid flow in a vertical porous channel with porous media and thermal radiation." *International Journal of Advances in Engineering Sciences and Applied Mathematics* 16.1 (2024): 1–12. https://doi.org/10.1007/s12572-023-00450-4
[18] Dadheech, Amit, Singh, G., Patel, R., Kumar, D., and Chamkha, A. J. "Second law analysis for MHD slip flow for Williamson fluid over a vertical plate with Cattaneo-Christov heat flux." *Case Studies in Thermal Engineering* 33 (2022): 101931. https://doi.org/10.1016/j.csite.2022.101931
[19] Shashikumar, N., S., Macha Madhu, S. Sindhu, B. J. Gireesha, and Naikoti Kishan. "Thermal analysis of MHD Williamson fluid flow through a microchannel." *International Communications in Heat and Mass Transfer* 127 (2021): 105582.
[20] Kothandapani, M., and J. Prakash. "Effects of thermal radiation parameter and magnetic field on the peristaltic motion of Williamson nanofluids in a tapered asymmetric channel." *International Journal of Heat and Mass Transfer* 81 (2015): 234–245.
[21] Shaheen, S., M. B. Arain, Kottakkaran Sooppy Nisar, Ashwag Albakri, M. D. Shamshuddin, and Fouad Othman Mallawi. "A case study of heat transmission in a Williamson fluid flow through a ciliated porous channel: A semi-numerical approach." *Case Studies in Thermal Engineering* 41 (2023): 102523.

[22] Manthesha, Mallikarjun PB, L. Kavitha, and K. C. Shobha. "Mixed convection of Williamson fluid along an inclined porous microchannel with chemical reaction by taking non-constant thermal conductivity: An entropy analysis." *Indian Journal of Science and Technology* 14.48 (2021): 3525–3536.

[23] Gireesha, B. J., Nagaraja, B., Srikantha, N., Rudraswamy, N. G., and Felicita, A. "Magnetically propelled Carreau fluid flow over penetrable sensor surface influenced by thermal radiation, Joule heating and heat generation." *Communications in Theoretical Physics* 74.2 (2022): 025002.

[24] Reddy, Y. Dharmendar, Fateh Mebarek-Oudina, B. Shankar Goud, and A. I. Ismail "Radiation, velocity and thermal slips effect toward MHD boundary layer flow through heat and mass transport of Williamson nanofluid with porous medium." *Arabian Journal for Science and Engineering* 47.12 (2022): 16355–16369.

[25] Khan, Masood, and Aamir Hamid. "Influence of non-linear thermal radiation on 2D unsteady flow of a Williamson fluid with heat source/sink." *Results in Physics* 7 (2017): 3968–3975.

[26] Hamid, M., M. Usman, Z. H. Khan, Rizwan Ul Haq, and W. Wang "Numerical study of unsteady MHD flow of Williamson nanofluid in a permeable channel with heat source/sink and thermal radiation." *The European Physical Journal Plus* 133.12 (2018): 527.

[27] Farooq, Umar, Hassan Waqas, Roa Makki, Mohamed R. Ali, Abdullah Alhushaybari, Taseer Muhammad, and Muhammad Imran. "Computation of Cattaneo-Christov heat and mass flux model in Williamson nanofluid flow with bioconvection and thermal radiation through a vertical slender cylinder." *Case Studies in Thermal Engineering* 42 (2023): 102736.

[28] Prakash, J., and Dharmendra Tripathi. "Electroosmotic flow of Williamson ionic nanoliquids in a tapered microfluidic channel in presence of thermal radiation and peristalsis." *Journal of Molecular Liquids* 256 (2018): 352–371.

[29] Hashim, Masood Khan, and Aamir Hamid. "Convective heat transfer during the flow of Williamson nanofluid with thermal radiation and magnetic effects." *The European Physical Journal Plus* 134 (2019): 1–12.

[30] Shobha, K. C., and B. Patil Mallikarjun. "Effect of nonlinear thermal radiation on flow of Williamson nanofluid in a vertical porous channel with heat source or sink by using Adomian decomposition method." *Journal of Nanofluids* 11.1 (2022): 39–47.

[31] Siddabasappa, C., G. Kalpana, and Babitha. "A numerical study on heat transfer and MHD flow of a Newtonian through a porous channel—Local thermal nonequilibrium model." *Heat Transfer* 52.1 (2023): 949–967.

[32] Kafoussias, N. G., and E. W. Williams. "Thermal-diffusion and diffusion-thermo effects on mixed free-forced convective and mass transfer boundary layer flow with temperature dependent viscosity." *International Journal of Engineering Science* 33.9 (1995): 1369–1384.

[33] Makinde, O. D., K. Zimba, and O. Anwar Bég. "Numerical study of chemically-reacting hydromagnetic boundary layer flow with Soret/Dufour effects and a convective surface boundary condition." *International Journal of Thermal and Environmental Engineering* 4.1 (2012): 89–98.

[34] Ahammad, N. Ameer, and M. Veera Krishna. "Numerical investigation of chemical reaction, Soret and Dufour impacts on MHD free convective gyrating flow through a vertical porous channel." *Case Studies in Thermal Engineering* 28 (2021): 101571.

[35] Sreedevi, Gandluru, D. R. V. Rao, Oluwole Daniel Makinde, and G. Reddy. "Soret and Dufour effects on MHD flow with heat and mass transfer past a permeable stretching sheet in presence of thermal radiation." *Frontiers in Heat and Mass Transfer (FHMT)* 8, (2017): 1–10. https://doi.org/10.5098/hmt.8.5

[36] Umadevi, K. B., and B. P. Mallikarjun. "Homotopy analysis method for an influence of DufourSoret and melting process on magnetohydrodynamic boundary-layer flow towards a wedge in an eyring powell fluid." *Indian Journal of Science and Technology* 15.15 (2022): 700–711.

[37] Beg, O. Anwar, Tasveer A. Beg, A. Y. Bakier, and V. R. Prasad. "Chemically-reacting mixed convective heat and mass transfer along inclined and vertical plates with Soret and Dufour effects: Numerical solutions." *International Journal of Applied Mathematics and Mechanics* 5.2 (2009): 39–57.

[38] Ou, Xueyu, Ruijin Wang, Tongwei Guo, Chun Shao, and Zefei Zhu. "Numerical investigation on the heat and mass transfer in microchannel with discrete heat sources considering the soret and dufour effects." *Micromachines* 13.11 (2022): 1848.

[39] Grosan, T., and I. Pop. "Fully developed mixed convection in a vertical channel filled by a nanofluid." 134.8 (2012): 082501. https://doi.org/10.1115/1.4006757

[40] Umavathi, J. C., Mikhail A. Sheremet, and Syed Mohiuddin. "Combined effect of variable viscosity and thermal conductivity on mixed convection flow of a viscous fluid in a vertical channel in the presence of first order chemical reaction." *European Journal of Mechanics-B/Fluids* 58 (2016): 98–108.

[41] Kierzenka, Jacek, and Lawrence F. Shampine. "A BVP solver based on residual control and the Maltab PSE." *ACM Transactions on Mathematical Software (TOMS)* 27.3 (2001): 299–316.

[42] Higham, Desmond J., and Nicholas J. Higham. *MATLAB guide* (3rd ed.). Philadelphia, PA: Society for Industrial and Applied Mathematics, 2016. https://doi.org/10.1137/1.9781611974652

[43] Shobha, K. C., and Mallikarjun B. Patil. "Irreversibility analysis of micropolar nanofluid flow in a vertical channel with the impact of inclined magnetic field and heat source or sink." *Heat Transfer* 51.3 (2022): 2723–2741.

[44] Chandan, K. G., B. Patil Mallikarjun, Soumya Ghosh, and Suvanjan Bhattacharyya. "Entropy analysis for two-dimensional double-diffusive mixed convective flow of a Williamson nanofluid through a porous medium." *Numerical Heat Transfer, Part B: Fundamentals* 85.1 (2024): 1–21. https://doi.org/10.1080/10407790.2024.2306021

[45] Patil, Mallikarjun B., K. C. Shobha, Suvanjan Bhattacharyya, and Zafar Said. "Soret and Dufour effects in the flow of Casson nanofluid in a vertical channel with thermal radiation: Entropy analysis." *Journal of Thermal Analysis and Calorimetry* 148.7 (2023): 2857–2867.

8 Thermal Analysis in MHD Ternary Nanofluid Models

Kandavkovi Mallikarjuna Nihaal, Ulavathi Shettar Mahabaleshwar, Gadhigeppa Myacher Sachin, Shravanakumari, Shanthaveerappa Mangala, Panchappa Mallikarjuna Kavana, and Basma Souayeh

8.1 INTRODUCTION

Ternary nanofluids are effective thermal fluids made up of three different components, which are a base fluid and three nanoparticles. Compared to typical fluids, these nanofluids have higher thermal conductivity and heat transmission properties. Ternary nanofluids are being extensively researched for their possible use in heat exchangers, cooling systems, and renewable energy technologies. They are useful for enhancing the efficacy of a variety of technical and industrial processes because of their capacity to boost heat transfer capabilities. Recently, many researchers have explored the influence of ternary nanofluid on Newtonian or non-Newtonian fluid flow on various geometries. Darvesh *et al.* [1] studied the effect of viscous dissipation, a non-uniform heat source sink, radiation, and infinite shear rate viscosity on bullet-shaped geometry in polymer-based ternary radiative Carreau nanofluid systems. The findings provide important insights into the predicting capabilities of increased heat transfer in ternary hybrid nanofluids utilizing ANN. Paul *et al.* [2] examined MHD Casson ternary nanofluid flow over a porous exponential stretching cylinder with the Darcy–Forchheimer model. The study shows that there is an improvement in heat and mass transfer rates by the ternary nanofluid model about 15.5% and 8%, respectively, compared to the nanofluid model. Nihaal *et al.* [3] explored ternary nanofluid over the bioconvective Darcy–Fochheimer flow model across a radially stretching disk under the influence of exponential heat source/sink and activation energy. Mahboobtosi *et al.* [4] examined parametric analysis and performance enhancement of ternary hybrid nanofluids and single nanofluids in convective flow around a stretched cylinder. According to this study, the ternary hybrid offers more benefits than the single nanofluid, including increased velocity and temperature. Farroq *et al.* [5] explored ternary nanofluid flow past a porous stretching cylinder in the presence of MHD, radiation, and chemical reaction effects with gyrostatic microorganisms.

DOI: 10.1201/9781003611820-10

Dusty fluid is a multiphase flow made up of solid particles scattered in a fluid (either gas or liquid). This fluid model has sparked widespread attention because of its numerous applications in domains, such as industrial operations, atmospheric science, and biomedical engineering. The presence of dust particles in the fluid can have a substantial impact on flow dynamics, heat transfer properties, and overall system performance. In recent years, various researchers have investigated fluid flow attributes with dust particle suspension. Das *et al.* [6] investigated the effects of MHD, radiation, and heat source/sink on Casson nanofluid flow across a vertically extended dusty porous medium. The results show that the magnetic parameter has a considerable impact on flow momentum, which in turn impacts dusty fluid dynamics. Ali *et al.* [7] investigated MHD dusty fluid flow across a stretched surface under the influence of Darcy–Fochheimer porosity and Brownian motion. Sachhin *et al.* [8] examined tetra dusty nanofluid flow past a stretching/shrinking sheet with the Darcy–Brinkman model under the influence of suction/injection. The results show that the tetra nanofluid phase outperforms the dusty phase in the case of heat and mass transfer. Mirza *et al.* [9] evaluated Levenberg–Marquardt neural networks for a two-phase flow of dusty trihybrid nanofluid with a C-C heat flux model. The results revealed that increasing the Marangoni convection factor improved the flow field of the dust and fluid phases. Koriko *et al.* [10] analyzed the impacts of partial slip and viscous dissipation on Carreau nanofluid and dusty fluid. The study's results show that increased viscous dissipation rate improves fluid velocity and temperature distribution across both fluid phase and dusty phase.

Magnetohydrodynamics (MHD) has a substantial influence on fluid flow and heat transfer capabilities since it incorporates magnetic fields into fluid dynamics. This interaction enables precise control of fluid behavior improving heat transfer efficiency and flow characteristics. MHD can be used to control the thickness and uniformity of coatings in industrial processes, such as wire coatings, by exploiting the interplay of magnetic fields with fluid dynamics. Rafique *et al.* [11] analyzed the impact of a magnetic field over ternary nanofluid flow over a stretched cylinder with velocity slip and thermal jump. It is observed that fluid velocity decreases for growing values of magnetic field parameter. Yanala *et al.* [12] studied the effect of MHD on heat transfer characteristics of an incompressible viscous fluid on a horizontal stretching cylinder submerged in a porous medium with internal heat production/sink. Galal *et al.* [13] examined nanofluid flow on a stretched cylinder with MHD and gyrostatic microorganisms. The findings show that as the magnetic parameter rises, the fluid velocity reduces. Raj *et al.* [14] investigated the influence of magnetic field, heat source, and thermal radiation on Maxwell fluid flow over a porous stretching cylinder. Nandi and Barman [15] examined the unsteady flow of hybrid nanofluid over a porous stretching cylinder with magnetic field, viscous dissipation radiation, and heat source. The study outcomes reveal that fluid velocity decreases due to the impact of magnetic field and unsteady parameter.

The radiation phenomenon is critical to fluid flow and heat transmission, especially in systems with large temperature gradients. It relates to the transport of heat by electromagnetic waves, which can change nonlinearly with temperature. This phenomenon is critical for enhancing thermal performance in areas such as solar energy systems and thermal insulation materials. Pashikanti and Priyadharshini [16] investigated the effects of radiation and chemical reaction on nanofluid flow over a stretched cylinder. Munazza and Farooq [17] examined the influence of radiation and heat generation/absorption on a convective MHD nanofluid flow over a stretched cylinder with variable thermal conductivity. Akbar *et al.* [18] investigated the flow of third grade MHD nanofluid past a non-linear stretching cylinder with thermal radiation. According to the study findings, the temperature curve increases due to radiation effects. Mahabaleshwar *et al.* [19] researched the impact of radiation on the Casson nanofluid flow on a stagnation point over a stretching plate and cylinder. The findings show that when radiation is amplified, the temperature field increases. Raj *et al.* [20] addressed the impact of heat generation and radiation on MHD Maxwell fluid flow in a stretched porous cylinder. According to quantitative data, the fluid s temperature profile rises with increased heat generation and radiation.

Boundary layer flow has drawn the attention of many researchers over various stretching surfaces because of its several engineering and industrial applications. Many of these industrial application processes feature boundary surfaces that resemble cylinders. The study of the flow of various fluids across a cylinder with numerous applications, such as tube exchangers, electronic cooling, and oil drilling, has drawn the interest of many researchers. The effect of a non-uniform heat source/sink on fluid flow on an inclined stretching cylinder with mixed convection was studied by Hayat *et al.* [21]. Rashid *et al.* [22] evaluated the impact of nanoparticle shape effect on heat transfer of a hybrid nanofluid flow over a horizontal shrinking and stretching cylinder. Mishra and Kumar [23] investigated MHD nanofluid flow via a stretching cylinder with slip effects, viscous dissipation, and Joule heating. Ganga *et al.* [24] explored the influence of machine learning on Maxwell hybrid nanofluid flow across an exponentially stretched cylinder in the presence of nonlinear radiation. The influence of radiation and viscous dissipation on MHD Maxwell fluid flow across a stretching cylinder with CNTs was analyzed by Faraz and Park [25].

According to the above-mentioned previous works of literature, no research has considered the impact of magnetic field and thermal radiation on dusty ternary nanofluid flow across a stretching cylinder. Given the numerous industrial uses of radiation MHD, and stretching cylinders, the current study addresses the dusty ternary nanofluid flow through stretching cylinder in an MHD while taking the radiation heat transfer phenomena into account. Employing an appropriate similarity approach, the flow governing and energy equations are transformed into non-linear ODEs and further numerical solutions are obtained via the shooting method and the RKF-45th method.

8.2 MATHEMATICAL MODELING

The present modeling showcases a two-dimensional, steady and incompressible fluid flow of dusty ternary fluid across a stretching cylinder with reference length L and positive stretching factor $\hat{a}$ is considered in the present study. The cylinder's horizontal direction is measured by x and radially r axis is considered in the direction perpendicular to the cylinder. Additionally, $U_w = x\hat{a}/L \ (\hat{a} > 0)$ denotes surface velocity measured parallel to the cylinder's surface. In the present modeling, the fluid flow and heat transfer are analyzed for both dusty and fluid phases under the impacts of a magnetic field and radiation. Taking the above considerations, the governing equations are given as follows (see [26, 27]):

Governing equations

Fluid phase

$$\frac{\partial (ur)}{\partial x}+\frac{\partial (vr)}{\partial r}=0, \tag{8.1}$$

$$u\frac{\partial u}{\partial x}+v\frac{\partial u}{\partial r}+\frac{KN}{\rho_{tnf}}\left(u-u_p\right)=\nu_{tnf}\left(\frac{\partial^2 u}{\partial r^2}+\frac{1}{r}\frac{\partial u}{\partial r}\right)-\frac{\sigma_{tnf}}{\rho_{tnf}}B^2u, \tag{8.2}$$

$$\begin{aligned}(\rho Cp)_{tnf}\left(u\frac{\partial T}{\partial x}+v\frac{\partial T}{\partial r}\right)=k_{tnf}\left(\frac{\partial^2 T}{\partial r^2}+\frac{1}{r}\frac{\partial T}{\partial r}\right)-\frac{1}{r}\frac{\partial (rq_r)}{\partial r},\\ +\frac{\rho_p (Cp)_f}{\tau_v}\left(T_p-T\right)\end{aligned} \tag{8.3}$$

Dusty phase

$$\frac{\partial (ru_p)}{\partial x}+\frac{\partial (rv_p)}{\partial r}=0, \tag{8.4}$$

$$u_p\frac{\partial u_p}{\partial x}+v_p\frac{\partial u_p}{\partial r}=\frac{K}{m}\left(u-u_p\right), \tag{8.5}$$

$$u_p\frac{\partial T_p}{\partial x}+v_p\frac{\partial T_p}{\partial r}=\frac{(Cp)_f}{\tau_t C_m}\left(T-T_p\right), \tag{8.6}$$

respective boundary constraints are

$$u=U_w, v=0, T=T_w, \text{at} r=R, \tag{8.7}$$

$$u\to 0,\ T\to T_\infty,\ u_p\to 0,\ v_p\to v,\ T_p\to T_\infty \text{ as } r\to\infty \tag{8.8}$$

Here, (u,v) are the velocity components along (x,r) coordinates and (u_p,v_p) are velocity components for the dust phase. (K,N) denotes Stoke's resistance and the number of density of particles, (ν,ρ,σ) indicates kinematic viscosity, density and electrical conductivity, B denotes the strength of the magnetic field, (T, T_w, T_∞, T_p) are temperature, wall temperature, ambient temperature, and temperature for the dust phase, (k, Cp) denotes thermal conductivity and specific heat capacity, the density of the dust particles is denoted by ρ_p and relaxation time for dust particles and thermal equilibrium for dusty particles are indicated by (τ_t, τ_v). The Rosseland approximation is used to calculate the radiation heat flux (q_r),

$$q_r = -\frac{4\sigma^*}{3k^*}\frac{\partial T^4}{\partial r} \tag{8.9}$$

Here, the Stefan–Boltzmann constant is σ^* and the mean absorption coefficient is denoted by k^*. The T^4 term in Eq. (8.9) is linearized with the help of the Taylor series about T_∞ with neglecting higher-order terms as shown in Eq. (8.10).

$$T^4 \cong 4TT_\infty^3 - 3T_\infty^4 \tag{8.10}$$

Thus using Eqs. (8.9) and (8.10), Eq. (8.3) can be transformed into

$$\begin{aligned}(\rho Cp)_{tnf}\left(u\frac{\partial T}{\partial x}+v\frac{\partial T}{\partial r}\right) = k_{tnf}\left(\frac{\partial^2 T}{\partial r^2}+\frac{1}{r}\frac{\partial T}{\partial r}\right)+\frac{16\sigma^* T_\infty^3}{3k^*}\left(\frac{\partial^2 T}{\partial r^2}+\frac{1}{r}\frac{\partial T}{\partial r}\right)\\ +\frac{\rho_p (Cp)_f}{\tau_v}(T_p - T)\end{aligned} \tag{8.11}$$

The proper similarities employed are

$$\eta = \sqrt{\frac{\hat{a}}{v_f L}}\left(\frac{r^2-R^2}{2R}\right) \tag{8.12}$$

Fluid phase

$$u = \frac{\hat{a}x}{L}f', v = -\frac{1}{r}\sqrt{\frac{\hat{a}v_f}{L}}Rf, \theta = \frac{T-T_\infty}{T_w - T_\infty} \tag{8.13}$$

Dusty phase

$$u_p = \frac{\hat{a}x}{L}F', v_p = -\frac{1}{r}\sqrt{\frac{\hat{a}v_f}{L}}RF, \theta_p = \frac{T_p - T_\infty}{T_w - T_\infty} \tag{8.14}$$

The thermophysical properties of a ternary nanofluid are as follows (see [28]):

$$\rho_{tnf} = (1-\phi_3)\left[(1-\phi_2)\left((1-\phi_1)\rho_f + \phi_1\rho_1\right)+\phi_2\rho_2\right]+\phi_2\rho_2,$$

$$\mu_{tnf}=\frac{\mu_f}{(1-\phi_3)^{2.5}(1-\phi_2)^{2.5}(1-\phi_1)^{2.5}},$$

$$k_{tnf}=k_f\left[\frac{k_3+2k_f-2\phi_3(k_f-k_3)}{k_3+2k_f+\phi_3(k_f-k_3)}\right],$$

$$k_{hnf}=k_{nf}\left[\frac{k_2+2k_{nf}-2\phi_2(k_{nf}-k_2)}{k_2+2k_{nf}+\phi_2(k_{nf}-k_2)}\right],$$

$$k_{tnf}=k_{hnf}\left[\frac{k_1+2k_{hnf}-2\phi_1(k_{hnf}-k_1)}{k_1+2k_{hnf}+\phi_1(k_{hnf}-k_1)}\right],$$

$$(\rho Cp)_{tnf}=(1-\phi_3)\left[(1-\phi_2)\left((1-\phi_1)(\rho Cp)_f+\phi_1(\rho Cp)_1\right)+\phi_2(\rho Cp)_2\right]+\phi_3(\rho Cp)_3$$

Upon substitution of similarities into Eqs. (8.1–8.8) and Eq. (8.11), the obtained reduced equations are

Fluid phase

$$\frac{1}{A_1A_2}(\{1+2\eta\lambda\}f'''+2\lambda f'')+\frac{l\beta_1}{A_2}(F'-f')-(f'^2-ff'')-\frac{A_3}{A_2}Mf'=0 \tag{8.15}$$

$$\left(A_4+\frac{4}{3}R\right)(\{1+2\eta\lambda\}\theta'')+\left(A_4+\frac{4}{3}R\right)2\theta'\lambda+\Pr A_5f\theta'+\Pr\beta_2 l(\theta_p-\theta)=0 \tag{8.16}$$

Dusty phase

$$FF''+\beta_1(f'-F')-F'^2=0, \tag{8.17}$$

$$F\theta_p'-\beta_2\gamma(\theta_p-\theta)=0, \tag{8.18}$$

and the corresponding boundary constraints are

$$f'(0)=1, f(0)=0, \theta(0)=1, \tag{8.19}$$

$$f'(\infty)\to 0, F'(\infty)\to 0, F(\infty)\to f(\infty), \theta(\infty)=0, \theta_p(\infty)=0 \tag{8.20}$$

where

$$A_1=\frac{\mu_{tnf}}{\mu_f}, A_2=\frac{\rho_{tnf}}{\rho_f}, A_3=\frac{\sigma_{tnf}}{\sigma_f}, A_4=\frac{k_{tnf}}{k_f}, A_5=\frac{(\rho Cp)_{tnf}}{(\rho Cp)_f} M=\frac{\sigma_f B^2 L}{\rho_f a}, l=\frac{\rho_p}{\rho_f}, \beta_1=\frac{L}{a\tau_v}, \tau_v=\frac{m}{K},$$

$R=\frac{4\sigma^* T_\infty^3}{k^* k_f}, \lambda=\sqrt{\frac{L\nu_f}{aR^2}}, \mathrm{Pr}=\frac{(\rho Cp)_f \nu_f}{k_f}, \beta_2=\frac{L}{a\tau_t}, \gamma=\frac{Cp_f}{C_m}$, represents thermophysical constants, magnetic field parameter, mass concentration of particles, interaction parameter for velocity, relaxation time for dust particles, radiation parameter, curvature parameter, Prandtl number, interaction parameter for temperature, and the ratio of specific heat.

The engineering factors and their respective reduced forms are obtained as follows:

$$Cf_x=\frac{\mu_{tnf}}{\rho_f U_w^2}\left.\frac{\partial u}{\partial r}\right|_{r=R} \Rightarrow \mathrm{Re}^{1/2} Cf_x = A_1 f''(0), \tag{8.21}$$

and

$$\begin{aligned} Nu_x &= \frac{x}{k_f(T_w-T_\infty)}\left(-\left\{k_{tnf}+\frac{16\sigma^* T_\infty^3}{3k^*}\right\}\right)\left.\frac{\partial T}{\partial r}\right|_{r=R} \\ &\Rightarrow \mathrm{Re}^{-1/2} Nu_x = -\left(A_4+\frac{4}{3}R\right)\theta'(0) \end{aligned} \tag{8.22}$$

where $\mathrm{Re}=\frac{ax^2}{\nu_f L}$ denotes the local Reynolds number.

8.3 NUMERICAL METHODOLOGY

The numerical solution enables the precise and efficient modeling of compressible and incompressible flows. A numerical method used in the current study is the RKF-45th method and shooting method. The shooting method converts a boundary value problem to an initial value problem. An initial guess for the unknown boundary is made, and RKF-45th integrates it. It is one of the numerical approximation methods for solving first and second order linear differential equations. It combines the fourth order and a fifth order Runge–Kutta algorithm to give an adaptive step size control procedure. Its wide stability interval and a great selection of stable formulas and step sizes make it particularly effective for solving stiff systems of ordinary differential equations. A new set of variables is employed in Eqs. (8.15–8.18) and boundary conditions (8.19–8.20), which examine the fluid flow and heat transfer models.

$f=a_1,\ f'=a_2,\ f''=a_3,$

$$f'''=-\frac{A_1A_2}{(1+2\eta\lambda)}\left(\frac{l\beta_1}{A_2}\{a_7-a_2\}-\{a_2^2-a_1a_3\}-\frac{A_3}{A_2}Ma_2+2\lambda a_3\right), \tag{8.23}$$

$\theta=a_4,\ \theta'=a_5,$

$$\theta''=-\frac{1}{(1+2\eta\lambda)\left(A_4+\frac{4}{3}R\right)}\left(\left\{A_4+\frac{4}{3}R\right\}\theta'\lambda+\mathrm{Pr}A_5 f\theta'+\mathrm{Pr}\beta_2 l\{\theta_p-\theta\}\right) \tag{8.24}$$

$F = a_6, F' = a_7,$

$$F'' = -\frac{1}{a_6}\left(\beta_1\left\{a_2 - a_7\right\} - a_7^2\right), \quad (8.25)$$

$$\theta_p = a_8.$$

$$\theta_p{}' = \frac{\beta_2\gamma}{a_6}\left(a_8 - a_4\right) \quad (8.26)$$

The algorithm of the RKF-45th order is given as

$$k_1 = hf\left(t_k, y_k\right),$$

$$k_2 = hf\left(\frac{h}{4} + t_k, \frac{k_1}{4} + y_k\right),$$

$$k_3 = hf\left(\frac{3h}{8} + t_k, \frac{3}{32}k_1 + \frac{9}{32}k_2 + y_k\right),$$

$$k_4 = hf\left(\frac{12h}{13} + t_k, \frac{1932}{2197}k_1 - \frac{7200}{2197}k_2 + \frac{7296}{2197}k_3 + y_k\right),$$

$$k_5 = hf\left(h + t_k, \frac{439}{216}k_1 - 8k_2 + \frac{3680}{513}k_3 - \frac{845}{4104}k_4 + y_k\right),$$

$$k_6 = hf\left(\frac{h}{2} + t_k, \frac{8}{27}k_1 + 2k_2 + \frac{3544}{2565}k_3 - \frac{1859}{4104}k_4 - \frac{11}{40}k_5 + y_k\right)$$

For the boundary conditions to be satisfied asymptotically, we have selected a distinct value for η_∞. In the present modeling, an appropriate finite value of η_∞ is taken as $\eta_\infty < 6$ so that both numerical solutions converge and boundary conditions specified at infinity obey asymptotically. 10^{-8} and $\Delta n = 0.001$ is the tolerance error and step size taken for convergence. Further, for computational purposes, the dimensionless parameter values are fixed such as $M = 0.5$, $l = 1$, $\lambda = 0.1$, $\beta_1 = 0.3$, $\beta_2 = 0.3$, $R = 0.5$, $\Pr = 6.3$, and $\gamma = 0.3$. Also, the values from the thermophysical properties from Table 8.1 are used for computational purposes.

8.4 RESULTS AND DISCUSSION

The current section addresses the outcomes of the impacts of magnetic field and radiation on dusty ternary nanofluid flow over a stretching cylinder. In the present study, the nanoparticles such as Fe_2O_4, ZnO, $CoFe_2O_4$ are suspended in H_2O. A numerical technique RKF-45th shooting method is employed to offer numerical solutions on the nature of fluid flow and heat transmission over-stretching cylinder with the help of graphs. Furthermore, engineering factors like skin friction and Nusselt number are estimated

TABLE 8.1
Thermo-Physical Properties of Nanoparticles and Base Fluid (see [29])

Properties	H_2O	Fe_2O_4	ZnO	$CoFe_2O_4$
Density	997.1	5180	5606	4907
Thermal conductivity	0.613	9.7	19	1.3×10^{-5}
Electrical conductivity	0.05	25000	0.01	1.1×10^{7}
Specific heat capacity	4179	670	544	700

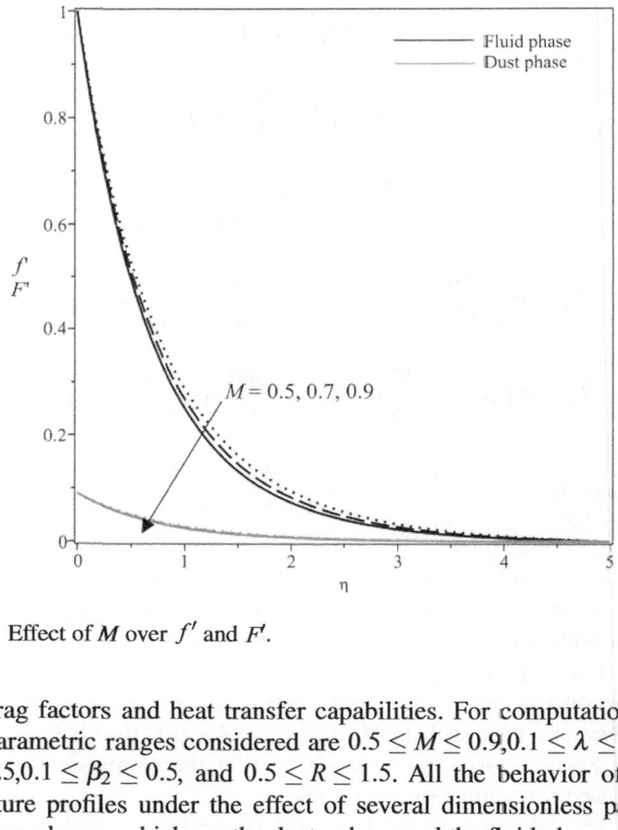

FIGURE 8.1 Effect of M over f' and F'.

to analyze drag factors and heat transfer capabilities. For computational purposes the default parametric ranges considered are $0.5 \le M \le 0.9, 0.1 \le \lambda \le 0.3, 1 \le l \le 3$ $0.1 \le \beta_1 \le 0.5, 0.1 \le \beta_2 \le 0.5$, and $0.5 \le R \le 1.5$. All the behavior of the velocity and temperature profiles under the effect of several dimensionless parameters are analyzed in two phases, which are the dusty phase and the fluid phase.

Figure 8.1 shows the impact of the magnetic field parameter M over fluid and dusty velocity profile (f', F'). As the values of M are increased, both profiles (f', F') decrease. Physically, the decrease in velocity profiles is attributed to the Lorentz force, which acts as a resistive factor generated due to growth in the magnetic field parameter, which slows down both fluid and dusty velocity profiles. Due to the larger size of dusty particles compared to fluid particles, the decrease is found more in the dusty profile.

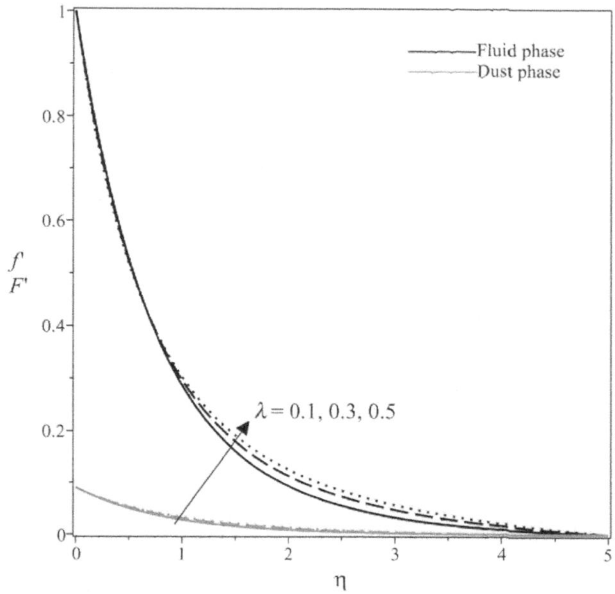

FIGURE 8.2 Effect of λ over f' and F'.

The effect of curvature parameter λ over fluid velocity f' and dusty velocity F' is depicted in Figure 8.2. It is evident from the figure that both profiles f' and F' increase for augmented values of λ. Also, it can be seen that fluid rise is the same as the rise in dusty velocity when $\lambda = 0.1$ and growth of λ increases dusty and fluid velocity within the boundary surface. For $\lambda = 0$, the problem is restricted to a flat surface. So due to the cylinder surface, the velocity inside the cylinder boundary layer is large compared to the flat surface. Physically, as the parameter of curvature rises, the cylindrical radius reduces, leading to fewer resistive force towards fluid motion thereby increasing fluid velocity. In the dusty phase, the velocity rises more quickly than in the fluid phase.

The influence of the particle's mass concentration l over (f', F') is demonstrated in Figure 8.3. As the mass concentration of particles rises, the mixture becomes viscous. This higher viscosity increases flow resistance, reducing both fluid and particle velocity. Physically, the higher the particle concentration, the more particle–particle interactions occur, which enhances drag forces inside the fluid and slows the flow. Meanwhile, the dusty phase's velocity is more directly impacted by particle interactions and mass than the fluid phase's, which shows that the decrease is more in the dusty phase compared to the fluid phase.

Figure 8.4 portrays the impact of velocity's interaction parameter β_1 over (f', F') profiles. It is clear from the figure that both (f', F') profiles increase for augmented values of β_1. An increase often signifies higher interactions between the fluid and the particles. This enhanced interaction results in more effective momentum transfer from

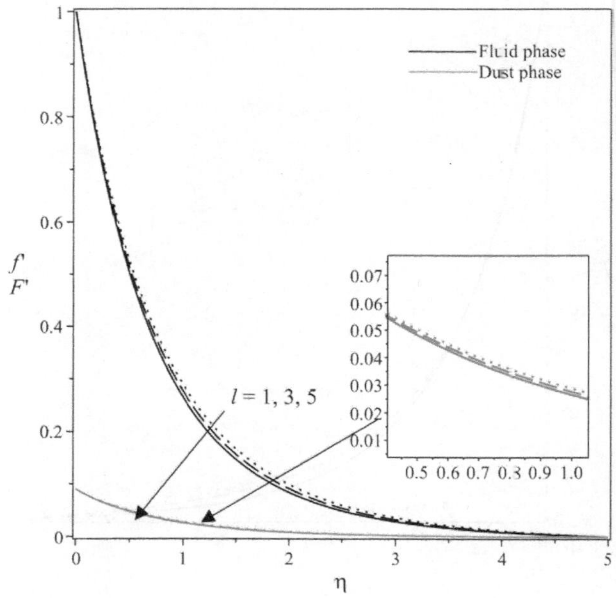

FIGURE 8.3 Effect of l over f' and F'.

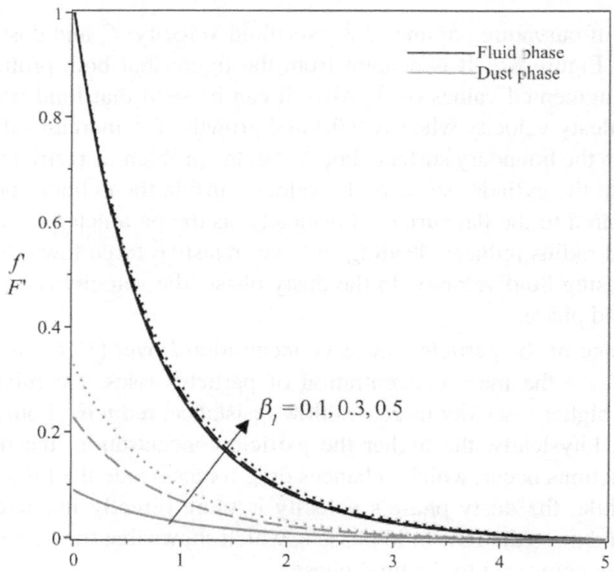

FIGURE 8.4 Effect of β_1 over f' and F'.

fluid to particles, thereby boosting the velocity of both phases. Physically the velocity interaction parameter depicts how efficiently the fluid can accelerate or decelerate the particles, with greater interactions resulting in higher velocities in both phases.

The effect of λ over temperature profiles (θ, θ_p) is displayed in Figure 8.5. It is seen from the figure that the profile (θ, θ_p) rises for augmented values of λ. Curvature offers a thinner boundary layer at the surface, which improves heat transmission. Physically, the curved surface can interrupt the boundary layer growth, resulting in areas with increased convective heat transfer coefficients. As a result, more heat is transmitted from the surface to the fluid, potentially boosting both the fluid and dusty temperature profiles along with raising the curved surface.

Consequences of the radiation parameter R on (θ, θ_p) is displayed in Figure 8.6. Physically, as the radiation parameter increases, so does the quantity of heat transmitted via radiation. Furthermore, this enhanced radiation parameter causes a thicker thermal boundary layer. This leads to a significant rise in temperature gradients in regions near surfaces raising overall temperature.

Increasing trend in (θ, θ_p) for both dusty and fluid phases for elevated values of mass concentration of particles l is depicted in Figure 8.7. Physically increased interaction of the particles with the fluid often generates extra heat, which in turn raises the overall temperature.

Variations in skin friction Cf_x for augmented values of interaction parameters for velocity β_1 and magnetic field parameter M is shown in Figure 8.8. Physically, the presence of a magnetic field gives rise to Lorentz force on electrically

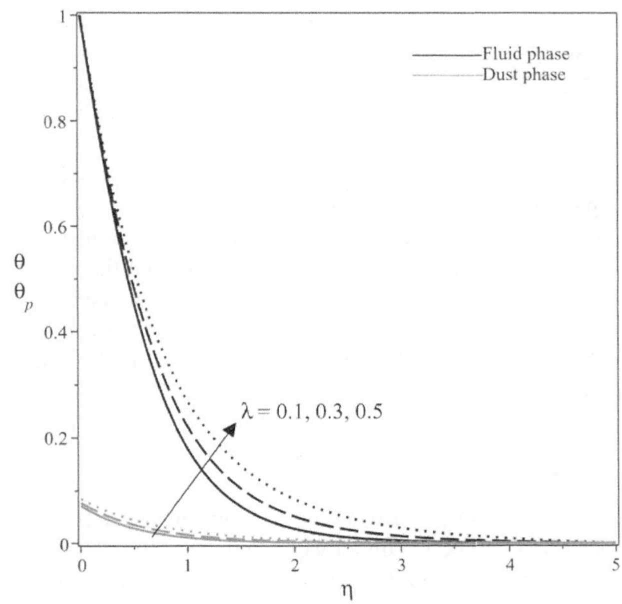

FIGURE 8.5 Impact of λ over θ and θ_p.

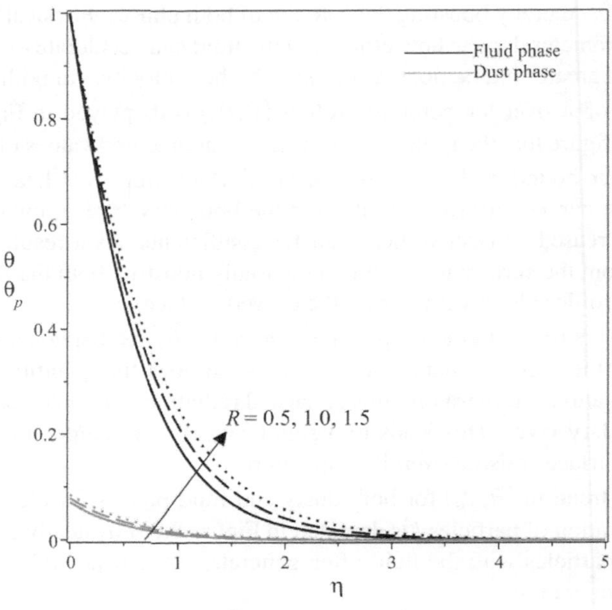

FIGURE 8.6 Impact of R over θ and θ_p.

conducting fluids, which opposes the fluid flow. This force increases the barrier to fluid motion at the surface, which causes upsurged shear stress and as a result, higher skin friction. Furthermore, a rise in the interaction parameter for velocity frequently indicates higher interactions between the fluid and particles or between various fluid layers, which increases friction with the accompanying of a magnetic field.

Variations in the Nusselt number Nu_x for elevated values of radiation parameter R and interaction parameter for temperature β_2 is displayed in Figure 8.9. Physically, an increase in the radiation parameter boosts thermal diffusion by adding radiation effects to the heat transfer process. This also enhances the thermal conductivity of the fluid leading to a more effective heat transmission mechanism thus increasing the Nusselt number. Further, elevated values of interaction parameters for temperature improve thermal contact between fluid particles, thus boosting heat transfer efficiency and Nusselt number.

8.5 CONCLUSION

In the current study, the impacts of magnetic field and radiation on dusty ternary nanofluid flow across a stretched cylinder were analyzed. The PDEs were transformed into ODEs with the help of suitable similarities. Further, numerical solutions were obtained using the RKF-45th method and shooting method. The impact of dimensionless parameters on velocity and momentum profile was studied in two phases: dusty and fluid phases. The main outcomes drawn from the current study are as follows:

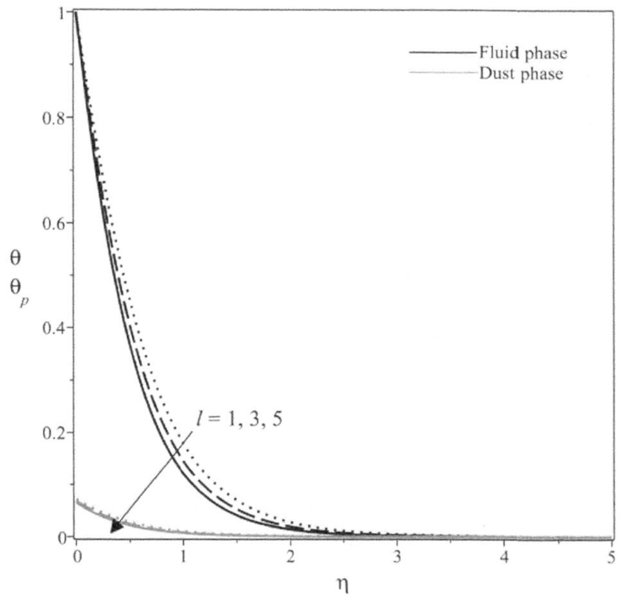

FIGURE 8.7 Impact of l over θ and θ_p.

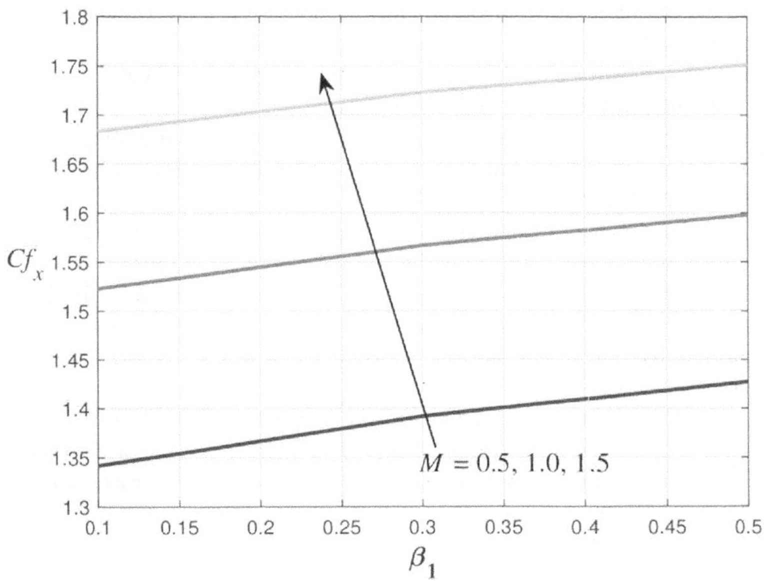

FIGURE 8.8 Behavioural change in Cf_x for varied values of β_1 and M.

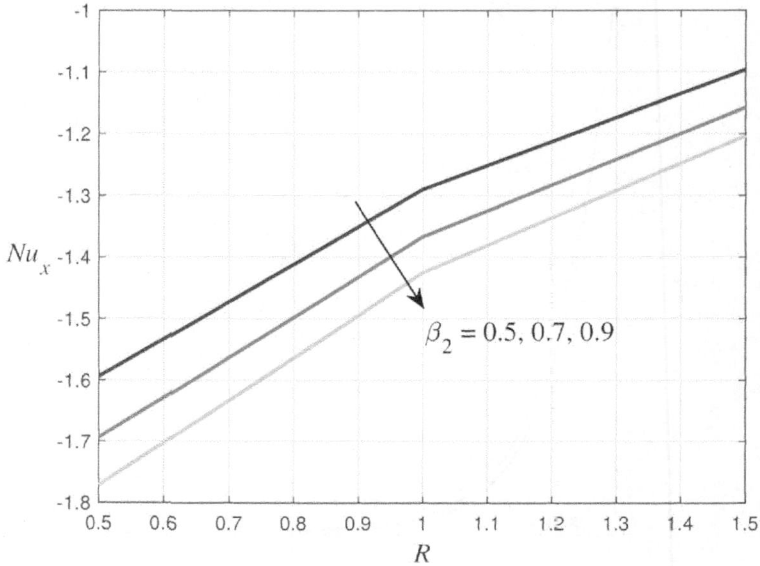

FIGURE 8.9 Behavioural change in Nu_x for varied values of R and β_2.

- Both the fluid velocity and dusty velocity profile are reduced for an elevated value of magnetic field parameter due to Lorentz force generated by magnetic field.
- Increasing values of the curvature parameter elevates velocity and thermal gradients for both fluid and dusty phases.
- A decreasing trend was found in the velocity profile and temperature profile for both cases of fluid and dusty phase for raising values of mass concentration of particles.
- Temperature profile increases for augmented values of radiation parameter due to an increase in the quantity of heat produced via radiation effect.
- Skin friction increases for improved values of magnetic field and interaction for velocity parameters. The rise is due to the Lorentz force accompanied by dense interaction between particles and fluid in various fluid layers.
- The rate of heat transfer enhances the growing values of radiation and interaction parameters for temperature. The increase is because more heat is supplied due to the radiation phenomenon and the interaction of fluid and particles also generates additional heat.

REFERENCES

[1] A. Darvesh, Maiz, Fethi Mohamed, Souayeh, Basma, Sánchez-Chero, Manuel, AL Garalleh, Hakim, Santisteban, Luis Jaime Collantes and Leonardo, Celso Nazario Purihuamán "Advanced ANN computational procedure for thermal transport prediction

in polymer-based ternary radiative Carreau nanofluid with extreme shear rates over bullet surface," *Applied Rheology* vol. 35, no. 1, 2025, doi: 10.1515/arh-2024-0029.

[2] A. Paul, N. Sarma, and B. Patgiri, "Numerical assessment of MHD thermo-mass flow of Casson ternary hybrid nanofluid around an exponentially stretching cylinder," *BioNanoScience*, vol. 14, no. 5, pp. 5230–5245, Dec. 2024, doi: 10.1007/s12668-024-01306-2.

[3] K. M. Nihaal, U. S. Mahabaleshwar, and S. W. Joo, "Darcy Forchhiemer imposed exponential heat source-sink and activation energy with the effects of bioconvection over radially stretching disc," *Sci. Rep.*, vol. 14, no. 1, p. 7910, Apr. 2024, doi: 10.1038/s41598-024-58051-5.

[4] M. Mahboobtosi, K. Hosseinzadeh, and D. D. Ganji, "Investigating the convective flow of ternary hybrid nanofluids and single nanofluids around a stretched cylinder: Parameter analysis and performance enhancement," *Int. J. Thermofluids*, vol. 23, p. 100752, Aug. 2024, doi: 10.1016/j.ijft.2024.100752.

[5] U. Farooq, T. Liu, A. Jan, U. Farooq, and S. Majeed, "Numerical investigation of magnetised bioconvection and heat transfer in a cross-ternary hybrid nanofluid over a stretching cylinder," *Multidiscip. Model. Mater. Struct.*, vol. 20, no. 6, pp. 913–936, Jan. 2024, doi: 10.1108/MMMS-03-2024-0079.

[6] S. Das, S. Sarkar, A. Ali, and R. Nath Jana, "Exploration of magnetically influenced flow dynamics of a dusty fluid induced by the ramped movement of a thermally active plate," *Arch. Appl. Mech.*, vol. 94, no. 2, pp. 407–433, Feb. 2024, doi: 10.1007/s00419-023-02531-z.

[7] L. Ali, P. Kumar, H. Poonia, S. Areekara, and R. Apsari, "The significant role of Darcy–Forchheimer and thermal radiation on Casson fluid flow subject to stretching surface: A case study of dusty fluid," *Mod. Phys. Lett. B*, vol. 38, no. 06, p. 2350215, Feb. 2024, doi: 10.1142/S0217984923502159.

[8] S. M. Sachhin, U. S. Mahabaleshwar, S. N. R. Nayakar, H. S. R. Patil, and B. Souayeh, "Darcy–Brinkman model for Boussinesq–Stokes suspension tetra dusty nanofluid flow over stretching/shrinking sheet with suction/injection," *Numer. Heat Transf. Part Appl.*, pp. 1–23, doi: 10.1080/10407782.2024.2359047.

[9] C. R. Mirza et al., "Intelligent computing technique to analyze the two-phase flow of dusty trihybrid nanofluid with Cattaneo-Christov heat flux model using Levenberg-Marquardt Neural-Networks," *Case Stud. Therm. Eng.*, vol. 68, p. 105891, Apr. 2025, doi: 10.1016/j.csite.2025.105891.

[10] O. K. Koriko, K. S. Adegbie, N. A. Shah, I. L. Animasaun, and M. A. Olotu, "Numerical solutions of the partial differential equations for investigating the significance of partial slip due to lateral velocity and viscous dissipation: The case of blood-gold Carreau nanofluid and dusty fluid," *Numer. Methods Partial Differ. Equ.*, vol. 40, no. 2, p. e22754, Mar. 2024, doi: 10.1002/num.22754.

[11] K. Rafique, Z. Mahmood, Usman, Adnan, U. Farooq, and W. Emam, "Numerical study of MHD flow over stretching cylinder with variable Prandtl number and viscous dissipation in ternary hybrid nanofluids with velocity and thermal slip conditions," *Mod. Phys. Lett. B*, vol. 39, p. 2450465, Jul. 2024, doi: 10.1142/S0217984924504657.

[12] D. Yanala, B. S. Goud, K. Nisar, B. Alshahrani, M. Mahmoud, and C. Park, "Heat absorption/generation effect on MHD heat transfer fluid flow along a stretching cylinder with a porous medium," *Alex. Eng. J.*, vol. 64, pp. 659–666 Sep. 2022, ISSN 1110-0168, doi: 10.1016/j.aej.2022.08.049.

[13] A. M. Galal et al., "Numerical exploration of bioconvection in optimizing nanofluid flow through heated stretched cylinder in existence of magnetic field," *Multidiscip. Model. Mater. Struct.*, vol. 21, no. 2, pp. 425–447, Jan. 2025, doi: 10.1108/MMMS-08-2024-0239.

[14] U. Raj, P. Sagadevan, and N. Alessa, "Effect of heat generation on MHD Maxwell fluid flow on a stretching cylinder embedded in a porous medium along with impact of radiation," *J. King Saud Univ. – Sci.*, vol. 36, no. 11, p. 103530, Dec. 2024, doi: 10.1016/j.jksus.2024.103530.

[15] S. Nandi and D. Barman, "Unsteady MHD hybrid nanofluid flow over a convectively heated linear stretching cylinder with velocity slip: A comparative study," *Int. J. Mod. Phys. B*, vol. 38, no. 21, p. 2450284, Aug. 2024, doi: 10.1142/S0217979224502849.

[16] J. Pashikanti and S. P. DR, "Influence of thermal radiation and chemical reaction due to graphene oxide nanofluid flow on a stretching cylinder," *Comput. Therm. Sci. Int. J.*, vol. 15, no. 5, pp. 17–32, 2023.

[17] F. Munazza and M. A. Farooq, "Numerical solution for cross nanofluid flow in porous medium over stretching cylinder under MHD and radiation effects," *ZAMM – J. Appl. Math. Mech. Z. Für Angew. Math. Mech.*, vol. 104, no. 3, p. e202300151, Mar. 2024, doi: 10.1002/zamm.202300151.

[18] A. A. Akbar et al., "Radiation-influenced magnetohydrodynamic third-grade nanofluid flow around non-linearly stretched cylinder," *J. Comput. Des. Eng.*, vol. 11, no. 3, pp. 72–90, Jun. 2024, doi: 10.1093/jcde/qwae038.

[19] U. S. Mahabaleshwar, T. Maranna, M. Mishra, M. Hatami, and B. Sunden, "Radiation effect on stagnation point flow of Casson nanofluid past a stretching plate/cylinder," *Sci. Rep.*, vol. 14, no. 1, p. 1387, Jan. 2024, doi: 10.1038/s41598-024-51963-2.

[20] U. Raj, P. Sagadevan, and N. Alessa, "Effect of heat generation on MHD Maxwell fluid flow on a stretching cylinder embedded in a porous medium along with impact of radiation," *J. King Saud Univ. – Sci.*, vol. 36, no. 11, p. 103530, Dec. 2024, doi: 10.1016/j.jksus.2024.103530.

[21] T. Hayat, S. Qayyum, A. Alsaedi, and S. Asghar, "Radiation effects on the mixed convection flow induced by an inclined stretching cylinder with non-uniform heat source/sink," *PLOS ONE*, vol. 12, no. 4, p. e0175584, Apr. 2017, doi: 10.1371/journal.pone.0175584.

[22] U. Rashid, H. Liang, H. Ahmad, M. Abbas, A. Iqbal, and Y. S. Hamed, "Study of (Ag and TiO2)/water nanoparticles shape effect on heat transfer and hybrid nanofluid flow toward stretching shrinking horizontal cylinder," *Results Phys.*, vol. 21, p. 103812, Feb. 2021, doi: 10.1016/j.rinp.2020.103812.

[23] A. Mishra and M. Kumar, "Velocity and thermal slip effects on MHD nanofluid flow past a stretching cylinder with viscous dissipation and Joule heating," *SN Appl. Sci.*, vol. 2, no. 8, p. 1350, Jul. 2020, doi: 10.1007/s42452-020-3156-7.

[24] S. Ganga, Z. Uddin, and R. Asthana, "Exploring swirling flow dynamics: Unsupervised machine learning in Maxwell hybrid nanofluid convection over an exponentially stretching cylinder with non-linear radiation effects," *Commun. Nonlinear Sci. Numer. Simul.*, vol. 140, p. 108378, Jan. 2025, doi: 10.1016/j.cnsns.2024.108378.

[25] M. Faraz and J. M. Park, "Thermal and MHD behavior of CNT Maxwell nanofluid over a stretchable cylinder," *J. Radiat. Res. Appl. Sci.*, vol. 18, no. 2, p. 101326, Jun. 2025, doi: 10.1016/j.jrras.2025.101326.

[26] S. U. Zaman, M. N. Aslam, M. B. Riaz, A. Akgul, and A. Hussan, "Williamson MHD nanofluid flow with radiation effects through slender cylinder," *Results Eng.*, vol. 22, p. 101966, Jun. 2024, doi: 10.1016/j.rineng.2024.101966.

[27] T. Hayat, N. Gull, M. Farooq, and B. Ahmad, "Thermal radiation effect in MHD flow of Powell—Eyring nanofluid induced by a stretching cylinder," *J. Aerosp. Eng.*, vol. 29, no. 1, p. 04015011, Jan. 2016, doi: 10.1061/(ASCE)AS.1943-5525.0000501.

[28] H. A. Madkhali, M. Ahmed, M. Nawaz, S. O. Alharbi, A. S. Alqahtani, and M. Y. Malik, "Numerical study on the role of ternary nanoparticles on heat transfer enhancement in MHD flow of cross-rheological-fluid," *Case Stud. Therm. Eng.*, vol. 51, p. 103579, Nov. 2023, doi: 10.1016/j.csite.2023.103579.

[29] Mahabaleshwar, U. S., K. M. Nihaal, L. M. Pérez, and H. F. Oztop, "An analysis of heat and mass transfer of ternary nanofluid flow over a Riga plate: Newtonian heating," *Numerical Heat Transfer, Part B: Fundamentals*, vol. 86, p. 1–16, 2023, doi: 10.1080/10407790.2023.2282165.

9 Trihybrid Nanofluids with Inclined Magnetic Fields

Liaqat Ali, Retna Apsari, and Pardeep Kumar

9.1 INTRODUCTION

Flows over stationary and moving surfaces have garnered substantial research attention owing to their wide-ranging engineering applications, including liquid film condensation, polymer sheet fabrication, and continuous fiber cooling in metalworking, and crystal growth processes. Since the inception of boundary layer theory, these phenomena have been rigorously analyzed. Nanofluids are now recognized as transformative agents across industrial manufacturing, scientific research, and engineering disciplines owing to their versatile applications. These nanoparticle-enhanced fluids optimize processes such as cryogenic cooling systems, bioengineered drug delivery, energy-efficient fuel synthesis, and thermal management in microelectronics. Their unique thermal properties have spurred global research into improving energy transport and hydrodynamic behavior. To enhance convective heat transfer, researchers have explored strategies such as dispersing metallic and non-metallic nanoparticles (e.g., Cu, Ag, SiO_2, Fe_3O_4, CNTs, graphene) into conventional base fluids like water, kerosene, ethylene glycol, or biological fluids. Recent studies [1–3] and Ahmad *et al.* [4] highlight breakthroughs in nanofluid dynamics, underscoring their potential to revolutionize fields from nanotechnology to renewable energy systems.

Marangoni convection, driven by surface tension gradients, has garnered significant scientific interest due to its diverse global applications. This phenomenon arises at liquid–liquid interfaces, where surface tension variations induce boundary layer formation. Beyond industrial uses such as crystal growth, welding, and semiconductor wafer drying, it plays a pivotal role in artistic techniques like fabric dyeing, where pigments are floated on a substrate's surface. Additionally, Marangoni effects are critical in stabilizing soap films, facilitating controlled convection processes, and enhancing precision in fields ranging from materials science to microfluidics. Hayat *et al.* [5] illustrated how chemical reactions affect the Marangoni convective flow of second-grade liquid. The activation energy-dependent chemical reaction flow of a nanoliquid was investigated by Ullah *et al.* [6]. Over a stretchable surface, Qayyum [7] investigated the Marangoni convective stream of hybrid nanofluid.

Magnetohydrodynamic (MHD) flows, characterized by their unique thermophysical properties and heat transfer efficiency, hold critical importance in both industrial and

DOI: 10.1201/9781003611820-11

scientific domains. These flows are integral to applications ranging from astrophysical phenomena (e.g., stellar formation) to biomedical innovations like magnetic drug delivery. Beyond fundamental physics and chemistry, MHD principles optimize industrial processes such as petroleum refinement, boundary layer regulation, and pump design [8, 9]. However, solving the non-linear convective terms inherent to MHD governing equations remains challenging. Advanced numerical methods, including finite difference, spectral element, and boundary element techniques, have been employed to address these complexities [10]. Among these, the finite element method (FEM) has emerged as a robust computational tool, widely adopted for resolving boundary value problems across diverse geometries [11–14].

The FEM traces its origins to foundational contributions in variational calculus by Lord Rayleigh (1870) and W. Ritz (1909), alongside B.G. Galerkin's weighted-residual techniques (1915). Concurrently, J. Argyris's independent research at the University of London advanced the method's theoretical framework, as detailed in his seminal work [15]. Historically, FEM's modern reinvention is closely tied to structural engineering, where decomposing complex systems into discrete, analyzable elements aligned naturally with traditional engineering practices. This approach, modeling intricate structures as assemblies of simpler components proved the revolutionary. Following its adoption in engineering, FEM attracted significant attention from applied mathematicians, particularly for its potential in solving partial differential equations numerically, bridging theoretical and practical advancements in computational science. In 1973, Strang and Fix [16] established the inaugural rigorous mathematical framework for the FEM, emphasizing its application to problems rooted in classical variational principles. FEM has since undergone extensive theoretical and applied investigation, generating a prolific volume of research, including thousands of academic papers and hundreds of specialized texts. By the early 1990s, FEM became the predominant numerical approach in structural analysis, solid mechanics, and related disciplines. Presently, it rivals the finite difference method in widespread adoption, particularly within heat transfer and fluid dynamics applications.

Trihybrid nanofluids are engineered by dispersing three distinct nanoparticle types into conventional base fluids, such as ethylene glycol, motor oil, or water, significantly augmenting their thermal performance. These advanced fluids are extensively employed in industrial and engineering contexts, including coolants for machinery, marine systems, pipelines, nuclear reactors, and heat exchangers. Within the energy sector, they enhance the efficiency of solar collectors and optimize thermal management in nuclear power systems. Notably, ternary nanofluids (TNFs) demonstrate substantially higher thermal conductivity and heat transfer efficiency compared to traditional hybrid nanofluids (HNFs). This superior thermal behavior underscores their potential to elevate energy efficiency and cooling capabilities in diverse applications, ranging from microfluidic devices and electric ovens to large-scale nuclear reactor operations. Some factors that influence the efficacy of ternary nanofluids include their sizes, shapes, ratios of nanoparticle mixtures, and the use of incompatible base fluids. The main topic of Mehmood *et al.* [17] has the flow of a three-hybrid Casson nanofluid in a conical enclosure with a rotating disc. Li and Leng [18] used sodium alginate to study how activation energy and an induced magnetic field affect the thermodynamic properties of two types of hybrid nanofluids: binary and ternary. Ali *et al.* [19] employed finite-element modeling to study the MHD effects of trihybrid nanoparticles in microgravity. Their studies demonstrated the significance of these particles in non-Newtonian fluids.

Existing literature primarily examines Marangoni convection in viscous non-Newtonian fluids, yet its application to trihybrid nanofluids containing cobalt ferrite ($CoFe_2O_4$), magnesium oxide (MgO), and silicon dioxide (SiO_2) nanoparticles remains unexplored. Thus the main objective of study is to investigates the thermal performance of a Marangoni-driven trihybrid nanofluid flow with the significant effect of an inclined magnetic field, thermal convection, and activation energy. Furthermore, the work introduces novel aspects, including Marangoni convection in trihybrid nanofluids and the combined effects of thermal radiation. Graphical analyses assess key parameters influencing velocity, temperature, concentration, Nusselt number, and skin friction.

Applications

Key practical applications of the flow model include the following:

1. Enhancing energy efficiency in systems involving heat and mass transfer, particularly in renewable energy technologies like solar panels.
2. Optimizing drug delivery systems and studying the motion of microorganisms for medical diagnostics and treatment.
3. Improving heat transfer mechanisms in industrial cooling systems, chemical reactors, and manufacturing processes involving hybrid nanofluids.
4. Designing advanced cooling systems for engines and aerospace equipment under rotating and magnetic field conditions.
5. Developing sustainable solutions for managing bio-convection processes in water treatment plants and environmental monitoring systems.

7.2 CONCEPTUAL FRAMEWORK AND ANALYTICAL FORMULATION

A trihybrid nanofluid is synthesized by dispersing cobalt ferrite $\left(CoFe_2O_4\right)$, magnesium oxide MgO, and silicon dioxide SiO_2 nanoparticles into a single base fluid. The study investigates Marangoni convection driven by temperature gradients between the fluid's free surface and its bulk. An applied magnetic field $B_0 \left(T = \text{kg s}^{-2}\ \text{A}^{-1}\right)$ exerts a notable influence on the fluid's velocity profile under inclined conditions. The analysis assumes a two-dimensional stretched surface, where velocity components govern the flow dynamics. Thermal radiation is integrated to enhance heat transfer analysis. Based on these assumptions, the governing equations for the system are formulated [12, 14, 20] as follows:

$$\frac{\partial u}{\partial x} + \frac{\partial v}{\partial y} = 0 \tag{9.1}$$

$$u\frac{\partial u}{\partial x}+v\frac{\partial u}{\partial y}=\nu_{thnf}\frac{\partial^2 u}{\partial y^2}-\frac{\sigma_{thnf}}{\rho_{thnf}}uB_0\sin^2\alpha \tag{9.2}$$

$$u\frac{\partial T}{\partial x}+v\frac{\partial T}{\partial y}=\frac{\kappa_{thnf}}{(\rho Cp)_{thnf}}\frac{\partial^2 T}{\partial y^2}-\frac{1}{(\rho Cp)_{thnf}}\frac{\partial q_r}{\partial y}-\frac{Q_0\,(T-T_\infty)}{(\rho Cp)_{thnf}}\exp\left(-nR_2x^{\frac{n-1}{3}}\right) \tag{9.3}$$

$$u\frac{\partial C}{\partial x}+v\frac{\partial C}{\partial y}=D\frac{\partial^2 C}{\partial y^2}-\kappa_0^2\,(C-C_\infty)\left(\frac{T}{T_\infty}\right)^n\exp\left(\frac{-E_a}{\kappa T}\right) \tag{9.4}$$

Here, the qr $\left(\mathrm{W\,m^{-2}}\right)$, $C_p\left(\mathrm{J\ kg^{-1}\ K^{-1}}\right)$, ρ_{nf}, μ_{nf} represents to thermal radiation, concentration of nanomaterials, density of nanofluid, respectively. The parameters ν_{nf}, σ, and E_a are kinematic viscosity, surface tension, and activation energy, respectively. Furthermore, α_{nf} represents thermal diffusivity, κ_{thnf} denotes the thermal conductivity of the trihybrid nanofluid, $D\left(\mathrm{m^2\ s^{-1}}\right)$ diffusivity coefficient, σ_{nf}, and $\gamma_T=\frac{-1}{\sigma_0}\frac{\partial\sigma}{\partial T}\Big|_T$, $\gamma_C=\frac{-1}{\sigma_0}\frac{\partial\sigma}{\partial C}\Big|_C$ are the surface tension's coefficients of temperature and concentration. Within the current framework, $(C,\ T)$ correspond to the volume fraction of nanoparticles and the fluid's temperature, respectively. The model's governing boundary conditions for velocity distribution, thermal gradients, and nanoparticle concentration dynamics are formulated as follows [12, 21]:

$$\left.\begin{aligned}
&\mu_{thnf}\frac{\partial u}{\partial y}=\frac{\partial\sigma}{\partial x}=\sigma_0\left(\gamma_C\frac{\partial C}{\partial x}+\gamma_T\frac{\partial T}{\partial x}\right),\quad v=0,\quad when\,y=0\\
&-\frac{\mu_{thnf}}{h_f}\frac{\partial T}{\partial y}=T_f-T,\quad C=C_w,\qquad when\,y=0\\
&u\to 0,\quad T\to T_\infty,\quad C\to C_\infty\qquad when\,y\to\infty
\end{aligned}\right\} \tag{9.5}$$

Here, $\sigma=\sigma_0\left(1-\gamma_T\left(T-T_\infty\right)-\gamma_C\left(C-C_\infty\right)\right)$ and at the surface particles concentration and temperature are C_w, T_w and ambient conditions C_∞, T_∞, respectively. We use the Rosseland approximation and an approximate representation for linear thermal radiation to quantify the radiative flux of heat $q_r=-\frac{4\alpha}{3\kappa}\frac{\partial T^4}{\partial y}$. Figure 9.1 illustrates the spatial arrangement of the investigated system. The temperature fluctuations are attributed to fluid flow dynamics and the application of a Taylor series expansion, with higher-order terms omitted. In this context, κ denotes the Rosseland mean absorption coefficient, while α corresponds to the Stefan–Boltzmann constant [20], as follows: $\frac{\partial q_r}{\partial y}=-\frac{16\alpha T_\infty^3}{3\kappa}\frac{\partial^2 T}{\partial y^2}$.

To non-dimensionalize the system of partial differential Eqs. (9.1–9.4) and their corresponding boundary conditions (5), a series of mathematical transformations has

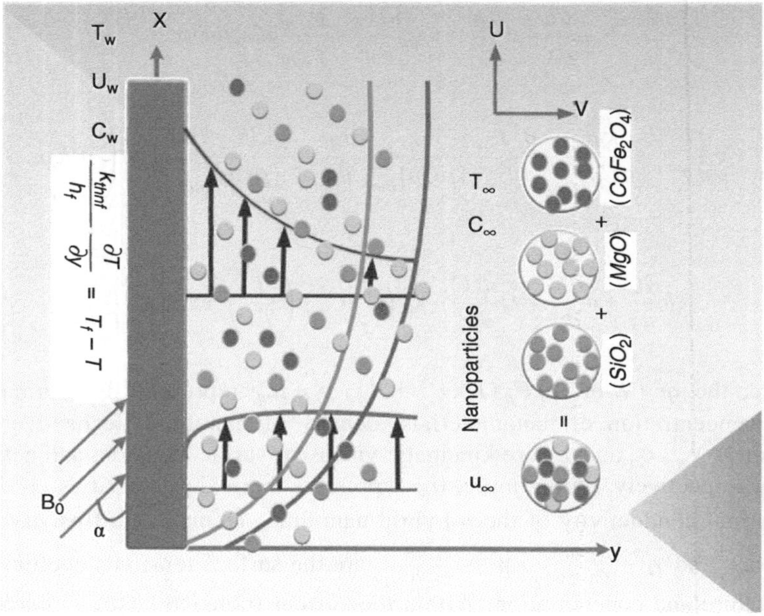

FIGURE 9.1 Visual depiction of the fluid flow configuration under investigation.

been systematically applied. These transformations reduced the governing PDEs to a set of ordinary differential equations, facilitating numerical or analytical solutions while preserving the physical integrity of the original problem [22].

$$\left.\begin{array}{l}[\xi,\ \psi,\ \theta,\ A,\ R_1, R_2, u,\ v] = yR_2x^{\frac{r-1}{3}}, R_1x^{\frac{2+r}{3}}f(\xi),\ \dfrac{T-T_\infty}{Ax^{(1+r)}},\ \dfrac{\Delta T}{L^{(1+r)}} \\ \left(\dfrac{\sigma_T A}{\rho_f^2}\mu_f\right)^{\frac{1}{3}},\ \left(\dfrac{\sigma_T A}{\mu_f^2}\rho_f\right)^{\frac{1}{3}},\ \dfrac{\partial\psi}{\partial y},\ -\dfrac{\partial\psi}{\partial x}\end{array}\right\} \quad (9.6)$$

Through similarity transformations (9.6) and the incorporation of Eq. (9.5) into the governing Eqs. (9.2–9.5), the system is reduced to coupled non-linear ordinary differential equations (ODEs). This reformulation preserves the essential dynamics of the original partial differential equations while enabling tractable analytical or numerical solutions. The resulting ODEs encapsulate the fundamental relationships governing the physical phenomena under investigation, as expressed below:

$$f'''(\xi) + \frac{A_2}{A_1}\left[\frac{r+2}{3}ff'' - \frac{1+2r}{3}f'^2\right] - \frac{A_3}{A_1}Mf'\sin^2\alpha = 0 \quad (9.7)$$

$$\left(A_4+\frac{4Rd}{3}\right)\theta''(\xi)+A_5\Pr\left(\frac{r+2}{3}f\theta'-(1+r)f'\theta\right)+Q_E.e^{-n\xi}=0 \tag{9.8}$$

$$\frac{1}{Sc}\phi''(\xi)+\left(\frac{r+2}{3}f\phi'-(1+r)\phi f'\right)-\mathrm{Kr}(1+\delta\theta)^m\exp\left(\frac{-E_1}{1+\delta\theta}\right)=0 \tag{9.9}$$

The transformed boundary conditions, derived through the outlined mathematical procedures, are explicitly defined as follows:

$$\left.\begin{array}{c} f(\xi)=0,\ A_1f''(\xi)=-2(1+M_a),\ A_4\theta'(\xi)=B_i(\theta-1),\ \phi(\xi)=1,\ at\ \xi=0 \\ f'(\xi)\to 0,\ \theta(\xi)\to 0,\ \ \phi(\xi)\to 0,\ \ at\ \ \xi\to\infty \end{array}\right\} \tag{9.10}$$

The emerging factors in Eqs. (9.7)–(9.9) are expressed as

$$A_1=\frac{\mu_{thnf}}{\mu_f},\ A_2=\frac{\rho_{thnf}}{\rho_f},\ A_3=\frac{\sigma_{thnf}}{\sigma_f},\ A_4=\frac{k_{thnf}}{k_f},\ A_4=\frac{(\rho c_p)_{thnf}}{(\rho c_p)_f},$$

$$M_a=\frac{C_0\gamma_C}{T_0\gamma_T},\ \Pr=\frac{\nu_f}{\alpha_f},\mathrm{Sc}=\frac{\nu_f}{D_B},\ \mathrm{Rd}=\frac{16\alpha T_\infty^3}{3\kappa_f\kappa}$$

In this framework, M denotes the magnetic body force parameter, which quantifies the influence of Lorentz forces on the fluid dynamics. The Prandtl number Pr, a dimensionless quantity characterizing the ratio of momentum diffusivity to thermal diffusivity, governs the relative thickness of the hydrodynamic and thermal boundary layers, the Eckert constant Ec, the Reynolds number Re_x, the heat source/sink coefficient Q_E, the activation energy E_1, the temperature difference is $\delta=\frac{T_0x^2}{T_\infty}$ the Marangoni number $M_a=\frac{C_0\gamma_C}{T_\infty\gamma_T}$, Schmidt number Sc, Biot number Bi, and the radiation parameter Rd.

The dimensionless parameters critical to this analysis: namely, the skin friction coefficient (quantifying surface shear stress), the local Nusselt number (characterizing convective heat transfer), and the Sherwood number (describing mass transfer rates) are mathematically defined as follows:

$$C_{fx}=\frac{2\tau_w}{\rho_f u_w^2},\ \mathrm{at}\tau_w=\mu_{thnf}+\left.\frac{\partial u}{\partial y}\right|_{y=0},N_{ux}=\frac{-xq_w}{\kappa(T_0-T_\infty)},\ q_w=-\left[\kappa_{thnf}+\frac{16\sigma T_\infty^3}{3\kappa}\right]\left.\frac{\partial T}{\partial y}\right|_{y=0}$$

And the skin friction tensor at wall is $\tau_w=\mu\left.\frac{\partial u}{\partial y}\right|_{y=0}$, the wall heat transfer is $q_w=-\kappa\left[1+\frac{16\sigma T_\infty^3}{3\kappa}\right]\left.\frac{\partial T}{\partial y}\right|_{y=0}$, and the mass flux from the sheet is $q_m=-\left(D_b\frac{\partial C}{\partial y}\right)_{y=0}$.

Using the similarity transformation Eq. (9.6), we acquired the dimensionless form of engineering quantities is expressed as below.

After utilizing Eq. (9.7), dimensionless form of engineering quantities is expressed as below:

$$Re_x^{1/2} C_{fx} = -A_1 \frac{d^2\mathbb{F}}{d\xi^2}(0), Re^{-1/2} Nux = -C_2 x^{\frac{2+r}{3}} \left[\frac{\kappa_{thnf}}{\kappa_f} + \frac{4}{3}Rd\right] \frac{d\theta}{d\xi}(0). \quad (9.11)$$

9.3 FINITE ELEMENT METHOD AND ITS BASIC CONFIGURATION

9.3.1 Linear Elements

A computational grid is established across the domain $0 \le x \le L$, comprising elements of varying dimensions. The spatial discretization is visually represented in Figure 9.2. Linear shape functions are defined over individual elements within the discretized domain to approximate the solution's behavior within each segment. These functions, typically first-order polynomials, facilitate interpolation of nodal values and ensure continuity across element boundaries. The governing equations for these shape functions are as follows:

$$A(x) = P_i(x) A_i + P_{i+1}(x) A_{i+1}, \quad \forall \; x_i \le x \le x_{i+1} \quad (9.12)$$

The global coordinate system defines the finite element mesh and associated shape functions, facilitating geometric representation of the problem. For a mesh comprising (n) elements, $(n+1)$ nodal points with coordinates $x_1, x_2, \ldots, x_{n+1}$ are established, as depicted in Figure 9.3. The spatial extents of each element are specified by

$$\eta_i : \{x | x_i \le x \le x_{i+1}\}, \quad \forall \; i = 1, 2, 3, 4, \ldots, n \quad (9.13)$$

Shape functions $P_i(x)$ adhere to the Kronecker delta property, ensuring $P_i(x_i)=1$ at node x_i and $P_i(x_{i+1})=0$ at adjacent nodes. Similarly, $P_{i+1}(x)=0$ satisfies $P_{i+1}(x_{i+1})=1$ and $P_{i+1}(x_i)=0$ as illustrated below.

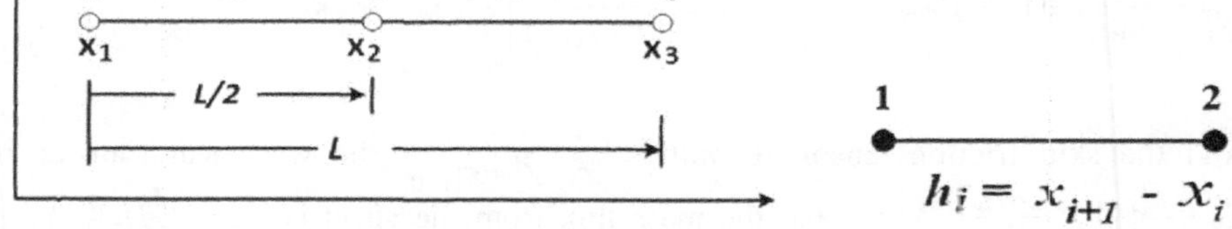

FIGURE 9.2 Segment discretization.

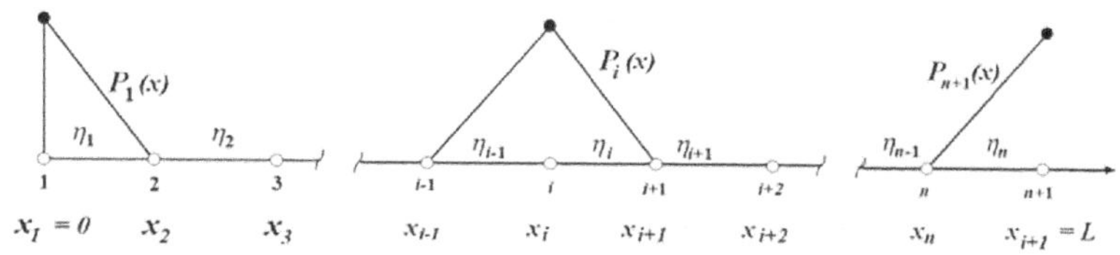

FIGURE 9.3 General form of pricewise linear elements of the shape function.

$$P_1(x) = \begin{cases} \dfrac{x_2 - x}{h_1} & \forall\ x_1 \le x \le x_2 \\ 0 & Otherwise \end{cases} \tag{9.14}$$

$$P_i(x) = \begin{cases} \dfrac{x_{i+1} - x}{h_i} & x_i \le x \le x_{i+1} \quad \forall\, i = 1, 2, 3, 4, \ldots, n \\ \dfrac{x - x_{i-1}}{h_{i-1}} & x_{i-1} \le x \le x_i \\ 0 & Otherwise \end{cases} \tag{9.15}$$

$$P_{n+1}(x) = \begin{cases} \dfrac{x - x_n}{h_n} & \forall\ x_n \le x \le x_{n+1} \\ 0 & Otherwise \end{cases} \tag{9.16}$$

where, $h_i = x_{i-1} - x_i$, $\forall\, i = 1, 2, 3, \ldots, n$. And the function $A(x)$ can be expressed as

$$A(x) = P_1(x) A_1 + P_2(x) A_2 + \ldots + P_{n+1}(x) A_{n+1} = \sum_{i=1}^{n+1} P_i(x) A_i \tag{9.17}$$

Here, A_i represents of A at x_i at node i and $A(x)$ is linear between nodes. For the meshing of a single element $(n = 1)$ and as the function $A(x)$ is linear between the two nodes, i.e. $A(x_2) = A_1$, and $A(x_2) = A_2$, then from the (Eq. (9.15)) we can get after rearranging the similar terms as

$$A(x) = \left(\frac{x_2 - x}{h_1}\right) A_1 + \left(\frac{x - x_1}{h_1}\right) A_2 \tag{9.18}$$

9.3.2 Quadratic Elements

Figure 9.4 illustrates how parabolic arcs, as element-wise quadratic basis functions, outperform linear segments in piecewise interpolation accuracy. The interpolating functions $A(x)$, expressed as second-degree polynomials per element, follow the general form:

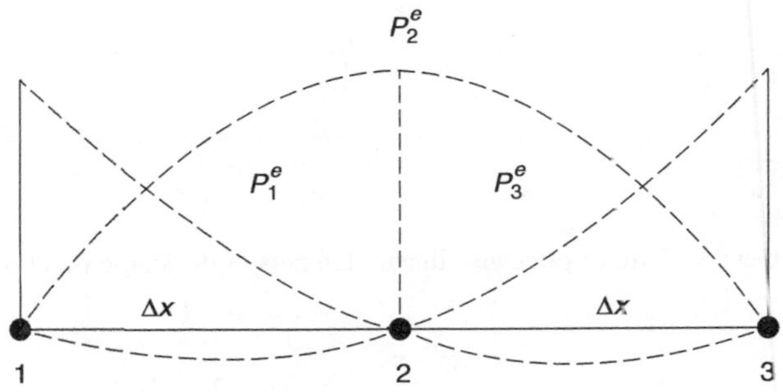

FIGURE 9.4 One-dimensional interpolation of quadratic function.

$$A(x) = a_1 + \alpha_2 x + a_3 x^2 \tag{9.19}$$

Quadratic interpolation requires three parameters, which endpoint interpolation alone cannot determine. By introducing a midpoint node within each element, this method generates an additional interpolation condition, providing the necessary third equation. Three-node elements (Figure 9.4) thus enable accurate quadratic interpolation across the discretized domain.

9.4 BASICS OF THE FINITE ELEMENT METHOD

Mathematical frameworks form the foundation for articulating physical principles. Core tenets such as the conservation of energy, mass, and momentum are rigorously formulated through partial differential equations (PDEs), which govern relationships between dependent variables (e.g., temperature, density, velocity, potential) and independent parameters like spatial coordinates x, y, z and time (t). These equations quantify incremental variations in dependent variables via derivatives, capturing their evolution as independent variables change. For instance, consider a solid exhibiting time-dependent temperature fluctuations with negligible spatial variation. In such cases, the conservation of internal energy (heat) under the influence of a heat source Q simplifies to an ODE, isolating temporal changes in temperature.

$$\rho C_p \frac{dT}{dt} = Q(T,t) \tag{9.20}$$

In this formulation, the density is denoted by ρ and C_p corresponds to the heat capacity. The dependent variable in this context is temperature T, while time t functions as the independent variable. A heat source, which may vary with both temperature and time, can be represented through a functional relationship. Equation (9.20) demonstrates that temporal changes in temperature must equilibrate with—or result from—a heat source. This relationship is encapsulated in a differential equation involving the derivative with respect to t, classifying it as an ODE. When an initial condition specifying the temperature at a particular time t_0 is provided, then Eq. (9.21) yields an analytical solution of the form.

$$T = h(t) \tag{9.21}$$

The temperature distribution within the solid is algebraically defined by Eq. (9.21), which establishes a direct relationship between a given time instant t_1 and its corresponding temperature value is T_1. Material characteristics frequently vary temporally and spatially. For instance, regions adjacent to a thermal source exhibit elevated temperatures compared to distal areas within a solid medium. Such thermal gradients induce conductive heat transfer between distinct material domains. In alignment with thermodynamic conservation principles, this energy redistribution process necessitates formulation of a PDE encompassing both temporal derivatives and spatial variables (e.g.,x), as demonstrated below:

$$\rho C_p \frac{dT}{dt} + \nabla . q = Q(T, t, x) \tag{9.22}$$

Within this formulation, temperature T serves as the dependent variable, while spatial coordinates $x\left(x = (x, y, z)\right)$ and t act as independent parameters. The heat flux vector $q = (q_x, q_y, q_z)$ quantifies directional thermal energy transfer, and its divergence $\nabla \cdot q$ captures the spatial variation of flux across the medium. In Cartesian coordinates, the divergence operator is mathematically defined as

$$\nabla . q = \frac{\partial q_x}{\partial x} + \frac{\partial q_y}{\partial y} + \frac{\partial q_z}{\partial z} \tag{9.23}$$

Equation (9.22) demonstrates that a non-zero divergence of the heat flux vector $\nabla \cdot q$, representing the net flux variation across spatial dimensions, necessitates either a temporal temperature shift or the presence of a heat source to maintain equilibrium or drive the process. The conductive heat flux within a solid is governed by Fourier's law, expressed as $q = -\kappa \nabla T$, where κ denotes thermal conductivity. Substituting this constitutive relationship into Eq. (9.22) yields the modified form:

$$\rho C_p \frac{dT}{dt} - \nabla . (-\kappa \nabla T) = Q(T, t, x) \tag{9.24}$$

In mathematical formalism, derivatives are associated with independent variables such as t, x, y and z. Differential equations involving derivatives with respect to multiple independent variables are classified as PDEs, since each partial derivative corresponds to variations along specific spatial or temporal dimensions. A key distinction between PDEs and ODEs lies in their notation: ODEs employ the symbol d (e.g.,$\frac{d}{dt}$) for single-variable derivatives, whereas PDEs use ∂ (e.g., $\frac{\partial}{\partial t}$) to denote partial derivatives in multivariable contexts. Analytical approaches are often bypassed in favor of numerical methods, where approximate solutions to PDEs are derived through computational models. Among these techniques, the FEM stands out as a robust computational framework for solving PDEs. Similar to the thermal energy conservation equations discussed earlier, PDE systems also arise from momentum and mass conservation principles—cornerstones of fluid dynamics. Furthermore, time-dependent electromagnetic field interactions and flux phenomena can be modeled using analogous PDE formulations, allowing for the analysis of complex multiphysics systems.

9.5 EVALUATION OF FEM ON THE PROPOSED PROBLEM

Finite element analysis (FEA) is a widely recognized computational method for deriving numerical solutions to ODEs and PDEs involving complex geometries or boundary conditions. In this study, the Galerkin finite element technique (GFET) was employed to computationally resolve the non-linear system of Eqs. (9.7)–(9.10), as analytical solutions proved intractable. GFET operates as a variational approach, utilizing trial functions that align with the test functions in the weak formulation, ensuring numerical stability. This method is particularly effective in addressing engineering challenges, especially in fluid dynamics, heat transfer, and biomechanics. Its versatility extends to diverse applications, including electromagnetic simulations, structural mechanics, and multiphysics systems, making it indispensable for solving real-world engineering problems [25, 26]. The researchers [27–29] provided an overview of the GFET and summarizes its major steps. This mathematical technique is widely used in engineering, particularly in the field of fluid dynamics. It is highly effective in addressing non-linear problems [21, 30]. The GFET solves certain problems efficiently, quickly, and precisely [31, 32]. To reduce the order of ODEs (9.7)–(9.9), we must first assume

$$f' = h \tag{9.25}$$

The simplified system of equations is expressed as follows:

$$h''(\xi) + \frac{A_2}{A_1}\left[\frac{r+2}{3} f h' - \frac{1+2r}{3} h^2\right] - \frac{A_3}{A_1} M h \sin^2\alpha = 0 \tag{9.26}$$

$$\left(A_4 + \frac{4Rd}{3}\right)\theta''(\xi) + A_5 Pr\left(\frac{r+2}{3} f\theta' - (1+r)\, h\theta\right) + Q_E . e^{-n\xi} = 0 \tag{9.27}$$

$$\frac{1}{Sc}\phi''(\xi) + \left(\frac{r+2}{3} f\phi' - (1+r)\,\phi h\right) - Kr(1+\delta\theta)^m \exp\left(\frac{-E_1}{1+\delta\theta}\right) = 0 \tag{9.28}$$

The boundary conditions governing the system are specified as follows:

$$\left.\begin{array}{c} f(\xi) = 0,\ A_1 h'(\xi) = -2(1+M_a),\ A_4\theta'(\xi) = B_i(\theta - 1),\ \phi(\xi) = 1,\ at\ \xi = 0 \\ h(\xi) \to 0,\ \theta(\xi) \to 0,\ \phi(\xi) \to 0,\ at\ \xi \to \infty \end{array}\right\} \tag{9.29}$$

9.5.1 Variational Formulations

Consider a scenario where the thermal profile of a radiator under investigation is governed by Eq. (9.25) under steady-state conditions, indicating that the temporal derivative of the temperature field in Eq. (9.25) is null. The governing equation characterizing the model domain Ω_c is defined as follows:

$$\nabla.\left(-\kappa\nabla T\right) = Q\left(T,x\right) \quad in\ \Omega_c \tag{9.30}$$

Subsequently, Eq. (9.25) is multiplied by a test function Φ and integrated across the computational domain Ω_c

$$\int_{\Omega\Omega_c} \{\nabla.\left(-\kappa\nabla T\right) - Q\} \emptyset dV = 0 \tag{9.31}$$

The test function Φ and solution T reside in a Hilbert space—an infinite-dimensional function space where functions exhibit properties akin to vectors in Euclidean space, such as forming linear combinations and possessing well-defined norms and inner products. The FEM systematically approximates these functions into finite-dimensional subspaces, enabling numerical computation via vector representations. The "weak form" (Eq. (9.25)) requires the governing equation to hold for all test functions in the Hilbert space, contrasting with the "strong form" (Eq. (9.25)), which demands point-wise validity across the domain Ω_c

$$\int_{\Omega\Omega_c} \{\kappa\nabla\mathrm{T}.\nabla\varnothing - \left(-\kappa\nabla T\right).\vec{n}Q\}\, dS = \int_{\Omega\Omega_c} \{Q\Phi\}\, dV \tag{9.32}$$

By enforcing the governing equation across all test functions within the Hilbert space, a weak formulation (or variational formulation) of Eq. (9.25) is derived. The discretization framework linked to Eqs. (9.25)–(9.28) is constructed over an individual element $\Omega_c = \left(\xi_c,\ \xi_{c+1}\right)$, as follows:

$$\int_{\xi_c}^{\xi_{c+1}} V_1 \left\{\frac{df}{d\xi} - h\right\} d\xi = 0 \tag{9.33}$$

$$\int_{\xi_c}^{\xi_{c+1}} V_2 \left\{h''\left(\xi\right) + \frac{A_2}{A_1}\left[\frac{r+2}{3}fh' - \frac{1+2r}{3}h^2\right] - \frac{A_3}{A_1}Mh\sin^2\alpha\right\} d\xi = 0 \tag{9.34}$$

$$\int_{\xi_c}^{\xi_{c+1}} V_3 \left\{\left(A_4 + \frac{4Rd}{3}\right)\theta''\left(\xi\right) + A_5\mathrm{Pr}\left(\frac{r+2}{3}f\theta' - \left(1+r\right)h\theta\right) + Q_E.e^{-n\xi}\right\} d\xi = 0 \tag{9.35}$$

$$\int_{\xi_c}^{\xi_{c+1}} V_4 \left\{\frac{1}{Sc}\phi''\left(\xi\right) + \left(\frac{r+2}{3}f\phi' - \left(1+r\right)\phi h\right) - \mathrm{Kr}(1+\delta\theta)^m \exp\left(\frac{-E_1}{1+\delta\theta}\right)\right\} d\xi = 0 \tag{9.36}$$

where V_1, V_2, V_3, and V_4 are arbitrary trial functions or form functions.

9.5.2 FEM Discretization

The spatial configuration of the finite element domain Ω_c is entirely defined by two distinct parameters. (i) The spatial coordinates of nodes within a discretized domain are critical for determining the geometric configuration of finite elements. (ii) The interpolation technique applied to the domain may align with the method used for interpolating the dependent variables in the system. Within the element domain Ω_c the spatial coordinates x of any point are mathematically represented through a combination of nodal position vectors x_n and predefined interpolation functions specific to the domain.

9.5.3 One-Dimensional

In the context of one-dimensional finite element analysis, the domains are modeled as line segments, which may exhibit straight or curved geometries, as demonstrated in Figure 9.5.

9.5.4 Two-Dimensional

In two-dimensional finite element analysis, domains are commonly modeled using triangular, rectangular, or quadrilateral elements with either linear or curvilinear boundaries. Geometrically complex elements are infrequently utilized in practical implementations. Figure 9.6 provides a representative taxonomy of standard element types, categorized by geometric form and polynomial order, such as the FEM approximations derived from Eqs. (9.33)–(9.36) are integrated into the computational framework allows one to generate the discretization of the FEM model:

$$\bar{f}=\sum_{n=1}^{p}\bar{f}_n\psi_n,\qquad \bar{h}=\sum_{n=1}^{p}\bar{h}_n\psi_n,\qquad \frac{d\bar{h}}{d\eta}=\sum_{n=1}^{p}\frac{d\bar{\theta}_n}{d\eta}\psi_n,\qquad \frac{d\bar{\Phi}}{d\eta}=\sum_{n=1}^{p}\frac{d\bar{\Phi}_n}{d\eta}\psi_n \quad (9.37)$$

With $V_1 = V_2 = V_3 = V_4 = \psi_n$, $\forall\,(n = 1, 2, 3)$, where the test functions ψ_n for a typical length element $\Omega_c = (\xi_c, \xi_{c+1})$ are specified.

The finite element modeled equationally is given as

FIGURE 9.5 Finite element mesh and grid.

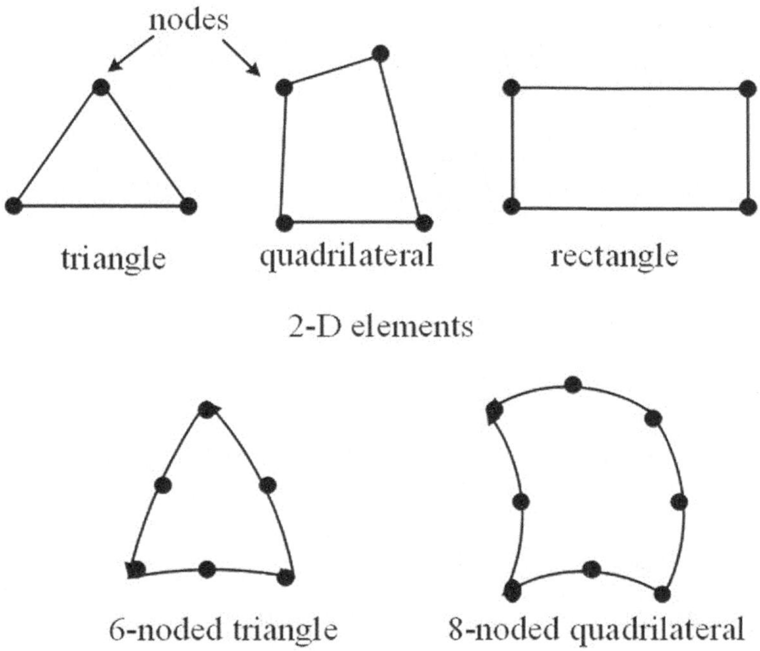

FIGURE 9.6 Finite element mesh and grid.

$$\begin{bmatrix} A^{11} & A^{12} & A^{13} & A^{14} \\ A^{21} & A^{22} & A^{23} & A^{24} \\ A^{31} & A^{32} & A^{33} & A^{34} \\ A^{41} & A^{42} & A^{43} & A^{44} \end{bmatrix} \begin{bmatrix} f \\ h \\ \theta \\ \Phi \end{bmatrix} = \begin{bmatrix} a_1 \\ a_2 \\ a_3 \\ a_4 \end{bmatrix} \tag{9.38}$$

Whereas the $\left[A^{mn}\right]$ and $\left[a_m\right]$, $\forall$ (m,n=1,2,3,4) are pretended below:

$$A_{ij}^{11} = \int_{\xi_c}^{\xi_{c+1}} \Psi_i \frac{d\Psi_j}{d\xi} d\xi, A_{ij}^{12} = -\int_{\xi_c}^{\xi_{c+1}} \Psi_i \Psi_j d\xi, A_{ij}^{13} = A_{ij}^{14} = A_{ij}^{21} = 0$$

$$A_{ij}^{22} = -\int_{\xi_c}^{\xi_{c+1}} \frac{d\Psi_i}{d\xi} \frac{d\Psi_j}{d\xi} d\xi + \int_{\xi_c}^{\xi_{c+1}} \bar{f}\Psi_i \frac{A_2(r+2)}{3A_1} \frac{d\Psi_j}{d\xi} d\xi - \frac{A_2(1+2r)}{3A_1} \int_{\xi_c}^{\xi_{c+1}} \bar{h}\Psi_i \Psi_j d$$

$$-\frac{MA_3}{A_1}\sin^2(\alpha)\int_{\xi_c}^{\xi_{c+1}}\Psi_i\Psi_j d, A_{ij}^{23}=0, A_{ij}^{24}=0, S_{ij}^{31}=0, A_{ij}^{32}=0$$

$$A_{ij}^{33}=\left(A_4-\frac{4R_d}{3}\right)\int_{\xi_c}^{\xi_{c+1}}\frac{d\Psi_i}{d\xi}\frac{d\Psi_j}{d\xi}d\xi+\frac{A_5P_r(r+2)}{3}\int_{\xi_c}^{\xi_{c+1}}\bar{f}\Psi_i\frac{d\Psi_j}{d\xi}d\xi$$

$$-A_5P_r(r+1)\int_{\xi_c}^{\xi_{c+1}}\bar{h}\Psi_i\Psi_j d\xi$$

$$+Q_E\int_{\xi_c}^{\xi_{c+1}}\exp(-n\xi)\Psi_i\Psi_j d\xi-\kappa_r\int_{\xi_c}^{\xi_{c+1}}(1+\delta\theta)^m\exp\left(\frac{-E_1}{1+\delta\theta}\right)\Psi_i\Psi_j d\xi$$

$$A_{ij}^{41}=A_{ij}^{42}=A_{ij}^{43}=0$$

$$A_{ij}^{44}=-\int_{\eta_e}^{\eta_{e+1}}\psi_i\frac{d\psi_j}{d\eta}d, S_{ij}^{44},$$

$$=-\frac{1}{Sc}\int_{\xi_c}^{\xi_{c+1}}\frac{d\Psi_i}{d\xi}\frac{d\Psi_j}{d\xi}d\xi+\frac{r+2}{3}\int_{\xi_c}^{\xi_{c+1}}\bar{f}\Psi_i\frac{d\Psi_j}{d\xi}d\xi$$

$$-(1+r)\int_{\xi_c}^{\xi_{c+1}}\Psi_i\Psi_j\bar{h}d\xi$$

$$a_i^1=0, a_i^2=-\left(\Psi\frac{dh}{d\xi}\right)_{\xi_c}^{\xi_{c+1}}, a_i^3=-\left(A_4-\frac{4}{3}Rd\right)\left(\Psi\frac{d\theta}{d\xi}\right)_{\xi_c}^{\xi_{c+1}},$$

$$a_i^4=-\frac{1}{Sc}\left(\Psi\frac{d\Phi}{d\xi}\right)_{\xi_c}^{\xi_{c+1}}$$

In this framework, the established parameters necessitate thorough consideration as $\bar{f}=\sum_{j=1}^{3}\bar{f}_n\psi_n$, $\bar{h}=\sum_{j=1}^{3}\bar{h}_n\psi_n$, $\frac{d\bar{\theta}}{d\eta}=\sum_{j=1}^{3}\frac{d\bar{\theta}_n}{d\eta}\psi_n$, $\frac{d\bar{\Phi}}{d\eta}=\sum_{j=1}^{3}\frac{d\bar{\Phi}_n}{d\eta}\psi_n$. It requires an iterative strategy to illuminate it for the efficient result when sufficient non-linear equality has been achieved through the accumulation of element equivalences. The relevant parameters in our problem have fixed values. The domain is kept at a constant length of six components and divided into 500 pieces because adding more pieces has not produced any noticeable difference.

9.6 RESULTS AND DISCUSSION

This investigation elucidates critical findings related to momentum transfer, thermal energy distribution, and nanomaterial concentration dynamics in fluid systems. The research employs an innovative methodology to analyze ternary hybrid nanofluids, integrating heat sources, activation energy, and thermal radiation to explore MHD flows influenced by nanoparticle diffusion. A key hypothesis posits that surface tension within the fluid exhibits linear dependencies on temperature and concentration gradients along an inclined boundary. The study advances significant insights by evaluating the system under diverse parametric variations, supported by comprehensive tabular data. Tables 9.1 and 9.2 delineate the thermophysical properties of nanoparticles, while Tables 9.3 and 9.4 validate the computational efficacy of the FEM through comparisons with established benchmarks. Specifically, Table 9.3 contrasts the proposed model's predictions with prior studies across varying Prandtl numbers (Pr), demonstrating robust alignment with referenced results.

Figures 9.7(a) and 9.7(b) depict the impact of the magnetic parameter M on the non-dimensional velocity profiles $F'(\xi)$. Observations reveal that the flow velocity $F'(\xi)$, along the stretching direction is constrained by the magnitude of M preventing it from

TABLE 9.1
Thermophysical Properties of Trihybrid Nanofluid [23, 24]

Properties	Ternary Hybrid Nanofluid
Viscosity	$\mu_{thnf}=\frac{\mu_f}{(1-\Phi_1)^{2.5}(1-\Phi_2)^{2.5}(1-\Phi_3)^{2.5}}$
Density	$\rho_{thnf}=(1-\Phi_1)\left[(1-\Phi_2)\left[(1-\Phi_3)+\Phi_3\frac{\rho_3}{\rho_f}\right]+\Phi_2\frac{\rho_2}{\rho_f}\right]+\Phi_1\frac{\rho_1}{\rho_f}$
Heat capacity	$(\rho c_p)_{thnf}=(1-\Phi_1)\left[(1-\Phi_2)\left[(1-\Phi_3)+\Phi_3\frac{(\rho c_p)_3}{(\rho c_p)_f}\right]+\Phi_2\frac{(\rho c_p)_2}{(\rho c_p)_f}\right]+\Phi_1\frac{(\rho c_p)_1}{(\rho c_p)_f}$
Thermal conductivity (Xue model)	$\frac{k_{nf}}{k_f}=\frac{1-\Phi_1+2\Phi_1\left(\frac{k_3}{k_3-k_f}\right)ln\frac{k_3+k_f}{2k_f}}{1-\Phi_1+2\Phi_1\left(\frac{k_f}{k_3-k_f}\right)ln\frac{k_3+k_f}{2k_f}};$ $\frac{k_{hnf}}{k_{nf}}=\frac{1-\Phi_2+2\Phi_2\left(\frac{k_2}{k_2-k_{nf}}\right)ln\frac{k_2+k_{nf}}{2k_{nf}}}{1-\Phi_2+2\Phi_2\left(\frac{k_{nf}}{k_2-k_{nf}}\right)ln\frac{k_2+k_{nf}}{2k_{nf}}};\frac{k_{thnf}}{k_{hnf}}=\frac{1-\Phi_3+2\Phi_3\left(\frac{k_1}{k_1-k_{hnf}}\right)ln\frac{k_1+k_{hnf}}{2k_{hnf}}}{1-\Phi_3+2\Phi_3\left(\frac{k_{hnf}}{k_1-k_{hnf}}\right)ln\frac{k_1+k_{hnf}}{2k_{hnf}}};$
Thermal conductivity (Yamada Ota model)	$\frac{k_{nf}}{k_f}=\frac{\frac{k_f}{k_3}+2\Phi_1{}^{0.2}\frac{L}{R}-2\Phi_1\Phi_1{}^{0.2}\frac{L}{R}\left(1-\frac{k_f}{k_3}\right)}{\frac{k_f}{k_3}+2\Phi_1{}^{0.2}\frac{L}{R}+\Phi_1\left(1-\frac{k_f}{k_3}\right)};$ $\frac{k_{hnf}}{k_{nf}}=\frac{\frac{nf}{k_2}+2\Phi_2{}^{0.2}\frac{L}{R}-2\Phi_2\Phi_2{}^{0.2}\frac{L}{R}\left(1-\frac{k_{nf}}{k_2}\right)}{\frac{k_{nf}}{k_2}+2\Phi_2{}^{0.2}\frac{L}{R}+\Phi_2\left(1-\frac{k_{nf}}{k_2}\right)};\frac{k_{thnf}}{k_{hnf}}=\frac{\frac{k_{hnf}}{k_1}+2\Phi_3{}^{0.2}\frac{L}{R}-2\Phi_3\Phi_3{}^{0.2}\frac{L}{R}\left(1-\frac{k_{hnf}}{k_1}\right)}{\frac{k_{hnf}}{k_1}+2\Phi_3{}^{0.2}\frac{L}{R}+\Phi_3\left(1-\frac{k_{hnf}}{k_1}\right)};$

TABLE 9.2
The Physical Properties of the Nanoparticles [33–35]

Physical Properties	ρ (kg m^{-1})	$C_p\left(\frac{J}{kg}K\right)$	$\kappa\left(\frac{W}{m}K\right)$	Σ (S/m)
$CoFe_2O_4$	4907.0	700.00	3.700	1.1×10^{7}
MgO	3580.0	937.00	54.90	2.6×10^{-6}
SiO_2	2650.0	730.00	1.500	5×10^{-4}
H_2O	0997.1	4179.0	0.613	$2.1*10^{-4}$

TABLE 9.3
Comparison of θ' (0) for Various Values of Pr when All Other Parameters are Held Constant at 0.1

Pr	Mabood *et al.* [36]	Haile *et al.* [37]	Pal [38]	Ishak *et al.* [39]	Ali *et al.* [40]	Results
00.72	0.8088	0.8086	0.80863134	0.80863134	0.80863325	0.808633
1.00	1.000	1.000	1.000	1.0004	1.00004	1.000020
3.00	1.9237	1.9236	1.9234	1.923682594	1.92367772	1.923674
10.00	3.7207	3.7207	3.7205	3.720673901	3.7206681685	3.720667

TABLE 9.4
A Comparative Analysis of $Nu/\sqrt{Re_x}$ is Conducted for Different Nanofluids Across Varying Values of M_a, Q, α, Bi with all Other Parameters Held Constant

Ma	α	Q_E	Bi	Mono Nanofluid $Nu/\sqrt{Re_x}$	Hybrid Nanofluid $Nu/\sqrt{Re_x}$	Ternary Nanofluid $Nu/\sqrt{Re_x}$
0.5	π/2	0.5	0.1	0.557888	0.938393	0.897955
1.0	-	-	-	0.560134	0.944837	0.918231
1.5	-	-	-	0.565516	0.957351	0.932724
-	0	-	-	0.565074	0.948970	0.926488
-	π/6	-	-	0.563841	0.947938	0.924422
-	π/3	-	-	0.561369	0.945871	0.920292
-	π/2	-	-	0.560134	0.944837	0.918231
-	-	0.3	-	0.565183	0.954293	0.928593
-	-	0.5	-	0.562659	0.949565	0.923412
-	-	0.7	-	0.557610	0.940109	0.913050
-	-	-	0.1	0.150074	0.264899	0.262049
-	-	-	0.4	0.557610	0.940109	0.913050
-	-	-	0.7	0.911036	1.478469	1.415350

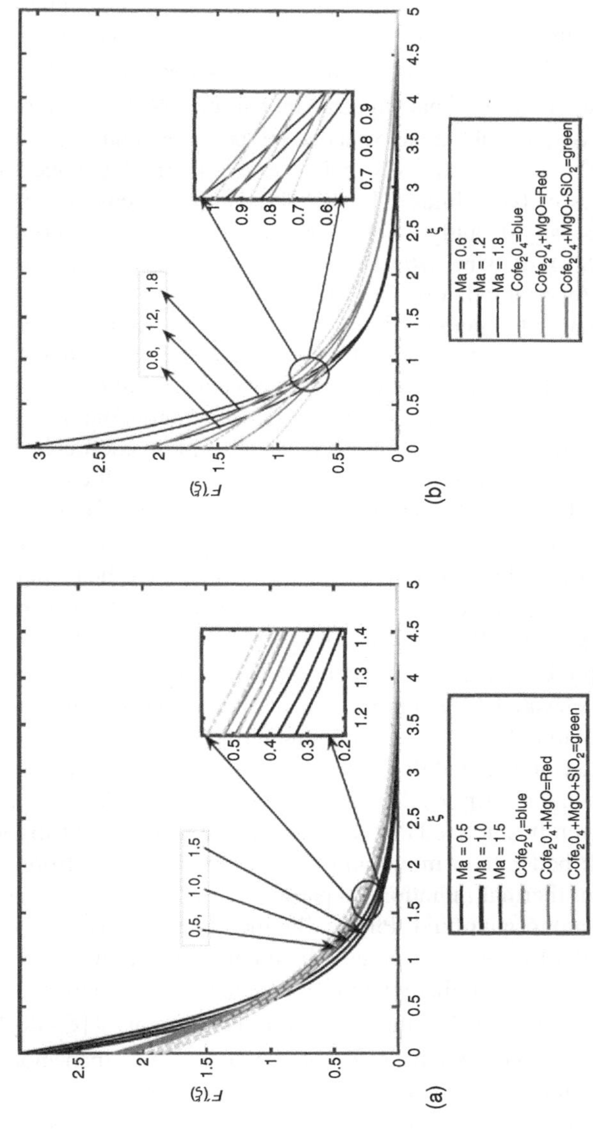

FIGURE 9.7 Fluctuation in $F'(\xi)$ under the effect of M (magnetic field) and M_a parameters.

exceeding this threshold. The velocity gradient diminishes progressively from its peak values to align with free stream conditions farther from the boundary layer, with this compression effect amplifying at elevated M. Conversely, increasing the Marangoni number M_a correlates with a rise in $F'(\xi)$. These velocity modulations stem from the Lorentz force, generated by the interplay of electric and magnetic fields. The negative sign associated with the magnetic field term in Eq. (9.2) reflects the Lorentz force's opposition to the primary flow direction, thereby suppressing $F'(\xi)$. Additionally, nanofluid behavior indicates that higher nanoparticle (NP) concentrations enhance thermal conductivity, leading to an expanded temperature boundary layer.

Three nanoparticles significantly enhance the thermal conductivity and heat transfer properties of base fluids. The characteristics, dimensions, and proportions of nanoparticles significantly affect the performance of ternary nanofluid. Figure 9.8(a),(b),(c),(d) demonstrates the influence of several parameters, namely Biot number Bi, thermal radiation Rd, Marangoni number M_a, and activation energy Q_E, on the fluctuating temperature gradient. This suggests that heat diffusion occurs more frequently than momentum diffusion. Furthermore, with the exception of the Marangoni number M_a, the temperature gradient exhibited a gradual increase influenced by the effects of thermal radiation Rd, the Marangoni number M_a, and the activation energy Q_E. The decrease in the mean absorption coefficient is attributed to an increase in the value of Rd. Radiative heat transfer is less efficient than conductive heat transfer, resulting in a reduction of the buoyancy force. High Rd enhances the transfer of heat to usable nanofluids, leading to an increase in temperature $\theta(\xi)$.

Radiative heat transfer is less efficient than conductive heat transfer, resulting in a reduction of the buoyancy force. Increasing the radiation parameter results in greater heating of the fluid. The result is an increase in the temperature profile. Physically, an uptick in Bi yields a rise in $\theta(\xi)$ caused by a raised thermal transfer coefficient. The elevated Biot number reveals enhanced heat transfer concerning the structure, as it quantifies the ratio of conduction to heat transfer compared to its interior to the exterior edges of the surface. As a result, the total energy of the entire system enhances and multiplies through the progressive thickening of the temperature boundary layer. It is noticed that the thermal conductance of trihybrid nanofluid (THNF) is much better than mono and hybrid nanofluid. As, from Figure 9.8(b) under the effect of thermal radiation Rd parameter the heat transfer rate of THNF is 33% better than mono-nanofluid. Physically, the boundary layer region's temperature is raised by the Biot number Bi and the intensity of thermal radiation.

Figure 9.9(a),(b) revealed that with the enhancing effect of temperature difference factor and Schmidt number Sc, the concentration gradient $\Phi(\xi)$ declined. And it is noticed that the concentration profile of THNF is better than mono and hybrid nanofluid but the impact is small. The selection process of values for parameters is crucial, as it ensures that every solution settles and is appropriate for physical occurrences. Figure 9.10 shows the trend of Nusselt number $\mathrm{Nu}/\sqrt{\mathrm{Re}_x}$ against the magnetic parameter and the Marangoni number Ma factors. It is observed that the Nusselt number $\mathrm{Nu}/\sqrt{\mathrm{Re}_x}$ declined with enhancing values of magnetic factor while inclined over the intensifying values of Marangoni number Ma factor. Physically, it

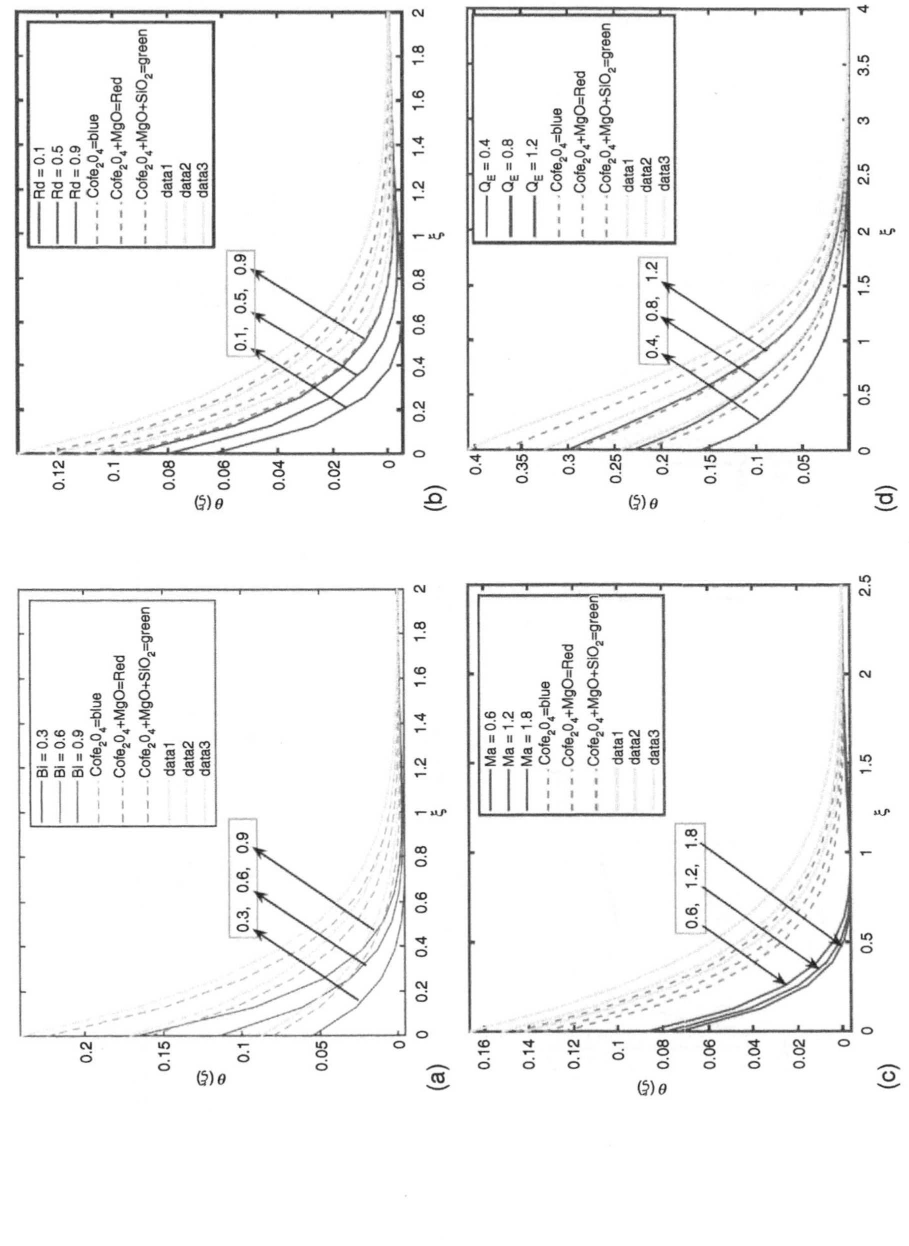

FIGURE 9.8 Fluctuation in $\theta\left(\xi\right)$ under the effect of Bi, Rd, Ma, and Q_E parameters.

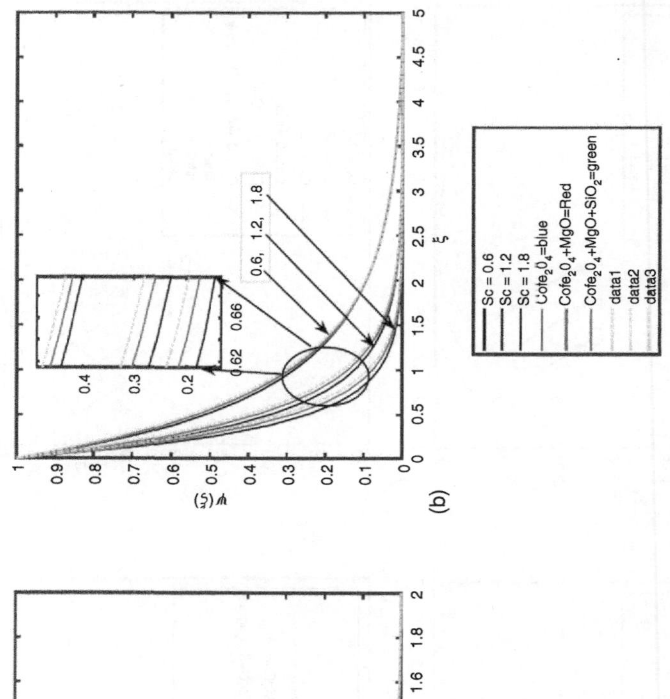

FIGURE 9.9 Fluctuation in concentration profile under the effect of δ and Sc parameters.

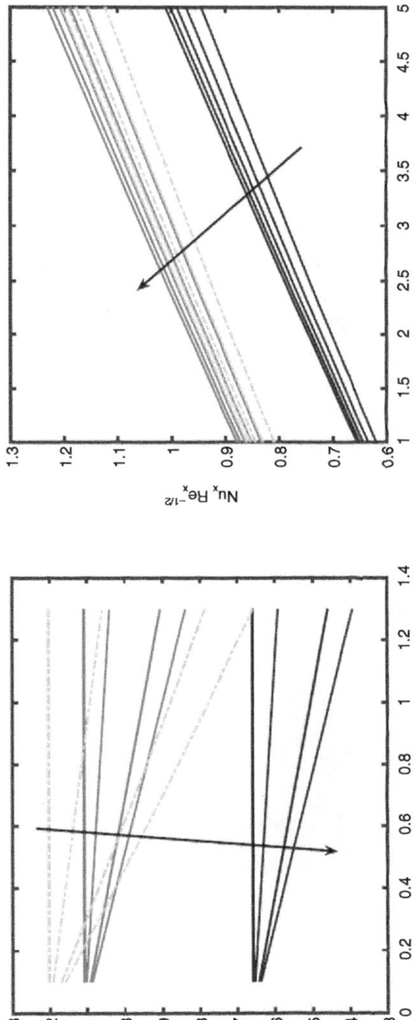

FIGURE 9.10 2D demonstration of Nusselt number over the effect of M and Ma parameters.

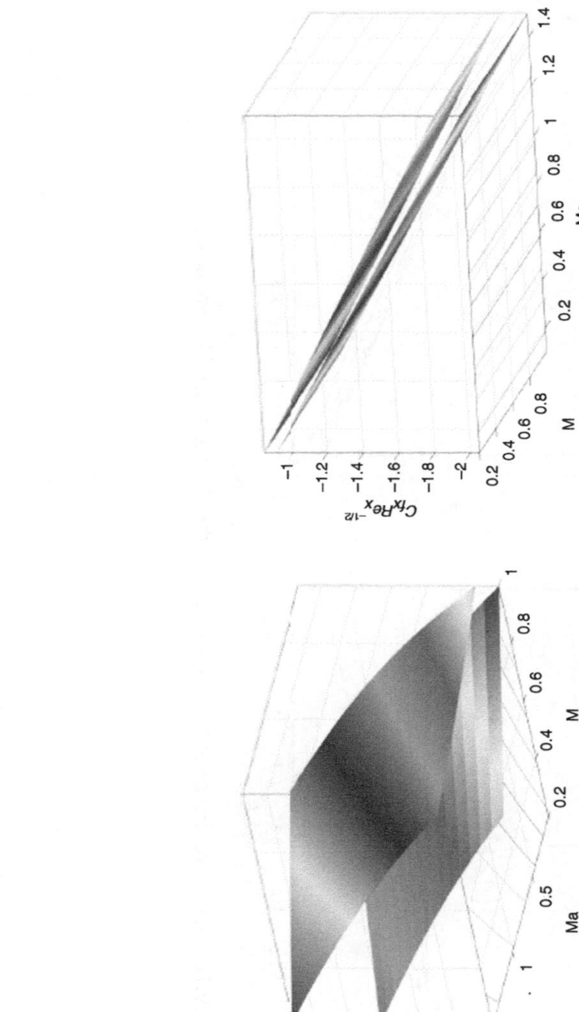

FIGURE 9.11 3D demonstration of Nusselt number and skin friction factor under influence of *M* and Ma.

can be explained by the fact that increasing magnetic field M values greatly improve conduction heat transfer, which lowers the local Nusselt number. Nusselt numbers are known to be the ratio of convection to conduction heat transfer. Figure 9.11 (a),(b) scrutinizing the three dimensional over view of Nusselt number $\mathrm{Nu}/\sqrt{\mathrm{Re}_x}$ and skin fraction $\sqrt{\mathrm{Re}_x}C_{fx}$ against the magnetic parameter M and the Marangoni number Ma factors. It revealed that with the enhancing impacts of parameters, both the Nusselt number and the skin friction coefficient were enhanced smoothly.

9.7 CONCLUSIONS

By modulating the nanoparticle volume fraction, researchers can enhance the characteristics of single-component nanofluids. But the approach has its limitations due to the compromise issue with the net negative of the viscosity rise. Researchers first used hybrid nanofluids and then trihybrid nanofluids to circumvent this conflict and improve heat transfer. The current findings demonstrate that trihybrid nanofluids outperform conventional fluids in terms of heat transfer efficiency. This analysis yielded the following key findings. Since cobalt ferrite $\left(CoFe_2O_4\right)$, magnesium oxide MgO, and silicon dioxide SiO_2 nanoparticles possess strong thermophysical properties, a comparative investigation found that physical limitations have a significant impact when employing this nanofluid. The notable findings are summarized below:

- With the increasing magnetic M impact reduces the velocity $F'\left(\xi\right)$ while the effect is particularly prominent in trihybrid nanofluids and the opposite trend is observed in the case for Marangoni number Ma.
- It is observed that with increasing values of Biot number Bi and thermal radiation parameter Rd the temperature profile $\theta\left(\xi\right)$ is intensified.
- As the temperature differential parameter becomes more intense, the concentration gradient decreases, which is an indicating a significant coupling effect.
- The activation energy Q_E and Marangoni number Ma impacting significantly on the temperature gradient but showing opposite effect and observing that the heat transfer rate in trihybrid nanofluid is significantly higher than mono-nanofluid.
- The Nusselt number $\mathrm{Nu}/\sqrt{\mathrm{Re}_x}$ decreased with intensifying the magnetic parameter M while the mass transfer rate was enhanced; meanwhile, the Nusselt number fluctuated on the basis of the nanofluid category with the enhanced values of the Marangoni number Ma with the influence of the magnetic field.
- Thermal radiation enhances heat transfer by 8.2% in trihybrid nanofluids; however, this improvement diminishes with increasing Marangoni number Ma. The observed enhancement highlights the superior efficacy of trihybrid nanofluids in boosting the thermal conductivity of base fluids.
- Further evidence suggests that trihybrid nanofluids could hold promise for a range of uses requiring for enhanced thermal performance.

ACKNOWLEDGEMENTS

The authors would like to acknowledge and express their gratitude to the School of Science for their support at Shaoyang University, China, and Universitas Airlangga (UNAIR), Surabaya, Indonesia.

REFERENCES

[1] A. A. Opanuga, S. O. Adesanya, H. I. Okagbue, O. O. Agboola, Impact of hall current on the entropy generation of radiative mhd mixed convection casson fluid, *International Journal of Applied and Computational Mathematics*, vol. 6, no. 2, 2020, pp. 1–18.

[2] L. Ali, X. Liu, B. Ali, S. Mujeed, S. Abdal, Finite element analysis of thermo-diffusion and multi-slip effects on mhd unsteady flow of casson nano-fluid over a shrinking/stretching sheet with radiation and heat source, *Applied Sciences,* vol. 9, no. 23, 2019, p. 5217.

[3] T. R. Mahapatra, A. Gupta, Heat transfer in stagnation-point flow towards a stretching sheet, *Heat and Mass Transfer*, vol. 38, no. 6, 2002, pp. 517–521.

[4] F. Ahmad, S. Abdal, H. Ayed, S. Hussain, S. Salim, A. O. Almatroud, The improved thermal efficiency of Maxwell hybrid nanofluid comprising of graphene oxide plus silver/kerosene oil over stretching sheet, *Case Studies in Thermal Engineering*, vol. 27, 2021, p. 101257.

[5] T. Hayat, S. A. Khan, A. Alsaedi, H. M. Fardoun, Marangoni forced convective flow of second grade fluid with irreversibility analysis and chemical reaction, *International Journal of Thermophysics*, vol. 42, 2021, pp. 1–21.

[6] I. Ullah, T. Hayat, A. Alsaedi, S. Asghar, Modeling for radiated marangoni convection flow of magnetonanoliquid subject to activation energy and chemical reaction, *Scientia Iranica*, vol. 27, no. 6, 2020, pp. 3390–3398.

[7] S. Qayyum, Dynamics of marangoni convection in hybrid nanofluid flow submerged in ethylene glycol and water base fluids, *International Communications in Heat and Mass Transfer*, vol. 119, 2020, p. 104962.

[8] J. L. Neuringer, Some viscous flows of a saturated ferro-fluid under the combined influence of thermal and magnetic field gradients, *International Journal of Non-Linear Mechanics*, vol. 1, no. 2, 1966, pp. 123–137.

[9] K. Raslan, S. Mohamadain, M. Abdel-wahed, E. Abedel-aal, Mhd steady/unsteady porous boundary layer of cu–water nanofluid with micropolar effect over a permeable surface, *Applied Sciences*, vol. 8, no. 5, 2018, p. 736.

[10] Y. S. Daniel, Z. A. Aziz, Z. Ismail, A. Bahar, F. Salah, Slip role for unsteady mhd mixed convection of nanofluid over stretching sheet with thermal radiation and electric field, *Indian Journal of Physics*, vol. 94, no. 2, 2020, pp. 195–207.

[11] L. Ali, X. Liu, B. Ali, S. Mujeed, S. Abdal, Finite element simulation of multi-slip effects on unsteady mhd bioconvective micropolar nanofluid flow over a sheet with solutal and thermal convective boundary conditions, *Coatings*, vol. 9, no. 12, 2019, p. 842.

[12] O. A. Beg, M. Ferdows, M. E. Karim, M. M. Hasan, T. A. B´eg, M. Shamshuddin, A. Kadir, Computation of non-isothermal thermo-convective micropolar fluid dynamics in a hall mhd generator system with non-linear distending wall, *International Journal of Applied and Computational Mathematics*, vol. 6, no. 2, 2020, pp. 1–44.

[13] B. Ali, M. B. Ghori, Melting effect on cattaneo–christov and thermal radiation features for aligned mhd nanofluid flow comprising microorganisms to leading edge: Fem approach, *Computers & Mathematics with Applications*, vol. 109, 2022, pp. 260–269.

[14] P. Kumar, H. Poonia, L. Ali, S. Areekara, The numerical simulation of nanoparticle size and thermal radiation with the magnetic field effect based on tangent hyperbolic nanofluid flow, *Case Studies in Thermal Engineering*, vol. 37, 2022, p. 102247.

[15] J. Argyris, H. Balmer, J. S. Doltsinis, P. Dunne, M. Haase, M. Kleiber, G. Malejannakis, H.-P. Mlejnek, M. M¨uller, D. Scharpf, Finite element method—the natural approach, *Computer Methods in Applied Mechanics and Engineering*, vol. 17, 1979, pp. 1–106.

[16] G. Strang, G. J. Fix, *An analysis of the finite element method(book- an analysis of the finite element method)*, Englewood Cliffs, NJ: Prentice-Hall, Inc, 1973, p. 318.

[17] Y. Mehmood, W. Abbasi, M. Alhussein, K. Aurangzeb, A. Rehman, M. S. Anwar, Computational behavior of trihybrid casson nanofluid blood flow occurring inside the conical gap between the rotating disk and the cone, *Contemporary Mathematics*, vol. 3, 2025, pp. 1–26.

[18] S. Li, Y. Leng, G. Atta, S. Ahmad, K. Ali, S. A. Idris, H. Ahmad, Thermal attributes of sodium alginate (na.c6h7o6) based binary and ternary hybrid nanofluids under activation energy and induced magnetic field environment, *Case Studies in Thermal Engineering*, vol. 59, 2024, p. 104449.

[19] B. Ali, I. Siddique, S. Majeed, Windarto, T. Lamoudan, S. A. Khan, Significance of trihybrid nanoparticles in nonnewtonian fluids: a finite-element simulation of magneto-hydrodynamic effects under microgravity conditions, *Mechanics of Time-Dependent Materials*, vol. 28, no. 3, 2024, pp. 1331–1348.

[20] L. Ali, B. Ali, X. Liu, T. Iqbal, R. M. Zulqarnain, M. Javid, A comparative study of unsteady mhd falkner–skan wedge flow for non-newtonian nanofluids considering thermal radiation and activation energy, *Chinese Journal of Physics*, vol. 77, 2022, pp. 1625–1638.

[21] L. Ali, Y.-J. Wu, B. Ali, S. Abdal, S. Hussain, The crucial features of aggregation in tio_2-water nanofluid aligned of chemically comprising microorganisms: a fem approach, *Computers & Mathematics with Applications*, vol. 123, 2022, pp. 241–251.

[22] A. H. Usman, Z. Shah, U. W. Humphries, P. Kumam, P. Thounthong, Soret, dufour, and activation energy effects on double diffusive convective couple stress micropolar nanofluid flow in a hall mhd generator system, *AIP Advances,* vol. 10, no. 7, 2020, p. 075010.

[23] L. Ali, V. Kumar, P. Kumar, A. Mahajan, B. Sharma, R. Apsari, The significant role of darcy–forchhiemer with integrated hybrid nanoparticles (graphene and tio_2) on dusty nanofluid flow subjected to heat conduction, *Numerical Heat Transfer, Part B: Fundamentals*, vol. 4, 2024, pp. 1–16.

[24] M. Z. Ullah, Irreversibility marangoni tri-hybrid nanoflow analysis for thermal enhancement applications, *Nanomaterials*, vol. 13, no. 3, 2023, p. 423.

[25] B. Ali, L. Ali, S. Abdal, M. I. Asjad, Significance of brownian motion and thermophoresis influence on dynamics of reiner-rivlin fluid over a disk with non-fourier heat flux theory and gyrotactic microorganisms: a numerical approach, *Physica Scripta*, vol. 96, no. 9, 2021, p. 094001.

[26] M. Uddin, P. Rana, O. A. B´eg, A. M. Ismail, Finite element simulation of magneto-hydrodynamic convective nanofluid slip flow in porous media with nonlinear radiation, *Alexandria Engineering Journal*, vol. 55, no. 2, 2016, pp. 1305–1319.

[27] J. N. Reddy, *Solutions manual for an introduction to the finite element method*, New York: McGraw-Hill, 1993.

[28] K. Jyothi, P. S. Reddy, M. S. Reddy, Carreau nanofluid heat and mass transfer flow through wedge with slip conditions and nonlinear thermal radiation, *Journal of the Brazilian Society of Mechanical Sciences and Engineering*, vol. 41, no. 10, 2019, p. 415.

[29] L. Ali, X. Liu, B. Ali, A. Din, Q. Al Mdallal, The function of nanoparticle's diameter and darcy-forchheimer flow over a cylinder with effect of magnetic field and thermal radiation, *Case Studies in Thermal Engineering*, vol. 9, 2021, p. 101392.

[30] G. Swapna, L. Kumar, P. Rana, B. Singh, Finite element modeling of a double-diffusive mixed convection flow of a chemically-reacting magneto-micropolar fluid with convective boundary condition, *Journal of the Taiwan Institute of Chemical Engineers*, vol. 47, 2015, pp. 18–27.

[31] L. Ali, X. Liu, B. Ali, S. Mujeed, S. Abdal, Finite element analysis of thermo-diffusion and multi-slip effects on mhd unsteady flow of casson nano-fluid over a shrinking/stretching sheet with radiation and heat source, *Applied Sciences,* vol. 9, no. 23, 2019, p. 5217.

[32] B. Ali, X. Yu, M. T. Sadiq, A. U. Rehman, L. Ali, A finite element simulation of the active and passive controls of the mhd effect on an axisymmetric nanofluid flow with thermo-diffusion over a radially stretched sheet, *Processes*, vol. 8, no. 2, 2020, p. 207.

[33] M. Gamal, M. Radwan, I. Elgizawy, M. Shedid, Thermophysical characterization on water and ethylene glycol/water-based mgo and zno nanofluids at elevated temperatures: an experimental investigation, *Journal of Molecular Liquids*, vol. 369, 2023, p. 120867.

[34] S. Elattar, U. Khan, A. Zaib, A. Ishak, N. Alwadai, A. M. Abed, Heat transfer characteristics of cobalt ferrite nanoparticles scattered in sodium alginate-based non-newtonian nanofluid over a stretching/shrinking horizontal plane surface, *Open Physics*, vol. 22, no. 1, 2024, p. 20230182.

[35] L. Ali, A. Manan, B. Ali, Maxwell nanofluids: Fem simulation of the effects of suction/injection on the dynamics of rotatory fluid subjected to bioconvection, lorentz, and coriolis forces, *Nanomaterials*, vol. 12, no. 19, 2022, p. 3453.

[36] F. Mabood, S. Shateyi, M. M. Rashidi, E. Momoniat, N. Freidoonimehr, MHD stagnation point flow heat andmass transfer of nanofluids in porous medium with radiation, viscous dissipation and chemical reaction, *Advanced Powder Technology*, vol. 27, no. 2, 2016, pp. 742–749.

[37] E. Haile, B. Shankar, Heat and mass transfer in the boundary layer of unsteady viscous nanofluid along a vertical stretching sheet, *Journal of Computational Engineering*, vol. 1, 2014, pp. 1–17.

[38] D. Pal, Combined effects of non-uniform heat source/sink and thermal radiation on heat transfer over an unsteady stretching permeable surface, *Communications in Nonlinear Science and Numerical Simulation*, vol. 16, no. 4, 2011, pp. 1890–1904.

[39] A. Ishak, R. Nazar, I. Pop, Boundary layer flow and heat transfer over an unsteady stretching vertical surface, *Meccanica,* vol. 44, 2009, pp. 369–375.

[40] B. Ali, R. A. Naqvi, L. Ali, S. Abdal, S. Hussain, A comparative description on time-dependent rotating magnetic transport of a water base liquid h2o with hybrid nano-materials al_2o_3-cu and al_2o_3-tio_2 over an extending sheet using buongiorno model: Finite element approach, *Chinese Journal of Physics*, vol. 70, no. 7, 2021, pp. 125–139.

Part III

Computational and Numerical Techniques

10 LES-RANS Hybrid Models in CFD

Ikram Ul Haq and Waqar Ul Hassan

10.1 INTRODUCTION

10.1.1 THE NEED FOR HYBRID MODELS IN CFD

The computational fluid dynamics (CFD) based on stand-alone large eddy simulation (LES) models provide highly accurate predictions of turbulence at large scales or the energy-containing scales, however, they are typically very expensive to run due to the high computational cost and therefore not suitable for complex optical applications in industry. In contrast, individual Reynolds-averaged Navier–Stokes (RANS) models, while computationally efficient for boundary layers, do not accurately replicate complex flows; these flows require the framework of time-averaged equations. Hybrid models solve these problems by leveraging the best of both models to create a compute-efficient and accurate approach [1]. This area of research is particularly relevant in overcoming the limitations of both standalone LES and RANS models and thus achieving a balance of computational efficiency and high precision for complex flows [2].

The characteristics of both techniques are described in unison and benefit from this coupling of LES models and RANS models. Also, LES is superior for large turbulence scale structures, which more profoundly unveil the complicated flow characteristics and can significantly capture most of the transient effect phenomena that are of great value in natural circulation systems. CFD, with its high fidelity, enables models to resolve the characteristics of the turbulent structures that can exist in these flows, which is an important contributor to understanding phenomena associated with flow reversals and thermal stratification [3].

On the other hand, the RANS models provide computational efficiency, especially over boundary layers, making it feasible and less costly to obtain averaged flow results. Hybrid models take advantage of the strengths of these approaches to improve the accuracy and efficiency of CFD simulations, which are crucial for numerous applications, including aerospace engineering and environmental systems [4].

DOI: 10.1201/9781003611820-13

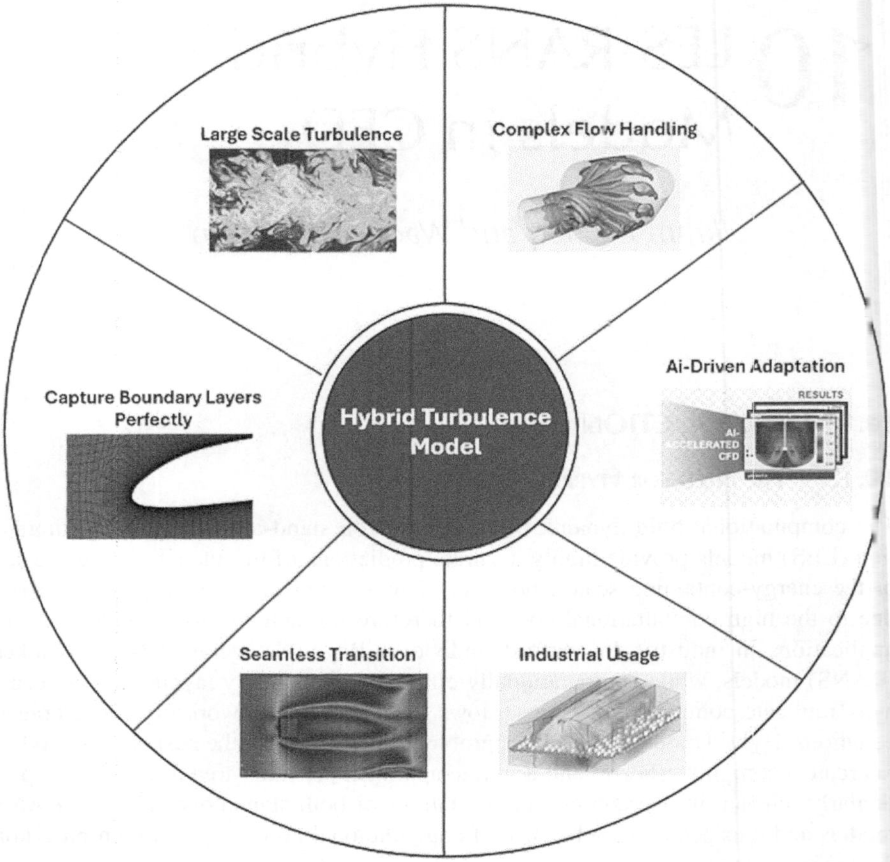

FIGURE 10.1 Applications for hybrid turbulence models.

10.1.2 Evolution of Hybrid Models

Hybrid models such as detached eddy simulation (DES) and scale-adaptive simulation (SAS) [5] have progressed significantly through this evolution. Specifically, DES was proposed by Spalart *et al.* in 1997, which also integrates RANS into wall models while using LES in the free stream, and between accuracy and computational efficiency [6]. This has become a standard practice for near-cavitation and heavily separated flows. A turbulence model more in line with the principles of SAS, or scale-adaptive simulation, conceived by Menter *et al.* in 2003, is an extension of the hybrid framework that is tuned according to the local flow dynamics (Menter *et al.*, 2003) [7]. Although hybrid modeling has already been preferred for many scientific engineering designs and dynamic DES (DDES) [10], further filtering hybrid modeling, with this option being more accurate and robust for numerous engineering

FIGURE 10.2 DES captures detailed flow structures and separation better than RANS [11].

designs ensuring seamless switching between RANS and LES modes as expressed in the following equations [8]:

$$\tilde{d} = \min\left(d_w, C_{DES}\Delta\right)$$

$$\Delta = \max\left(\Delta_1, \Delta_2, \Delta_3\right)$$

$$\tilde{d} = d_w - f_d \max\left(0, d_w - C_{DES}\Delta\right)$$

Hybrid models are important for bridging the gap between high-fidelity simulations and real industrial implementations. Such models combine the advantages of LES, which achieves a high spatial detail resolving the turbulence structures, with the results of a time-averaged approach using RANS equations reached at relatively small computational costs. This makes it possible to resolve complex flow phenomena like transient behaviors and local turbulence while achieving the computational efficiency needed in the context of large-scale industrial problems. Consequently, such hybrid models are progressively being implemented in numerous domains like aerospace, automotive, and energy to attain the best design and enhance performance [8,9].

10.2 THEORETICAL FOUNDATIONS

10.2.1 Governing Equations

10.2.1.1 LES (Large Eddy Simulation)

LES resolves large scales of turbulence and models smaller scales using a sub-grid scale (SGS) model. It is a method of solving the equations using LES by filtering velocity and pressure fields with spatial-filter approaches, this filtering leads to governing equations derived from Navier–Stokes. Upon inserting this filter, the Navier–Stokes equations take the form:

$$\frac{\partial \overline{u_i}}{\partial t} + \frac{\partial \overline{u_i' u_j}}{\partial x_j} = -\frac{\partial \bar{p}}{\partial x_i} + \frac{\partial}{\partial x_j}\left(\mu \frac{\partial \overline{u_i}}{\partial x_j}\right) - \frac{\partial \tau_{ij}}{\partial x_j}$$

$$\frac{\partial \overline{\mathrm{u_i}}}{\partial \mathrm{x_i}} = 0$$

10.2.1.2 RANS (Reynolds-Averaged Navier–Stokes)

RANS models use time-averaged Navier–Stokes equations to describe the mean flow and turbulence. The governing equations for RANS are obtained by taking the time average of the Navier–Stokes equations. The Reynolds-averaged Navier–Stokes equations are

$$\frac{\partial \langle u_i \rangle}{\partial t} + \frac{\partial \langle u_i u_j \rangle}{\partial x_j} = -\frac{\partial \langle p \rangle}{\partial x_i} + \frac{\partial}{\partial x_j}\left(\mu \frac{\partial \langle u_i \rangle}{\partial x_j}\right) + \frac{\partial \left\langle u_i^{\prime} u_j^{\prime} \right\rangle}{\partial x_j}$$

$$\frac{\partial \langle \mathrm{u_i} \rangle}{\partial \mathrm{x_i}} = 0$$

Here, $\langle u_i \rangle$ is the time-averaged velocity, $\langle p \rangle$ is the time-averaged pressure, μ is the dynamic viscosity, and $\left\langle u_i^{\prime} u_j^{\prime} \right\rangle$ is the Reynolds stress tensor, which represents the effect of the turbulent fluctuations on the mean flow.

10.2.1.3 Turbulence Closure Requirements

In both cases, turbulence closure is needed to represent the influence of the small-scale turbulence on the large scales (LES), or the mean flow (RANS). In LES, the sub-grid scale stress tensor: τ_{ij} is modeled using the SGS model, for example, the Smagorinsky model or the WALE model. In RANS, the ⟨ui′ uj′ ⟩ is modeled by the turbulence model, e.g. k-ε model, and the k-ω model [10–12].

10.2.2 Hybridization Strategies

10.2.2.1 Zonal vs. Non-Zonal Method

I. **Zonal Methods: Domain Decomposition**

These approaches consist of separating the computational or flow domain into different areas in which either LES or RANS is used independently. This technique is especially effective when certain portions of the flow can be identified as favoring the use of either LES or RANS [13]. LES, for instance, may be applied in terms of complex turbulence regions while the RANS is used in some boundary layers where the flow is [slightly] well-behaved [14].

- **Advantages:**
  RANS is computationally inexpensive, and can be used in areas where resolving the flow in detail is unnecessary. LES could be used in areas that require high-fidelity turbulence modeling.

- **Disadvantages:**
  However, it can be difficult to ensure continuity and consistency at the LES-RANS interface. Usually at the cost of careful domain decomposition and interface management.

II. **Non-Zonal Methods: Blended Formulations**

Non-zonal techniques combine the LES and RANS formulations in the same domain, which enables a seamless transition between LES and RANS. To do this, a series of hybridized models like non-linear LES (NLDE) and partially-averaged Navier–Stokes (PANS) are employed.

- **Advantages:**
  Offers a smoother coupling between LES and RANS, thus minimizing interfacing problems. Trained on data extending until October 2023.

- **Disadvantages:**

It is proposed that you become a sort of master blender and learn the secrets behind the art of blending. Can be computationally expensive for the requirement to solve both the LES and RANS components

III. **Examples of Blended Formulations**

Non-Linear LES (NLDE)

NLDE, as a non-linear model, is a combination of LES and RANS and it is a type of non-linear eddy viscosity model. This approach can provide smoother transitions to turbulent/laminar regions; hence it is suitable for arbitrary turbulence levels. 3. Non-linear eddy viscosity/ The non-linear eddy viscosity method aids in better capturing the complex interactions between various scales of turbulence [15].

Partially-Averaged Navier-Stokes (PANS)

PANS is a hybrid that partially averages the Navier–Stokes equations and thus can provide a compromise between the computational expense of LES and the fidelity offered by RANS. PANS is capable of using both LES and RANS simultaneously, depending on the complexity of the flow, making it an attractive option for many complex flows. The smoothness adaptivity of meshes is particularly beneficial for simulations where flow characteristics in different regions of the domain are very different [16].

10.2.2.2 Dynamic Coupling

In hybrid homogeneous models, dynamic coupling is defined as the adaptive activation of the RANS model with the knowledge of the local turbulence intensity, the calculation of the Van Driest integrated length scale in the inlet, and the dimensions of the computational grid for the near-wall RANS zone. This enables the model to intelligently adapt the turbulence resolution level depending on the region of interest, providing a balance between computational efficiency and accuracy where it matters most in the simulation space [17].

I. **Turbulence Intensity**
For adaptive switching, the intensity of turbulence is an important parameter. With the potential to resolve large-scale turbulence structures, LES would generally be more suited to these areas of high turbulence intensity. On the other hand, in regions where the intensity of the turbulence is less, we can use RANS to reduce the use of computational resources [18], where I is turbulence intensity defined as

$$I = \frac{\sqrt{\frac{2}{3}k}}{U}$$

where k is the turbulent kinetic energy and U is the mean flow velocity. By monitoring the turbulence intensity, the model can switch between LES and RANS dynamically, ensuring that LES is used only where necessary.

II. **Grid Resolution**
The adaptive switching criterion is another widely suggested switching method that is based on the grid resolution [19]. Whereas in regions with fine grid resolution, LES is better suited to resolve smaller turbulent scales. In the coarser region, RANS can capture relatively high computational costs. The grid resolution, which can be expressed in terms of the grid spacing, Δx and the local Reynolds number, Re_Δ, is defined as

$$\mathrm{Re}_\Delta = \frac{U\Delta x}{\nu}$$

where (ν) is the kinematic viscosity. By evaluating (Re_Δ), the model can determine whether the grid resolution is sufficient for LES or if RANS should be used instead.

III. **Adaptive Switching Mechanisms**
Adaptive switching mechanisms consist of various approaches, including the following:

1. **Sensor-Based Switching:** Turbulence intensity or mesh resolution-based LES-RANS switch [20]. So, for example I could trigger LES in regions where I exceed some threshold (say, based on a sensor monitoring the local turbulence intensity).
2. **Hybrid Models with Adaptive Criteria:** Several hybrid models, e.g. the SAS, inherently adapt to local flow conditions. The SAS model dynamically switches between LES and RANS based on local flow conditions.
3. **Machine Learning Approaches:** Preferring ML algorithms to maximize the switching criteria is the latest development in the past few

years. Frequent switching has been modeled using machine learning algorithms capable of inferring optimal switching characteristics from historical data.

10.2.3 Key Hybrid Frameworks

10.2.3.1 Improved DES Variants (IDDES, DDES)

IDDES (improved delayed detached eddy simulation) is a DES (detached eddy simulation) that has the same approach as DDES (delayed detached eddy simulation). So these techniques take care of the few limitations of DES to make it more accurate and robust. IDDES and DDES have been particularly productive in these areas, with their application in complex flows, where flow separation and reattachment occur, for example, flow around bluff bodies and in problems related to aerodynamics. DES models can balance the advantages of RANS and LES, improving the performance of turbulence prediction without loss of computation efficiency [21].

10.2.3.2 Partially-Averaged Navier–Stokes (PANS)

PANS is a hybrid method that aims to bridge the gap from RANS to LES. It enables fully resolved or averaged description at any section of the flow, depending on its complexity and computational potential. PANS can be very effective in capturing flows with a large range of turbulence scales, including transitional flows and flows in complex geometries. This offers a versatile means to trade-off accuracy with computational cost [22].

10.2.3.3 Embedded LES (ELES)

Like the RANS/LES hybrid methods, ELES treats part of the domain as LES, while still using RANS in the portion of the domain where the flow can be considered attached. The latter method can provide high-resolution large eddy simulation in regions of interest (e.g., flow separation regions or vortex motions) while premiering the remaining domain to RANS, thereby saving computational resources. In terms of mesh handling, the main complication in ELES arises at the interface between RANS and LES, where the turbulence must not be resolved but must be adequately formed at their junction and turbulent RANS flow be also hashed when it overpasses with the LES one.

10.2.3.4 Machine Learning-Enhanced Hybridization

Like the RANS/LES hybrid methods, ELES treats part of the domain as LES, while still using RANS in the portion of the domain where the flow can be considered attached. The latter method can provide high-resolution large eddy simulation in regions of interest (e.g., flow separation regions or vortex motions) while premiering the remaining domain to RANS, thereby saving computational resources. In terms of mesh handling, the main complication in ELES arises at the interface between RANS and LES, where the turbulence must not be resolved but must be adequately formed at their junction and turbulent RANS flow be also hashed when it overpasses with the LES one [23].

10.3 METHODOLOGICAL INNOVATIONS

A major difficulty in hybrid LES-RANS computations is a smooth transition between the two separate approaches. In this section, we look at advanced methods that have been implemented to alleviate what has been termed "gray-area" issues that are produced at the interface between LES and RANS regions.

10.3.1 Seamless Transition Techniques

103.3.1 Strategies to Mitigate Gray-Area Problems at LES-RANS Interfaces

The gray-area problem arises from the inability to accurately resolve the flow physics at the interface of LES and RANS regions. In this respect, it is especially acute since LES and RANS have very different approaches both in terms of resolution and modeling. To overcome this issue, various methods have been proposed [24]:

I. **Blending Functions:**
One down-scaling approach you will see used for this is to make use of "blending functions" to smoothly transition between the LES and RANS models. These can be functions of many different flow parameters, including distance from the wall, turbulence intensity, and local gradients in the flow. The smoothing functions are a method to smoothly go from one model to another to stabilize the simulation and limit numerical artifacts at the interface.

II. **Buffer Zones:**
Or simply, a buffer region can be created where the LES and RANS equations are solved simultaneously. This dual-solution approach leads to a calmer, steadier, incremental transition between communities. The buffer region helps to smooth the transition in flow properties between the high-fidelity LES part and the low-fidelity RANS part.

III. **Consistent Turbulence Modeling:**
It is critical to maintain consistency in turbulence modeling between LES and RANS across the LES-RANS interface to preserve the accuracy of the simulation. In LES this can be done by utilizing compatible subgrid-scale models, while in RANS this is possible using eddy viscosity models. The discontinuities are markedly reduced, and numerical instabilities are mitigated by corresponding the two mesh regions.

IV. **Dynamic Coupling:**
Dynamic coupling techniques provide a more flexible solution for handling the LES-RANS interface. These approaches resize the interface location in (very) real-time as flow characteristics change, e.g. turbulence intensity or vorticity. Dynamic coupling accommodates changing conditions of flows concerning each other, as the interface can move and then allow you the capability of simulation to stabilize.

10.3.3.2 Sensor-Based Adaptive Models

Adaptive models based on sensors have also been proposed to make the transition between the LES and RANS well-grounded. In these models, real-time flow information is used to activate LES or RANS depending on set criteria. Also, with the significant computational expense of high-fidelity modeling, it provides a way to more intelligently allocate computational resources based on the relevance of the model, by applying high-order LES in regions of relative importance in the flow and switching to busier, more robust RANS-like models away from those regions.

- **Vorticity-Based Sensors:**
  Vorticity sensors order most of them, which tend to be the indicators of the turbulent structure. The model toggles between using LES to deal with these features but reverts to RANS in various areas, such that the vorticity levels can trigger an initiation of the LES switch. This adaptive strategy guarantees that the usage of computational resources is efficient in a gentle exchange for fidelity accuracy where it matters [25].
- **Strain-Rate Criteria:**
  Another useful tool for adaptive modeling is strain-rate sensors. These sensors can find areas of the domain that are experiencing high strain rates, and in those areas, switch their functionality to LES, which can solve features at a greater resolution where necessary. This strategy serves to enhance the efficiency of computational resources by encapsulating areas with substantial flow gradients and turbulence [26].
- **Adaptive Mesh Refinement:**
  Moreover, particle-based models in the existing hybrid simulations can use the adaptive mesh refinement (AMR) technique to improve the accuracy even further for absolute concentration near the cellular membrane in sensor-based models. AIMR corrects the mesh resolution automatically, depending on the flow properties, making sure that a high-resolution LES is computed only where it is needed. Such an adaptive approach helps to increase the accuracy of the simulation with minimal computation effort since no unnecessary high-resolution calculations will be done in the less critical areas.
- **Hybrid Sensors:**
  Hybrid sensors that combine more than one characteristic can be used to obtain a stronger and more flexible adaptive model. These sensors combine data from multiple flow parameters including vorticity, strain rate, and wall distance to characterize the overall flow conditions.

10.3.2 HPC Integration

Hybrid models need to integrate with high-performance computing (HPC) platforms based on supercomputers to make them efficient at running complex simulations. In this section, we explore the scalability of hybrid models on GPU/CPU architectures, as well as the use of reduced-order modeling for hybrid frameworks.

10.3.2.1 Scalability of Hybrid Models on GPU/CPU Architectures

Scalability of hybrid models on modern HPC architectures (especially on heterogeneous architectures, e.g., involving GPUs and CPUs [27]). is one of the major aspects of achieving efficient, realistic, and timely simulations. To tantalize you further, the following techniques are critical in achieving models capable of leveraging the parallel-processing capacity of these architectures:

I. **Parallelization Strategies:**
These parallelization strategies are very important to distribute the computation on different/cores from this or multiple. This is possible via domain discretization, by breaking the computational domain into increasingly many subdomains that can be computed independently in parallel across cores or GPUs. A key aspect of parallel processing is a balanced workload in terms of computation since they will be parallel then they will be idle some of the time, and so on.

$$\Omega = \bigcup_{i=1}^{N} \Omega_i$$

where Ω is the total computational domain and Ω_i are the subdomains.

II. **GPU Acceleration:**
Because GPUs can carry out numerous calculations simultaneously, they provide strong computational benefits. The hybrid models can take advantage of GPU acceleration by delegating the heavy calculations to GPU, for instance, solving the LES equations. This enables faster computation times and utilizing the device correctly. But a consideration of the transfer of data between the CPU and GPU is necessary to prevent overhead and a stutter in operation.

$$\frac{\partial \boldsymbol{u}}{\partial t} + \boldsymbol{u} \cdot \nabla \boldsymbol{u} = -\nabla p + \nu \nabla^2 \boldsymbol{u} + \boldsymbol{f}$$

where u is the velocity field, p is the pressure, ν is the kinematic viscosity and f represents external forces.

III. **Hybrid Parallelism:**
To scale a hybrid model on nodes are often combined the MPI (message passing interface) for inter-node communication with OpenMP (open multi-processing) or CUDA (compute unified device architecture) for intra-node parallelism. By integrating both CPUs and GPUs in this manner, it is possible to parallelize computation in an efficient manner utilizing the available resources of the hardware.

$$\text{Hybrid Parellelism: MPI} \times (\text{OpenMP or CUDA})$$

10.3.2.2 Reduced-Order Modeling for Hybrid Frameworks

Reduced-order modeling (ROM) considered as powerful way for minimizing the computational complexity of hybrid frameworks while maintaining a high level of accuracy. ROM involves the construction of a lower-dimensional representation of the original system, which can be solved more efficiently [28].

I. **Proper Orthogonal Decomposition (POD):**
POD a technique widely used in reduced order model construction. It is based on the decomposition of the original system into a set of orthogonal basis functions that describe the dominant flow features.

$$u(x,t) \approx \sum_{i=1}^{r} \alpha_i(t)\,\phi_i(x)$$

where $u(x,t)$ is the velocity field, $\alpha_i(t)$ are the time-dependent coefficients and $\phi_i(x)$ are the spatial basis functions.

II. **Galerkin Projection:**
This formulation is typically achieved by Galerkin projection of the governing equations onto the reduced basis functions, producing a (lower-dimensional) system of equations.

$$\frac{d\alpha}{dt} = A\alpha + B$$

where α is the vector of coefficients, A is the reduced system matrix and B represents the forcing terms.

10.3.3 Multiphysics Extensions

Hybrid models are crucial to consider since a significant real-world engineering problem consists of many Multiphysics phenomena. We will discuss hybrid frameworks and their coupling with combustion, acoustics, and multiphase flows in this section.

10.3.3.1 Integration with Combustion

If gas at different temperatures is mixed in the combustion process, the resolve to a specific thermal state becomes a multi-stage physical process characterized by complex interactions of fluid dynamics, chemical reactions, and heat transfer. Coupling hybrid models with combustion simulations [29] requires special attention to the coupling between flow field and chemical reaction.

- **Combustion Models:**
For example, combustion models like EDC or FGM can be coupled with hybrid LES-RANS frameworks to model the underlying chemistry-turbulence interaction in a detailed manner.

$$\frac{\partial Y_k}{\partial t} + u \cdot \nabla Y_k = \frac{\partial}{\partial x_i}\left(\Gamma_k \frac{\partial Y_k}{\partial x_i}\right) + \omega_k$$

where Y_k is the mass fraction of species k, Γ_k is the diffusion coefficient and ω_k is the chemical reaction rate.

10.3.3.2 Integration with Acoustics

The acoustic phenomena are often coupled to fluid dynamics, for example in aero acoustics and combustion noise applications. Hybrid models are meant to be combined with acoustic simulations and require the application of corresponding acoustic models and coupling methods [30].

- **Acoustic Analogies:**
  For example, acoustic analogies like light hill are used to describe the acoustic sources of the flow field.

$$\frac{\partial^2 p'}{\partial t^2} - c^2 \nabla^2 p' = \rho_0 \frac{\partial^2 u \cdot u}{\partial t^2}$$

where p' is the acoustic pressure perturbation, c is the speed of sound and ρ_0 is the reference density.

10.3.3.3 Integration with Multiphase Flows

The interaction between different phases is complex, as such flows are multiphase flows gas liquid or gas-solid, for example. Multiphase models and coupling methods must be combined with hybrid models as well [31].

- **Volume of Fluid (VOF) Method:**
  The volume of fluid (VOF) method is commonly used for tracking the interface which is the boundary between two or more immiscible fluids in a multiphase flow. This can be used in combination with hybrid LES-RANS frameworks that resolve the internal dynamics of the interface.

$$\frac{\partial \alpha}{\partial t} + u \cdot \nabla \alpha = 0$$

where α is the volume fraction of one of the phases

10.4 VALIDATION AND VERIFICATION

10.4.1 Benchmark Cases

10.4.1.1 Canonical Flows for Validation

In CFD, canonical flows, such as the backward-facing step, and periodic hill, are commonly used validation cases. These flows are selected because of their well-established features and the existence of abundant experimental and computational data.

I. **Backward-Facing Step:**
   This canonical flow is characterized by an abrupt channel enlargement and consequent flow separation and reattachment at the downstream side. The

characteristics of the flow include recirculation zones and complicated turbulence structures. This case is commonly used for turbulence models and CFD solver validation.

II. **Periodic Hill:**
Periodic hill is a standard example that is built to include the complexity of the flow features due to flow separation and reattachment in a series of hills. Flow over periodic hills is a challenging case study for CFD validation, wherein separation and reattachment zones are observed

104.1.2 Comparison with DNS and Experimental Data

Direct numerical simulation (DNS) is the ultimate fluid flow simulation method for resolving all scales of turbulence, which means that no turbulence model is needed. Such high-level validation can be achieved through comparisons between CFD predictions and DNS data. However, DNS is computationally expensive and generally restricted to simpler geometries and lower Reynolds numbers [32].

I. **DNS Data:**
In the literature, DNS data are available for canonical flows such as the backward-facing step and the periodic hill. These datasets include turbulence statistics, mean flow profiles, and various other flow characteristics.

II. **Experimental Data:**
Validation requires experimental data as well, especially for these canonical flows. Detailed velocity fields and turbulence statistics can be obtained through techniques such as particle image velocimetry (PIV) and laser Doppler anemometry (LDA). The agreement of CFD results with experimental data is used to evaluate the validity of numerical simulations [33].

- **Pressure Gradient Matching**
  In a study comparing DNS and experimental data for a transitional boundary layer, the pressure gradient was adjusted to match the DNS conditions. The pressure coefficient C_p is calculated as

$$C_p = 1 - \left(\frac{U(x)}{U_\infty} \right)^2$$

where $U(x)$ is the mean velocity at a given streamwise location x and U_∞ is the free-stream velocity.

- **Free-Stream Turbulence (FST) Matching**
  The turbulence intensity Tu and the streamwise integral length scale Λ_x are important parameters for matching FST conditions. The turbulence intensity is defined as

$$\mathrm{Tu} = \frac{\sqrt{u^2}}{U_\infty} = \frac{u_{rms}}{U_\infty}$$

where u_{rms} is the root-mean-square value of the streamwise velocity component.

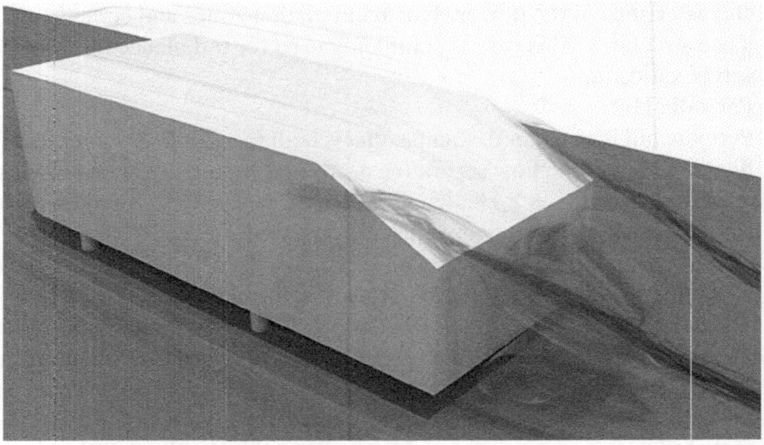

FIGURE 10.3 Ahmed body bench mark [50].

The integral length scale Λ_x is calculated using Taylor's hypothesis:

$$\Lambda_x = U_\infty \int_0^{\tau^*} R_{uu}(\tau)\, d\tau$$

where τ^* is the first zero crossing of the autocorrelation R_{uu} of the velocity signal.

10.4.2 Industrial Validation

10.4.2.1 Automotive (Aerodynamics)

I. **Ahmed Body Benchmark**
The Ahmed body is a standard benchmark case used in the validation of CFD simulations in automotive aerodynamics. It is a generic car model created to replicate the key characteristics of the flow around modern cars. This case is suitable for validation studies as we have high-quality experimental data available for this case [34].

- **Geometry and Setup**
The Ahmed body is placed in a wind tunnel with dimensions of $15 \times 1.4 \times 1.87$ m in the X, Y and Z directions, respectively. The slant angle is set to $25°$ and the Reynolds number is $\text{Re} = 2{,}784{,}000$.
- **Validation Parameters**
Key parameters for validation include the drag coefficient, velocity profiles along the body, and flow structures behind the body. Simulations should match the experimental drag coefficient $(C_d = 0.299)$ and lift coefficient $(C_l = 0.345)$ with reasonable accuracy.

10.4.2.2 Aerospace (Wing-Body Juncture Flows)

I. **Onera M6 Wing Benchmark**
The Onera M6 wing is a common case for the validation of CFD codes for aerospace applications. Especially the tendency for wing-body juncture flows and complicated aerodynamic interactions [35].
- **Geometry and Setup**:
The Onera M6 wing is a transonic airfoil shape with a complex geometry that includes a wing-body juncture. It is also a challenging case for CFD validation, as the flow conditions are typically high-Reynolds numbers and transonic Mach numbers.
- **Validation Parameters**:
Validation key parameters: pressure distributions, lift and drag coefficients, and flow separation patterns. According to the copious experimental data on wind tunnel tests, the results can be compared with the CFD results.
Wake Velocity Deficit:

For wind turbine wakes, the velocity deficit (ΔU) in the wake is calculated as:

$$\Delta U = U_\infty - U_w$$

where (U_w) is the velocity in the wake.

10.4.2.3 Energy (Wind Turbine Wakes)

I. **Wind Turbine Wake Benchmark**
Wind turbine wakes play a crucial role in the characteristics of wind turbines and their performance. CFD simulations should be validated against experimental data to make accurate predictions of the wake and turbine performance in the examined flow environment [36].
- **Geometry and Setup**:
The typical procedure includes placing the wind turbine model in a wind tunnel or an open-field test site. The flow conditions can be at steady or turbulent inflow, with the overall aim to reproduce the wake structure and velocity deficit of the turbine far downstream of the turbine.
- **Validation Parameters**:
The main parameters for validation are the wake velocity deficit, turbulence intensity and power from the turbine. Verification of the CFD results is performed through a comparison of the results obtained from the PIV measurements and the power output tests.

10.5 CUTTING-EDGE APPLICATIONS

10.5.1 Aerospace Engineering

10.5.1.1 Shock/Boundary-Layer Interaction in Supersonic Flows

Shock/boundary layer interaction (SBLI) is one of the significant phenomena in supersonic and hypersonic flows, which is manifested through the interaction of shock waves with the boundary layer, giving rise to complex flow structures and leading to significant changes in aerodynamic performance. The prediction of SBLI is critical to the design and optimization of aerospace vehicles, especially for their flight in the high-speed regime [37].

A. **Benchmark Cases and Validation**
   I. **ONERA M6 Wing Benchmark:**
      - **Geometry and Setup**:
        The ONERA M6 wing is an extensively used test case for transonic and supersonic flows. With a relatively modest sweep angle, the model is capable of operating at a range of Mach numbers and angles of attack making it a great candidate for shockwave and boundary layer interaction studies.
      - **Validation Parameters**:
        Parameters such as pressure coefficients (Cp), lift & drag coefficients, and flow separation patterns. Wind tunnel test experimental data provides a good basis for validating numerical results
      - **Turbulence Models:**
        The application of various turbulence models including Spalart-Allmaras (SA) and *k*-omega SST were compared for their ability to capture the complex characteristics of flow. The results showed that the SA model was more appropriate for lower angles of attack, however, the *k*-omega SST model predicted splitter plate flows more reliably.
   II. **Supersonic Flow Benchmark:**
      - **Geometry and Setup:**
        Although there are many CFD codes in use, supersonic flow over a flat plate or a wedge has long been a classical benchmark case to study the SBL. They feature a high Mach number and strong shockwave interaction with the boundary layer.
      - **Validation Parameters:**
        Parameters such as the shock location, pressure distribution, and boundary layer thickness are being studied. Validation is performed on experimental data of shock tube and wind tunnel tests.
      - **Turbulence Models:**
        Updated: Turbulence is a significant factor in considering turbulence calculations, traditional turbulence models are commonly used, such as the RANS equations with *k*-omega SST or LES to consider complex flow structures.

B. **Cutting-Edge Applications**
   I. **Exascale Computing:**
      - **SCALABLE Project**:
      - SCALABLE managed to harness the benefits of the best existing CFD tooling to provide state-of-the-art modeling technology for exascale computing. The technology was tested in benchmark problems with

convection and turbulence modeling, which is key for the design of airplane engines and wind turbines.

- **Prototype Software**:
  The specific test case applied to a few airplanes landing gear verified whether CFD prototype software is performant and precise. The academic and industrial partners will build on the successes and breakthroughs made during the SCALABLE project.

II. **Advanced Turbulence Models:**

- **LES and DDES**:
  High-fidelity simulations of SBLI are being performed using LES, and its variant, DDES with increasing frequency. These models give excellent local insight into flow field structures and work especially well for boundary-layer-type interactions at supersonic speeds.
- **Computational Efficiency**:
  Different turbulence models can have a major impact on computational efficiency. Both LES and DDES are more accurate than RANS but demand considerable computational resources. The Spalart-Allmaras model, in contrast, offers a good compromise in accuracy versus computational cost for some applications.

10.5.2 Automotive Design

10.5.2.1 Aerodynamic Optimization with Transient Separation Control

Aerodynamic optimization is crucial in automotive design to improve fuel efficiency, reduce drag, and enhance vehicle stability. Transient separation control involves manipulating the flow separation patterns around the vehicle to achieve these goals. This can be achieved through active flow control techniques, such as using synthetic jets or vortex generators [38].

A. Benchmark Cases and Validation

I. **Ahmed Body Benchmark:**

- **Geometry and Setup**:
  The Ahmed body is a generic car model used to study flow separation and drag reduction. It is placed in a wind tunnel with dimensions of 15 × 1.4 × 1.87 m in the *X*, *Y*, and *Z* directions, respectively. The slant angle is set to 25°, and the Reynolds number is Re=2,784,000.
- **Validation Parameters**:
  Key parameters for validation include the drag coefficient, velocity profiles along the body, and flow structures behind the body. Simulations should match the experimental drag coefficient (Cd = 0.299) and lift coefficient (Cl = 0.345) with reasonable accuracy.

II. **Active Flow Control:**

- **Techniques**:
  Apply synthetic jets, vortex generators, and plasma actuators to induce flow separation. By employing these techniques, it is possible to greatly lower drag and enhance aerodynamic performance.

- **Validation Parameters**:
  Some key parameters are how efficient the flow control devices are in the reduction of drag, the factor that improves the lift-to-drag ratio and the change of flow topology. Validation is performed using wind tunnel test and PIV measurements experimental data.

B. Cutting-Edge Applications

I. **Advanced CFD Simulations:**

- **High-Fidelity Models**:
  The use of advanced turbulence models (LES and DDES) provides accurate resolution of transient flow features. All of these models feature detailed information regarding the separation and reattachment patterns.
- **Computational Efficiency**:
  In order to computationally fast develop this code, a turbulence model was selected. LES and DDES, providing high accuracy, comes with a high computational cost. On the other hand, the Spalart-Allmaras model serves as a good trade-off between accuracy and computational efficiency in some cases.

II. **Optimization Algorithms:**

- **Genetic Algorithms**:
  Flow control device placement and operation optimization by genetic algorithms. Such algorithms iteratively modify the design variables until the optimal aerodynamic performance is desired.
- **Machine Learning**:
  Increasingly, machine learning methods are being developed to help predict flow patterns and optimize control strategies. By applying these techniques, the computational cost can be significantly reduced and the analysis accuracy can be improved for aerodynamic optimization.

10.5.3 Environmental and Energy Systems

10.5.3.1 Urban Airflow Prediction, Wind Farm Optimization, and Nuclear Reactor Thermal Hydraulics

Urban airflow prediction, wind farm optimization, and nuclear reactor thermal hydraulics are critical areas where CFD simulations play a significant role. Accurate predictions in these areas are essential for improving energy efficiency, reducing environmental impact, and ensuring safety [39].

A. Benchmark Cases and Validation

I. **Urban Airflow Prediction:**

- **Geometry and Setup**:
  Aerodynamics of buildings and urban structures: Predicting airflow in the urban environment. The setup simulates the urban area in which the game is played, with buildings and streets accurately modeled.

- **Validation Parameters**:
  The main validation parameters are wind speed and direction, turbulence intensity, and pollutant dispersion. The validation uses experimental data from wind tunnel tests and field measurements.

II. **Wind Farm Optimization:**

- **Geometry and Setup**:
  One of the use cases is wind farm optimization, where the flow around the wind turbines is simulated and placement is optimized to maximize energy production. A multi-rotor design, which is generally in a wind farm formation of several turbines.
- **Validation Parameters**:
  Turbine power output, wake velocity deficit, and turbulence intensity are integral validation parameters. Wind tunnel tests and field measurements validate the experimental data.

III. **Nuclear Reactor Thermal Hydraulics:**

- **Geometry and Setup**:
  Nuclear reactor thermal hydraulics involves simulating the flow and heat transfer within the reactor core [40]. The setup typically includes a detailed representation of the reactor core, with fuel rods and coolant channels modeled accurately.
- **Validation Parameters**:
  Key parameters for validation include coolant temperature, pressure drop, and heat transfer coefficients. Experimental data from reactor tests and computational models are used for validation.

B. Cutting-Edge Applications

I. **Advanced CFD Simulations:**

- **High-Fidelity Models**:
  They contain fewer assumptions than their previous versions to accurately capture the complex flow features, advanced turbulence models such as large eddy sloshing (LES) and DDES can be applied. The models allow a detailed understanding of flow structures and heat transfer phenomena.
- **Computational Efficiency**:
  Hence, the selection of the turbulence model is a very crucial aspect of determining computational efficiency. Although LES and DDES are highly accurate, they are very resource-intensive. For some applications, the Spalart-Allmaras model is a good compromise between accuracy and computational cost.

II. **Optimization Algorithms:**

- **Genetic Algorithms**:
  Related work consideration of wind turbine design optimization including genetic algorithms. These algorithms search through design parameter space and refine iteratively to achieve the highest energy output.

- **Machine Learning**:
  In recent times, machine learning-based techniques have made it easy to predict flow patterns and optimize control strategies. Such methods can help aerodynamic optimization become computationally cheap and highly accurate.

10.6 CHALLENGES AND CRITICAL PERSPECTIVES

10.6.1 Persistent Limitations

Simulating unsteady flows presents several persistent limitations that researchers and engineers must navigate. These challenges are rooted in the complex nature of fluid dynamics and the computational resources required to accurately model them [41].

10.6.1.1 Grid Sensitivity

A central issue in unsteady flow simulations is the grid sensitivity. The grid generated is also important because the numerical solutions heavily depend upon the quality and resolution of the grid. However, this can also cause a significant increase in computational overhead, as fine grids are usually needed to resolve the complex nature of the unsteady flows. High gradients or complex geometries in the flow field can make it sensitive to grid changes, and also contribute to grid sensitivity issues in such flows.

10.6.1.2 Transition Modeling

One of the major limitations in unsteady flow simulation is transition modeling. Correctly predicting the transition from laminar to turbulent flow is key to many applications, but still a difficult problem. This can lead to significant uncertainty in the predicted sediment transport rates and gross/net mass and energy exchange across the sediment-water interface. In this regard, this may cause errors in the simulation results, particularly where the transition process happens in areas containing complex flow interactions.

10.6.1.3 Computational Overhead

Simulating unsteady flows involves considerable computational overhead. The necessity to be able to describe the time-dependent characteristics of the flow involves small time steps and high spatial resolutions, which can result in a very high computational cost. For example, this overhead may become a serious obstacle to big simulation or real-time applications, which have a limited amount of computational resources.

10.6.2 Best Practices for Implementation

10.6.2.1 Guidelines for Mesh Design

Effective mesh design is crucial for accurate and efficient simulations of unsteady flows. The following guidelines can help optimize mesh quality and performance [42]:

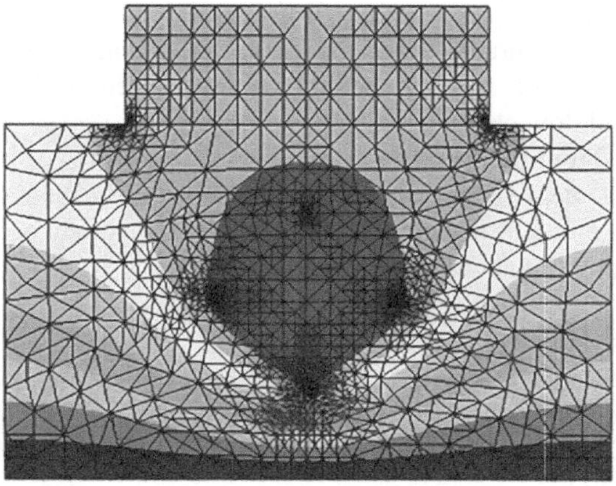

FIGURE 10.4 Adaptive mesh refinement representation [43].

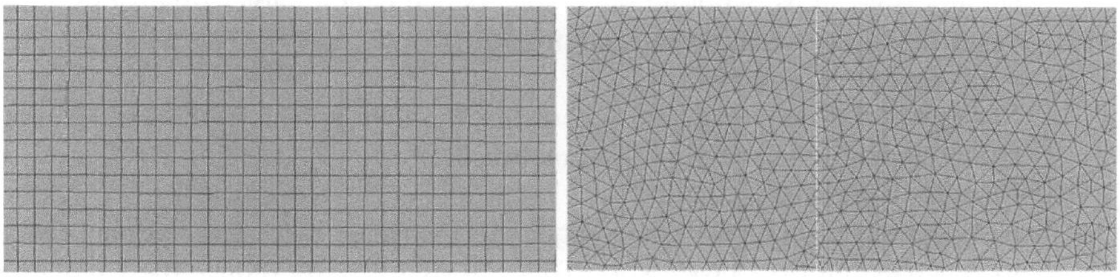

FIGURE 10.5 Detailed view of structured vs unstructured mesh.

I. **Adaptive Mesh Refinement (AMR):**
Use AMR techniques to dynamically adjust the mesh resolution based on the flow features. This allows for finer resolution in regions of interest while maintaining coarser resolution elsewhere, reducing computational cost.

II. **Structured vs. Unstructured Meshes:**
Choose between structured and unstructured meshes based on the specific requirements of the problem. Structured meshes are generally more efficient for regular geometries, while unstructured meshes offer greater flexibility for complex geometries.

III. **Boundary Layer Resolution:**
Ensure sufficient resolution in the boundary layer to capture the near-wall flow behavior accurately. This is particularly important for simulations involving turbulent flows.

10.6.2.2 Time-Step Selection

Selecting an appropriate time step is essential for balancing accuracy and computational efficiency. The following guidelines can help:

I. **CFL Condition:**
Adhere to the Courant–Friedrichs–Lewy (CFL) condition to ensure numerical stability. The CFL number should be chosen based on the specific numerical scheme and flow characteristics.

II. **Adaptive Time Stepping:**
Implement adaptive time-stepping algorithms to adjust the time step dynamically based on the flow behavior. This can help maintain accuracy while reducing computational cost.

10.6.2.3 Solver Settings

Optimizing solver settings is critical for achieving accurate and efficient simulations. Consider the following best practices [44]:

I. **Solver Type:**
Choose between explicit and implicit solvers based on the problem requirements. Explicit solvers are generally simpler and faster but may require smaller time steps, while implicit solvers can handle larger time steps at the cost of increased computational complexity.

II. **Convergence Criteria:**
Set appropriate convergence criteria for iterative solvers to ensure accurate solutions without excessive computational effort.

III. **Parallel Computing:**
Utilize parallel computing techniques to distribute the computational workload across multiple processors. This can significantly reduce the simulation time for large-scale problems.

10.7 FUTURE DIRECTIONS

10.7.1 AI-Driven Hybridization

10.7.1.1 Reinforcement Learning for Adaptive LES/RANS Domain Partitioning

This work proposes adaptive large time simulation (LTS) and Reynolds-Averaged Navier–Stokes domain partitioning via reinforcement learning (RL). RL algorithms can help in this regard and adaptively partition the domain within the simulation based on the flow characteristics and computational requirements. Such an adaptive strategy enables computational resources to be used more efficiently, applying LES of higher fidelity in regions where precision in eddy structures is important and reporting to RANS in flow structures where predictability is adequate. This hybrid approach can effectively minimize the computational burden while providing a balance of in accuracy across regions of interest [45].

10.7.1.2 Physics-Informed Neural Networks (PINNs) for Closure Modeling

PINNs are a novel approach for closure modeling in fluid dynamics. PINNs use physical laws, like the Navier–Stokes equations, as part of the neural network architecture

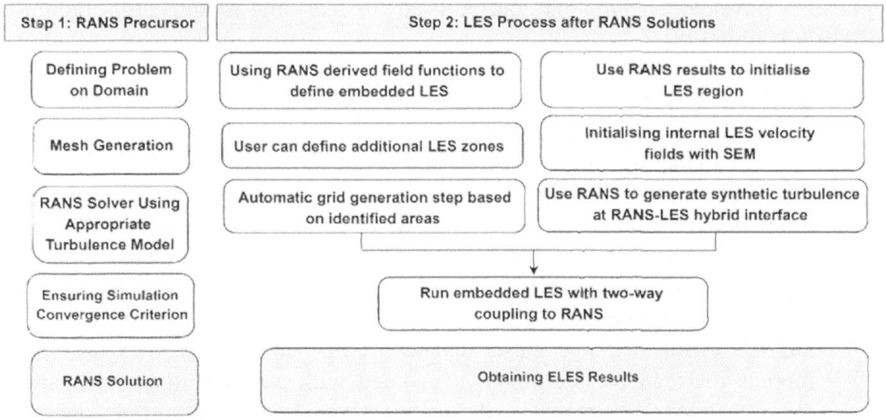

FIGURE 10.6 Hybridization of the LES and RANS. A flow mapping [46].

so that the solutions follow known physical rules. It has been used in a couple of works to simulate turbulent flows that can capture some of the relevant features, such as the formation of eddies, and energy dissipation, in complex systems. Given that turbulence modeling classical CFD relies on either simplifications of physics or empirical correlations, PINNs can address their shortcomings particularly well [47].

Future directions for PINNs in fluid dynamics include the following:

I. **Extension to Three-Dimensional Flows:**
Most of the recent applications of PINNs have been on two-dimensional flows. Extending these methods to three-dimensional flows is critical for real-world applications.

II. **Adaptive Collocation Point Sampling:**
Investigating adaptive sampling techniques to improve resolution in high-gradient regions can enhance the accuracy and efficiency of PINNs.

III. **Data-Driven Turbulence Modeling:**
Extending the use of data-driven techniques to inform turbulence models at extreme Reynolds numbers may further improve the robustness and fidelity of both PINN and its scientific application.

10.7.2 Exascale Computing and Digital Twins

10.7.2.1 Real-Time Hybrid CFD for Digital Twin Applications

The integration of exascale computing and digital twin technology is poised to revolutionize the field of CFD. Exascale computing, which refers to computing systems capable of at least one exaflop (10^{18} floating-point operations per second), provides the necessary computational power to perform highly detailed and real-time simulations. This capability is essential for the development of digital twins, which are virtual replicas of physical systems that can be used for real-time monitoring, prediction and optimization [48].

- **Real-Time Hybrid CFD** is a key component of this integration. By combining the strengths of different CFD methods, such as LES and RANS, hybrid CFD approaches can provide more accurate and efficient simulations. For instance, LES can be used in regions where high-fidelity turbulence modeling is required, while RANS can be employed in less critical areas to reduce computational cost. This adaptive approach ensures that the simulation remains computationally feasible while maintaining the necessary accuracy for real-time applications.
- **Digital Twins** leverage these advanced CFD simulations to create virtual models that can be continuously updated with real-time data. This allows for dynamic adjustments and optimizations based on current operating conditions. For example, in the coke making process, a digital twin can predict heat transfer and optimize energy consumption in real-time, leading to significant cost savings and improved efficiency.
- **AI-Driven Enhancements** further enhance the capabilities of hybrid CFD for digital twin applications. Machine learning algorithms can analyze vast amounts of data to identify patterns and correlations that inform design decisions, reducing the need for physical prototypes and lowering costs. Additionally, adaptive learning techniques allow AI systems to continuously refine their models, improving accuracy over time.
- **Hardware Advancements** are also crucial for supporting these real-time simulations. High-performance computing (HPC) systems with powerful CPUs, GPUs, and ample RAM are essential for handling the computational demands of exascale CFD. Cloud-based solutions offer scalability and cost efficiency, enabling access to large computational resources on demand.

10.7.3 Toward Universal Hybrid Models

10.7.3.1 Cross-Disciplinary Standardization and Open-Source Frameworks

The development of universal hybrid models in the field of CFD is a promising area under research, driven by the need for more efficient and accurate simulations. One key aspect of this development is the push for cross-disciplinary standardization and the use of open-source frameworks, such as OpenFOAM extensions [49].

- **Cross-disciplinary standardization** makes hybrid model integration across fields and usage possible. This requires the establishment of shared standards and protocols that enable smooth communication and data interchange among different simulation tools and platforms. As an example, as our AI-driven CFD models are integrated with traditional numerical methods, there must be a common approach for using standardized methods that ensure compatibility and interoperability.

- **Open-Source Frameworks.** The major part of this process is open source Open-Source frameworks like OpenFOAM. OpenFOAM is an open-source CFD solver package with various solvers for many applications. Such a capability makes OpenFOAM a perfect tool to couple with AI-based tools along with new simulation technologies.
- **OpenFOAMGPT** is a project showing how OpenFOAM can be augmented with AI capabilities, creating an interactive agent that guides users through simulation setup, execution, and analysis. This integration utilizes AI to enhance the speed and accuracy of CFD simulations.
- **Universal Hybrid Models** are the next evolution in this journey. The goal of these hybrid models to take advantage of the best aspects of different simulation methods in combination with AI-based approaches such as reinforcement learning and physics-informed neural networks (PINN). All of these steps would greatly enhance the accuracy and computational cost of the simulations, making them appropriate for expanding stream applications along with complex flow conditions.

10.8 CONCLUSION

They also provide cheap high-fidelity simulations of complex systems, which is going to be huge in the aerospace, automotive, and energy sectors. Grid and modeling challenges will need to be tackled collaboratively.

Nevertheless, the confluence of high-resolution and very efficient turbulence models has afforded novel ways of simulating fluid flows balancing accuracy with feasibility. Because the approach tackles such big swirling eddies but cheaply models smooth, near-wall areas, these hybrid techniques are a must for practical problems ranging from aircraft design to urban wind studies. This journey, from initial efforts in detached eddy simulation to sophisticated adaptive frameworks augmented by machine learning, demonstrates an unrelenting determination to deliver high-fidelity simulations to a user base that extends far beyond academic institutions.

Essentially, these hybrid models employ smart methods to seamlessly stitch together various approaches to simulation that smoothly traverse between systems with low discontinuities and high fidelity. But distant prospects for studying everything from unsteady aerodynamics in autos to turbulent airflow around wind turbines have cramped up as innovations like adaptive sensors and high-performance computing have accelerated their potential. It has been validated through extensive testing with actual, real-world experiments and super-high-definition simulations, and industries are flocking to it in increasingly large numbers to save money and speed innovation.

Yet hurdles remain. Particularly as industries trend to real-time digital simulations and AI-led design, fine-tuning grids, modeling erratic flow switches, and coping with heavy computational loads remain to be solved. However, the advent of artificial intelligence introduces interesting solutions in the form of adaptive algorithms that will contour the simulation regions automatically, and neural networks could be trained on the rules of physics to bridge the gap on the modeling side. But progress depends on collaboration: shared standards, open-source tools, and partnerships across fields will be crucial to adapting these tools for widespread use.

This is not a simple simulation game, solving the ordinary differential equations of a fluid-filled Turing machine, but rather an early version of what will become a Reality (R) game played with hyper-realistic simulations of digital twins. Hybrid turbulence models represent a hybrid zone where theory gets real. But their future is not just about sharper algorithms, it is about a cooperative notching-up of expectations. The way forward is simple: keep iterating, keep validating, and keep bridging disciplines. As a result, such models will continue to unravel the depths of fluid dynamics, enabling engineers and scientists to innovate faster and cleverer and one simulation at a time.

REFERENCES

[1] A. Battistini, A. Cammi, S. Lorenzi, M. Colombo, and M. Fairweather, "Development of a CFD – LES model for the dynamic analysis of the DYNASTY natural circulation loop," *Chem. Eng. Sci.*, vol. 237, p. 116520, Jun. 2021, doi: 10.1016/j.ces.2021.116520.

[2] "A hybrid RANS/LES simulation of turbulent channel flow," *ResearchGate*, Oct. 2024, doi: 10.1007/s00162-003-0089-x.

[3] "RechAerosp_1994_SpalartAllmaras.pdf." Accessed: Feb. 14, 2025. [Online]. Available: https://turbmodels.larc.nasa.gov/Papers/RechAerosp_1994_SpalartAllmaras.pdf

[4] J. Smagorinsky, "General circulation experiments with the primitive equations: I. the basic experiment*," *Mon. Weather Rev.*, vol. 91, no. 3, pp. 99–164, Mar. 1963, doi: 10.1175/1520-0493(1963)091<0099:GCEWTP>2.3.CO;2.

[5] "A dynamic hybrid RANS/LES modeling methodology for turbulent/transitional flow field prediction – ProQuest." Accessed: Feb. 14, 2025. [Online]. Available: www.proquest.com/openview/5ba938ee363167832138e37fc81106d5/1?pq-origsite=gscholar&cbl=18750

[6] "(PDF) Comments on the feasibility of LES for wings, and on a hybrid RANS/LES approach," in *ResearchGate*, Accessed: Feb. 14, 2025. [Online]. Available: www.researchgate.net/publication/236888805_Comments_on_the_Feasibility_of_LES_for_Wings_and_on_a_Hybrid_RANSLES_Approach

[7] "A scale-adaptive simulation model for turbulent flow predictions | Request PDF," in *ResearchGate*, doi: 10.2514/6.2003-767.

[8] B. Chaouat, "The state of the art of hybrid RANS/LES modeling for the simulation of turbulent flows," *Flow Turbul. Combust.*, vol. 99, no. 2, pp. 279–327, Sep. 2017, doi: 10.1007/s10494-017-9828-8.

[9] "Highly resolved experimental results of the separated flow in a channel with streamwise periodic constrictions," *J. Fluid Mech.*, Accessed: Feb. 14, 2025. [Online]. Available: www.cambridge.org/core/journals/journal-of-fluid-mechanics/article/abs/highly-resolved-experimental-results-of-the-separated-flow-in-a-channel-with-streamwise-periodic-constrictions/28DB7EE6898D25824EF8AAC343A8FF63

[10] "(PDF) Partially-averaged navier stokes model for turbulence: Implementation and validation," in *ResearchGate*. Accessed: Feb. 14, 2025. [Online]. Available: www.resea

rchgate.net/publication/24336418_Partially-Averaged_Navier_Stokes_Model_for_Turbulence_Implementation_and_Validation

[11] "Hybrid RANS / LES modeling in CFD: An introduction." Accessed: Feb. 18, 2025. [Online]. Available: www.linkedin.com/pulse/hybrid-rans-les-modeling-cfd-introduction-mohamed-aly-sayed

[12] L. Temmerman, M. Hadžiabdić, M. A. Leschziner, and K. Hanjalić, "A hybrid two-layer URANS–LES approach for large eddy simulation at high Reynolds numbers," *Int. J. Heat Fluid Flow*, vol. 26, no. 2, pp. 173–190, 2005.

[13] M. Terracol. "ELES. – Google 搜索" *AIAA J.*, 2001. Accessed: Feb. 14, 2025. [Online]. Available: www.google.com.hk/search?q=12.%09Terracol%2C+M.+%282001%29.+AIAA+J.+%E2%80%93+ELES.&newwindow=1&sca_esv=568a1eef6a07acf0&ei=jhiwZ9ucDIGqur8PwYmIuA0&ved=0ahUKEwjbm7-V7MSLAxUB1e4BHcEEAtcQ4dUDCBA&uact=5&oq=12.%09Terracol%2C+M.+%282001%29.+AIAA+J.+%E2%80%93+ELES.&gs_lp=Egxnd3Mtd2l6LXNlcnAiKjEyLglUZXJyYWNvbCwgTS4gKDIwMDEpLiBBSUFBIEouIOKAkyBFTEVTLjIFEAAY7wVI5ApQ5QZY5QZwAXgAkAEAmAG8A6ABvAOqAQM0LTG4AQPIAQD4AQH4AQKYAgGgAr8DqAIAmAMC8QXT-SPIGNzCKJIHAzQtMaAHqAI&sclient=gws-wiz-serp

[14] "(PDF) hybrid LES — review and assessment." Accessed: Feb. 14, 2025. [Online]. Available: www.researchgate.net/publication/225330234_Hybrid_LES_-_Review_and_assessment

[15] "International journal of numerical methods for heat & fluid flow | Emerald Publishing." Accessed: Feb. 14, 2025. [Online]. Available: www.emeraldgrouppublishing.com/journal/hff

[16] "The spectrum of turbulence" in *Proceedings of the Royal Society of London. Series A – Mathematical and Physical Sciences*. Accessed: Feb. 14, 2025. [Online]. Available: https://royalsocietypublishing.org/doi/10.1098/rspa.1938.0032

[17] P. R. Spalart, S. Deck, M. L. Shur, K. D. Squires, M. Kh. Strelets, and A. Travin, "A new version of Detached-eddy simulation, resistant to ambiguous grid densities," *Theor. Comput. Fluid Dyn.*, vol. 20, no. 3, pp. 181–195, Jul. 2006, doi: 10.1007/s00162-006-0015-0.

[18] S.-H. Peng, "Hybrid Rans-les modelling: Some recent development and aerodynamic applications".

[19] "(PDF) Ten years of industrial experience with the SST turbulence model." Accessed: Feb. 14, 2025. [Online]. Available: www.researchgate.net/publication/228742295_Ten_years_of_industrial_experience_with_the_SST_turbulence_model

[20] "(PDF) A review of physics-informed machine learning in fluid mechanics," in *ResearchGate*, Dec. 2024, doi: 10.3390/en16052343.

[21] M. Strelets, "Detached eddy simulation of massively separated flows," in *39th Aerospace Sciences Meeting and Exhibit*, Reno, NV, U.S.A.: American Institute of Aeronautics and Astronautics, Jan. 2001. doi: 10.2514/6.2001-879.

[22] S. Lakshmipathy and S. S. Girimaji, "Partially averaged Navier–Stokes (PANS) method for turbulence simulations: Flow past a circular cylinder," 2010, Accessed: Feb. 15, 2025. [Online]. Available: https://asmedigitalcollection.asme.org/fluidsengineering/article-abstract/132/12/121202/451184

[23] K. Duraisamy, "Perspectives on machine learning-augmented Reynolds-averaged and large eddy simulation models of turbulence," *Phys. Rev. Fluids*, vol. 6, no. 5, p. 050504, May 2021, doi: 10.1103/PhysRevFluids.6.050504.

[24] L. Hao, L. Wei, and W. Shengye, "Numerical investigation of fixed and non-fixed separation with shear layer adapted delayed detached Eddy simulation," *Int. J. Aeronaut. Space Sci.*, vol. 22, no. 3, pp. 531–546, Jun. 2021, doi: 10.1007/s42405-020-00342-y.

[25] J.-B. Chapelier, C. Scalo, and B. Wasistho, "Development of adaptive subgrid scale models based on a large-scale vorticity sensor," in *8th AIAA Theoretical Fluid Mechanics Conference*, Denver, Colorado: American Institute of Aeronautics and Astronautics, Jun. 2017. doi: 10.2514/6.2017-4168.

[26] J. Larsson. "Strain-rate criteria. – Google 搜索," *AIAA J.*, 2015. Accessed: Feb. 15, 2025. [Online]. Available: www.google.com.hk/search?q=19.%09Larsson%2C+J.+%282015%29.+AIAA+J.+%E2%80%93+Strain-rate+criteria.&newwindow=1&sca_esv=7104c08f085ecc62&ei=xnOxZ6fYCZjQkPIPi-DtkA8&ved=0ahUKEwin08mmt8eLAxUYKEQIHQtwG_IQ4dUDCBA&uact=5&oq=19.%09Larsson%2C+J.+%282015%29.+AIAA+J.+%E2%80%93+Strain-rate+criteria.&gs_lp=Egxnd3Mtd2l6LXNlcnAiOTE5LglMYXJzc29uLCBKLiAoMjAxNSkuIEFJQUEgSi4g4oCTIFN0cmFpbi1yYXRlIGNyaXRlcmlhLki1ClCDBliDBnABeACQAQCYAZgCoAGYAqoBAzItMbgBA8gBAPgBAfgBApgCAKACAKgCAJgDAfEFtbWsv9wH_f6SBwCgB3c&client=gws-wiz-serp

[27] D. A. Jacobsen. "GPU solvers. – Google 搜索," *Comput. Phys. Commun.*, 2015. Accessed: Feb. 15, 2025. [Online]. Available: www.google.com.hk/search?q=21.%09Jacobsen%2C+D.A.+%282015%29.+Comput.+Phys.+Commun.+%E2%80%93+GPU+solvers.&newwindow=1&sca_esv=7104c08f085ecc62&ei=AHSxZ8nPIqj8kPIP6uWQ2A0&ved=0ahUKEwjJz7bCt8eLAxUoPkQIHeoyBNsQ4dUDCBA&uact=5&oq=21.%09Jacobsen%2C+D.A.+%282015%29.+Comput.+Phys.+Commun.+%E2%80%93+GPU+solvers.&gs_lp=Egxnd3Mtd2l6LXNlcnAiQTIxLglKYWNvYnNlbiwgRC5BLiAoMjAxNSkuIENvbXB1dC4gUGh5cy4gQ29tbXVuLiDigJMgR1BVIHNvbHZlcnMuSLMKUPsFWPsFcAF4AJABAJgB9wKgAfcCqgEDMy0xuAEDyAEA-AEB-AECmAIAoAIAqAIAmAMB8QW5UJyFczfL8ZIHAKAHfw&client=gws-wiz-serp

[28] C. W. Rowley. "ROMs for CFD. – Google 搜索," *J. Fluid Mech* ., 2004. Accessed: Feb. 15, 2025. [Online]. Available: www.google.com.hk/search?q=1.%09Rowley%2C+C.W.+%282004%29.+J.+Fluid+Mech.+%E2%80%93+ROMs+for+CFD.&newwindow=1&sca_esv=7104c08f085ecc62&ei=DHixZ-m7Crmjur8Ps6Xb8Q0&ved=0ahUKEwjp8Z6wu8eLAxW5ke4BHbPSNt4Q4dUDCBA&uact=5&oq=1.%09Rowley%2C+C.W.+%282004%29.+J.+Fluid+Mech.+%E2%80%93+ROMs+for+CFD.&gs_lp=Egxnd3Mtd2l6LXNlcnAiODEuCVJvd2xleSwgQy5XLiAoMjAwNCkuIEouIEZsdWlkIE1lY2guIOKAkyBST01zIGZvciBDRkQuSNwKUMYGWMYGcAF4AJABAJgBAKABAKoBALgBA8gBAPgBAfgBApgCAKACAKgCAJgDAfEFsGRkUuSR-4SSBwCgBwA&client=gws-wiz-serp

[29] W. P. Jones. "Hybrid... – Google 学术搜索," *Combust. Flame*, 2012. Accessed: Feb. 15, 2025. [Online]. Available: https://scholar.google.com/scholar?q=23.+Jones,+W.P.+(2012).+Combust.+Flame+%E2%80%93+Hybrid+combustion.&hl=zh-CN&as_sdt=0&as_vis=1&oi=scholart

[30] D. J. Bodony. "Acoustics coupling. – Google 搜索," *J. Comput. Phys.*, 2005. Accessed: Feb. 15, 2025. [Online]. Available: www.google.com.hk/search?q=24.%09Bodony%2C+D.J.+%282005%29.+J.+Comput.+Phys.+%E2%80%93+Acoustics+coupling.&newwindow=1&sca_esv=1b49f95c0843824b&ei=_XyxZ4HpA7HIkPIPm4Cu8Ak&ved=0ahUKEwjB27GLwMeLAxUxJEQIHRuAC54Q4dUDCBA&uact=5&oq=24.%09Bodony%2C+D.J.+%282005%29.+J.+Comput.+Phys.+%E2%80%93+Acoustics+coupling.&gs_lp=Egxnd3Mtd2l6LXNlcnAiQTI0LglCb2RvbnksIEQuSi4gKDIwMDUpLiBKLiBDb21wdXQuIFBoeXMuIOKAkyBBY291c3RpY3MgY291cGxpbmcuSKQKUP4FWP4FcAF4AJABAJgB5wGgAecBqgEDMi0xuAEDyAEA-AEB-AECmAIAoAIAqAIAmAMA8QXkrKr2vEcy65IHAKAHfw&client=gws-wiz-serp

[31] T. Höhne. "Reactor hydraulics. – Google 搜索," *Nucl. Eng. Des.*, 2018. Accessed: Feb. 15, 2025. [Online]. Available: https://www.google.com.hk/search?q=25.%09H%C3%B6hne%2C+T.+%282018%29.+Nucl.+Eng.+Des.+%E2%80%93+Reactor+hydraulics.&newwindow=1&sca_esv=1b49f95c0843824b&ei=_XyxZ4HpA7HIkPIPm4Cu8Ak&ved=0ahUKEwjB27GLwMeLAxUxJEQIHRuAC54Q4dUDCBA&uact=5&oq=25.%09H%C3%B6hne%2C+T.+%282018%29.+Nucl.+Eng.+Des.+%E2%80%93+Reactor+hydraulics.&gs_lp=Egxnd3Mtd2l6LXNlcnAiPjI1LglIw7ZobmUsIFQuICgyMDE4KS4gTnVjbC4gRW5nLiBEZXMuIOKAkyBSZWFjdG9yIGh5ZHJhdWxpY3MuMgUQABjvBTIIEAAYgAQYogQyBRAAGO8FMgUQABjvBTIFEAAY7wVIrT9QjztYjztwAngAkAEAmAGnAqABpwKqAQMyLTG4AQPIAQD4AQH4AQKYAgGgAqsCqAIAmAMB8QUHEnT2cknJxZIHAzItMaAH7gI&sclient=gws-wiz-serp

[32] C. Wenzel, B. Selent, M. Kloker, and U. Rist, "DNS of compressible turbulent boundary layers and assessment of data/scaling-law quality," *J. Fluid Mech.*, vol. 842, pp. 428–468, 2018.

[33] L. Hakim *et al.*, "Large eddy simulations of multiple transcritical coaxial flames submitted to a high-frequency transverse acoustic modulation," *Proc. Combust. Inst.*, vol. 35, no. 2, pp. 1461–1468, 2015.

[34] A. Rao, G. Minelli, B. Basara, and S. Krajnović, "On the two flow states in the wake of a hatchback Ahmed body," *J. Wind Eng. Ind. Aerodyn.*, vol. 173, pp. 262–278, Feb. 2018, doi: 10.1016/j.jweia.2017.10.021.

[35] P. Panagiotou, S. Fotiadis-Karras, and K. Yakinthos, "Conceptual design of a blended wing body male UAV," *Aerosp. Sci. Technol.*, vol. 73, pp. 32–47, Feb. 2018, doi: 10.1016/j.ast.2017.11.032.

[36] "Analytical modeling of wind farms: A new approach for power prediction." Accessed: Feb. 16, 2025. [Online]. Available: www.mdpi.com/1996-1073/9/9/741

[37] "Numerical simulations of shock/boundary layer interactions using time-dependent modeling techniques: A survey of recent results – ScienceDirect." Accessed: Feb. 16, 2025. [Online]. Available: / www.sciencedirect.com/science/article/pii/S0376042108000535

[38] "Complete_body_aerodynamic_study_of_three_vehicles-2017(1)."

[39] Y. Tominaga and B. Blocken, "Wind tunnel experiments on cross-ventilation flow of a generic building with contaminant dispersion in unsheltered and sheltered conditions," *Build. Environ.*, vol. 92, pp. 452–461, Oct. 2015, doi: 10.1016/j.buildenv.2015.05.026.

[40] N. Mérigoux, "Multiphase Eulerian-Eulerian CFD supporting the nuclear safety demonstration," *Nucl. Eng. Des.*, vol. 397, p. 111914, Oct. 2022, doi: 10.1016/j.nucengdes.2022.111914.

[41] I. Celik, M. Klein, and J. Janicka, "Assessment measures for engineering LES applications," 2009, Accessed: Feb. 17, 2025. [Online]. Available: https://asmedigitalcollection.asme.org/fluidsengineering/article-abstract/131/3/031102/467161

[42] L. Turton *et al.*, "Guidelines for best practice in the audiological management of adults with severe and profound hearing loss," *Semin. Hear.*, vol. 41, no. 3, pp. 141–246, Aug. 2020, doi: 10.1055/s-0040-1714744.

[43] "Adaptive mesh refinement — NGS-Py 6.2.2406-142-ge678c9e40 documentation." Accessed: Feb. 17, 2025. [Online]. Available: https://docu.ngsolve.org/nightly/whetting_the_appetite/adaptive.html

[44] "(PDF) A review of guidelines and best practices for subsonic aerodynamic simulations using RANS CFD." Accessed: Feb. 17, 2025. [Online]. Available: www.researchgate.net/publication/337756101_A_review_of_guidelines_and_best_practices_for_subsonic_aerodynamic_simulations_using_RANS_CFD

[45] C. Vignon, J. Rabault, J. Vasanth, F. Alcántara-Ávila, M. Mortensen, and R. Vinuesa, "Effective control of two-dimensional Rayleigh–Bénard convection: Invariant multi-agent reinforcement learning is all you need," *Phys. Fluids*, vol. 35, no. 6, 2023. Accessed: Feb. 17, 2025. [Online]. Available: https://pubs.aip.org/aip/pof/article/35/6/065146/2900730

[46] J. Holgate, A. Skillen, T. Craft, and A. Revell, "A Review of Embedded Large Eddy Simulation for Internal Flows," *Arch. Comput. Methods Eng.*, vol. 26, no. 4, pp. 865–882, Sep. 2019, doi: 10.1007/s11831-018-9272-5.

[47] M. Raissi, P. Perdikaris, and G. E. Karniadakis, "Physics-informed neural networks: A deep learning framework for solving forward and inverse problems involving nonlinear partial differential equations," *J. Comput. Phys.*, vol. 378, pp. 686–707, 2019.

[48] "digital-twin-institute-position-paper-(december-2020).pdf." Accessed: Feb. 17, 2025. [Online]. Available: www.aiaa.org/docs/default-source/uploadedfiles/issues-and-advocacy/policy-papers/digital-twin-institute-position-paper-(december-2020).pdf

[49] L. Orgogozo, T. Xavier, H. Oulbani, and C. Grenier, "Permafrost modelling with OpenFOAM®: New advancements of the permaFoam solver," *Comput. Phys. Commun.*, vol. 282, p. 108541, 2023.

[50] "Flow around the Ahmed body – CFD simulation." Available: https://help.sim-flow.com/validation/ahmed-body

11 Laplace Homotopy Method for Fuzzy Fractional KdV Equations

Bellagubbala Bala Sai Sankar and Perumandla Karunakar

11.1 INTRODUCTION

Tsunami waves are one of the catastrophic waves in nature. These waves are different from regular ocean waves, which are caused by tides and winds. Tsunami waves are caused generally due to natural phenomena, such as volcanic eruptions, large earthquakes, and submarine landslides. These phenomena lead to the development of long wavelengths, waves having high speed and energy. Tsunami waves are more damaging than tidal waves, due to their long wavelength [1]. In shallow waters, the speed of tsunami waves is slow compared to deep water, indicating that the speed of the tsunami waves depends on the fluid depth.

Nowadays, the waves mechanism may be imagined either through mathematical formulation or by using existing differential equations for modeling. In shallow waters, the tsunami waves may be analyzed using geophysical Korteweg de-Vries equation (gKdVE). The KdV equation is a nonlinear wave equation that explains the propagation of solitary waves in shallow water and dispersive medium. It was first derived in 1895 by Dutch mathematicians Diederik Korteweg and Gustav de Vries [2], the equation governs the balance between dispersion and nonlinearity in wave motion.

Forms of the KdV equations are listed below [3].

KdV equation	$z_t + azz_x + bz_{xxx} = 0,$
Modified KdV equation	$z_t + az^2z_x + bz_{xxx} = 0,$
Generalized KdV equation	$z_t + az^nz_x + bz_{xxx} = 0,\ n \geq 3.$

The KdV equation and its modified form play a vital role in understanding the solitary wave theory [4]. Numerous researches have contributed to the KdV equation and its modified form [5, 6].

In geophysical fluid dynamics, the KdV equation finds applications in modeling wave dynamics in atmosphere and oceans. In oceans, the KdV equation captures the behavior of tsunamis, internal waves in stratified fluids, and shallow water waves. The

DOI: 10.1201/9781003611820-14

atmospheric applications of this equation encompass modeling gravity waves and Rossby waves, which play crucial roles in climate and weather patterns.

The geophysical KdV equation proposed by Geyer and Quirchmayr [7, 8] reads

$$u_t - \omega_0 u_x + 1.5uu_x + \frac{1}{6}u_{xxx} = 0 \tag{11.1}$$

subject to

$$u(x,0) = 2(c+\omega_0)\operatorname{sech}^2\left(\sqrt{1.5(c+\omega_0)}\ x\right). \tag{11.2}$$

The exact solution [7] of Eq. (11.1) is

$$u(x,t) = 2(c+\omega_0)\operatorname{sech}^2\left(\sqrt{1.5(c+\omega_0)}\ (x-ct)\right). \tag{11.3}$$

Recognizing that natural systems often exhibit behaviors not adequately captured by classical integer-order differential equations. To study all such characteristics, researchers have extended the KdV equation to incorporate fractional derivatives [8, 9].

The time fractional geophysical KdV equation (tfgKdVE) of fractional order α is

$$D_t^{\alpha}\eta - \omega_0\eta_x + 1.5\eta\eta_x + \frac{1}{6}\eta_{xxx} = 0, 0 < \alpha \leq 1 \tag{11.4}$$

here $\eta(x,t)$ represents the free surface elevation, ω_0 is the parameter pertaining to the Coriolis effect and α denotes the order of fractional derivative. The term $\omega_0\eta_x$ is mandatory to notice the effect of Earth's spinning on the wave propagation. The parameter ω_0 relies on the depth of the water [7]. Across the ocean, this depth differs from one place to the other. The value of ω_0 consists of measuring errors while estimating the depth of the basin. This leads to uncertainty in analyzing the dynamics of the gKdVE. This uncertainty may be handled properly by incorporating a fuzzy parameter. Many authors used fuzzy parameters in place of crisp parameters to examine the wave dynamics. Verma *et al.* [10] solved the fuzzy fractional KdV equation using q-homotopy analysis Sumudu transform method (q-HASTM). Parthkumar *et al.* solved the fuzzy fractional differential equations system representing the gravity wave model [11] using q-HASTM. Saima *et al.* [12] studied nonlinear shallow water models with generalized Homotopy perturbation transform method. To get a greater idea on handling uncertain parameters, the interested readers are encouraged to refer to [13–15]. In this work, we consider ω_0 as triangular fuzzy number (TFN) as it is uncertain in nature.

So, the fuzzy tfgKdVE in Caputo sense is of the form

$$D_t^{\alpha}\tilde{\eta} - \tilde{\omega}_0\tilde{\eta}_x + 1.5\widetilde{\eta\eta}_x + \frac{1}{6}\tilde{\eta}_{xxx} = 0, \quad 0 < \alpha \leq 1, \tag{11.5}$$

Subject to

$$\tilde{\eta}\left(x,0\right)=2\left(c+\tilde{\omega}_0\right)\operatorname{sech}^2\left(\sqrt{1.5\left(c+\tilde{\omega}_0\right)}\ x\right).$$

Here, fuzzy free surface elevation is denoted by $\tilde{\eta}$, the fuzzy Coriolis parameter by $\tilde{\omega}_0$, and the wave speed by c. Handling the nonlinear fuzzy differential equations is quite tricky and many researchers used different methods to solve them. Behera *et al.* [16] used the Sine-Cosine method to obtain solitary wave solutions of gKdVE. Hosseini *et al.* [3] studied gKdVE with the source term using the Painleve test. Aljahdaly [17] studied the solutions of gKdVE with a damping term. Karunakar and Chakraverty [18] studied the effect of the Coriolis constant on gKdVE using Homotopy perturbation method. Jhangeer *et al.* [19] applied Lie symmetry to study the solution of gKdV. Sayed *et al.* [20] obtained the solutions of generalized gKdVE under the time-space conformable operator. Sahoo and Chakraverty [21] examined the approximate solution of the gKdV equation in fuzzy environment using Adomian decomposition method (ADM) and ADM-Pade approximation. Naowarat *et al.* [22] computed the solution of generalized gKdVE using the tanh-coth method. Hirota bilinear technique is used to generalize the perturbed KdV equation by Arshad *et al.* [23]. Alharbi *et al.* [24] obtained the solitary wave solution to gKDVE by improved expansion technique. Rizvi *et al.* [25] studied lump and interaction solutions of gKDVE through the Hirota bilinear method. Turgut *et al.* [26] studied the bifurcation analysis of gKdVE using the finite element method. In [27] tfgKdVE was studied using the hybrid methods. In the existing literature, it may be seen that gKdVE is studied by treating the Coriolis parameter, $\tilde{\omega}_0$ as the crisp parameter. Less work has been performed by treating $\tilde{\omega}_0$ as the uncertain parameter. This work incorporates HPM with Laplace transform to study the fractional gKdVE in fuzzy uncertain nature by treating $\tilde{\omega}_0$ as TFN to understand the dynamic behavior much better.

The structure of the article is as follows. Section 11.2 provides the fundamental definitions and concepts related to the fractional derivatives and fuzzy arithmetic. Section 11.3 presents the methodology utilized to analyze the proposed model. Section 11.4 presents the Laplace Homotopy perturbation method (LHPM) solution of tfgKdVE. Section 11.5 gives the findings and the analysis. Section 11.6 gives the conclusion of our work.

11.2 PRELIMINARIES

The basic concepts related to the methodology are discussed in this section.

A. Riemann–Liouville (RL) Fractional Integral [28]

For the function $q\left(t\right)\in C_\mu$, and $\mu\geq -1$, the RL integral of arbitrary order $\alpha>0$ is given as

$$J^\alpha\, q\left(t\right)=\frac{1}{\Gamma\left(\alpha\right)}\int_0^t\left(t-x\right)^{\alpha-1}q\left(y\right)dy \tag{11.6}$$

B. Caputo Fractional Derivative (CFD) [28]

In Caputo sense, the fractional derivative of $q(t)$, is given as

$$D_t^{\alpha} q(t) = \frac{1}{\Gamma(k-\alpha)} \int_0^t (t-x)^{k-\alpha-1} q(x)\, dx \tag{11.7}$$

for $k-1 \langle\, \alpha \leq k,\ k \in N,\ t \rangle 0$.

C. Laplace Transform (LT) [29]

The LT of a function $q(t)$ is given by

$$L\left[q(t)\right] = \int_0^{\infty} e^{-st} q(t)\, dt = Q(s), \tag{11.8}$$

where s is a parameter.

The LT of fractional derivative of a function $q(t)$ in Caputo sense [27] is defined as

$$L\left[q(t)\right] = s^{\alpha} L\left[q(t)\right] - \sum_{n=0}^{k-1} s^{\alpha-1-n} q^{(n)}(0+) \tag{11.9}$$

for $k-1 < \alpha \leq k,\ k \in N$.

D. Interval Operations [30, 31]

An interval is a set of real numbers $\mathbb{R}$ in a way such that $\left[S_L, S_U\right] = \{x/S_L \leq x \leq S_U,\ \ S_L, S_U \in R\}$.

If $A = \left[A_1, A_2\right]$ and $B = \left[B_1, B_2\right]$ are two intervals where A_1, A_2 and B_1, B_2 are lower and upper bounds of the intervals A and B, respectively, then addition and subtraction operations are as follows:

$$A + B = \left[A_1, A_2\right] + \left[B_1, B_2\right] = \left[A_1 + B_1,\ A_2 + B_2\right],$$

$$A - B = \left[A_1, A_2\right] - \left[B_1, B_2\right] = \left[A_1 - B_2,\ B_1 - A_2\right].$$

E. Triangular Fuzzy Number (TFN) [32]

A triplet denoted by $\tilde{A} = \left[A_L, A_M, A_U\right]$ is known as TFN, having a membership function as below.

$$\mu_{\tilde{A}} = \begin{cases} 0 & \text{if } x < A_L, \\ \dfrac{x - A_L}{A_M - A_L} & \text{if } x \in [A_L, A_M), \\ \dfrac{A_U - x}{A_U - A_M} & \text{if} x \in [A_M, A_U), \\ 0 & \text{if} x \geq A_U. \end{cases}$$

F. Parametric Representation [33]

The r-cut rule is that used for converting a fuzzy number in parametric form. TFN $\mu_{\tilde{A}}$ may be expressed as

$$\tilde{A} = [A_L + (A_M - A_L) r, A_U - (A_U - A_M) r], \quad r \in [0, 1].$$

An interval can be written as a crisp parameter as below.

$$\tilde{A} = [\underline{A}, \bar{A}] = \underline{A} + b[\bar{A} - \underline{A}], \text{ where } b \in [0, 1].$$

11.3 METHODOLOGY OF LHPM

This section extends the LHPM [34] for solving the considered fuzzy fractional equation.

Consider a nonlinear fuzzy time-fractional partial differential equation of the form

$$D_t^{\alpha} \tilde{\eta} + M\tilde{\eta} + N\tilde{\eta} = p \tag{11.10}$$

Subject to

$$\tilde{\eta}(x,0) = h(x),$$

here, $\tilde{\eta}$ is the function of x and t, D_t^{α} is the CFD, N and M are nonlinear and linear operators, respectively, and the inhomogeneous term is denoted by p.

Employing LT to either side of the Eq. (11.10), we get

$$L[D_t^{\alpha} \tilde{\eta}] + L[\widetilde{M\eta}] + L[\widetilde{N\eta}] = L[p] \tag{11.11}$$

Proceeding the LT, we have

$$L[\tilde{\eta}] = s^{-1} h(x) + s^{-\alpha} L[p] - s^{-\alpha} L[M\tilde{\eta} + \widetilde{N\eta}] \tag{11.12}$$

Applying inverse LT on both sides of Eq. (11.12) leads to

$$\tilde{\eta} = h(x) + L^{-1}[s^{-\alpha} L[p - M\tilde{\eta} - N\tilde{\eta}]] \tag{11.13}$$

Next, suppose solution as series that holds an embedding parameter $p \in [0, 1]$ as

$$\tilde{\eta} = \sum_{k=0}^{\infty} p^k \tilde{\eta}_k \tag{11.14}$$

The nonlinear terms are handled by using He's polynomials as

$$N\tilde{\eta} = \sum_{k=0}^{\infty} p^k \tilde{H}_k \tag{11.15}$$

Here $H_k(x)$ represents He's polynomials which are given below [18, 35]

$$H_k(\tilde{\eta}_0, \tilde{\eta}_1, \ldots \tilde{\eta}_k) = \frac{1}{k!}\left[\frac{\partial^k}{\partial p^k} N\left(\sum_{i=0}^{\infty} p^i \tilde{\eta}_i\right)\right]. \tag{11.16}$$

Now by placing Eq. (11.14) and Eq. (11.15) in Eq. (11.13) and incorporating LT with HPM, we get

$$\sum_{k=0}^{\infty} p^k \tilde{\eta}_k = h(x) + p\left(L^{-1}\left[s^{-\alpha} L\left[p - M\sum_{k=0}^{\infty} p^k \tilde{\eta}_k - \sum_{k=0}^{\infty} p^k H_k(\tilde{\eta})\right]\right]\right) \tag{11.17}$$

On comparing the same power coefficients on the both sides of "p" of Eq. (11.17) results

$$\begin{aligned}
p^0 &: \tilde{\eta}_0 = h(x), \\
p^1 &: \tilde{\eta}_1 = L^{-1}[s^{-\alpha} L[p - M\tilde{\eta}_0 - H_0(\tilde{\eta})]], \\
p^2 &: \tilde{\eta}_2 = L^{-1}[s^{-\alpha} L[p - M\tilde{\eta}_1 - H_1(\tilde{\eta})]], \\
&\vdots \\
p^k &: \tilde{\eta}_k = L^{-1}[s^{-\alpha} L[p - M\tilde{\eta}_k - H_k(\tilde{\eta})]], \\
&\vdots
\end{aligned}$$

One can get the series solution of Eq. (11.10) as

$$\tilde{\eta} = \lim_{N \to \infty} \sum_{n=0}^{N} \tilde{\eta}_n$$

11.4 SOLUTION OF FUZZY TFGKDVE

In this segment, we employ the methodology discussed in the previous section to get the fuzzy solution of the tfgKdVE.

Applying LT to Eq. (11.5) leads to

$$L\left[D_t^{\alpha}\tilde{\eta}\right]=L\left[\tilde{\omega}_0\tilde{\eta}_x\right]-L\left[1.5\tilde{\eta}\tilde{\eta}_x\right]-L\left[\frac{1}{6}\tilde{\eta}_{xxx}\right]=0 \tag{11.18}$$

Proceeding further we get

$$\tilde{\eta}\left(x,t\right)=\tilde{\eta}\left(x,0\right)+L^{-1}\left[\frac{1}{s^{\alpha}}L\left[\tilde{\omega}_0\tilde{\eta}_x-1.5\tilde{\eta}\tilde{\eta}_x-\frac{1}{6}\tilde{\eta}_{xxx}\right]\right] \tag{11.19}$$

Now proceeding with LHPM given in section 11.3, we get

$$\sum_{k=0}^{\infty}p^k\tilde{\eta}_k=\tilde{\eta}_0+p\Bigg(L^{-1}\bigg[s^{-\alpha}L\bigg[\tilde{\omega}_0\Big[\sum_{k=0}^{\infty}p^k\tilde{\eta}_k\Big]_x -\frac{1}{6}\Big[\sum_{k=0}^{\infty}p^k\tilde{\eta}_k\Big]-1.5\Big[\sum_{k=0}^{\infty}p^kH_k(x,t)\Big]\bigg]\bigg]\Bigg) \tag{11.20}$$

here H_k represents He's polynomial.

Some of He's polynomials for $\tilde{\eta}\tilde{\eta}_x$ are

$$H_0\left(\tilde{\eta}\right)=\tilde{\eta}_0\tilde{\eta}_{0x},$$

$$H_1\left(\tilde{\eta}\right)=\tilde{\eta}_0\tilde{\eta}_{1x}+\tilde{\eta}_1\tilde{\eta}_{0x},$$

$$H_2\left(\tilde{\eta}\right)=\tilde{\eta}_0\tilde{\eta}_{2x}+\tilde{\eta}_1\tilde{\eta}_{1x}+\tilde{\eta}_2\tilde{\eta}_{0x},$$

$$\vdots$$

Now by correlating the coefficients of alike powers of "p" on the both sides of the Eq. (11.20), we get

$$p^0:\ \tilde{\eta}_0=\tilde{\eta}\left(x,0\right),$$

$$p^1:\ \tilde{\eta}_1=L^{-1}\left[\frac{1}{s^{\alpha}}L\left[\tilde{\omega}_0\tilde{\eta}_{0x}-1.5\tilde{\eta}_0\tilde{\eta}_{0x}-\frac{1}{6}\tilde{\eta}_{0xxx}\right]\right],$$

$$p^2:\ \tilde{\eta}_2=L^{-1}\left[\frac{1}{s^{\alpha}}L\left[\tilde{\omega}_0\tilde{\eta}_{1x}-1.5(\tilde{\eta}_0\tilde{\eta}_{1x}+\tilde{\eta}_0\tilde{\eta}_{1x})-\frac{1}{6}\tilde{\eta}_{1xxx}\right]\right],$$

$$p^3:\ \tilde{\eta}_3=L^{-1}\left[\frac{1}{s^{\alpha}}L\left[\tilde{\omega}_0\tilde{\eta}_{2x}-1.5(\tilde{\eta}_0\tilde{\eta}_{2x}+\tilde{\eta}_1\tilde{\eta}_{1x}+\tilde{\eta}_2\tilde{\eta}_{0x})-\frac{1}{6}\tilde{\eta}_{2xxx}\right]\right],$$

$$\vdots$$

On computing and simplifying, we get the successive approximations as

$$\tilde{\eta}_1 = \frac{2\sqrt{6}ct^{\alpha}\sinh\left(\frac{\sqrt{6}x\sqrt{c+\tilde{\omega}_0}}{2}\right)(c+\tilde{\omega}_0)^{3/2}}{\cosh\left(\frac{\sqrt{6}x\sqrt{c+\tilde{\omega}_0}}{2}\right)^3\Gamma(\alpha+1)},$$

$$\tilde{\eta}_2 = \frac{6c^2\alpha t^{\alpha+2}(c+\tilde{\omega}_0)^2\left(2\cosh\left(\frac{\sqrt{6}x\sqrt{c+\tilde{\omega}_0}}{2}\right)^2-3\right)}{\cosh\left(\frac{\sqrt{6}x\sqrt{c+\tilde{\omega}_0}}{2}\right)^4\Gamma(\alpha+1)\Gamma(2\alpha+1)},$$

$$\vdots$$

The series solution of the tfgKdVE is

$$\tilde{\eta} = \tilde{\eta}_0+\tilde{\eta}_1+\tilde{\eta}_2+\ldots \tag{11.21}$$

11.5 RESULT ANALYSIS AND DISCUSSIONS

This section presents numerical solution of fuzzy tfgKdVE obtained using LHPM. In this study, the uncertain Coriolis parameter ω_0 has been taken as a fuzzy number, i.e., $\omega_0 = [0.6,\ 1,\ 1.4]$.

Table 11.1, provides a comparison of the fuzzy solution of tfgKdVE at $\alpha = 1$ using LHPM with the exact solution for $r = 0$ and $b = 0.5$. Table 11.2, shows the term wise fuzzy solution of tfgKdVE at $\alpha = 1$ for $r = 0$, $b = 0.5$. It is noted that as the number of terms increases, the fuzzy LHPM solution is converging to the fuzzy exact solution, demonstrating the efficiency of the method.

TABLE 11.1
Comparison of Fuzzy Solution of tfgKdVE Using LHPM with its Exact Solution in Integer Order Case for $t = 1$, $c = 0.01$ at $r = 0$ and $b = 0.5$.

x	LHPM	Exact	Absolute Error
0.1	1.995413527911981	1.995412962907464	1.0e-05*0.056500451628416
0.2	1.913430915210523	1.913429890882983	1.0e-05*0.102432753945614
0.3	1.783011554499525	1.783010297420468	1.0e-05*0.125707905707095
0.4	1.617704492645299	1.617703238947418	1.0e-05*0.125369788084306
0.5	1.432571136488859	1.432571136488859	1.0e-05*0.107082945666193
0.6	1.241665196313246	1.241664402335935	1.0e-05*0.079397731145292
0.7	1.056332469156842	1.056331967277013	1.0e-05*0.050187982858496
0.8	0.884520531785863	0.884520284599264	1.0e-05*0.024718659974443
0.9	0.730875043353829	0.730874989443702	1.0e-05*0.005391012714018
1.0	0.597318774703674	0.597318850136853	1.0e-05*0.007543317936953

TABLE 11.2

Term Wise Fuzzy Solution of tfgKdVE in Integer Order Case for $t = 1$, $c = 0.01$ at $r = 0$ and $b = 0.5$

x	1st term	2nd term	3rd term	4th term	Exact
0.1	1.989703457548856	1.995701996952712	1.995414120134323	1.995413527911981	1.995412962907464
0.2	1.902368823768384	1.913669815762838	1.913431956909565	1.913430915210523	1.913429890882983
0.3	1.767803592873315	1.783180326361386	1.783012816702852	1.783011554499525	1.783010297420468
0.4	1.599835244061606	1.617796872059344	1.617705740756466	1.617704492645299	1.617703238947418
0.5	1.413528055227335	1.432594530637092	1.432573265786183	1.432571136488859	1.432571136488859
0.6	1.222721802749776	1.241631876716788	1.241665975486450	1.241665196313246	1.241664402335935
0.7	1.038441227525017	1.056260940464917	1.056332957115678	1.056332469156842	1.056331967277013
0.8	0.868287580264999	0.884427309384697	0.884520767774014	0.884520531785863	0.884520284599264
0.9	0.716603243990787	0.730773500603964	0.730875089339764	0.730875043353829	0.730874989443702
1.0	0.585079258640797	0.597218436829672	0.597318694353781	0.597318774703674	0.597318850136853

Figure 11.1, shows the fuzzy free surface elevation at $\alpha = 1$. It is noted that the center solution of the fuzzy LHPM matches with exact solution. Figure 11.2 shows the fuzzy free surface elevation at $\alpha = 0.7$.

Figure 11.3, displays the TFN solution for different α. It is observed that the arbitrary order has a crucial impact in the dynamics of the uncertain geophysical waves modeled by the KdV equation. It is noted that as α increases both solution bounds increase. Also, it may be seen that as the fractional order α increases from 0.1 to 0.7, the distance between the fuzzy solution bounds widens. This suggests that fractional order α is proportional to the width of the interval solution.

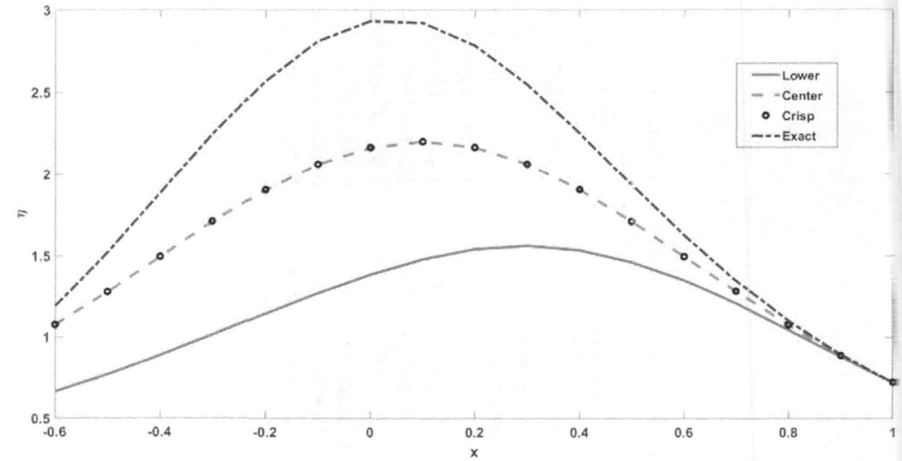

FIGURE 11.1 Fuzzy free surface elevation at $\alpha = 1$.

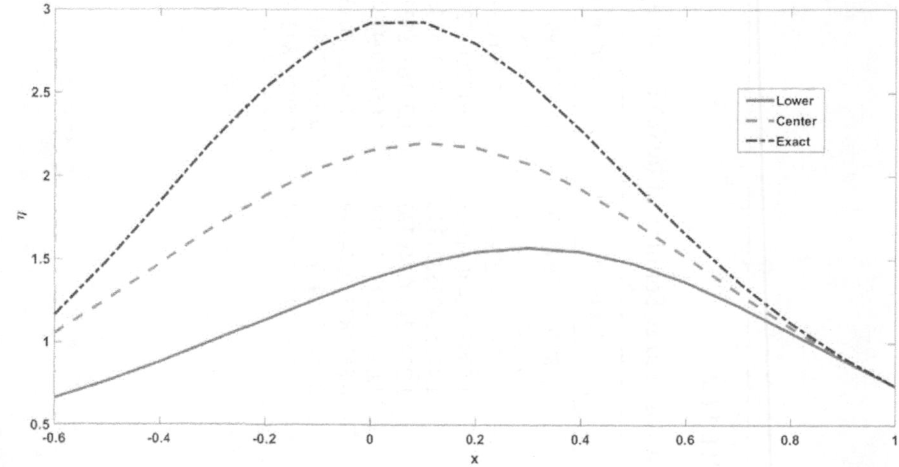

FIGURE 11.2 Fuzzy free surface elevation at $\alpha = 0.7$.

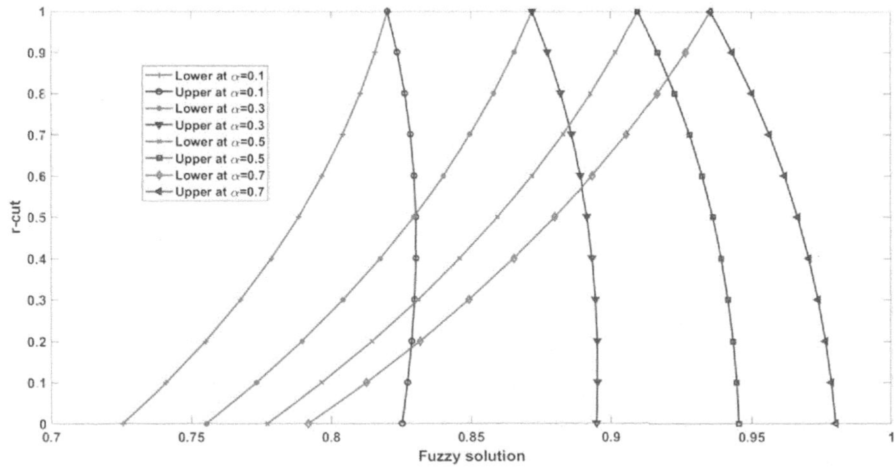

FIGURE 11.3 TFN plots for different fractional orders.

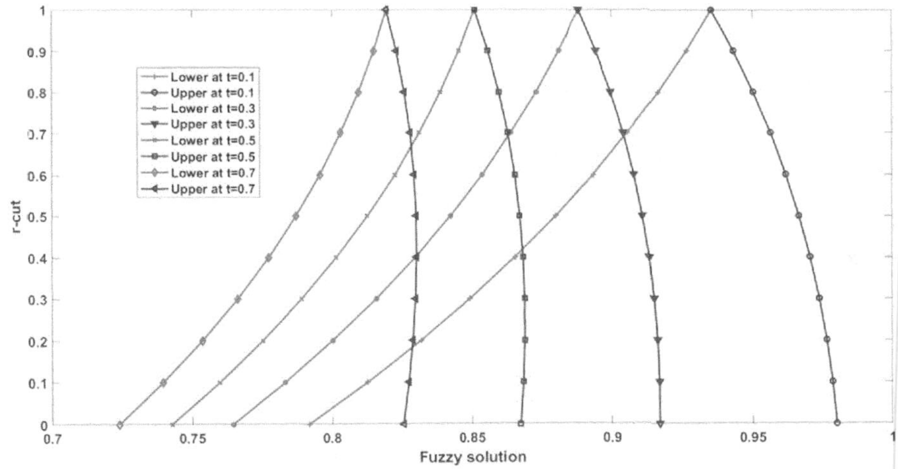

FIGURE 11.4 TFN plots for different t at $\alpha = 0.7$.

Figure 11.4 displays the TFN solution for different values of t at $\alpha = 0.7$. Both solution bounds increase as time t decreases from 0.7 to 0.1. It may be noted that the gap between solution interval decreases as time t increases.

Figure 11.5 shows the TFN solution for different spatial values x at $\alpha = 0.7$. It is noted that as x increases from 0.1 to 0.7, both solution bounds decrease. The width of the solution interval increases as x decreases.

From Figures 11.4 and 11.5, the inverse relation between the width of the solution interval and spatial and temporal coordinates x and t is observed.

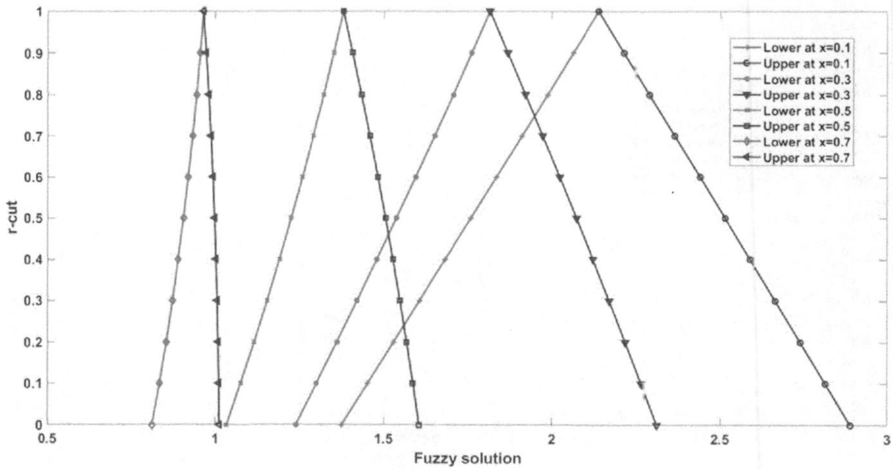

FIGURE 11.5 TFN plots for different x at $\alpha = 0.7$.

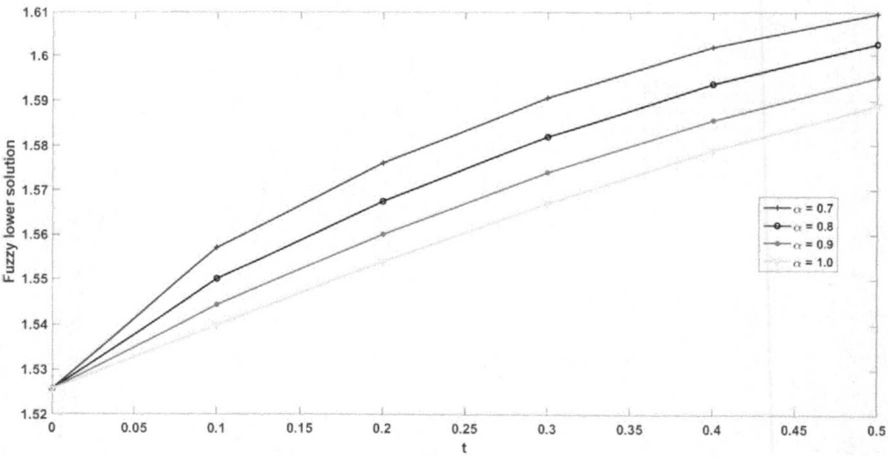

FIGURE 11.6 Fuzzy lower solution against t.

Figure 11.6 presents the fuzzy lower solution against the t-coordinate for varying α. The fuzzy upper solution against the t-coordinate for different α is presented in Figure 11.7. From Figures 11.6 and 11.7, we may see the decreasing behavior of the fuzzy solution at α increasing from 0.7 to 1. Figure 11.8 displays the fuzzy lower solution against the x-coordinate for varying α. Figure 11.9 shows the fuzzy upper solution against the x-coordinate for different α. As similar to Figures 11.8 and 11.9, the decreasing behavior of the fuzzy solution is observed as α varies from 0.7 to 1.

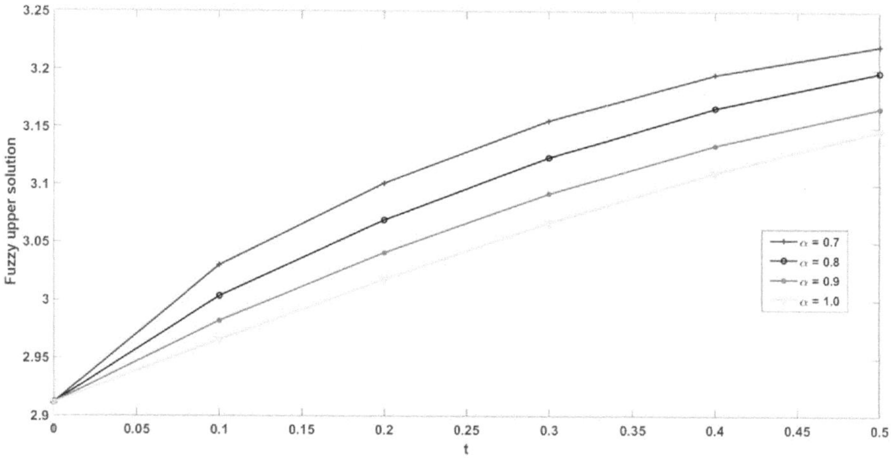

FIGURE 11.7 Fuzzy upper solution against *t*.

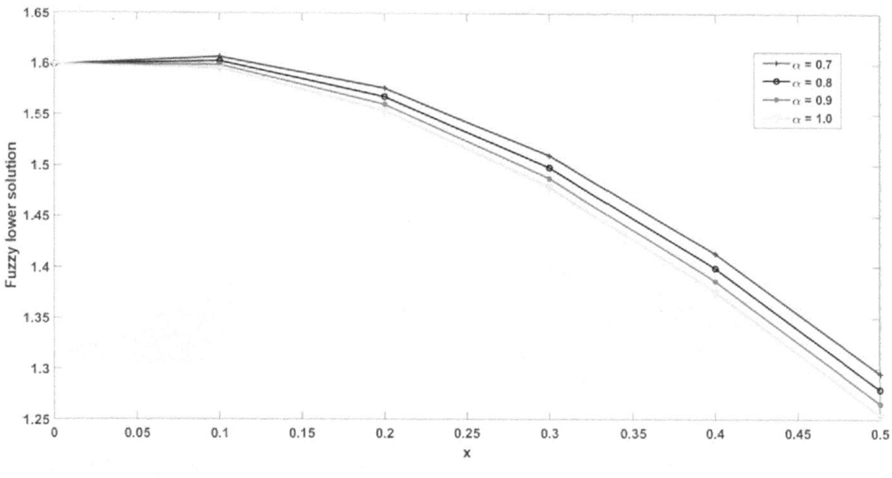

FIGURE 11.8 Fuzzy lower solution against *x*.

Figure 11.10 shows the 3D surface plot of the lower, center, and upper solutions for $\alpha = 1$. Figure 11.11 depicts the surface plot of the lower, center, and upper solutions for $\alpha = 0.5$. It is noted that as the Coriolis parameter ω_0 increases from 0.6 to 1.4, the fuzzy solution becomes sharper, particularly for small values of c. Figure 11.12, illustrates the 3D fuzzy solution at t for $\alpha = 0.1$ and $\alpha =$ 0.3. Figure 11.13, represents the 3D fuzzy solution at $t = 1$ for $\alpha = 0.5$ and $\alpha = 0.7$. Additionally, Figure 11.14 shows the 3D fuzzy solution at $x = 0.1$ for fractional orders $\alpha = 0.1$ and $\alpha = 0.3$, while

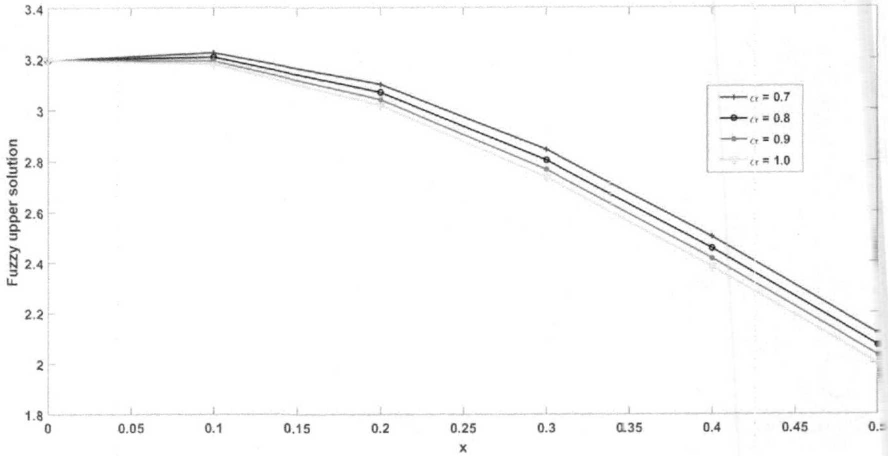

FIGURE 11.9 Fuzzy upper solution against x.

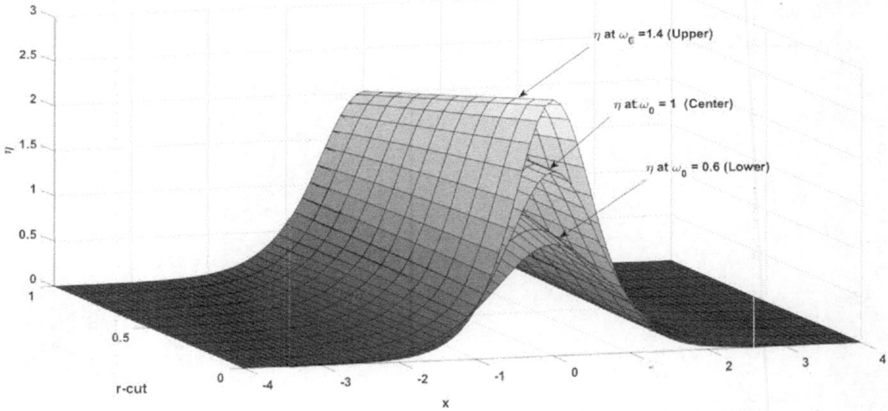

FIGURE 11.10 Surface plots for lower, center, and upper solution for $\omega_0 = [0.6, 1, 1.4]$ at $\alpha = 1$.

Figure 11.15 demonstrates the 3D fuzzy solution at $x = 0.1$ for $\alpha = 0.5$ and $\alpha = 0.7$. From Figures 11.12 and 11.13, it is observed that lower and upper fuzzy bounds interchanges due to different values of x. The analysis of 2D and 3D plots reveals important findings about the behavior of the lower and upper bound solutions as they relate to fractional order. This indicates that the uncertain Coriolis parameter significantly influences the shape of the fuzzy solution.

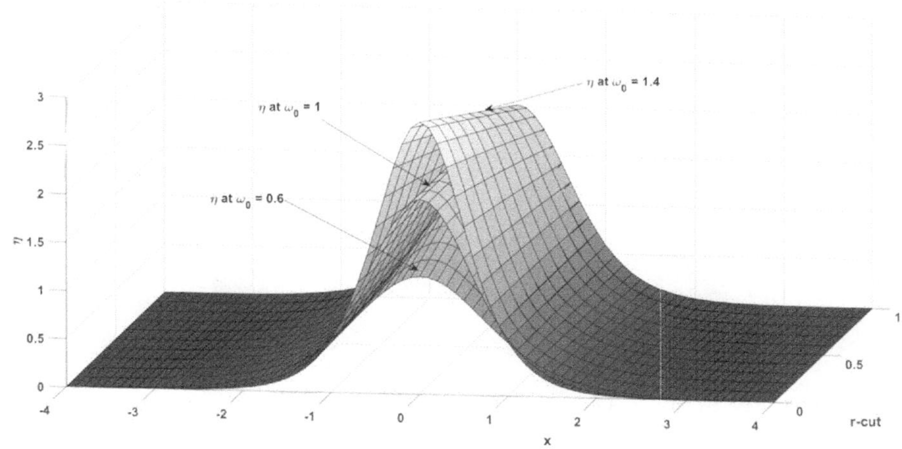

FIGURE 11.11 Surface plots for lower, center, and upper solution for $\omega_0 = [0.6, 1, 1.4]$ at $\alpha = 0.5$.

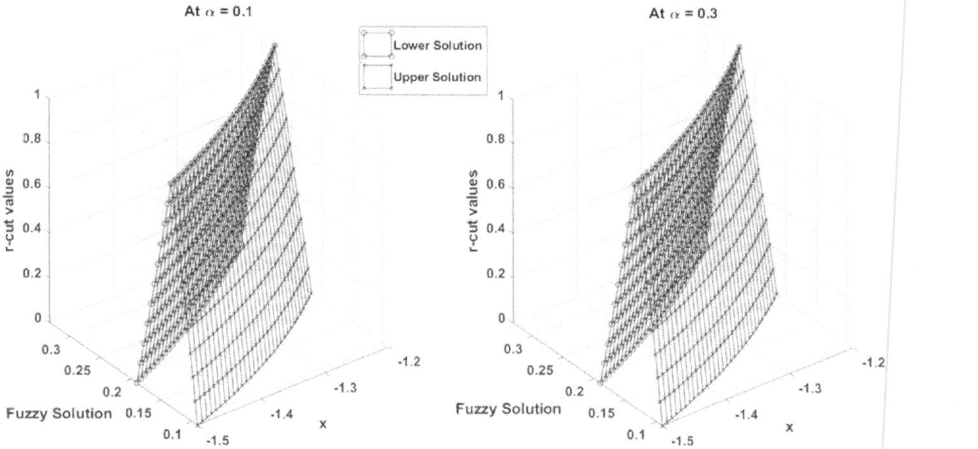

FIGURE 11.12 3D fuzzy solution for $t = 1$ at $\alpha = 0.1$ and $\alpha = 0.3$ (red-lower, blue-upper).

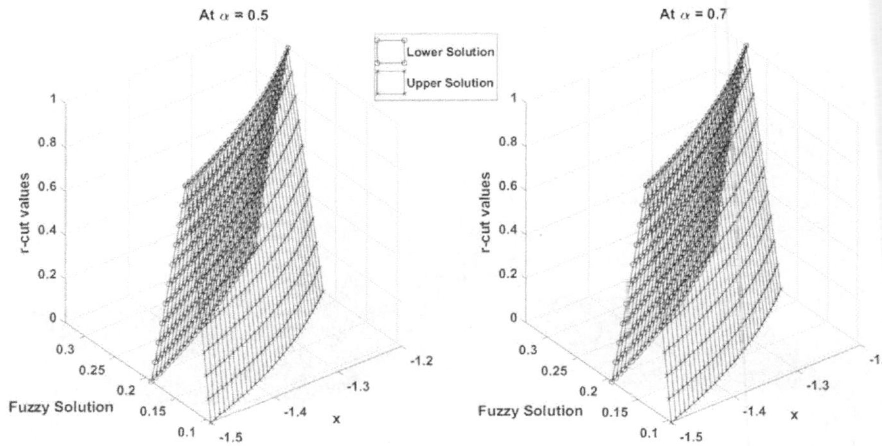

FIGURE 11.13 3D fuzzy solution for $t = 1$ at $\alpha = 0.5$ and $\alpha = 0.7$ (red-lower, blue-upper).

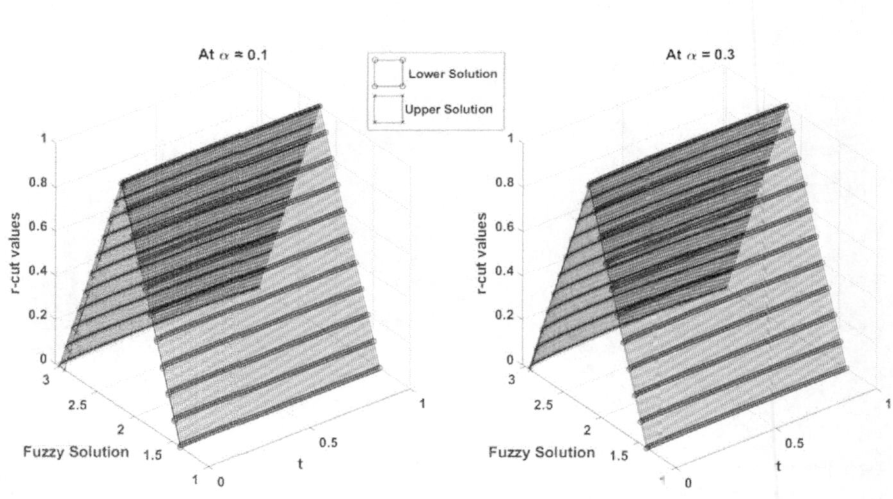

FIGURE 11.14 3D fuzzy solution for $x = 0.1$ at $\alpha = 0.1$ and $\alpha = 0.3$ (red-lower, blue-upper).

11.6 CONCLUSIONS

LHPM is applied successfully to study the tfgKdVE in fuzzy environment. All the simulations and calculations are done using MATLAB®. Our analysis reveals significant insights into the dynamics of geophysical waves as influenced by fractional orders and the fuzzy parameter. The analysis of the fuzzy tfgKdVE reveals that increasing the fractional order leads to an increase in the width of the solution interval. Also, at a particular fractional order ($\alpha = 0.7$), the width of the solution interval decreases with an increase in the spatial and temporal coordinates. The influence of uncertain

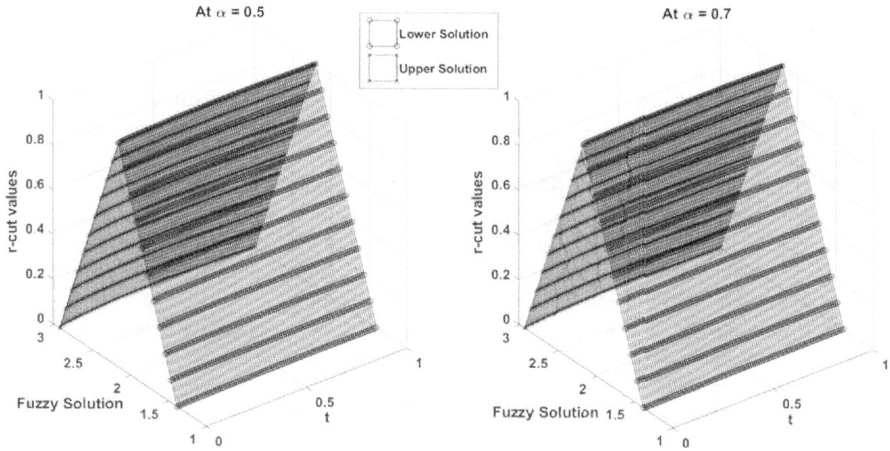

FIGURE 11.15 3D fuzzy solution for $x = 0.1$ at $\alpha = 0.5$ and $\alpha = 0.7$ (red-lower, blue-upper).

Coriolis parameter on the shape of the fuzzy solution is studied. It is also seen how the fuzzy bound solutions are affected by the different values of the parameters. This works highlights the significance of encompassing uncertainty in modeling geophysical waves to account for variability and complexity, offering valuable insights on wave behavior in fractional order. Further, the present work may be continued by treating the speed of the wave c, which depends on the fluid depth [17], with a TFN to model the dynamics much better.

REFERENCES

[1] T. Y. Wu, Long waves in ocean and coastal waters, *Journal of the Engineering Mechanics Division* 107 (3) (1981) 501–522.

[2] D. J. Korteweg, G. De Vries, Xli. on the change of form of long waves advancing in a rectangular canal, and on a new type of long stationary waves, *The London, Edinburgh, and Dublin Philosophical Magazine and Journal of Science* 39 (240) (1895) 422–443.

[3] K. Hosseini, D. Baleanu, E. Hincal, S. Manukure, S. Salahshour, B. Kaymakam zade, Painlevé analysis and kink-type solitary waves of the geophysical KdV equation involving a source, *International Journal of Applied and Computational Mathematics* 10 (2) (2024) 74.

[4] A.-M. Wazwaz, Two wave mode higher-order modified KdV equations: Essential conditions for multiple soliton solutions to exist, *International Journal of Numerical Methods for Heat & Fluid Flow* 27 (10) (2017) 2223–2230.

[5] P. Karunakar, S. Chakraverty, Differential quadrature method for solving fifth-order KdV equations, in: *Recent Trends in Wave Mechanics and Vibrations: Select Proceedings of WMVC 2018*, Springer, 2019, pp. 361–369.

[6] M. Iqbal, D. Lu, A. R. Seadawy, Z. Zhang, Nonlinear behaviour of dust acoustic periodic soliton structures of nonlinear damped modified Korteweg–de Vries equation in dusty plasma, *Results in Physics* 59 (2024) 107533.

[7] A. Geyer, R. Quirchmayr, Shallow water equations for equatorial tsunami waves, *Philosophical Transactions of the Royal Society A: Mathematical, Physical and Engineering Sciences* 376 (2111) (2018) 20170100.

[8] P. Karunakar, S. Chakraverty, Solutions of time-fractional third-and fifth-order Korteweg–de-Vries equations using homotopy perturbation transform method, *Engineering Computations* 36 (7) (2019) 2309–2326.

[9] A. H. Ganie, F. Mofarreh, A. Khan, On new computations of the time-fractional nonlinear KdV-burgers equation with exponential memory, *Physica Scripta* 9 (2024) 045217.

[10] L. Verma, R. Meher, Z. Avazzadeh, O. Nikan, Solution for generalized fuzzy fractional Korteweg-de Varies equation using a robust fuzzy double parametric approach, *Journal of Ocean Engineering and Science* 8 (6) (2023) 602–622.

[11] P. P. Sartanpara, R. Meher, A robust fuzzy-fractional approach for the atmospheric internal wave model, *Journal of Ocean Engineering and Science* 8 (3) (2023) 308–322.

[12] S. Rashid, R. Ashraf, Z. Hammouch, New generalized fuzzy transform computations for solving fractional partial differential equations arising in oceanography, *Journal of Ocean Engineering and Science* 8 (1) (2023) 55–78.

[13] P. Karunakar, S. Chakraverty, Solving shallow water equations with crisp and uncertain initial conditions, *International Journal of Numerical Methods for Heat & Fluid Flow* 28 (12) (2018) 2801–2815.

[14] P. Karunakar, S. Chakraverty, Homotopy perturbation method for predicting tsunami wave propagation with crisp and uncertain parameters, *International Journal of Numerical Methods for Heat & Fluid Flow* 31 (1) (2021) 92–105.

[15] R. Vana, K. Perumandla, Computational approach and convergence analysis for interval-based solution of the Benjamin–Bona-Mahony equation with imprecise parameters, *Engineering Computations* 41 (4) (2024) 1067–1085.

[16] S. Behera, N. H. Aljahdaly, Soliton solutions of nonlinear geophysical KdV equation via two analytical methods, *International Journal of Theoretical Physics* 63 (5) (2024) 107.

[17] N. H. Aljahdaly, Study tsunamis through approximate solution of damped geophysical Korteweg-de Vries equation, *AIMS Mathematics* 8 (5) (2024) 10926–10934.

[18] P. Karunakar, S. Chakraverty, Effect of Coriolis constant on geophysical Korteweg-de Vries equation, *Journal of Ocean Engineering and Science* 4 (2) (2019) 113–121.

[19] A. Jhangeer, T. Jamal, M. Z. Hussain, M. Imran, A lie symmetry approach to travelling wave solutions, bifurcation, chaos and sensitivity analysis of the geophysical Korteweg–de Vries equation, *Partial Differential Equations in Applied Mathematics* 10 (2024) 100734.

[20] S. Saifullah, N. Fatima, S. A. Abdelmohsen, M. M. Alanazi, S. Ahmad, D. Baleanu, Analysis of a conformable generalized geophysical KdV equation with Coriolis effect, *Alexandria Engineering Journal* 73 (2023) 651–663.

[21] M. Sahoo, S. Chakraverty, Influence of uncertain Coriolis parameter on wave solution of Korteweg-de Vries equation, *GEM-International Journal on Geomathematics* 15 (1) (2024) 10.

[22] S. Naowarat, S. Saifullah, S. Ahmad, M. De la Sen, Periodic, singular and dark solitons of a generalized geophysical KdV equation by using the tanh-coth method, *Symmetry* 15 (1) (2023) 135.

[23] A. Khan, S. Saifullah, S. Ahmad, J. Khan, D. Baleanu, Multiple bifurcation solitons, lumps and rogue waves solutions of a generalized perturbed KdV equation, *Nonlinear Dynamics* 111 (6) (2023) 5743–5756.

[24] A. R. Alharbi, M. Almatrafi, Exact solitary wave and numerical solutions for geophysical KdV equation, *Journal of King Saud University-Science* 34 (6) (2022) 102087.

[25] S. Rizvi, A. R. Seadawy, F. Ashraf, M. Younis, H. Iqbal, D. Baleanu, Lump and interaction solutions of a geophysical Korteweg–de Vries equation, *Results in Physics* 19 (2020) 103661.

[26] T. Ak, A. Saha, S. Dhawan, A. H. Kara, Investigation of Coriolis effect on oceanic flows and its bifurcation via geophysical Korteweg–de Vries equation, *Numerical Methods for Partial Differential Equations* 36 (6) (2020) 1234–1253.

[27] B. B. S. Sankar, P. Karunakar, Solution characteristics of time-fractional geophysical kdv equation with coriolis parameter through hybrid methods, *Journal of Vibration Engineering & Technologies* 13 (1) (2025) 103.

[28] I. Podlubny, *Fractional differential equations: An introduction to fractional derivatives, fractional differential equations, to methods of their solution and some of their applications*, Elsevier, 1998.

[29] M. R. Spiegel, *Laplace transforms*, McGraw-Hill, New York, 1965.

[30] R. E. Moore, C. Yang, *Interval analysis i, Technical Document LMSD-285875*, Lockheed Missiles and Space Division, Sunnyvale, CA, USA, 1959.

[31] P. Karunakar, S. Chakraverty, Solving fully interval linear systems of equations using tolerable solution criteria, *Soft Computing* 22 (2018) 4811–4818.

[32] M. C. J. Anand, J. Bharatraj, Theory of triangular fuzzy number, *Proceedings of NCATM* 80 (2017) 80–83.

[33] D. Behera, S. Chakraverty, New approach to solve fully fuzzy system of linear equations using single and double parametric form of fuzzy numbers, *Sadhana* 40 (2015) 35–49.

[34] M. Qayyum, A. Tahir, S. Acharya, New solutions of fuzzy-fractional fisher models via optimal he–laplace algorithm, *International Journal of Intelligent Systems* 2023 (1) (2023) 7084316.

[35] P. Karunakar, S. Chakraverty, Solution of interval-modified Kawahara differential equations using homotopy perturbation transform method, In *Wave Dynamics*, pp. 193–202. 2022.

12 Machine Learning for Lid-Driven Cavity Flow

A.J.D. Nanthakumar

12.1 INTRODUCTION

12.1.1 Fluid Mechanics

The flow of water in a river and the lift of air planes are governed by the same laws of fluid mechanics – the study of fluids in motion. Many of the systems humanity encounters, either natural or engineered, involve fluid flow starting from the flow of blood in human arteries to the movement of climatic currents resulting in extreme weather patterns. The applications of fluid mechanics is versatile and omnipresent thus generating many sub disciplines including aerodynamics, hydrodynamics, rheology, etc. The scope of fluid mechanics include the analysis of the behavior of fluids under motion, subjected to force in the form of pressure and also involving energy transfer. The history of fluid mechanics can be traced back to ancient civilizations where the major implications could be found in irrigation, sailing, and architectural feats. The formalization of such features of fluid mechanics started with Archimedes who pioneered the concepts of buoyancy and hydrostatics, which stand relevant to date. The next formal advancements was made in the seventeenth century wherein Blaise Pascal and Evangelista Torricelli explored the concepts of fluid pressure and vacuum, setting the stage for future breakthroughs. Daniel Bernoulli in the eighteenth century introduced the energy conservation principle, which remains central to the study of fluid mechanics. The fundamental mathematical framework for the analysis of fluid flow dynamics was carried over by Claude-Louis Navier and George Gabriel Stokes in the nineteenth century resulting in the formulation of Navier–Stokes equations, as informed by Temam (2024). With the rapid increase in computational power in the twentieth century, scientists can tackle complex fluid dynamic equations numerically, transforming theoretical possibilities into practical tools. Such an increase in the capability of the computing hardware has enabled accurate simulations possible in the domains including aerospace engineering, automotive engineering, climate modeling, etc. The domain of CFD has enhanced the analysis capabilities in modern engineering so that design testing and optimization is carried over well before the build of physical prototypes.

 DOI: 10.1201/9781003611820-15

12.1.2 Artificial Intelligence (AI) and Machine Learning (ML)

The domain of artificial intelligence (AI) involves emulation of machines and systems capable of simulating human cognizance for varied tasks. Such a cognitive function can be applied for problem solving tasks, language understanding, recognition of logical patterns coupled with data analysis. This has resulted in autonomous implementation of complex tasks.

Machine learning can be considered as an inclusive subset of AI that could create systems capable of learning from data and improvise over time without specific programming for such tasks, as discussed by Samuel (1959). Conventional programming includes providing specific instructions whereas the ML approach identifies patterns in the datasets, and make decisions based on such patterns. ML results from the intersection of algorithm creation, statistical functions and utilizing working logic observed in natural systems including human brain and living organisms.

12.1.2.1 Supervised Learning

In supervised learning, an algorithm is formulated and training of the algorithm carried over using a well-informed dataset. The algorithm utilizes the training data set to create a mapping function relating inputs and outputs. Such a technique is predominantly employed for classification tasks (such as email classification), regression tasks (such as predicting temperature). Algorithms include linear regression, K – nearest neighbor regression, random forests, decision trees, least squares regression, artificial neural networks.

12.1.2.2 Unsupervised Learning

In supervised learning, the formulated algorithm needs to find patterns over datasets that are not complete with input and output parameters. Such approaches find a place in clustering tasks, dimensionality reduction tasks, etc. A practical example involves segmentation/classification of customer base so that customers with similar purchasing behaviors are segregated. Algorithms include K-means clustering, principal component analysis (PCA), hierarchical clustering, etc. Unsupervised learning has found applications in market analysis and image compression domains.

12.1.2.3 Semi-Supervised Learning (SSL)

When unlabeled data is available in abundance when compared with labeled data, Semi-Supervised Learning technique is employed. It is suitable for fields where obtaining labeled data is unavailable or time consuming and costly. SSL techniques improves the performance of machine learning models without necessity of a very large labeled dataset.

12.1.2.4 Reinforcement Learning

Reinforcement learning is a subcategory of machine learning that allows an agent to learn through trial and error in an uncertain environment. The agent interacts with the environment by taking actions from a set A in a given state S, receiving feedback through a reward function R. The agent's goal is to learn a policy, a strategy that defines the best action to take for a particular state s that maximizes cumulative rewards over time. This process can be represented by the expression $S \times A \rightarrow R$. Common reinforcement learning algorithms include Q-learning, $TD(\lambda)$-learning.

12.1.2.5 Deep Learning

Neural networks containing numerous layers—known as deep neural networks—form the foundation of deep learning, a specialized branch of machine learning. Such deep neural networks are capable of discovering intricate patterns as well as independent detection of such features with minimal manual intervention and input.

12.2 LITERATURE REVIEW

12.2.1 Lid-Driven Cavity Problem

The lid-driven cavity problem can be considered as a fundamental problem in fluid dynamics analysis, as discussed by Kuhlmann and Romano (2019) and is used by researchers as a benchmark for testing numerical analysis procedures. It can be deployed as a simple as well as fundamental model to study the fundamental aspects of dynamic flow of fluids. The significant aspect of the model is its inherent characteristic of possessing flow patterns similar to industrial processes including mixing tanks and lubrication systems. Behind its simple settings for fluid flow, it exhibits many of the fundamental fluid flow structures including vortex mixing, separation of flow corner complexities, boundary layer flow, chaotic mixing, and turbulent behavior It has rendered a side and broad scope of fluid dynamic analysis making it a live problem for ongoing research.

The dynamics of vortex formation was investigated by Zhu et al. (2020) for varying Reynolds and Stokes number values. Based on the vortex formulations, four flow patterns were identified and studied. Turkyilmazoglu (2022) analyzed the streamline patterns for the recirculating primary vortex and secondary vortices for a double lid-driven cavity. Migeon et al. (2000) discussed the effect of the shape of the cavity on the flow dynamics. Approximately similar vortex generation had been observed for all the examined cavities, with differences appearing over time. The stability of the lid-driven cavity flow was undertaken by Ramanan and Homsy (1994) who numerically computed 2D flow for varying Reynolds numbers. The mechanism of instability of flow was discussed through energy calculations. Sheu and Tsai (2002) analyzed the lid-driven cavity as a three-dimensional cube with the top surface moving. The topological analysis was carried over to understand the vortex formations in the case study. The spiraling nature of the flow in the ineer regions was also analyzed. The fundamental issue of Hopf bifurcation was studied by Bruneau and Saad (2006) who shed light on the bifurcation for Reynolds numbers up to Re = 10,000.

The turbulent flow inside a cubic lid-driven cavity was studied by Ferrari et al. (2024) who observed that the flow regime change occurred between Reynolds numbers 20000 and 25000. The turbulent dissipation moves from the bottom of the cavity towards the downstream wall. For Reynolds numbers higher than 50000, the dissipation was localized near the moving lid. The thickness of the boundary layer in a cavity was studied by Alleborn *et al.* (1999). The effect of the wall velocity on the heat transfer rate and mass transfer rate as well as on the concentration of the species alongside the boundaries were studied. The convection nature of the fluid flow along with the heat transfer performance for a circular body located in the case of a lid-driven cavity were analyzed by Khanafer and Aithal (2013). It was found that major velocity gradients occurred inside the boundary layer. The characteristics of the turbulence in a 3D lid-driven cavity cube was studied by Samantaray and Das (2019) in which large eddy simulation techniques were applied to analyze fluid behavior within a cubic enclosure featuring a moving boundary and revealed that turbulent characteristics exhibited spatial variation even at relatively modest Reynolds values (3200). Turbulence was anisotropic, especially at near wall and corner regions. Gnanasekaran and Satheesh (2024) investigated the two-dimensional steady incompressible turbulent flow characteristics in an enclosed cavity. The contours of turbulence kinetic energy, dissipation, and viscosity were examined and it was found that the average intensity of turbulent quantities was decreased by increasing the speed ratio. The transition from laminar to chaotic flow in the 2D lid-driven cavity was studied by Peng *et al.* (2003) in which interesting phenomenon were found. The transition from the laminar to the periodic flow was through supercritical Hopf bifurcation. The incompressible form of the Navier–Stokes equation were solved for varying Reynolds number.

12.2.2 Machine Learning in Fluid Dynamics

Nearly every research discipline is being transformed by the Fourth Industrial Revolution, driven by advances in big data, machine learning, and artificial intelligence. Traditionally, fluid mechanics has relied on massive datasets generated from experiments, field measurements, and large-scale numerical simulations to understand complex flow dynamics. Although fluid experiments and simulations have generated a large amount of data, machine learning in this field still struggles with data sparsity. This is due to the high-dimensional nature of the flow state, the numerous geometrical parameters and boundary conditions, and the inherently complex, cross-scale, random, and non-linear characteristics of fluid mechanics. The interface between machine learning and fluid dynamics has a long, and perhaps surprising, history. Machine learning (ML) algorithms offer promising solutions to some of these challenges. CFD calculations involve high computational requirements that has made the technique for systems with less computational prowess such as mobile computing and real-time monitoring. In such situations ML can replace complex fluid dynamics calculations eliminating fine grid computations thereby reducing computational requirements and speeding up simulations.

In the 1940s, Kolmogrov (1991) understood that turbulence can be considered as an application for his theories on random variables and prediction. Norbert Weiner (Wiener, 2019), worked on expert systems for emulating human brains behavior. The concept of perceptron (Rosenblatt, 1958) was able to automate tasks like classification and mathematical regression. However it was informed that single layer perceptrons were not able to solve problems like XOR function (Minsky and Papert, 1969).

Gholami *et al.* (2019) utilized machine learning algorithms for estimating the parameters of fluid flow depth and flow velocity. Colvert *et al.* (2018) used neural networks for understanding flow patterns from local vorticity measurements. An oscillating aerofoil was analyzed for the formation of vortices behind its wake and the neural networks were able to undermine the salient features of the wake formations. Costa Rocha *et al.* (2023) solved the partial differential equation for the RANS turbulence model using a Fourier neural operator inside a lid-driven cavity. OpenFOAM was utilized for the study. In a lid-driven cavity, the phenomenon of advection-diffusion, viscous shock and stoeks flow characteristics was captured by Boull'e et al. (2022) who utilized ratinal neural networks for the approximation of Green function of linear partial differential equations.

Ranade *et al.* (2021) utilized the ML technique for solving partial differential equations. The developed technique was able to handle a steady flow in incompressible form of Navier–Stokes equation for the lid-driven cavity case. Kashir *et al.* (2021) extracted fluid flow features using neural networks. The flow features were extracted at the highest possible resolution but transforming each computational cell into a detection pixel. Fukami *et al.* (2020) utilized convolutional neural networks (CNNs) for analyzing two-dimension wakes, isotropic turbulence studies, turbulent flow in a 3D channel. The direct numerical simulation (DNS) data for a turbulent channel flow was used and analyzed by Pandey *et al.* (2020) for turbulent channel flow in which isotropic turbulence and Rayleigh–Bernard convections were observed.

Kokash *et al.* (2024) conducted regression modeling for assessing the fluid flow inside a square cavity with the flow initiated by one of the moving walls. Physics-informed neural networks (PINNs) were deployed for the modeling of flow and thermal behavior at both low and high Reynolds numbers. The results were compared among actual data from computational fluid dynamics (CFD) simulations. As numerical simulations used in engineering disciplines takes tens of hours for solving simple problems, smart proxy models (Aboaba *et al.*, 2020) are capable of replicating numerical simulations with considerable accuracy. Such an approach significantly reduced the typical long time-to-solution of CFD simulations, maintaining traditional CFD-level accuracy for the specific model under study. Drakoulas *et al.* (2023) proposed a new machine learning (ML) platform for devel oping reduced-order models (ROMs) to address large-scale, computationally intensive problems involving transient, non-linear partial differential equations. The proposed framework demonstrated its efficiency and versatility across several challenging benchmarks, including 2D linear convection–diffusion, fluid flow around a cylinder, the 2D lid-driven cavity at high Reynolds numbers, and 3D blood flow in an arterial segment. These examples highlight the framework's ability to efficiently solve both low- and high-dimensional flow problems across diverse applications.

However, fluid dynamics phenomena are inherently complex, and no single ML or neural network model can handle all conditions. Fluid behavior varies based on environmental conditions and the geometry of the objects involved, which limits the applicability of a single model. As a result, hybrid approaches that combine traditional mathematical models with ML algorithms have emerged to address this complexity, offering tailored solutions for different types of flow scenarios.

12.3 PROBLEM DESCRIPTION

12.3.1 Lid-Driven Cavity Problem

Rectangular and cubic containers are among the most fundamental geometries for studying fluid motion in confined spaces. A simple yet effective way to drive fluid movement within such containers is through the tangential motion of one of the walls. When one of the solid walls of a cuboid moves parallel to itself, the resulting setup is known as a lid-driven cavity. Such a chamber of fluid with enclosed walls can be considered as a standard test case in fluid dynamics study and research. The upper boundary is assumed to move at a specific tangential speed thereby generating a distinct circular flow structure, including a large and prominent central rotating pattern along with smaller rotational patterns developed in the corners. The Reynolds number, which represents the ratio of inertial to viscous forces, strongly influences the flow structure: as the Reynolds number increases, the vortices become more numerous and complex, often leading to turbulent behavior. Such a simple yet fundamental case study can be utilized for validating CFD models.

There have been many variations considered by researchers based on the problem. A 3D version introduces span-wise effects and secondary flows, adding three-dimensional complexity. Multi-lid configurations, where more than one wall moves (sometimes in opposite directions), create interacting vortices and complex shear layers. In thermally-driven cavities, a temperature gradient introduces buoyancy forces, creating a mixed convection scenario that combines forced and natural convection. High Reynolds numbers make the flow turbulent, requiring advanced turbulence models. Inclined or rotating cavities introduce asymmetric or swirling flows, relevant for applications like natural ventilation or mixers. Porous cavities include a porous medium, which dampens vortex structures and adds flow resistance, useful for studying filtration and groundwater flow. Additionally, unsteady lid motion introduces time-dependent changes, and adding internal obstacles, like cylinders, simulates flow around objects within confined spaces, applicable to engineering scenarios such as electronics cooling. Each variation presents unique challenges, allowing researchers to validate and benchmark CFD solvers across a range of flow regimes.

12.3.2 Problem Definition

The geometry of the problem is shown in Figure 12.1.

The geometry depicts a square cavity with three rigid walls that impose no-slip conditions. The fourth wall, or "lid" moves tangentially across the top with a unit velocity, creating shear flow within the cavity. The lower left corner of the cavity is imposed with a reference static pressure of 0.

12.3.3 CFD Analyses

The lid-driven cavity problem can be modeled using a non-dimensional form of the Navier–Stokes equations, which simplifies analysis and enhances generalization by reducing the number of parameters. In COMSOL Multiphysics version 6, the laminar

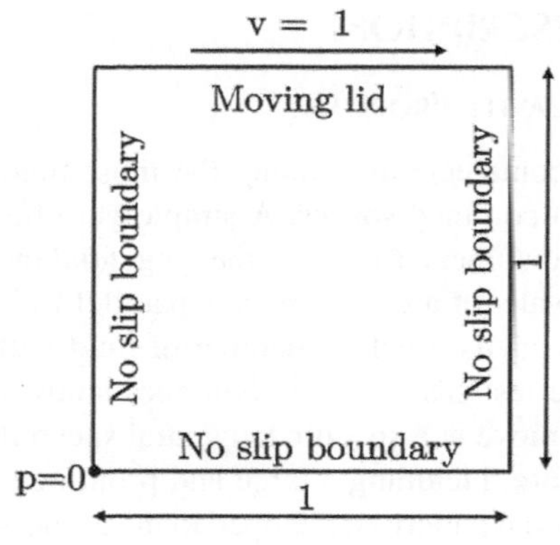

FIGURE 12.1 Lid-driven cavity – geometry.

flow physics interface allows users to solve the fluid dynamics equations of Navier–Stokes and given by

$$\rho\left(u\cdot\nabla\right) = -\nabla p + \nabla^2 \mathbf{u} \tag{12.1}$$

The classical Reynolds number is given by

$$\mathrm{Re} = \frac{\rho v L}{\mu} \tag{12.2}$$

where ρ is the density of the fluid, v is the dynamic viscosity of the fluid, L is the characteristic length, and μ is the kinematic viscosity. It is an advantageous method to solve the flow equation after non-dimensionalizing the terms. When expressed in the non-dimensional form, the equations offer the advantage of consolidating multiple physical variables into a unitary dimensionless value of the Reynolds number that correlates the inertial forces and the viscous forces in fluid flow. By working with this simplified form, the flow behavior can be analyzed across different regimes simply by varying the Reynolds number, without needing to adjust other variables. Such a non-dimensional form can be solved effectively by choosing appropriate values for density and viscosity, scaled to normalize the equations for the desired Reynolds number. Such a procedure involves nondimensionalizing the velocity term, pressure term and the length scale as depicted below:

$$u^* = \frac{u}{v};\quad p^* = \frac{p}{\rho v^2};\quad r^* = \frac{r}{L};\quad \nabla^* = L \tag{12.3}$$

The resulting Navier–Stokes equation is given by

$$\left(u^*\cdot\nabla^*\right)u^* = -\nabla p^* + \frac{1}{Re}\nabla^{*^2}u^* \tag{12.4}$$

The computational domain is configured as a square geometry measuring 1 unit on each of the sides with the fluid density value set to unitary value and the dynamic viscosity represented as the reciprocal of Reynolds number (1/Re). For boundary conditions, the upper wall of the cavity acts as a moving wall with a constant horizontal velocity of 1, which generates shear driven flow within the cavity. The remaining boundaries (left, right, and bottom walls) are considered as walls with no slipping and having a velocity value of zero. A unique solution is achieved when the constraint on the pressure at a point is assumed. In steady-state analyses of closed systems, where there are no inlets or outlets, none of the boundary conditions naturally anchor the pressure. Without this constraint, the pressure field would be undefined, leading to possible numerical drift or instability in the solution. By specifying the pressure at a single point in the domain, we establish a reference that makes the problem well-posed. This can be implemented by selecting a single node or point in the domain and setting its pressure to zero (or another fixed value) to serve as the reference pressure. The problem is solved for a range of Reynolds number from 100 to 10000.

12.3.4 Streamline Pattern

Streamline plots have been plotted for the lid-driven cavity problem at various Reynolds numbers, ranging from Re = 100 to 10,000. Figure 12.2 depicts the velocity streamlines for Reynolds numbers Re =100, 200, 400, 800, 1200, 1600. Figure 12.3 depicts the velocity streamlines for Reynolds numbers Re = 2000, 2400, 4000, 6000, 8000, 10000.

For Reynolds number ranging from Re = 100 to 400, only a single vortex formation is evident that occupies a major portion of the cavity space. Moreover, the vortex shape can be approximated to be near circular with symmetric nature. Boundary layers are more evident near the moving lid. For Reynolds number ranging from Re = 800 to 1600, very small vortices were evident in the bottom portion of the cavity. This is due to the interactions between the primary vortex and the stationary wall boundaries. Flow separation occurs at locations near the stationary walls. The loss of symmetry starts at these Reynolds number values. The recirculation effect has started for such conditions. For Reynolds number ranging from Re = 2000 to 4000, establishment of smaller vortices that have already formed occurs. The symmetric nature of the streamline patterns gets further deteriorated. The center of the primary vortex has started shifting towards the right (direction of velocity). The vortices at the bottom portion show an increase in its size and strength. The stretching effect of the primary vortex has occurred and the primary vortex undergoes a higher shear rate as seen from denser streamlines. For Reynolds number ranging from Re = 6000 to 10000, the expansion of secondary vortices is seen in the streamline plot. As Reynolds number increases towards turbulent regime, the flow structure becomes more irregular along with appearance of more recirculating regions throughout the flow.

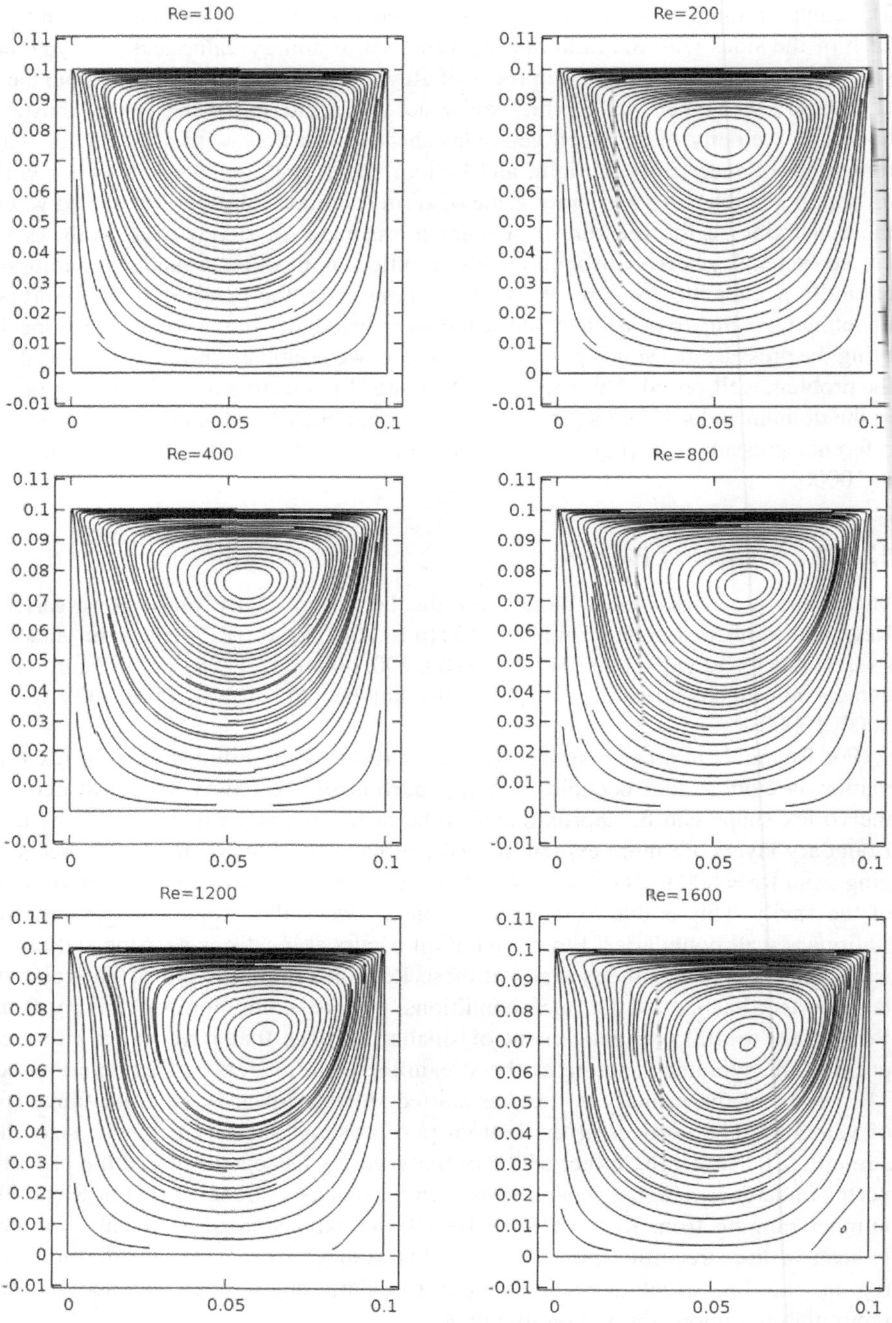

FIGURE 12.2 Velocity streamlines for Re = 100, 200, 400, 800, 1200, 1600.

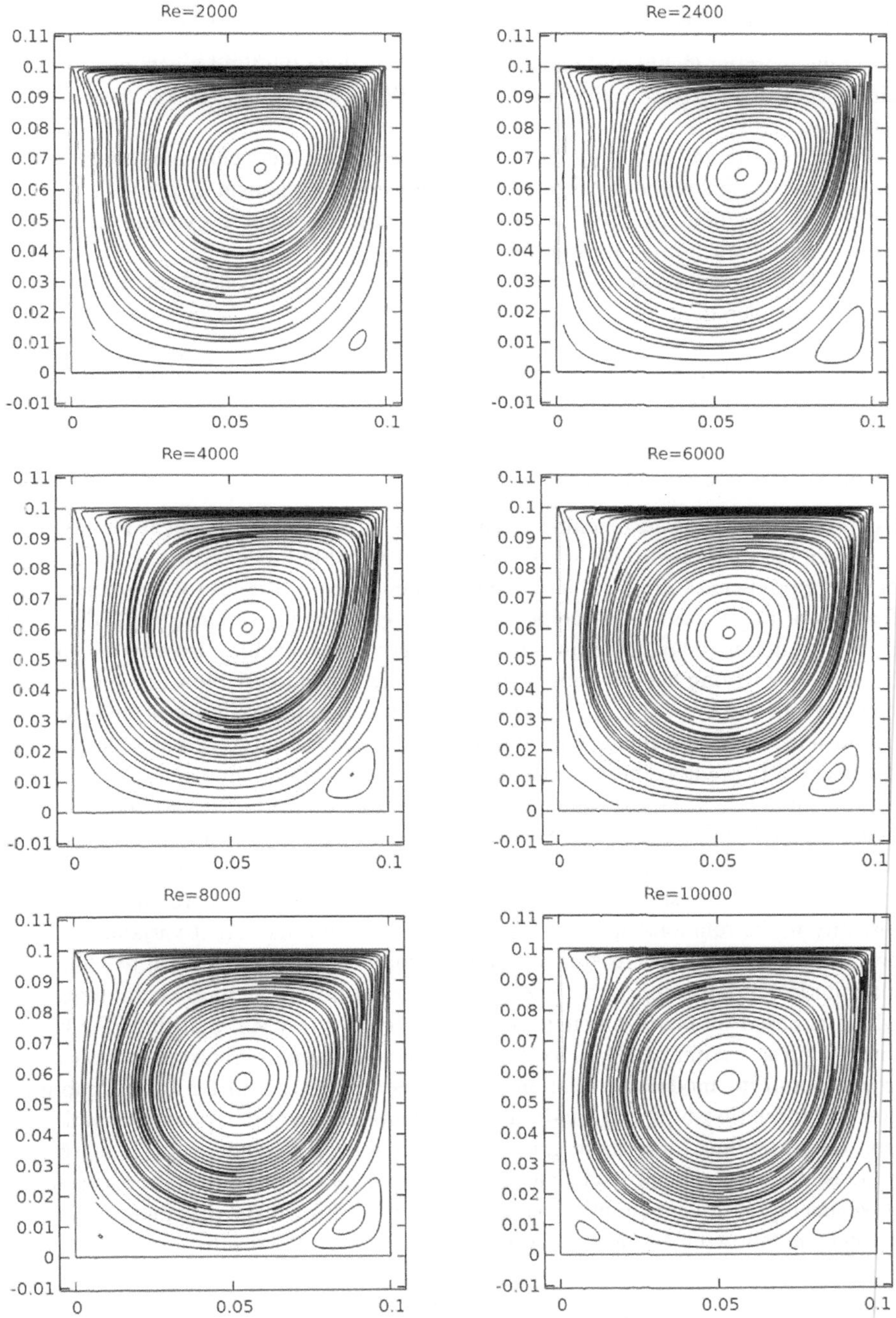

FIGURE 12.3 Velocity streamlines for Re = 2000, 2400, 4000, 6000, 8000, 10000.

12.4 REGRESSION MODELING OF THE FLOW VELOCITY

Machine learning techniques can be applied to analyze the fluid mechanics problem similar to the lid-driven cavity in three ways (Wang and Wang, 2021):

1. Data-driven modeling to obtain the relation between input and output parameters without considering the physical mechanism involved.
2. Employment of physical models for optimization of existing models by machine learning algorithms.
3. Hybrid approach in which both the data and physical mechanisms are considered.

In the presented research work, data-driven modeling based on regression analysis has been carried over.

12.4.1 Regression Modeling

Regression modeling in machine learning is a fundamental statistical method used to model the relationship between a dependent variable (also known as the target or output or predicted value) and one or more independent variables (often referred to as features or predictors). In machine learning, regression is primarily applied for predictive modeling, where the aim is to forecast continuous values. Regression analysis involves identifying the relationship between a dependent variable y and one or more independent variables $x_1, x_2, x_3, \ldots x_n$ by fitting a function that minimizes the difference between the predicted and actual values of y. Regression analysis creates a model that can accurately predict continuous outcomes based on input features, uncover patterns in data, and understand the influence of each feature on the target variable. There are many methods in regression modeling including linear regression wherein a linear relationship between the dependent and independent variables is formulated.

Polynomial regression wherein polynomial terms are added for allowing non-linearity in the relationship, logistic regression wherein the target variable is binary or categorical, Lasso regression wherein a penalty term is introduced to reduce the magnitude of coefficients, which helps prevent over fitting, Support vector regression which is based on based on support vector machines (SVMs) that finds the best-fit line within a margin of tolerance (epsilon), decision tree regression, which uses a tree structure to model the relationship between features and the target, capturing non-linear relationships and handling both categorical and numerical data well, decision trees combined into aggregated models characterize random forest regression, an ensemble-based approach that leverages multiple independent tree predictors, Bayesian regression that applies Bayes' theorem to estimate the distribution of model parameters, allowing for probabilistic interpretations of the results.

12.4.2 Polynomial Regression

The usability of linear regression can be extended by utilizing polynomial expressions with the polynomial expression specified by its degree n. The polynomial expressions are useful in cases where non-linear relationships exist and curves are necessary rather than fit lines. In addition, multiple polynomial regression includes multiple independent variables with each term consisting of inner polynomial terms. Thus, it can capture complex and non-linear interactions among the variables. In multiple polynomial regression, the dependent variable y is represented as a polynomial function of several independent variables $x_1, x_2, \ldots x_n$ and is given by

$$y = \beta_0 + \beta_1 x_1 + \beta_2 x_2 + \ldots + \beta_{11} x_1^2 + \beta_{12} x_1 x_2 + \ldots + \beta_{nn} x_n^2 + \tag{12.5}$$

The interaction terms also include cross product terms and squared terms for modeling complex relationships. The selection of polynomial regression as the primary machine learning approach rather than other approaches can be justified due to the following reasons:

- The polynomial regression model possesses an explicit mathematical form enabling direct analysis and mapping of coefficients with the physical phenomenon whereas in other techniques such as neural networks the relationship is encoded across multiple layers and weight coefficients.
- Polynomial regression model is computationally efficient as it involves closed form least squares solutions instead of iterative optimization.
- Computational time is less for polynomial regression model as it does not require multiple forward passes as in neural networks.
- The memory storage requirements for a polynomial regression model involves only the polynomial coefficients rather than multiple weight parameters as in neural networks and support vectors as in support vector regression.

12.4.2.1 Polynomial Regression Model Fitting

The accuracy of the polynomial regression depends upon degree of the polynomial, type of the feature scaler, normalization techniques, dataset size and quality. Higher degree polynomials can capture more complex patterns for dynamic phenomenon whereas lower degree polynomials with fewer parameters lend a simpler analysis thereby avoiding overfitting. Underfitting is evident when the regression model is very simplistic and is unable to capture the inner patterns. Overfitting is evident when the regression model apart from capturing the flow details also captures the noise data. Overfitting can be understood when the training error is low but the testing error is very high. In the presented research work, polynomial features are generated for each specified degree of the polynomial $n = 3, 6, 9, 12$ and models are trained accordingly.

Normalizing input variables through different scaling techniques enhances polynomial regression model stability and effectiveness. This normalization step is crucial for polynomial regression since the values of higher-degree terms can vary dramatically, potentially causing computational problems and convergence failures.

The research evaluated several normalization approaches to determine their effects on model outcomes. MinMax scaling compresses variables into a specific range—typically between zero and one—which functions well with evenly distributed datasets but struggles with extreme values. Standard scaling centers data around zero with unit variance, assuming normal distribution characteristics and accommodating diverse variable ranges.

Robust scaling utilizes median values and interquartile measurements for transformation, offering greater resilience against outliers than both MinMax and Standard techniques. This method proves particularly valuable for skewed distributions containing atypical values. Normalizer transformation adjusts individual data points to uniform magnitude, beneficial for distance-dependent applications like clustering or when comparing differently scaled magnitude values.

The selected scaling methodology substantially influences feature transformation, directly affecting polynomial regression performance. Proper normalization prevents coefficient inflation that might otherwise cause numerical instability and reduced model generalization. The investigation implemented MinMax, Standard, Robust, and Normalizer scaling techniques to process the input variables.

For each of the Reynolds number, the linear regression modeling is carried over for the polynomial degrees of 3, 6, 9, and 12 and all the four scalers mentioned above.

12.4.2.2 Framework for using Polynomial Regression in Fluid Dynamics

The mathematical expression for the Navier–Stokes equation for fluid flow in non-dimensionalized form is given in Eq. (12.4). The machine learning exercise is a mapping F from the spatial coordinates and Reynolds number to the velocity variable.

$$\mathscr{F} : (x, y, \mathrm{Re}) \mapsto \text{velocity} \tag{12.6}$$

For each Reynolds number, a polynomial expression is constructed for the degrees of 3, 6, 9, and 12. The logic for utilizing polynomial expression for velocity can be understood through Taylors series expansion of the velocity values in each coordinate (x, y) given by

$$\text{velocity}(x, y) = \sum_{i=0}^{\infty}\sum_{j=0}^{\infty} \frac{1}{i!j!} \frac{\partial^{i+j} u(0,0)}{\partial x^i \partial y^j} x^i y^j \tag{12.7}$$

when the above series is truncated at degree n, a polynomial approximation of the velocity filed. As n increases the polynomial expression converges to the true solution.

12.4.2.3 Evaluation Metrics

To assess polynomial regression fit quality, this study employed three key performance metrics. The coefficient of determination (R^2) quantifies how well independent variables explain variance in the dependent variable, ranging from 0-1, with higher values indicating better explanatory power. Root mean squared error (RMSE) calculates the square root of averaged squared differences between predictions and actual values. Lower RMSE values signal smaller prediction deviations and superior

TABLE 12.1
Best Combination of Polynomial Degree and Scaler for Each Reynolds Number

Reynolds Number	Polynomial Degree	Scaler	R^2 Score	RMSE	MAE
100	12	MinMax	0.97907838	0.03060052	0.015709325
200	12	MinMax	0.97910334	0.03059635	0.015733002
400	12	MinMax	0.97918855	0.03058123	0.015804995
800	12	Robust	0.97937445	0.03053670	0.016033472
1200	12	MinMax	0.97947321	0.03045064	0.016280945
1600	12	Robust	0.97952084	0.03028414	0.016328455
2000	12	Robust	0.97947344	0.03012181	0.016225521
2400	12	Standard	0.97928166	0.03004413	0.016113959
4000	12	MinMax	0.97752424	0.03049513	0.016441443
6000	12	Robust	0.97469139	0.03168872	0.017478620
8000	12	MinMax	0.97174376	0.03306506	0.018677467
10000	12	Standard	0.96863413	0.03455274	0.020071788

model performance, while elevated values indicate significant discrepancies and reduced accuracy.

12.4.2.4 Determination of the Best Fit

Mean absolute error (MAE) measures the average magnitude of prediction errors in absolute terms. Lower MAE values demonstrate predictions closer to actual observations on average, suggesting stronger model performance, while higher values reflect larger prediction errors and poorer accuracy. The model selection process considered all three metrics simultaneously, prioritizing R^2 maximization while minimizing both RMSE and MAE to balance goodness-of-fit with precision. For each Reynolds number condition, optimal models were identified based on achieving the highest R^2 alongside minimal RMSE and MAE values. Table 12.1 presents the selected polynomial degrees and scaling methods that satisfied these criteria across different flow conditions.

12.5 RESULTS AND DISCUSSION

In fluid dynamics computational studies, velocity contour plots function as essential visualization instruments for examining fluid flow patterns. These plots provide clear depictions of velocity variations throughout the analyzed region, helping researchers comprehend intricate flow characteristics. This visualization approach delivers valuable understanding of flow behaviors, supporting better-informed analysis and design decisions.

The visualization data comes from numerical flow simulations performed using computational fluid dynamics methods. After calculating the velocity field, researchers create contour representations to showcase specific flow aspects, from localized variations to broader patterns. These visualizations excel at identifying key flow features such as turbulent and laminar regions, zero-velocity stagnation points, and circular flow patterns. These insights prove crucial when optimizing aerodynamic structures, fluid transport networks, or industrial processing systems.

For this investigation, velocity contour line visualizations were generated using both traditional CFD simulation results and the polynomial regression machine learning approach. Figure 12.4 depicts the contour line plot of fluid flow velocity obtained from CFD simulations and machine learning model for Reynolds number = 100, 200, 400, 800. Figure 12.5 depicts the contour line plot of fluid flow velocity obtained from CFD simulations and machine learning model for Reynolds number = 1200, 1600, 2000, 2400. Figure 12.6 depicts the contour line plot of fluid flow velocity obtained from CFD simulations and machine learning model for Reynolds number = 4000, 6000, 8000, 10000. The actual velocity represents the fluid flow velocity obtained from CFD simulation and predicted velocity represents the fluid flow velocity obtained from the machine learning regression model.

The overall flow patterns depicted in the actual velocity contour plots and the predicted velocity contour plots demonstrate a significant degree of similarity. This resemblance suggests that the machine learning model has successfully learned and generalized the flow characteristics across different Reynolds numbers to a considerable extent. For lower Reynolds numbers (e.g., Re = 100, 200), where the flow is predominantly laminar, the predicted velocity contours closely align with the actual contours in terms of the shape and positioning of velocity zones. These flow patterns are smooth and streamlined, a fundamental characteristic of laminar flow, and the machine learning model effectively captures these characteristics with high accuracy.

As the Reynolds number increases (e.g., Re = 4000, 6000, 10,000), the flow patterns become more complex, with more intricate velocity distributions, sharper gradients, and the emergence of pronounced vortices. While the predicted velocity contours still successfully capture the overall distribution of the flow field, few deviations are visible at these higher Reynolds numbers. For instance, the actual velocity contours exhibit tighter clustering of contour lines in regions with steep velocity gradients, reflecting rapid changes in velocity. In contrast, the predicted contours tend to smooth out these regions, resulting in less sharpness and precision. One common difference is the placement of contour lines representing specific velocity magnitudes (e.g., 0.1, 0.2, etc.). While the machine learning model captures the overall distribution of velocity magnitudes across the domain, slight shifts in the contour lines are evident, particularly in regions with rapid changes in velocity. Overall, the machine learning model performs well in capturing the broader flow characteristics and velocity distributions across a range of Reynolds numbers. Its accuracy is particularly high at lower Reynolds numbers, where the flow is laminar, and velocity gradients are mild. However, as the Reynolds number increases and the flow becomes more turbulent, the model's predictions show some deviation in capturing the sharpness of velocity gradients, boundary layer behavior, and the detailed structure of turbulent features.

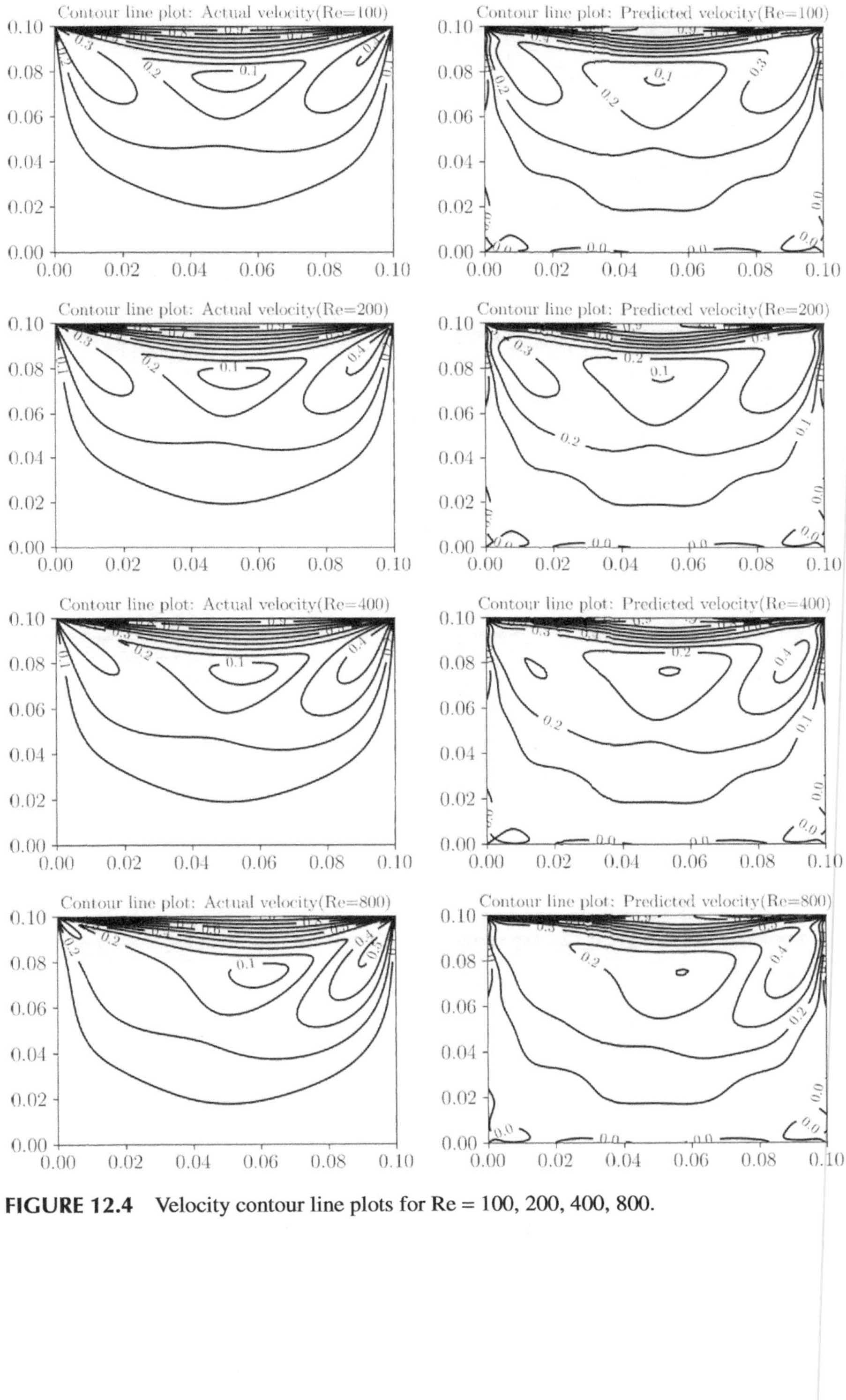

FIGURE 12.4 Velocity contour line plots for Re = 100, 200, 400, 800.

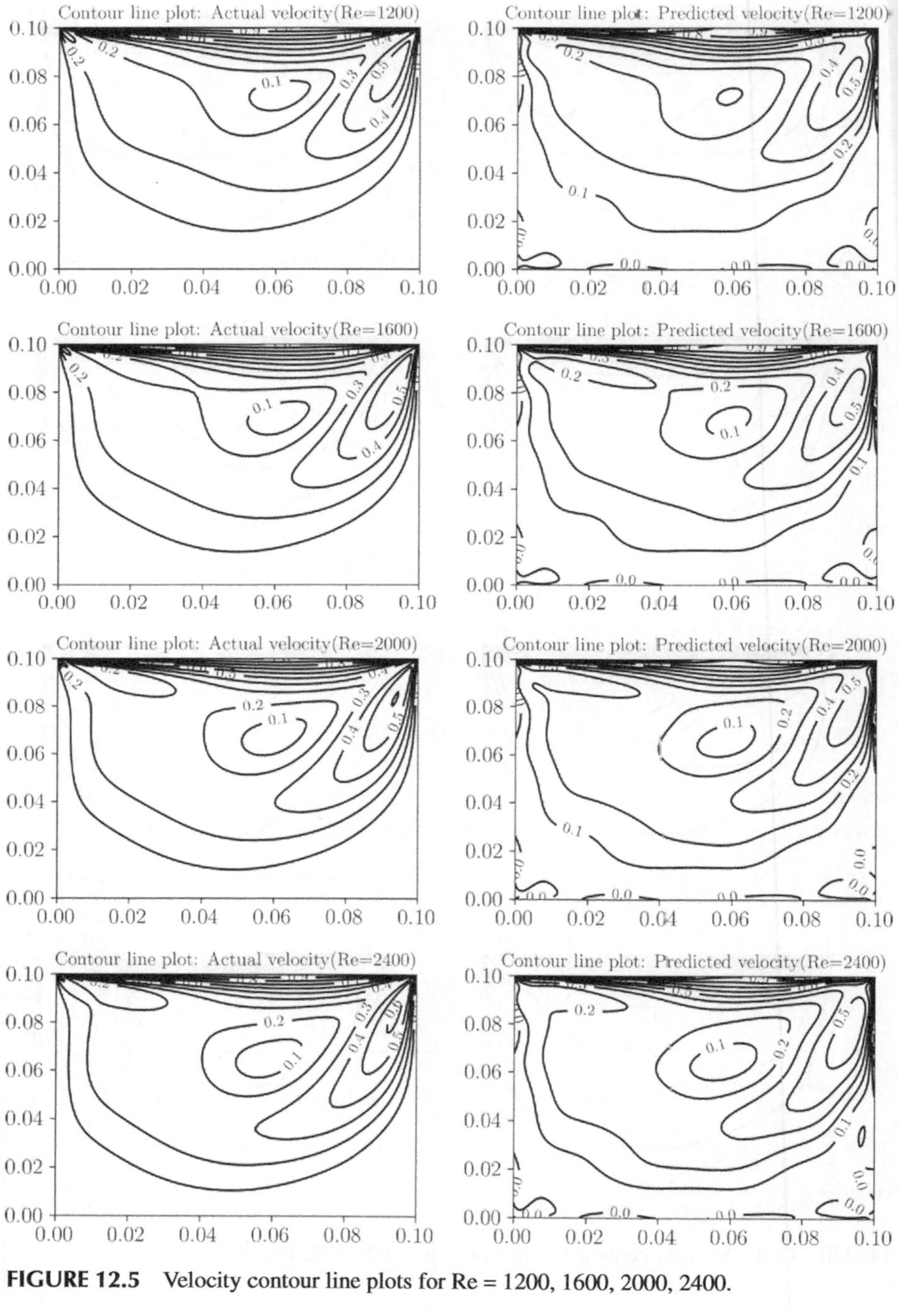

FIGURE 12.5 Velocity contour line plots for Re = 1200, 1600, 2000, 2400.

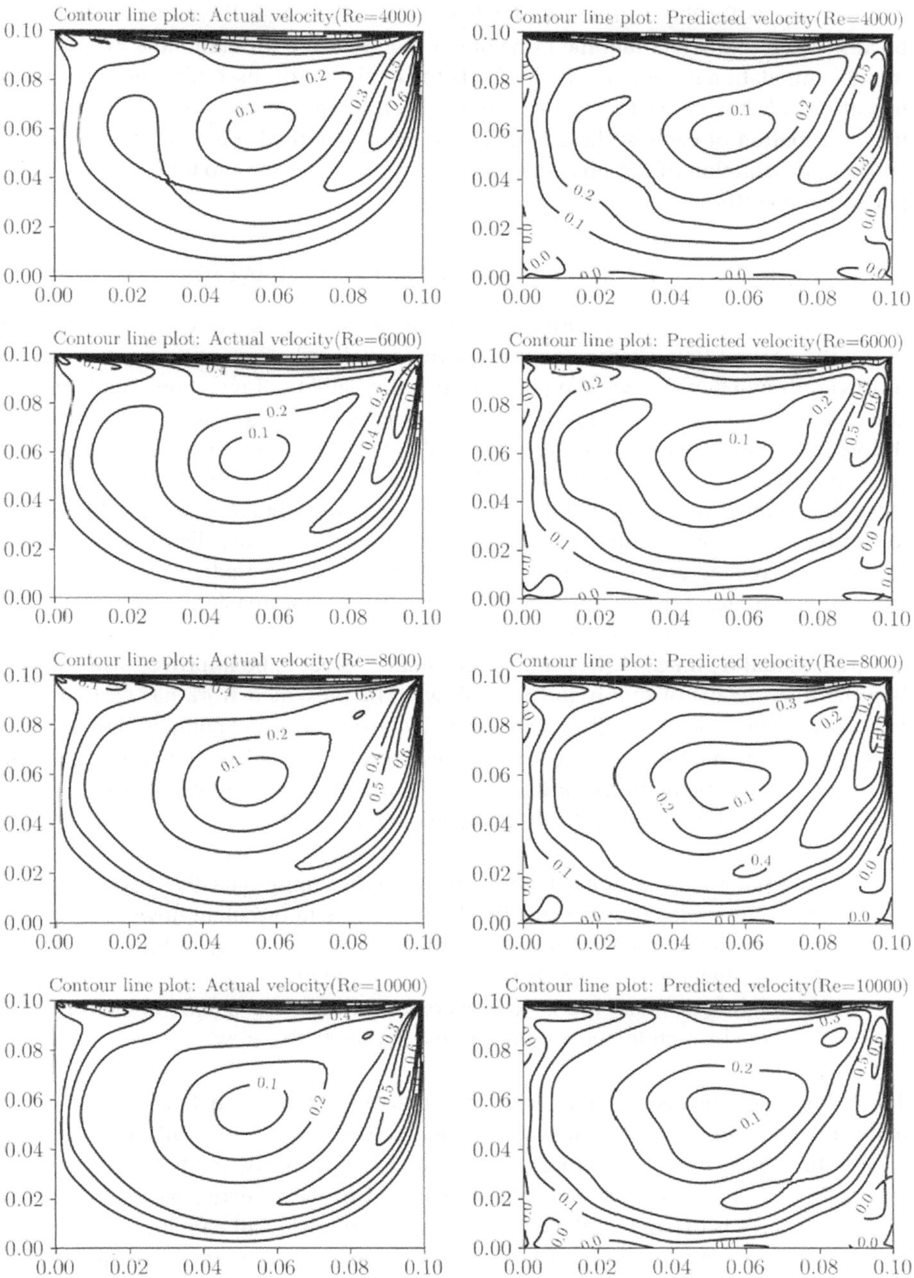

FIGURE 12.6 Velocity contour line plots for Re = 4000, 6000, 8000, 10000.

The reasons for the decline in model performance at high Reynolds number can be at tributed to the fact that the polynomial functions are differentiable everywhere in the computational domain which affects the depiction of discontinuities and steep transitions of flow characteristics existing in high Reynolds number flow. Moreover, the contribution of each of the coefficient in the polynomial function influences the entire representation of the physical phenomenon making the modeling of localized phenomenon difficult.

12.6 CONCLUSION AND SCOPE FOR FUTURE WORK

In conclusion, the machine learning model demonstrates strong capability in capturing the general flow characteristics and velocity distributions across a wide range of Reynolds numbers, showcasing high accuracy in laminar flow regimes and reasonably effective predictions in more turbulent scenarios. At lower Reynolds numbers, the predicted velocity contours closely resemble the actual ones, accurately reflecting smooth, streamlined flow patterns and mild velocity gradients. However, as the Reynolds number increases and the flow transitions to turbulence, the model begins to exhibit limitations. While it successfully captures the broader distribution of flow features, discrepancies emerge in regions with sharp velocity gradients, localized vortices, and complex boundary layer behavior. Overall, the machine learning model proves to be a valuable tool for approximating flow behavior, especially in laminar and transitional regimes. However, its performance is yet to be improved when handling the finer details of highly turbulent flows. The presented work has carried over polynomial regression modeling, which relatively is simple and computationally less expensive. A plethora of machine learning models are available that could be utilized for more accurate predictions for fluid flow conditions with steep variations.

The work can be extended that may include the following:

- Incorporating additional features or refined training data that better capture the high gradient regions and localized vortices in turbulent flows.
- Expanding the training dataset to include a greater variety of high-Reynolds-number flow cases for predicting intricate structures.
- Embedding physical laws, such as conservation of mass and momentum into the model's architecture could improve its predictive power.

The polynomial regression under consideration considered the spatial features as the main input features. Additional features such as distance to the wall, angular position with respect to cavity corners can definitely improve the predictive results for the study and the corresponding dataset analyzed in the research paper. When the training data set is expanded, the machine learning model shall capture the phenomenon found in the current study such as secondary and tertiary vortices, steep velocity gradients evident in thin boundary layers, points of flow separation and transitional flow structures. The embedding of physical laws in the current study would enable the machine learning model to give results incorporating the fluid dynamics principles that can improve the predictive capability of the machine learning model at higher

Reynolds number. Thus, the above specific suggestions result in the feasibility of improvement of the predicted output specific to the considered study.

REFERENCES

Aboaba, A., Y. Martinez, S. Mohaghegh, M. Shahnam, C. Guenther, and Y. Liu (2020). Smart proxy modeling application of artificial intelligence & machine learning in CFD. Technical report, National Energy Technology Laboratory (NETL), Pittsburgh.

Alleborn, N., H. Raszillier, and F. Durst (1999). Lid-driven cavity with heat and mass transport. *International Journal of Heat and Mass Transfer 42* (5), 833–853.

Boull´e, N., C. J. Earls, and A. Townsend (2022). Data-driven discovery of green's functions with human-understandable deep learning. *Scientific Reports 12* (1), 4824.

Bruneau, C.-H. and M. Saad (2006). The 2d lid-driven cavity problem revisited. *Computers & Fluids 35* (3), 326–348.

Colvert, B., M. Alsalman, and E. Kanso (2018). Classifying vortex wakes using neural networks. *Bioinspiration & Biomimetics 13* (2), 025003.

Costa Rocha, P. A., S. J. Johnston, V. Oliveira Santos, A. A. Aliabadi, J. V. G. The, and B. Gharabaghi (2023). Deep neural network modeling for CFD simulations: Benchmarking the Fourier neural operator on the lid-driven cavity case. *Applied Sciences 13* (5), 3165.

Drakoulas, G., T. V. Gortsas, G. C. Bourantas, V. N. Burganos, and D. Polyzos (2023). A reduced-order modeling framework based on machine learning for real-time applications. *Computer Methods in Applied Mechanics and Engineering 414*, 116155.

Ferrari, M., A. Franco, and L. Hegele (2024). Turbulent flow inside a cubic lid-driven cavity using moment representation lattice Boltzmann method. *Physics of Fluids 36* (7), 075104.

Fukami, K., K. Fukagata, and K. Taira (2020). Assessment of supervised machine learning methods for fluid flows. *Theoretical and Computational Fluid Dynamics 34* (4), 497–519.

Gholami, A., H. Bonakdari, A. A. Akhtari, and I. Ebtehaj (2019). A combination of computational fluid dynamics, artificial neural network, and support vectors machines models to predict flow variables in curved channel. *Scientia Iranica 26* (2), 726–741.

Gnanasekaran, M. and A. Satheesh (2024). Numerical analysis of turbulent flow characteristics with the influence of speed ratio in a double-sided cavity. *MethodsX 12*, 102594.

Kashir, B., M. Ragone, A. Ramasubramanian, V. Yurkiv, and F. Mashayek (2021). Application of fully convolutional neural networks for feature extraction in fluid flow. *Journal of Visualization 24*, 771–785.

Khanafer, K. and S. Aithal (2013). Laminar mixed convection flow and heat transfer characteristics in a lid driven cavity with a circular cylinder. *International Journal of Heat and Mass Transfer 66*, 200–209.

Kokash, H., K. Khanafer, and M. Burzo (2024). Machine learning based predictions of flow and heat transfer characteristics in a lid -driven cavity with a rotating cylinder. *Energies 17* (20), 5220.

Kolmogorov, A. N. (1991). The local structure of turbulence in incompressible viscous fluid for very large reynolds numbers. *Proceedings of the Royal Society of London. Series A: Mathematical and Physical Sciences 434* (1890), 9–13.

Kuhlmann, H. C. and F. Romano (2019). The lid-driven cavity. In Alexander Gelfgat (Ed.), *Computational Modelling of Bifurcations and Instabilities in Fluid Dynamics,* Springer Cham. 233–309.

Migeon, C., A. Texier, and G. Pineau (2000). Effects of lid-driven cavity shape on the flow establishment phase. *Journal of Fluids and Structures 14* (4), 469–488.

Minsky, M. and S. Papert (1969). An introduction to computational geometry. *Cambridge Tiass., HIT 479* (480), 104.

Pandey, S., J. Schumacher, and K. R. Sreenivasan (2020). A perspective on machine learning in turbulent flows. *Journal of Turbulence 21* (9–10), 567–584.

Peng, Y.-F., Y.-H. Shiau, and R. R. Hwang (2003). Transition in a 2-d lid-driven cavity flow. *Computers & Fluids 32* (3), 337–352.

Ramanan, N. and G. M. Homsy (1994). Linear stability of lid-driven cavity flow. *Physics of Fluids 6* (8), 2690–2701.

Ranade, R., C. Hill, and J. Pathak (2021). Discretization net: A machine-learning based solver for Navier–stokes equations using finite volume discretization. *Computer Methods in Applied Mechanics and Engineering 378*, 113722.

Rosenblatt, F. (1958). The perceptron: A probabilistic model for information storage and organization in the brain. *Psychological Review 65* (6), 386.

Samantaray, D. and M. K. Das (2019). Nature of turbulence inside a cubical lid-driven cavity: Effect of Reynolds number. *International Journal of Heat and Fluid Flow 80*, 108498.

Samuel, A. L. (1959). Some studies in machine learning using the game of checkers. *IBM Journal of Research and Development 3* (3), 210–229.

Sheu, T. W. and S. Tsai (2002). Flow topology in a steady three-dimensional lid-driven cavity. *Computers & Fluids 31* (8), 911–934.

Temam, R. (2024). *Navier–Stokes equations: Theory and numerical analysis*, Volume 343. American Mathematical Society.

Turkyilmazoglu, M. (2022). Driven flow motion by a dually moving lid of a square cavity. *European Journal of Mechanics-B/Fluids 94*, 17–28.

Wang, B. and J. Wang (2021). Application of artificial intelligence in computational fluid dynamics. *Industrial & Engineering Chemistry Research 60* (7), 2772–2790.

Wiener, N. (2019). *Cybernetics or Control and Communication in the Animal and the Machine.* MIT Press.

Zhu, J., L. E. Holmedal, H. Wang, and D. Myrhaug (2020). Vortex dynamics and flow patterns in a two-dimensional oscillatory lid-driven rectangular cavity. *European Journal of Mechanics-B/Fluids 79*, 255–269.

13 Analyzing Heat and Mass Transfer in an Unsteady MHD Williamson Nanofluid Flow within a Porous Matrix

Numerical and Statistical Investigation

Ragavi Moorthy, Tamalapakula Poornima, and Pandikunta Sreenivasulu

13.1 INTRODUCTION

The investigation of non-Newtonian fluid behavior has been gaining much attention because of its numerous applications in industrial processes like polymer extrusion, food processing, biomedical engineering, and coating technologies. Among the various non-Newtonian models, the Williamson fluid model is especially significant in modeling pseudo-plastic (shear-thinning) fluids, which are characterized by a reduction in apparent viscosity with an increase in shear rate. In contrast to Newtonian fluids of constant viscosity, Williamson fluids give a more realistic description of complex materials like blood, paint, and polymeric solutions. Crane [1] made a pioneering contribution by analyzing boundary layer flow over a linearly stretching sheet, which became the basis for subsequent research in this area. In the context of non-Newtonian fluids, Andersson *et al.* [2] examined the flow of power-law fluids over a stretching sheet, emphasizing how shear-thinning properties influence boundary layer thickness. Ariel [3] focused on the axisymmetric flow of a second-grade fluid over a surface.

Several studies have incorporated additional physical effects. For instance, Vajravelu and Nayfeh [4] examined MHD effects on non-Newtonian flow over a stretching sheet. Later, Srinivas *et al.* [5] included the influence of chemical reactions and thermal radiation, while Nadeem *et al.* [6] analyzed slip boundary conditions. Maxwell nanofluid flow subject to the convective boundary condition was investigated by Hayat *et al.* [7]. Forced convective heat transmission in an electrically conducting Sisko fluid across a nonlinear radially extending sheet was studied analytically by

DOI: 10.1201/9781003611820-16

Khan *et al.* [8]. Shahzad *et al.* [9] assessed heat transfer and boundary layer movement over an extended surface, offering a meticulous and insightful evaluation of numerical and analytical approaches. Using chemical repercussions power and multi-slip factor, Faraz *et al.* [10] investigated the consequences of thermo-diffusion on unstable axisymmetric Casson movement across a time-varying radially extensible sheet.

Rheological fluid behavior with ohmic heating and viscous dissipation is examined in Refs. [11–14]. Zhu *et al.* [15] investigate the Williamson nanofluid mass and heat transfer characteristics flowing on an expanding/contracting surface. The analysis incorporates the repercussions regarding Brownian motion and thermophoresis, making it relevant to various engineering applications involving heat transfer enhancement. Asjad *et al.* [16] studied an unstable Williamson fluid flow with additional effects such as thermal radiation, thermal dissipation, a uniform magnetic field, and chemical reactions. Qayyum *et al.* [17] examined the simulation and evaluation of an unstable Carreau fluid with a magnetohydrodynamical consequence for an elastic sheet. Islam *et al.* [18] focused on the issues of 2D unstable border thickness of thermal and mass transmission movement along the sheet that expands exponentially placed within a permeable medium under a consistent magnetic field. The study included additional factors like suction/blowing, thermal generation/imbibition, and heat radiation.

Mathematical evaluation of heat imbibition along with thermal radiative consequences on 3D MHD Williamson nano liquid flow via extendable sheet is analyzed by Shiva and Reddy [19]. Ramesh *et al.* [20] explored how velocity slip and Newtonian heating parameters affect Williamson nanofluid on a stretching surface with variable conditions. The results of this research have practical applications in industries like petroleum, heat exchangers, conjugate heat transfer around fins, and solar radiation. Saini *et al.* [21] scrutinized the axisymmetric flow characteristics of a non-Newtonian Williamson liquid via an absorbent medium and a radially extending surface that contains nanoparticles. The flow profiles are examined with the combined effects of viscous dissipation and Joule heating. In addition, numerical solutions for porous and non-porous media are compared in the study. Sarfraz *et al.* [22] examined the nanofluid thermal transmission and flow within the zone of stagnation region around a whirling spheroid cylinder under the influence of magnetohydrodynamics (MHD). The Buongiorno model was implemented to look at the thermal characteristics of nanoscale particles, taking into account thermophoresis and Brownian diffusion on concentration and temperature profiles.

According to the preceding literature, there has been no research done on unstable Williamson fluid circulation across a radial extendable surface via a porous medium, including convection using the regression analysis and Keller box method. This chapter examines the dynamic behavior of unsteady Williamson nanofluid flow over a radially extendable sheet, providing a detailed analysis of its complex motion and characteristics. We also consider the consequence of viscous dissipation, ohmic heating, and chemical reactions. By employing similarity variables, the governing nonlinear PDEs are reformulated into a system of ODEs. The numerical solutions of converted equations are obtained through MATLAB® by utilizing the Keller box method. The influence of various pertinent factors on concentration, temperature, and velocity is thoroughly investigated using extensive tables and graphical representations. Furthermore, the skin friction coefficient, Sherwood number, and Nusselt number are computed and presented in a tabular format. This configuration is applicable for investigating heat transfer processes in devices such as heat exchangers, coating processes, and the manufacturing of composite materials.

13.2 MATHEMATICAL FORMULATION

A 2D unsteady laminar incompressible Williamson nanofluid flow over a radial stretching surface is scrutinized. This intricate analysis takes into account the influential presence of porous media, viscous dissipation, chemical reaction, Joule heating, and convective boundary conditions. The sheet coincides with the plane $z = 0$. This structure greatly improves the flow of the system. To provide mathematical information, we utilized the cylindrical polar coordinate scheme (r, θ, z) with a non-uniform transverse magnetic field strength $B = \frac{B_0}{\sqrt{1-\varepsilon t}}$ is applied in the z-direction and the velocity field is $v = [u(r,z,t), 0, w(r,z,t)]$. Flow occurs along the radial direction due to stretching with velocity $U_w = \frac{ar}{1-\varepsilon t}$. The axial and radial components of the velocity are denoted by w and u, respectively. The temperature T of the fluid and the concentration C of particles at the surface change utilizing time 't' in the manner $T_w = T_\infty + \frac{ar}{1-\varepsilon t}$ is the wall temperature and the distance 'r' toward the surface. Figure 13.1 shows a schematic depiction of the physical model under discussion.

Under the mentioned conditions, the boundary layer equation can be expressed as follows [10,23,24]:

$$\frac{\partial u}{\partial r} + \frac{u}{r} + \frac{\partial w}{\partial z} = 0 \tag{13.1}$$

$$\frac{\partial u}{\partial t} + u\frac{\partial u}{\partial r} + w\frac{\partial u}{\partial z} = \nu\left(\frac{\partial^2 u}{\partial z^2} + \sqrt{2}\Gamma\frac{\partial^2 u}{\partial z^2}\frac{\partial u}{\partial z}\right) - \frac{\sigma B^2 u}{\rho} - \frac{\nu u}{K^*} \tag{13.2}$$

$$\frac{\partial T}{\partial t} + u\frac{\partial T}{\partial r} + w\frac{\partial T}{\partial z} = \frac{k}{\rho C_p}\frac{\partial^2 T}{\partial z^2} + \frac{\nu}{\rho C_p}\left(\frac{\partial u}{\partial z}\right)^2 + \frac{\sigma B^2 u^2}{\rho C_p} \tag{13.3}$$

$$\frac{\partial C}{\partial t} + u\frac{\partial C}{\partial r} + w\frac{\partial C}{\partial z} = D_b\frac{\partial^2 C}{\partial z^2} + \frac{D_t}{T_\infty}\frac{\partial^2 T}{\partial z^2} - K(C - C_\infty) \tag{13.4}$$

and the boundary conditions are [25–27]

$$u = U_w,\ \ w = 0,\ -k_f\frac{\partial T}{\partial z} = h_s(T_w - T),\ \ C = C_w \text{ at } z = 0$$

$$u \to u_\infty,\ T \to T_\infty,\ C \to C_\infty \quad \text{as} \quad z \to \infty \tag{13.5}$$

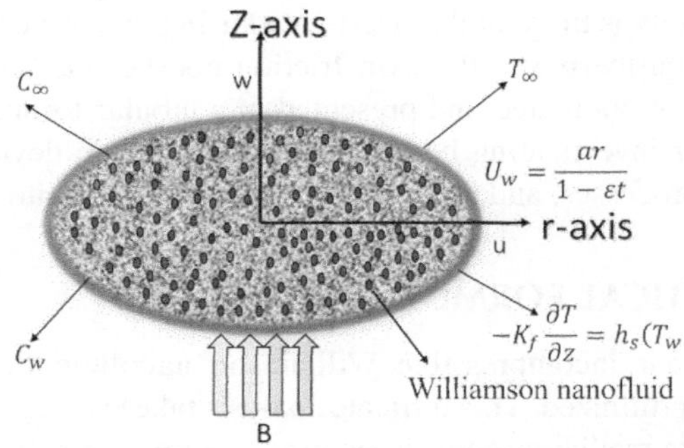

FIGURE 13.1 Physical model of the problem.

The similarity transformations utilized in this study are presented as [28],

$$\psi = \frac{-r^2\sqrt{a\nu_f}}{\sqrt{1-\varepsilon t}} f(\eta),$$

$$\eta = z\sqrt{\frac{a}{\nu_f(1-t\varepsilon)}},$$

$$\theta(\eta) = \frac{T-T_\infty}{T_w-T_\infty}, \tag{13.6}$$

$$\phi(\eta) = \frac{C-C_\infty}{C_w-C_\infty}.$$

The following describes the velocity transverse and longitudinal components are

$$u = \frac{1}{r}\frac{\partial\psi}{\partial z} = U_w f'(\eta),\; w = \frac{-1}{r}\frac{\partial\psi}{\partial r} = -2\sqrt{\frac{a\vartheta}{1-\varepsilon t}} f(\eta) \tag{13.7}$$

The PDEs in Eq. (13.2)–(13.4) are modified as a set of ODEs, which are presented in Eqs. (13.8)–(13.10).

$$f''' + Wef''f''' + 2ff'' - (f')^2 - J\left(f' + \frac{\eta}{2}f''\right) - Mf' - Df' = 0 \tag{13.8}$$

$$\frac{\theta''}{Pr} + 2f\theta' - f'\theta + E_t(f'')^2 - J\left(\theta + \frac{\eta}{2}\theta'\right) + ME_t(f')^2 = 0 \tag{13.9}$$

$$\phi'' + \frac{Nt}{Nb}\theta'' - L_e\left[J\left(\phi + \frac{\eta}{2}\phi'\right) + 2f'\phi - 2f\phi'\right] - N_r L_e\phi = 0 \tag{13.10}$$

Now, the boundary condition (13.5) transforms into

$$\text{at } \eta = 0, \quad f(\eta) = 0, \; f'(\eta) = 1, \quad \theta'(\eta) = -B_t\left(1 - \theta(\eta)\right), \; \phi(\eta) = 1,$$

$$\text{as } \eta \to \infty, \qquad f'(\eta) \to 0, \theta(\eta) \to 0, \phi(\eta) \to 0 \tag{13.11}$$

Here are a few constants that do not have dimensions: $E_t = \frac{a\nu_f}{C_p(T_w - T_\infty)}$ is the Eckert number; $D = \frac{(1-\varepsilon t)\mu_f}{\rho_f a K_1^*}$ is the porosity variable, $J = \frac{\varepsilon}{a}$ refers unsteadiness parameter, $\Pr = \frac{\rho C_p \nu_f}{k_f}$ is Prandtl number, $M = \frac{B_0^2 \sigma_f}{a\rho_f}$ is the magnetic variable, $B_t = \frac{h_0}{k}\sqrt{\frac{\nu_f}{a}}$ is Biot number, $\mathrm{Re} = \frac{rU_w}{\nu_f}$ represent Reynolds number, $N_b = \frac{D_b(c_w - c_\infty)}{\nu_f}$ represent Brownian motion parameter, $N_r = \frac{K(1-\varepsilon t)}{a}$ is the chemical reaction parameter, $N_t = \frac{D_t(T_w - T_\infty)}{\nu_f T_\infty}$ is thermophoresis parameter, $L_e = \frac{\nu_f}{D_b}$ represent Lewis number, $\mathrm{We} = \sqrt{2}\Gamma\sqrt{\frac{a}{\nu_f(1-\varepsilon t)}}\frac{ar}{(1-\varepsilon t)}$ is the Williamson parameter.

Problem physical quantities:

The important physical parameters that influence the governing flow problem consist of the skin friction (Cf), the Sherwood number (Sh), and the Nusselt number (Nu), defined as follows:

$$C_f = \frac{\tau_w}{\rho_f U_w^{\,2}}, \quad \mathrm{Sh} = \frac{rq_m}{D_B\left(C_w - C_\infty\right)}, \quad \mathrm{Nu} = \frac{rq_w}{k_f\left(T_w - T_\infty\right)} \tag{13.12}$$

where $\tau_w = \mu_f\left[\frac{\partial u}{\partial z} + \sqrt{2}\Gamma\left(\frac{\partial u}{\partial z}\right)^2\right]$, $q_m = -D_B\frac{\partial C}{\partial z}$ and $q_w = -k_f\frac{\partial T}{\partial z}$ at $z = 0$.

The dimensionless expressions of Eq. (13.12) are obtained in the following format.

$$C_f\sqrt{Re} = f''(0)\left[1 + \mathrm{We} f''(0)\right],$$

$$\mathrm{Re}^{-1}/2Nu = -\theta'(0),] \tag{13.13}$$

$$\mathrm{Re}^{-1}/2Sh = -\phi'(0).$$

13.3 NUMERICAL PROCEDURE

A non-dimensional momentum and energy Eqs. (13.8–13.10) along with the corresponding boundary conditions (13.11) are solved numerically using the Keller box method via MATLAB®. This approach exhibits a level of accuracy that is of second order. The solution can be reached using the following procedures:

- Convert the system of differential equations to a set of first-order equations.
- Utilizing centered-difference approximations to write equations in finite difference form.
- Linearize the equations using Newton's linearization approach (if they are nonlinear).
- Compute the solution of the reduced system by employing the block tridiagonal elimination method.

New dependent variables are introduced such as l, s, q, and p to rewrite the equations into a first-order system.

$$f' = l \tag{13.14}$$

$$l' = s \tag{13.15}$$

$$\theta' = q \tag{13.16}$$

$$\phi' = p \tag{13.17}$$

Then,

$$s' + Wess' + 2fs - (l)^2 - J\left(l + \frac{\eta}{2}s\right) - l(M + D) = 0 \tag{13.18}$$

$$\frac{q'}{Pr} + 2fq - l\theta + Ec(s)^2 - J\left(\theta + \frac{\eta}{2}q\right) + ME_t l^2 = 0 \tag{13.19}$$

$$p' + \frac{Nt}{Nb}q' - L_e\left[J\left(\phi + \frac{\eta}{2}q\right) + l\phi - 2fp\right] - L_e N_r \phi = 0 \tag{13.20}$$

The boundary condition becomes

$$f(\eta) = 0,\ l(\eta) = 1,\ v(\eta) = -B_t\left(1 - \theta(\eta)\right),\ \phi(\eta) = 1 \quad \text{at} \quad \eta = 0$$

$$l(\eta) = 0,\ \theta(\eta) = 0,\ \phi(\eta) = 0 \quad \text{as} \quad \to \infty \tag{13.21}$$

Finite Difference Technique:

Discretization of the governing Eqs. (13.14–13.20) via the finite difference approximation yields a system of algebraic relationships

$$f_j - f_{j-1} - \left(l_j - l_{j-1}\right)\frac{h_j}{2} = 0 \tag{13.22}$$

$$l_j - l_{j-1} - \left(s_j - s_{j-1}\right)\frac{h_j}{2} = 0 \tag{13.23}$$

$$\theta_j - \theta_{j-1} - \left(q_j - q_{j-1}\right)\frac{h_j}{2} = 0 \tag{13.24}$$

$$\phi_j-\phi_{j-1}-\left(p_j-p_{j-1}\right)\frac{h_j}{2}=0 \tag{13.25}$$

$$\begin{aligned}&s_j-s_{j-1}+\frac{We}{2}\left(s_j+s_{j-1}\right)\left(s_j-s_{j-1}\right)+\frac{h_j}{4}\left(f_j+f_{j-1}\right)\left(s_j+s_{j-1}\right)-\frac{h_j}{4}\left(l_j+l_{j-1}\right)^2\\&-\frac{h_j}{4}J\left[\left(l_j+l_{j-1}\right)+\frac{\eta}{2}\left(s_j+s_{j-1}\right)\right]-\frac{h_j}{2}\left(l_j+l_{j-1}\right)\left(M+D\right)=0\end{aligned} \tag{13.26}$$

$$\begin{aligned}&\frac{1}{Pr}\left[q_j-q_{j-1}\right]-\frac{h_j}{4}\left(l_j+l_{j-1}\right)\left(\theta_j+\theta_{j-1}\right)+E_t\frac{h_j}{4}\left(s_j+s_{j-1}\right)^2-\frac{J}{2}\frac{h_j}{2}\left[\left(\theta_j+\theta_{j-1}\right)+\frac{\eta}{2}\left(q_j+q_{j-1}\right)\right]\\&+ME_t\frac{h_j}{4}\left(l_j+l_{j-1}\right)^2+\left(f_j+f_{j-1}\right)\left(q_j+q_{j-1}\right)2\frac{h_j}{4}=0\end{aligned} \tag{13.27}$$

$$\begin{aligned}&p_j-p_{j-1}+\frac{Nt}{Nb}\left(q_j-q_{j-1}\right)-\frac{h_j}{4}JL_e\left[\left(\phi_j+\phi_{j-1}\right)+\frac{\eta}{2}\left(p_j+p_{j-1}\right)\right]\\&-\frac{h_j}{4}L_e(l_j+l_{j-1})(\phi_j+\phi_{j-1})+L_e\cdot 2\cdot\frac{h_j}{4}(f_j+f_{j-1})(p_j+p_{j-1})\\&-\frac{h_j}{2}L_eK_r(\phi_j+\phi_{j-1})=0\end{aligned} \tag{13.28}$$

These equations are imposed for j = 1, 2, 3, … … – 1 and at j=0, j=J, the boundary conditions are

$$\begin{gathered}f_0=0,\ \ l_0=1,\ \ q_0=-Bi\left(1-\theta_0\right),\ \ \phi_0=1\\ l_J=0,\ \ \theta_J=0,\ \ \phi_J=0\end{gathered} \tag{13.29}$$

Newton's Technique of Linearization:

To address the challenge of the nonlinear Eqs. (13.22–13.28), we utilized Newton's method and introduced a subsequent $\left[f_j^i,\ l_j^i,\ s_j^i,\ \theta_j^i,\ q_j^i,\ \phi_j^i,\ p_j^i\right]$ where i = 0, 1, 2, ... For subsequent iterates, we established the following:

$$\begin{gathered}f_j^{i+1}=f_j^i+\delta f_j^i,\ \ l_j^{i+1}=l_j^i+\delta l_j^i,\ \ s_j^{i+1}=s_j^i+\delta s_j^i,\\ \theta_j^{i+1}=\theta_j^i+\delta\theta_j^i,\qquad q_j^{i+1}=q_j^i+\delta q_j^i\\ \phi_j^{i+1}=\phi_j^i+\delta\phi_j^i,\qquad p_j^{i+1}=p_j^i+\delta p_j^i.\end{gathered} \tag{13.30}$$

Following the substitution of the right-hand side expressions from Eqs. (13.22–13.28) and subsequent linearization through the neglect of higher-order terms in the δ. The derived linear equations are (i has been excluded):

$$\delta f_j-\delta f_{j-1}-\frac{h_j}{2}\left(\delta l_j-\delta l_{j-1}\right)=\left(r_1\right)_j,$$

$$\delta l_j-\delta l_{j-1}-\frac{h_j}{2}\left(\delta s_j-\delta s_{j-1}\right)=\left(r_2\right)_j,$$

$$\delta\theta_j-\delta\theta_{j-1}-\frac{h_j}{2}\left(\delta q_j-\delta q_{j-1}\right)=\left(r_3\right)_j,$$

$$\delta\phi_j-\delta\phi_{j-1}-\frac{h_j}{2}\left(\delta p_j-\delta p_{j-1}\right)=\left(r_4\right)_j \tag{13.31}$$

$$\left(a_1\right)_j\delta s_j+\left(a_3\right)_j\delta l_j+\left(a_5\right)_j\delta f_j+\left(a_2\right)_j\delta s_{j-1}+\left(a_4\right)_j\delta l_{j-1}+\left(a_6\right)_j\delta f_{j-1}=\left(r_5\right)_j$$

$$\begin{aligned}\left(b_1\right)_j\delta q_j+\left(b_3\right)_j\delta l_j+\left(b_5\right)_j\delta f_j+\left(b_7\right)_j\delta\theta_j+\left(b_9\right)_j\delta s_j+\left(b_2\right)_j\delta q_{j-1}+\left(b_4\right)_j\delta l_{j-1}\\+\left(b_6\right)_j\delta f_{j-1}+\left(b_8\right)_j\delta\theta_{j-1}+\left(b_{10}\right)_j\delta s_{j-1}=\left(r_6\right)_j,\end{aligned}$$

$$\begin{aligned}(c_1)_j\delta p_j+(c_3)_j\delta\phi_j+(c_5)_j\delta q_j+(c_7)_j\delta f_j+(c_9)_j\delta l_j\\+(c_2)_j\delta p_{j-1}+(c_4)_j\delta\phi_{j-1}+(c_6)_j\delta q_{j-1}+(c_8)_j\delta f_{j-1}+(c_{10})_j\delta l_{j-1}=(r_7)_j\end{aligned}$$

where,

$$\left(a_1\right)_j=1+Wes_{j-\frac{1}{2}}+h_jf_{j-\frac{1}{2}}-\tfrac{h_j}{2}\tfrac{\eta}{2}J,$$

$$\left(a_2\right)_j=-1-Wes_{j-\frac{1}{2}}+h_jf_{j-\frac{1}{2}}-\tfrac{h_j}{2}\tfrac{\eta}{2}J,$$

$$\left(a_3\right)_j=-h_jl_{j-\frac{1}{2}}-\tfrac{h_j}{2}J-(M+D)\tfrac{h_j}{2}=\left(a_4\right)_j,$$

$$\left(a_5\right)_j=h_js_{j-\frac{1}{2}}=\left(a_6\right)_j,$$

$$\left(b_1\right)_j=\tfrac{1}{Pr}+\tfrac{2h_j}{2}f_{j-\frac{1}{2}}-\tfrac{h_j}{2}\tfrac{\eta}{2}J,$$

$$\left(b_2\right)_j=-\tfrac{1}{Pr}+\tfrac{2h_j}{2}f_{j-\frac{1}{2}}-\tfrac{h_j}{2}\tfrac{\eta}{2}J,$$

$$\left(b_3\right)_j=-2\tfrac{h_j}{4}\theta_{j-\frac{1}{2}}+ME_t\tfrac{h_j}{2}l_{j-\frac{1}{2}}=\left(b_4\right)_j,$$

$$\left(b_5\right)_j=\tfrac{2h_j}{2}q_{j-\frac{1}{2}}=\left(b_6\right)_j,$$

$$\left(b_7\right)_j=-J\tfrac{h_j}{2}-2\tfrac{h_j}{4}l_{j-\frac{1}{2}}=\left(b_8\right)_j,$$

$$\left(b_9\right)_j=E_t\tfrac{h_j}{2}s_{j-\frac{1}{2}}=\left(b_{10}\right)_j,$$

$$\left(c_1\right)_j=1-\tfrac{\eta}{2}\tfrac{h_j}{2}L_eJ+2L_e\tfrac{h_j}{2}f_{j-\frac{1}{2}},$$

$$(c_2)_j = -1 - \frac{\eta}{2}\frac{h_j}{2}L_e J + 2L_e\frac{h_j}{2}f_{j-\frac{1}{2}},$$

$$(c_3)_j = -J\frac{h_j}{2}L_e - L_e\frac{h_j}{2}l_{j-\frac{1}{2}} - \frac{h_j}{2}L_e N = (-c_4)_j,$$

$$(c_5)_j = \frac{Nt}{Nb} = (-c_6)_j,$$

$$(c_7)_j = 2L_e\frac{h_j}{2}p_{j-\frac{1}{2}} = (c_8)_j,$$

$$(c_9)_j = -2L_e\frac{h_j}{2}\phi_{j-\frac{1}{2}} = (c_{10})_j,$$

$$(r_1)_j = f_{j-1} - f_j + h_j l_{j-\frac{1}{2}},$$

$$(r_2)_j = l_{j-1} - l_j + h_j s_{j-\frac{1}{2}},$$

$$(r_3)_j = \theta_{j-1} - \theta_j + h_j q_{j-\frac{1}{2}},$$

$$(r_4)_j = \phi_{j-1} - \phi_j + h_j p_{j-\frac{1}{2}},$$

$$(r_5)_j = s_{j-1} - s_j + \frac{We}{2}\left(s_{j-1}^2 - s_j^2\right) - 2h_j f_{j-\frac{1}{2}} s_{j-\frac{1}{2}} + h_j\left(l_{j-\frac{1}{2}}\right)^2 + Jh_j\left[l_{j-\frac{1}{2}} + \frac{\eta}{2}s_{j-\frac{1}{2}}\right] + h_j l_{j-\frac{1}{2}}(M+D),$$

$$(r_6)_j = \frac{1}{Pr}\left(q_{j-1} - q_j\right) - 2h_j f_{j-\frac{1}{2}} q_{j-\frac{1}{2}} + h_j l_{j-\frac{1}{2}}\theta_{j-\frac{1}{2}} - E_t h_j\left(s_{j-\frac{1}{2}}\right)^2 - ME_t\left(l_{j-\frac{1}{2}}\right)^2 h_j + J\left[\theta_{j-\frac{1}{2}} + \frac{\eta}{2}q_{j-\frac{1}{2}}\right]h_j,$$

$$(r_7)_j = (p_{j-1} - p_j) + \frac{Nt}{Nb}(q_{j-1} - q_j) + L_e h_j J\left(\phi_{j-\frac{1}{2}} + \frac{\eta}{2}p_{j-\frac{1}{2}}\right) + h_j L_e l_{j-\frac{1}{2}}\phi_{j-\frac{1}{2}} - 2L_e h_j f_{j-\frac{1}{2}} p_{j-\frac{1}{2}} + L_e h_j N\phi_{j-\frac{1}{2}}.$$

Structure of Block Tridiagonal Matrix:
The system of linearized Eqs. (13.31) can be written in matrix notation as follows:

$$\begin{bmatrix} [A_1] & [C_1] & & & & & \\ [B_2] & [A_2] & [C_2] & & & & \\ & [B_3] & [A_3] & [C_3] & & & \\ & & & \ddots & \cdots & & \\ & & & & [B_{j-1}] & [A_{j-1}] & [C_{j-1}] \\ & & & & & [B_j] & [A_j] \end{bmatrix} \begin{bmatrix} [\delta_1] \\ [\delta_2] \\ [\delta_3] \\ \vdots \\ [\delta_{j-1}] \\ [\delta_J] \end{bmatrix} = \begin{bmatrix} [r_1] \\ [r_2] \\ [r_3] \\ \vdots \\ [r_{j-1}] \\ [r_j] \end{bmatrix}$$

or

$$[A][\delta]=[r] \tag{13.32}$$

$$[A_1]=\begin{bmatrix} 0 & 0 & 0 & 1 & 0 & 0 & 0 \\ -h/2 & 0 & 0 & 0 & -h/2 & 0 & 0 \\ 0 & -h/2 & 0 & 0 & 0 & -h/2 & 0 \\ 0 & 0 & -h/2 & 0 & 0 & 0 & -h/2 \\ (a_2)_1 & 0 & 0 & (a_5)_1 & (a_1)_1 & 0 & 0 \\ (b_{10})_1 & (b_2)_1 & 0 & (b_5)_1 & (b_9)_1 & (b_1)_1 & 0 \\ 0 & (c_6)_1 & (c_2)_1 & (c_7)_1 & 0 & (c_5)_1 & (c_1)_1 \end{bmatrix} \tag{13.33}$$

for $2 \le j \le J$,

$$[A_j]=\begin{bmatrix} -h/2 & 0 & 0 & 1 & 0 & 0 & 0 \\ -1 & 0 & 0 & 0 & -1/2 & 0 & 0 \\ 0 & -1 & 0 & 0 & 0 & -1/2 & 0 \\ 0 & 0 & -1 & 0 & 0 & 0 & -1/2 \\ (a_4)_j & 0 & 0 & (a_5)_j & (a_1)_j & 0 & 0 \\ (b_4)_j & (b_8)_j & 0 & (b_5)_j & (b_9)_j & (b_1)_j & 0 \\ (c_{10})_j & 0 & (c_4)_j & (c_7)_j & 0 & (c_5)_j & (c_1)_j \end{bmatrix} \tag{13.34}$$

$$[B_j]=\begin{bmatrix} 0 & 0 & 0 & -1 & 0 & 0 & 0 \\ 0 & 0 & 0 & 0 & -1/2 & 0 & 0 \\ 0 & 0 & 0 & 0 & 0 & -1/2 & 0 \\ 0 & 0 & 0 & 0 & 0 & 0 & -1/2 \\ 0 & 0 & 0 & (a_6)_j & (a_2)_j & 0 & 0 \\ 0 & 0 & 0 & (b_6)_j & (b_{10})_j & (b_2)_j & 0 \\ 0 & 0 & 0 & (c_8)_j & 0 & (c_6)_j & (c_2)_j \end{bmatrix} \tag{13.35}$$

$$[C_j] = \begin{bmatrix} -h/2 & 0 & 0 & 0 & 0 & 0 & 0 \\ 1 & 0 & 0 & 0 & 0 & 0 & 0 \\ 0 & 1 & 0 & 0 & 0 & 0 & 0 \\ 0 & 0 & 1 & 0 & 0 & 0 & 0 \\ (a_3)_j & 0 & 0 & 0 & 0 & 0 & 0 \\ (b_3)_j & (b_7)_j & 0 & 0 & 0 & 0 & 0 \\ (c_9)_j & 0 & (c_3)_j & 0 & 0 & 0 & 0 \end{bmatrix} \tag{13.36}$$

We will proceed to solve Eq. (13.32) using LU decomposition, under the assumption that matrix A is non-singular and can be resolved into

$$[A] = [L]\,[U] \tag{13.37}$$

where,

$$[L] = \begin{bmatrix} [\alpha_1] & & & & \\ [\beta_2] & [\alpha_2] & & & \\ & & \ddots & & \\ & & & [\alpha_{J-1}] & \\ & & & [\beta_J] & [\alpha_J] \end{bmatrix}, \quad [U] = \begin{bmatrix} [I_1] & [\gamma_1] & & & \\ & [I_2] & [\gamma_2] & & \\ & & \ddots & & \\ & & & [I_{J-1}] & [\gamma_{J-1}] \\ & & & & [I_J] \end{bmatrix} \tag{13.38}$$

These $[\alpha_j]$, $[\beta_j]$, $[\gamma_j]$ are organized in 7×7 matrices, the resultant equations are provided as follows:

By plugging Eq. (13.37) into Eq. (13.32), we obtain

$$[L]\,[U]\,[\delta] = [r] \tag{13.39}$$

Let

$$[U]\,[\delta] = [S] \tag{13.40}$$

Then,

$$[L]\,[S] = [r] \tag{13.41}$$

where,

$$[S] = \begin{bmatrix} [S_1] \\ [S_2] \\ \vdots \\ [S_{J-1}] \\ [S_J] \end{bmatrix} \tag{13.42}$$

In this scenario, $[S_j]$ represents 7×1 column matrix structure with components obtained through the solution of Eq. (13.37) such as

$$[\alpha_1]\,[S_1] = [r_1],\quad [\alpha_j]\,[S_j] = [r_j - \beta_j]\,[S_{j-1}],\quad 2 \le j \le J \tag{13.43}$$

By using a forward sweep, the values of γ_j, α_j, and S_j are determined. In addition, backward sweeps are utilized to obtain the δ elements through Eq. (13.40), and the obtained values are

$$[\delta_J] = [S_J],\quad [\delta_j] = [S_j] - [\gamma_j]\,[\delta_{j+1}],\quad 1 \le j \le J-1 \tag{13.44}$$

The iteration processes are continuously performed until the desired level of accuracy, represented by $\left|\delta v_o^{(i)}\right| < \varepsilon$ is achieved.

13.4 OUTCOMES AND ANALYSIS

The interpretation of increasing physical parameters, such as thermophoresis parameter (N_t), Eckert number (E_t), porosity parameter (D), Biot number (B_t), unsteadiness parameter (J), Brownian motion (N_b), magnetic parameter (M), Lewis number (L_e), chemical reaction parameter (N_r) on spatial distributions of temperature, velocity, and concentration are discussed in this section. Parameters Pr = 0.7, $M = J = 0.2$, $N_b = B_t = N_t = 0.5$, $E_t = N_r = 0.3$, $L_e = 1$, and We = 0.2 are held constant in this investigation.

13.4.1 Effect of Physical Parameter Variation on the Fluid Velocity Field

The flow pattern for changing magnetic variables is displayed in Figure 13.2. Raising the magnetic parameter results in a decline in the flow distribution due to generating a resistive force (Lorentz force), which hinders fluid motion. Figure 13.3 highlights the consequences of the unsteadiness parameter (J). Increases J, the velocity profile decreases because the fluid flow changes over time. The flow may become less steady and more oscillatory as the unsteadiness variable is elevated. The phenomenon results from increased unsteadiness affecting the fluid flow, causing fluctuations in

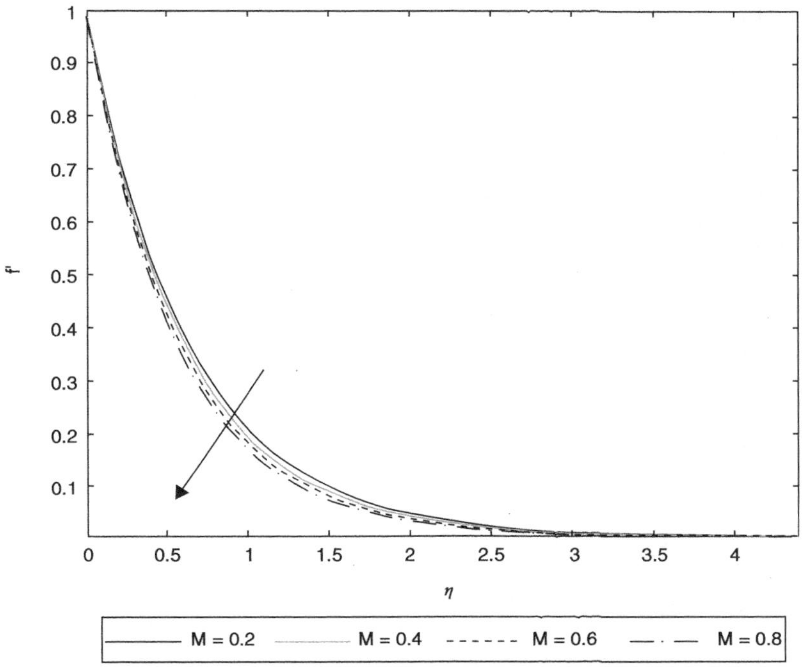

FIGURE 13.2 Modulation of velocity profile by magnetic parameter (M).

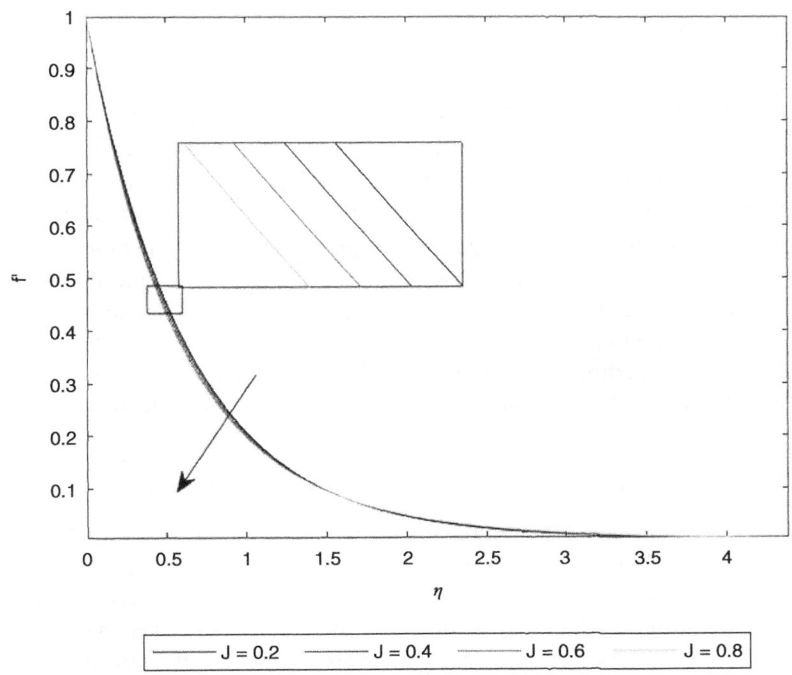

FIGURE 13.3 Modulation of velocity profile by unsteadiness parameter (J).

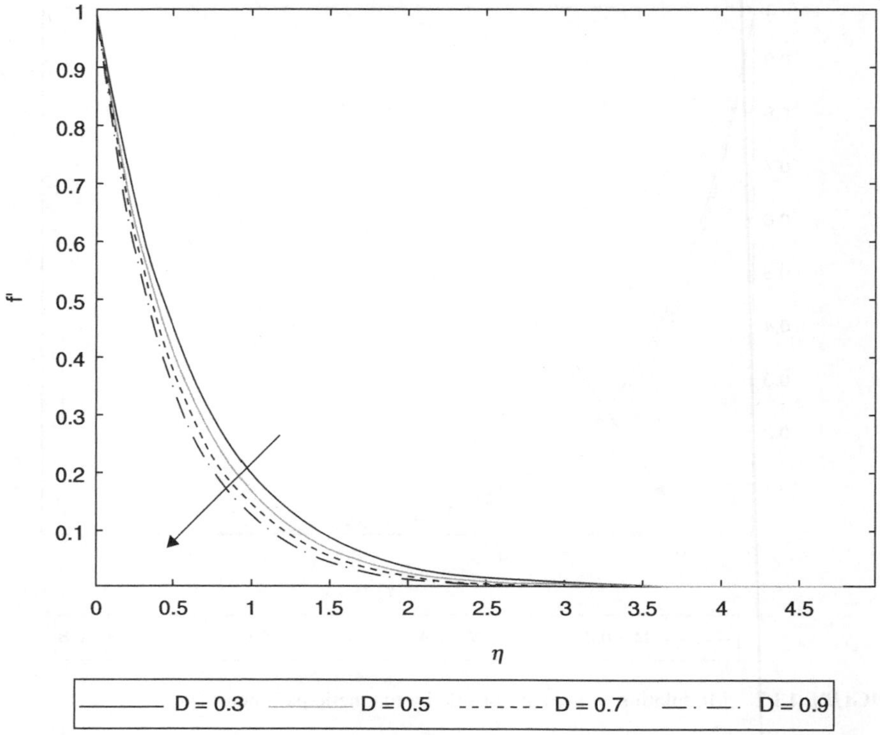

FIGURE 13.4 Modulation of velocity profile by porosity parameter (D).

velocity components. Figure 13.4 shows the impact of the porosity parameter (D) on $f'(\eta)$. The velocity profile decreases as the porosity parameter is increased. This phenomenon occurs because of the porous medium's capacity to resist the movement of the fluid, resulting in a rapid decrease in velocity towards the surface of the sheet. Figure 13.5 shows the Williamson parameter impact on $f'(\eta)$. The ratio of the relaxation time (time for the fluid to return to equilibrium after stress) to the retardation time (time for the internal structure of the fluid to break down under stress). If the viscosity increases, the fluid provides a greater barrier to flow. This resistance functions as a drag force, impeding the motion of fluid particles and therefore decreasing the total velocity. As the Williamson variable rises, the velocity profile diminishes.

13.4.2 Effect of Parameter Variation on the Fluid Energy

Figure 13.6 illustrates the effect of the magnetic parameter on the thermal profile. An increase in magnetic parameter tends to augment the thermal relocation amount due to the suppression of boundary layer thickness, which results in a higher temperature profile along the sheet. Figure 13.7 illustrates the changes in Eckert number (E_t) on $\theta'(\eta)$. It has been observed that raising the Eckert number has the potential to lead to an elevated temperature. The dimensionless Eckert number characterizes the effect of

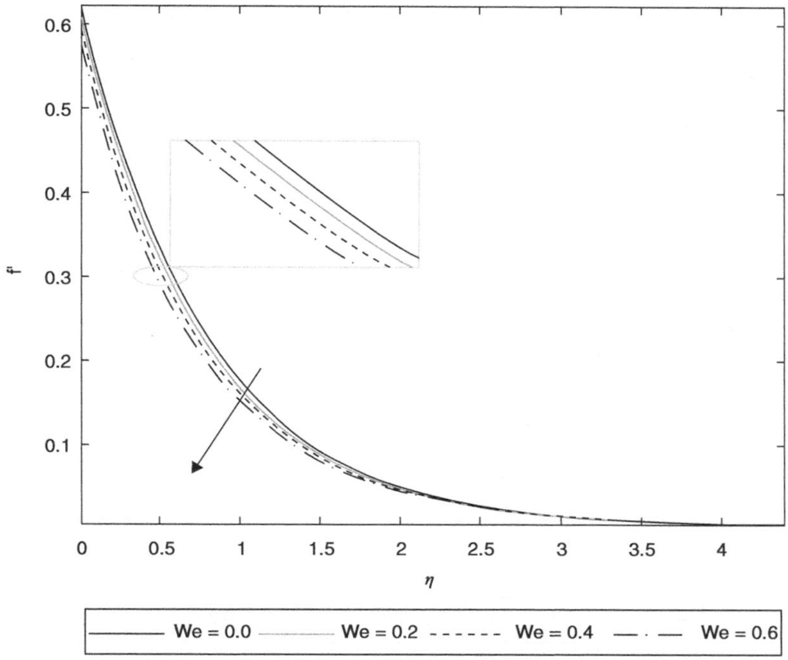

FIGURE 13.5 Modulation of velocity profile by Williamson parameter (We).

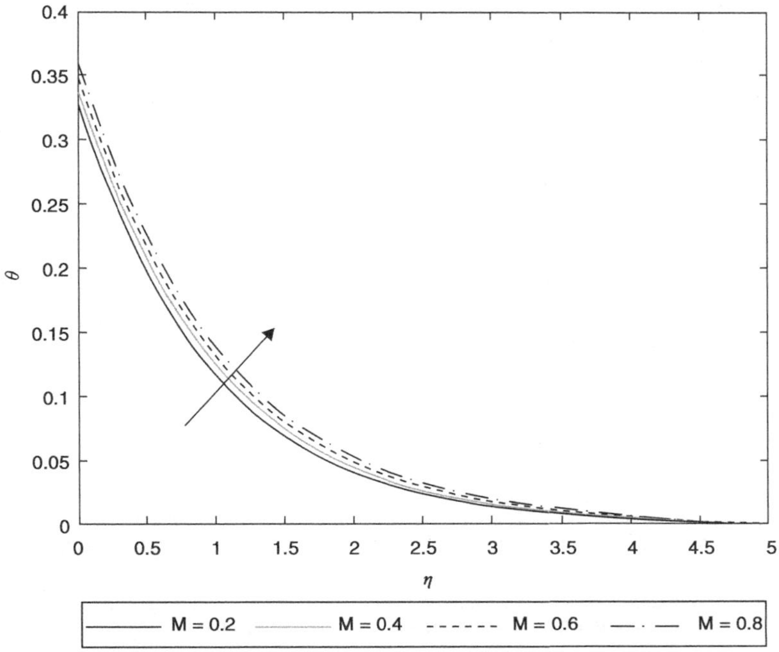

FIGURE 13.6 Modulation of temperature profile by magnetic parameter (M).

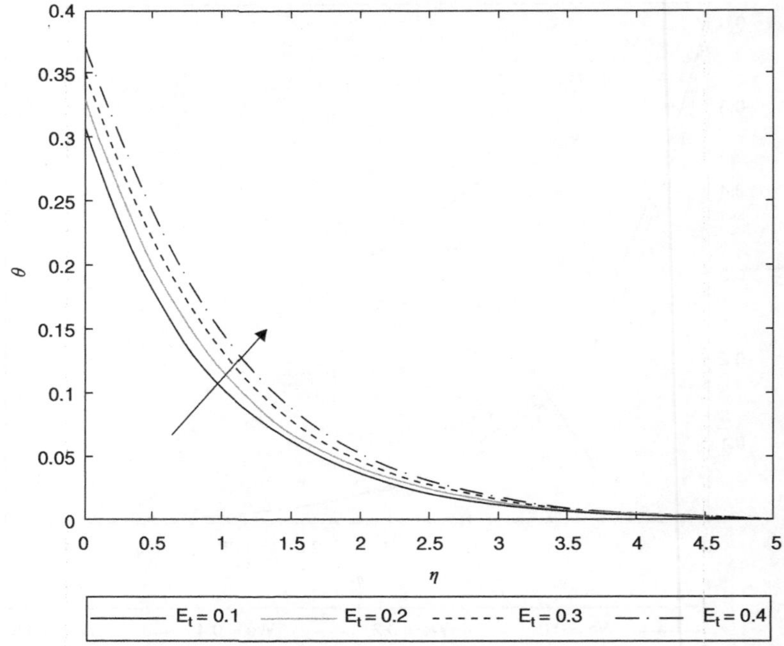

FIGURE 13.7 Modulation of temperature profile by Eckert number (E_t).

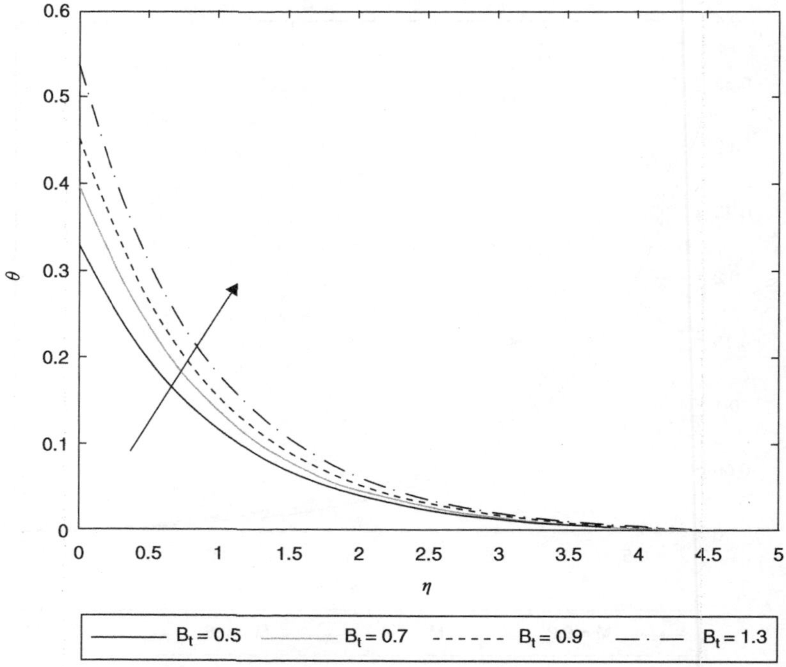

FIGURE 13.8 Modulation of temperature profile by Biot number (B_t).

thermal dissipation on fluids. It describes the influence of fluid auto-heating caused by energy dissipation phenomena. A consequence of B_t is shown in Figure 13.8. A Biot number is a dimensionless group that compares the relative transport resistance between external and internal heat transfer. Increasing the Biot number can enhance convective cooling and potentially lead to an increase in temperature near the surface.

13.4.3 Effect of Parameter Variation on Species Diffusion

The implications of the chemical reaction parameter (N_r) are emphasized in Figure 13.9. Concentration profiles tend to get flatter as the chemical reaction parameter is increased. This is due to the accelerated consumption of reactants at a higher reaction rate, leading to a reduction in their concentration within the system. Figure 13.10 shows the consequence of the Lewis number (L_e) on (Φ'). It signifies the relationship between thermal diffusivity and mass diffusivity. A rise in the Lewis number leads to a notable decrease in the concentration profile. This occurs because a higher Lewis number signifies faster heat diffusion relative to mass diffusion, resulting in a decline in particle concentration. Figure 13.11 displays the Brownian motion's impact on concentration. The microscopic movement of particles due to their collisions with

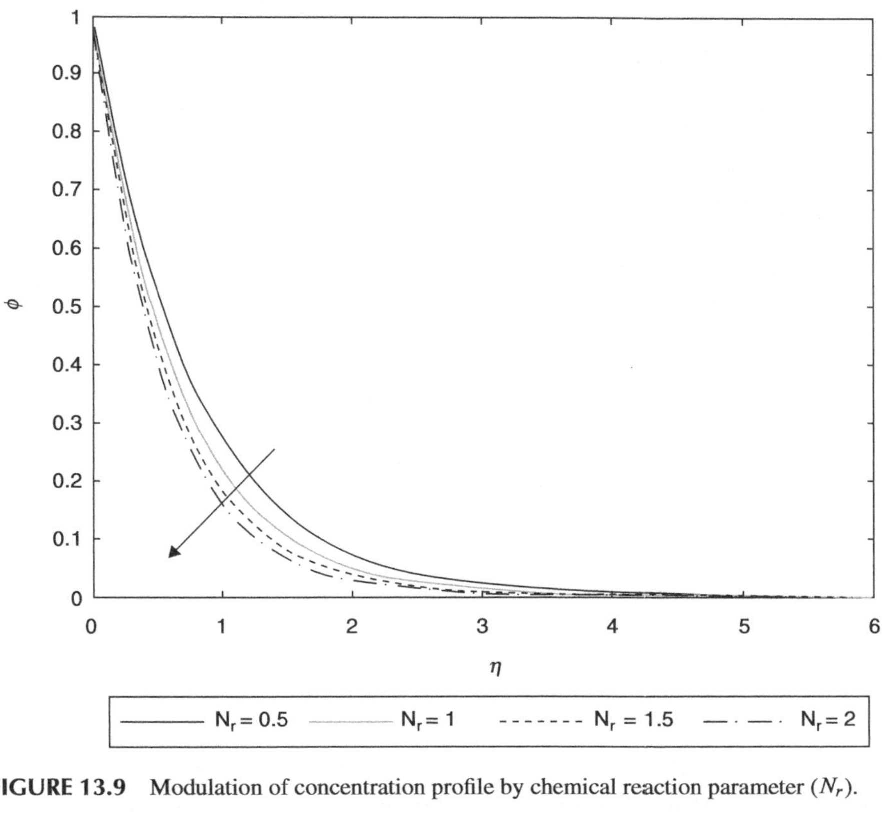

FIGURE 13.9 Modulation of concentration profile by chemical reaction parameter (N_r).

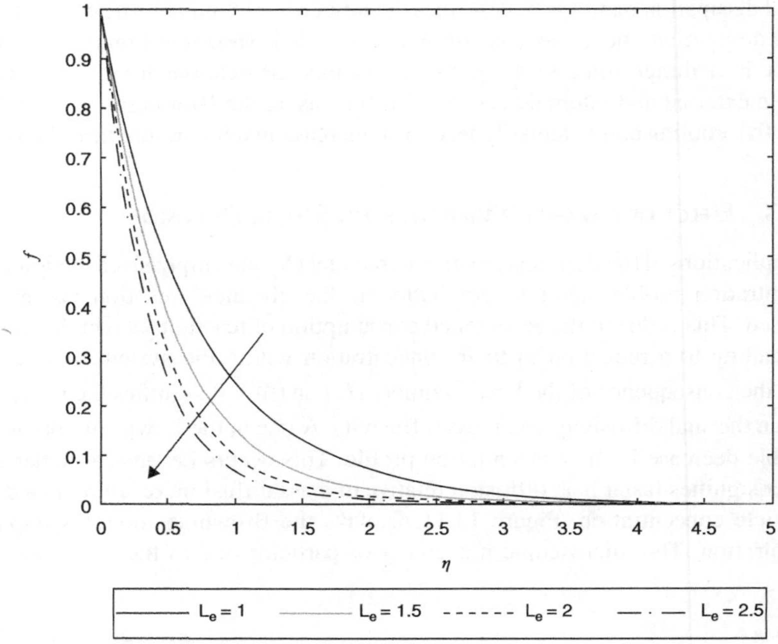

FIGURE 13.10 Modulation of concentration profile by Lewis number (L_e).

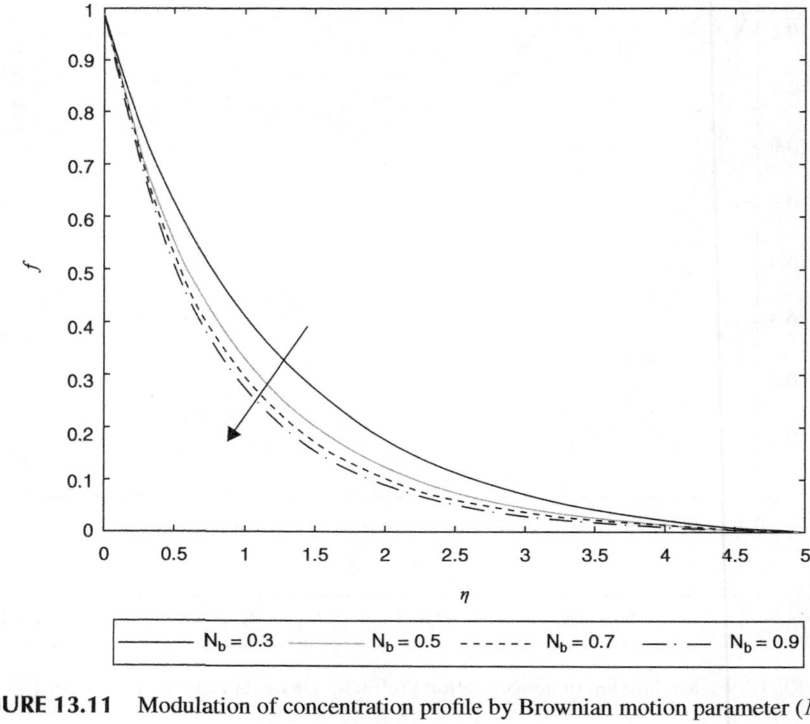

FIGURE 13.11 Modulation of concentration profile by Brownian motion parameter (N_b).

surrounding fluid molecules is known as Brownian motion. A higher N_b indicates stronger Brownian motion, meaning particles experience more frequent and intense collisions. Increasing the Brownian motion variable generally leads to a decrease in the concentration profile due to enhanced diffusion. The impact of the thermophoresis parameter is shown in Figure 13.12. An increase in N_t can lead to a gradual increase in the concentration profile. This phenomenon arises because stronger thermophoresis compels particles to migrate toward zones with higher temperatures.

13.4.4 Effect of Parameter Variation on the Engineering Quantities

Table 13.1 displays the impact of regulating factors on the skin friction coefficient. According to the table's data, the friction factor decreases as magnetic, unsteadiness, and porosity variables are increased. The results of the Nusselt number are displayed in Table 13.2. Heat transfer is enhanced by optimizing the magnet, Biot, and Eckert numbers. The effect of several parameters on the Sherwood number is displayed in Table 13.3. For higher values of the Brownian motion parameter (N_b), Lewis number (L_e), and unsteadiness parameter (J), the local Sherwood number drops, whereas it increases for the thermophoresis parameter (N_t) levels.

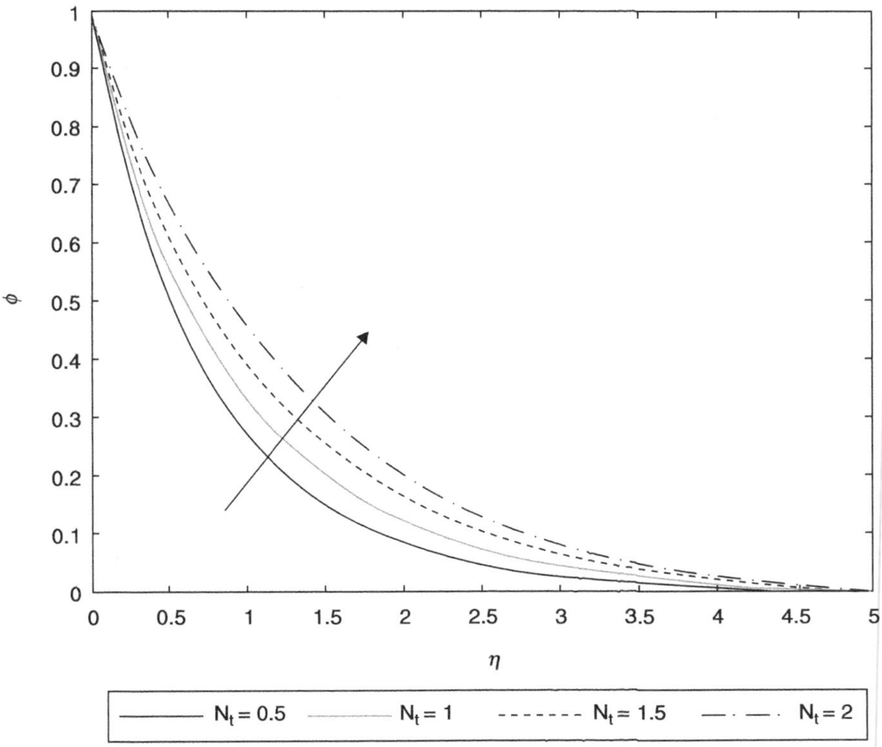

FIGURE 13.12 Modulation of concentration profile by thermophoresis parameter (N_t).

TABLE 13.1
Numerical Values of Skin Friction Coefficient

M	*D*	*J*	We	$C_f\sqrt{Re}$	Predicted
0.2	0.2	0.2	0.2	-1.06508	-1.04421
0.4	0.2	0.2	0.2	-1.09844	-1.0834
0.6	0.2	0.2	0.2	-1.12773	-1.12258
0.8	0.2	0.2	0.2	-1.15332	-1.16176
0.2	0.3	0.2	0.2	-1.08229	-1.06211
0.2	0.5	0.2	0.2	-1.11357	-1.09789
0.2	0.7	0.2	0.2	-1.14097	-1.13368
0.2	0.9	0.2	0.2	-1.16481	-1.16946
0.2	0.3	0.2	0.2	-1.08229	-1.06211
0.2	0.3	0.4	0.2	-1.10506	-1.09221
0.2	0.3	0.6	0.2	-1.12617	-1.12232
0.2	0.3	0.8	0.2	-1.14557	-1.15242
0.2	0.3	0.2	0.1	-1.26596	-1.37416
0.2	0.3	0.2	0.2	-1.08229	-1.06211
0.2	0.3	0.2	0.3	-0.83201	-0.75005
0.2	0.3	0.2	0.4	-0.34291	-0.43799

TABLE 13.2
Numerical Values of Nusselt Number

E_t	B_t	*J*	*M*	Re	Predicted
0.1	0.5	0.2	0.2	0.307673	0.309039
0.2	0.5	0.2	0.2	0.290458	0.29319
0.3	0.5	0.2	0.2	0.273243	0.27734
0.4	0.5	0.2	0.2	0.256027	0.261491
0.3	0.5	0.2	0.2	0.273243	0.27734
0.3	1	0.2	0.2	0.404737	0.370467
0.3	1.5	0.2	0.2	0.482067	0.463594
0.3	2	0.2	0.2	0.532983	0.556721
0.3	0.5	0.2	0.2	0.273243	0.27734
0.3	0.5	0.4	0.2	0.28145	0.283047
0.3	0.5	0.6	0.2	0.288677	0.288754
0.3	0.5	0.8	0.2	0.295044	0.294461
0.3	0.5	0.2	0.2	0.273243	0.27734
0.3	0.5	0.2	0.4	0.264471	0.267045
0.3	0.5	0.2	0.6	0.256005	0.256751
0.3	0.5	0.2	0.8	0.247811	0.246456

TABLE 13.3
Numerical Values of Sherwood Number

L_e	J	N_b	N_t	N_r	Re	Predicted
1	0.2	0.5	0.5	0.3	1.212895	1.211494
1.5	0.2	0.5	0.5	0.3	1.585482	1.543121
2	0.2	0.5	0.5	0.3	1.899593	1.874748
2.5	0.2	0.5	0.5	0.3	2.175692	2.206376
1	0.2	0.5	0.5	0.3	1.212895	1.211494
1	0.4	0.5	0.5	0.3	1.265704	1.26413
1	0.6	0.5	0.5	0.3	1.317528	1.316767
1	0.8	0.5	0.5	0.3	1.368371	1.369403
1	0.2	0.3	0.5	0.3	1.145968	1.180219
1	0.2	0.5	0.5	0.3	1.212895	1.211494
1	0.2	0.7	0.5	0.3	1.241578	1.242768
1	0.2	0.9	0.5	0.3	1.257513	1.274043
1	0.2	0.5	0.5	0.3	1.212895	1.211494
1	0.2	0.5	1	0.3	1.112505	1.111704
1	0.2	0.5	1.5	0.3	1.012115	1.011914
1	0.2	0.5	2	0.3	0.911724	0.912125
1	0.2	0.5	0.5	0.3	1.212895	1.211494
1	0.2	0.5	0.5	0.5	1.310047	1.302486
1	0.2	0.5	0.5	0.7	1.397973	1.393478
1	0.2	0.5	0.5	0.9	1.478954	1.484471

13.4.5 Interpretation on Particle Flow Path

Flow paths are constructed for different parameter values across the model to demonstrate the characteristics of the fluid motion. Figures 13.13–13.16 display the streamline patterns. It provides a graphical representation of fluid motion patterns. Figure 13.13 illustrates the effects of streamlines at the magnetic parameter ($M = 0$). Figure 13.14 illustrates the streamline effects at $M = 0.2$. As seen in Figure 13.15, the streamline pattern is present at $D = 0$. Figure 13.16 displays the elegant streamline effects at $D = 0.3$.

13.5 MULTIPLE LINEAR REGRESSION

It is a statistical technique used to predict a continuous outcome variable based on multiple input variables. This helps to understand how the dependent variable changes when any one of the independent variables is varied while the other independent variables are held constant. The equation takes the form of [29–32]:

$$y = \delta_0 + \delta_1 x_1 + \delta_2 x_2 + \delta_3 x_3 \ldots\ldots\ldots\ldots + \delta_n x_n \tag{13.42}$$

The least squares method determines the coefficients by reducing the sum of squared differences between the actual and predicted values of the dependent variable. This technique ensures that the model aligns well with the data, leading to more precise predictions. We employ a multivariate regression method to estimate skin friction, the

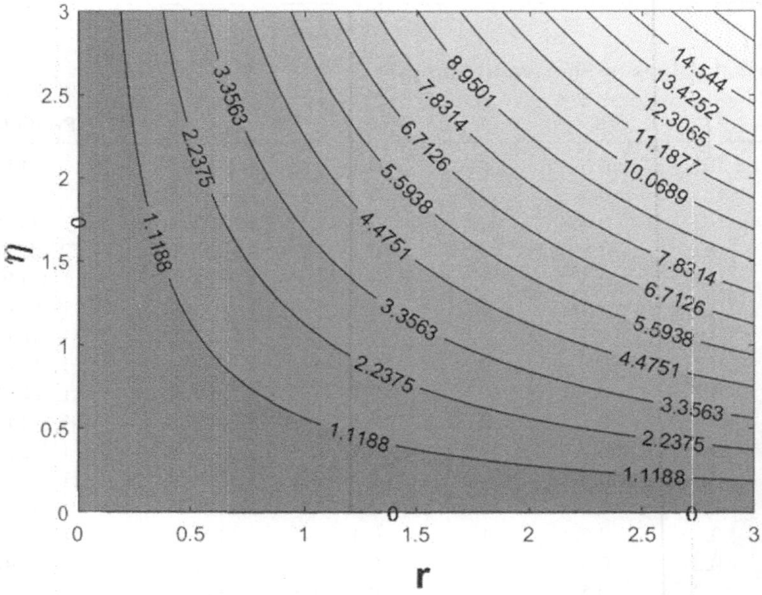

FIGURE 13.13 Streamline effects at $M = 0$.

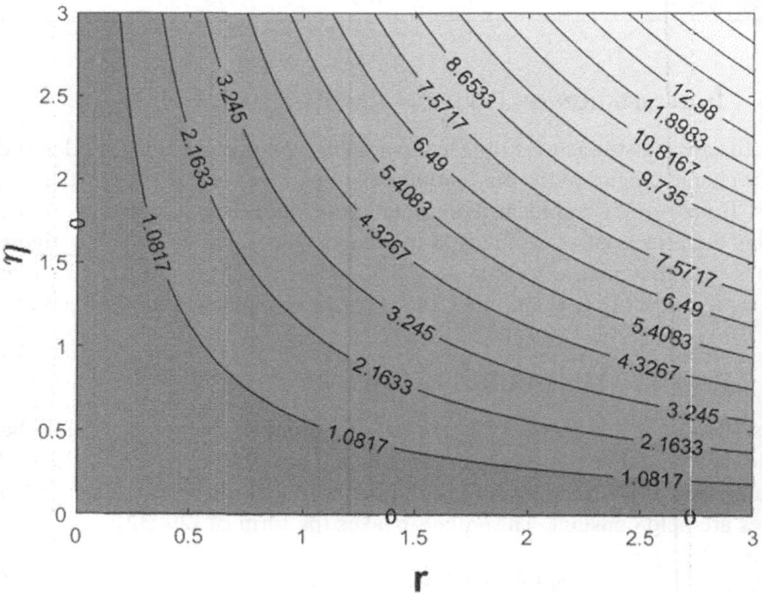

FIGURE 13.14 Streamline effects at $M = 1$.

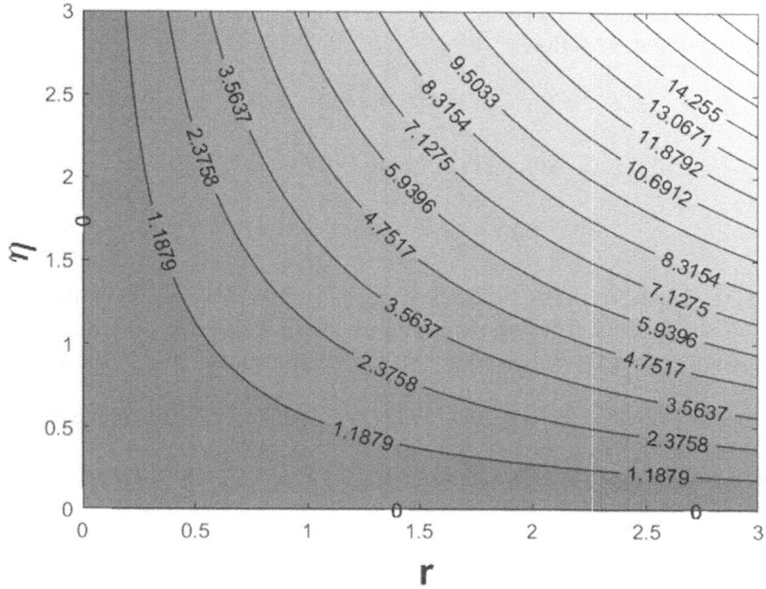

FIGURE 13.15 Streamline effects at $D = 0$.

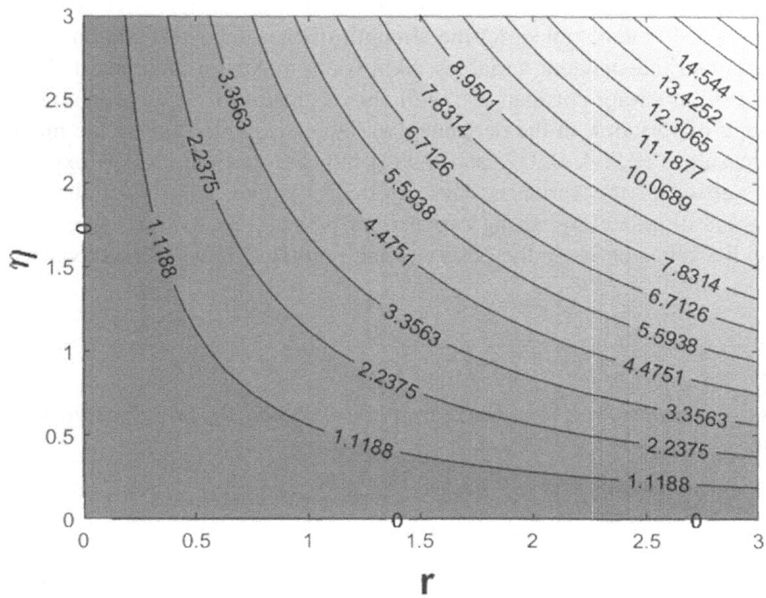

FIGURE 13.16 Streamline effects at $D = 1$.

Nusselt number, and the Sherwood number. The predicted models can be represented in a general format, as follows:

$$Cf_{est} = \alpha_1 M + \alpha_2 D + \alpha_3 J + \alpha_4 We + \alpha \tag{13.43}$$

$$\mathrm{Nu}_{est} = \beta_1 E_t + n_2 B_t + n_3 J + n_4 M + \beta \tag{13.44}$$

$$\mathrm{Sh}_{est} = \gamma_1 L_e + \gamma_2 J + \gamma_3 N_b + \gamma_4 N_t + \gamma_5 N_r + \gamma \tag{13.45}$$

Here, $\alpha_1, \alpha_2, \alpha_3, \alpha_4, \alpha, \beta_1, \beta_2, \beta_3, \beta_4, \beta, \gamma_1, \gamma_2, \gamma_3, \gamma_4, \gamma_5$ and γ represents the calculated regression coefficients. The following regression models have been developed, providing a comprehensive representation of the relationships between the variables:

$$Cf_{est} = -1.563249 - 0.195916M - 0.178928D - 0.150526J + 3.120553\mathrm{We} \tag{13.46}$$

$$\mathrm{Nu}_{est} = 0.236349 - 0.158494E_t + 0.186254B_t + 0.028534J - 0.051474M \tag{13.47}$$

$$\begin{aligned}\mathrm{Sh}_{est} = {} & 0.380717 + 0.663254L_e + 0.263182J + 0.156373N_b \\ & - 0.199579N_t + 0.454961N_r\end{aligned} \tag{13.48}$$

Table 13.4 presents the regression statistics for the skin friction coefficient (Cf), Nusselt number (Nu), and Sherwood number (Sh). The multiple R value serves as the correlation coefficient, reflecting the strength of the linear association between the dependent and independent variables. Skin friction, Nusselt number, and Sherwood number exhibit strong correlations, with values close to 1. R squared represents the percentage of variation in the dependent variable accounted for by the model. High R-square values (0.954, 0.972, and 0.996) indicate that the models explain a substantial portion of the variance. The adjusted R square adjusts for the number of predictors, still indicating strong explanatory power. The standard error reflects the average difference between the observed and predicted values. Lower values (0.052

TABLE 13.4
Regression Analysis for Skin Friction, Nusselt Number, and Sherwood Number

	Regression Statistics		
	Re	Re	Re
Multiple R	0.976976531	0.989703187	0.998437264
R Square	0.954483143	0.979512398	0.99687697
Adjusted R Square	0.937931559	0.972062361	0.995761602
Standard Error	0.052218636	0.014133813	0.018753239
Observations	60	60	75

0.014, and 0.019) imply that the predictions closely match the actual data points, indicating high precision.

The correlation matrix displayed in Figures 13.17, 13.18, and 13.19 illustrates the relationships between variables and physical quantities. Each cell in the matrix shows the correlation coefficient between two variables, with values ranging from -1 to 1. From Figure 13.17, we observe that the variable Williamson parameter (We) and skin friction (Cf) have a strong positive correlation (0.9577). In Figure 13.18, the Biot

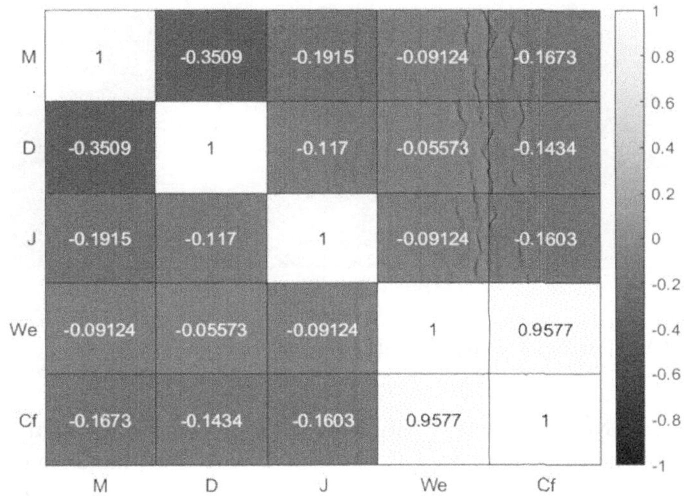

FIGURE 13.17 Correlation analysis of skin friction coefficient.

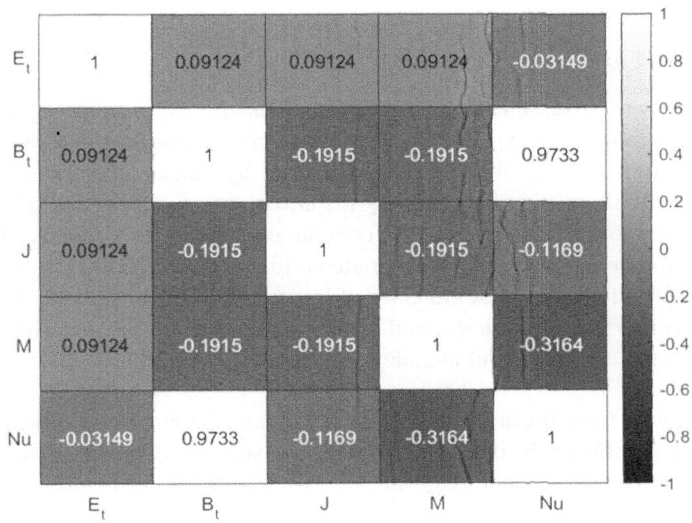

FIGURE 13.18 Correlation analysis of Nusselt number.

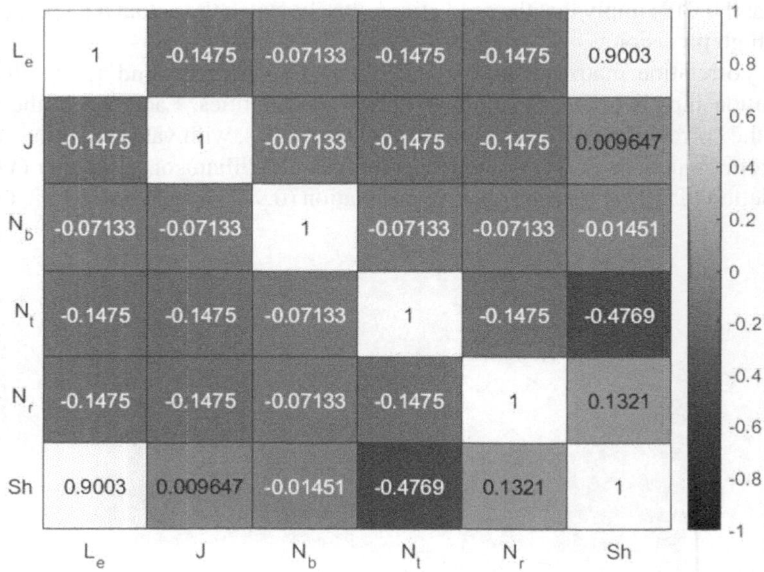

FIGURE 13.19 Correlation analysis of Sherwood number.

number (B_t) and Nusselt number (Nu) show a strong positive correlation (0.9733). Other variables have weaker correlations, such as E_t and M, indicating a moderate inverse relationship. The correlation matrix in Figure 13.19 shows the relationships among L_e, J, N_b, N_t, N_r, and Sh. The strongest positive correlation, 0.9003, is observed between the Lewis number (L_e) and Sherwood number (Sh). Other correlations are relatively weak, suggesting minimal linear association among the remaining variables.

13.6 CONCLUSION

This chapter scrutinizes the thermal and flow behavior of an unsteady Williamson nanofluid over a radially expanding surface through a permeable substrate with convective heat transfer. The flow is also influenced by viscous dissipation, chemical reaction, and Joule heating. To simplify the equations, we utilized similarity transformations to convert the partial differential equation into ordinary differential equations. Numerical solutions are computed utilizing the KBM (Keller box method) in MATLAB® software. In addition, we utilize multiple linear regression (MLR) to predict the skin friction coefficient and Nusselt number based on numerical data. Key highlights of the current model include the following:

- Elevating the influence of both the porosity and the non-Newtonian Williamson variable effectively diminishes the fluid's velocity distribution. The porosity affects the fluid's flow speed, while the Williamson parameter affects its viscosity.
- The velocity experiences a flattening trend with increasing magnetic parameters but the temperature profile exhibits an opposite phenomenon. The velocity and temperature distribution diminish as unsteadiness parameters increase.

- Temperature distribution improves with a higher range of the Eckert number (E_t) and Biot number B_t. As a result, the heat transfer is more efficient.
- A rise in the Brownian motion parameter results in a reduction of $\phi(\eta)$, whereas an increase in the thermophoresis parameter amplifies the concentration profiles.
- The analysis revealed a decline in concentration distribution $\phi(\eta)$ with increased Lewis number and chemical reaction parameter values.
- The regression model predicting skin friction is formulated as follows:

$$Cf_{est} = -1.563249 - 0.195916M - 0.178928D - 0.150526J + 3.120553\mathrm{We}.$$

- The predicted Nusselt number values are estimated using the regression model:

$$\mathrm{Nu}_{est} = 0.236349 - 0.158494E_t + 0.186254B_t + 0.028534J - 0.051474M$$

- The Sherwood number regression model takes the following form:

$$\mathrm{Sh}_{est} = 0.380717 + 0.663254L_e + 0.263182J + 0.156373N_b - 0.199579N_t + 0.454961N_r$$

Future research can build upon this by investigating various hybrid nanofluid combinations, considering additional factors, and verifying the accuracy of numerical and multiple linear regression predictions through experimental studies. Applying advanced machine learning models, such as neural networks, could further improve predictive capabilities. The findings provide valuable insights for enhancing heat transfer in industrial processes.

NOMENCLATURE

a – Positive constant	M – Magnetic Parameter
k – Thermal conductivity	Nu – Nusselt number
μ – Viscosity coefficient	C_f – Skin friction coefficient
α – Thermal diffusivity	Sh – Sherwood number
C_p – Specific heat capacity	J – Unsteadiness parameter
ε – Rate of stretching	L_e – Lewis number
σ – Electrical conductivity	E_t – Eckert number
C_w – Concentration of sheet	D – Porosity parameter
ρ – Fluid density	Pr – Prandtl number
T_w – Temperature near the surface	B_t – Biot number
D_b – Brownian diffusion coefficient	N_r – Chemical reaction parameter
T_∞ – Ambient temperature	N_t – Thermophoresis parameter
D_t – Thermophoretic diffusion coefficient	N_b – Brownian motion parameter
U, w – Velocity components along r and z direction	$\phi(\eta)$ – Dimensionless concentration
B_0 – Magnetic field strength	$f'(\eta)$ – Dimensionless velocity
U_w – Velocity near the sheet	$\theta(\eta)$ – Dimensionless temperature

AVAILABILITY OF DATA AND MATERIALS

On reasonable request, the corresponding author will provide access to the datasets used and analyzed during the present work.

ACKNOWLEDGMENTS

The authors express their gratitude to VIT, Vellore, for providing the necessary facilities for this study.

COMPETING INTERESTS

The authors declare no conflict of interest.

REFERENCES

1. Crane, L.J., 1970. Flow past a stretching plate. *Zeitschrift für angewandte Mathematik und Physic ZAMP*, *21*, pp.645–647. https://doi.org/10.1007/BF01587695
2. Andersson, H.I., Bech, K.H. and Dandapat, B.S., 1992. Magnetohydrodynamic flow of a power-law fluid over a stretching sheet. *International Journal of Non-Linear Mechanics*, *27*(6), pp.929–936. https://doi.org/10.1016/0020-7462(92)90045-9
3. Ariel, P.D., 2001. Axisymmetric flow of a second-grade fluid past a stretching sheet. *International Journal of Engineering Science*, *39*(5), pp.529–553. https://doi.org/10.1016/S0020-7225(00)00052-5Get rights and content
4. Vajravelu, K. and Nayfeh, J., 1992. Hydromagnetic flow of a dusty fluid over a stretching sheet. *International Journal of Non-Linear Mechanics*, *27*(6), pp.937–945. https://doi.org/10.1016/0020-7462(92)90046-A
5. Srinivas, S., Reddy, P.B.A. and Prasad, B.S.R.V., 2015. Non-Darcian unsteady flow of a micropolar fluid over a porous stretching sheet with thermal radiation and chemical reaction. *Heat Transfer—Asian Research*, *44*(2), pp.172–187. https://doi.org/10.1002/htj.21090
6. Nadeem, S., Rashid, M. and Noreen, S.A., 2015. Partial slip effect on non-aligned stagnation point nanofluid over a stretching convective surface. *Chinese Physics B*, *24*(1), p.014702. 10.1088/1674-1056/24/1/014702
7. Hayat, T., Muhammad, T., Shehzad, S.A., Chen, G.Q. and Abbas, I.A., 2015. Interaction of magnetic field in flow of Maxwell nanofluid with convective effect. *Journal of Magnetism and Magnetic Materials*, *389*, pp.48–55. 10.1016/j.jmmm.2015.04.019
8. Khan, M., Malik, R., Munir, A. and Shahzad, A., 2016. MHD flow and heat transfer of Sisko fluid over a radially stretching sheet with convective boundary conditions. *Journal of the Brazilian Society of Mechanical Sciences and Engineering*, *38*, pp.1279–1289. https://doi.org/10.1007/s40430-015-0437-y
9. Shahzad, A., Ali, R., Hussain, M., and Kamran, M., 2017. Unsteady axisymmetric flow and heat transfer over a time-dependent radially stretching sheet. *Alexandria Engineering Journal*, *56*(1), pp.35–41. https://doi.org/10.1016/j.aej.2016.08.030
10. Faraz, F., Haider, S., Imran, S.M., 2020. Study of magneto-hydrodynamics (MHD) impacts on an axisymmetric Casson nanofluid flow and heat transfer over an unsteady radially stretching sheet. *SN Applied Sciences*, *2*, pp.1–17. https://doi.org/10.1007/S42452-019-1785-5

11. Reza, M. and Rana, A., 2020. Analysis of exact solutions of electromagnetohydrodynamic flow and heat transfer of non-Newtonian Casson fluid in a microchannel with viscous dissipation and joule heating. In *Advances in Fluid Mechanics and Solid Mechanics: Proceedings of the 63rd Congress of ISTAM 2018* (pp. 117–127). Springer Singapore. 10.1007/978-981-15-0772-4_11
12. Kazemi, M.A., Javanmard, M., Taheri, M.H. and Askari, N., 2020. Heat transfer investigation of the fourth-grade non-Newtonian MHD fluid flow in a plane duct considering the viscous dissipation, joule heating, and forced convection on the walls. *SN Applied Sciences*, *2*(10), p.1752. https://doi.org/10.1007/s42452-020-03567-4
13. Gorfie, E.H., Zergaw, G.A. and Assres, H.D., 2022. Thermally radiant Williamson nanofluid flow over a permeable stretching sheet with viscous dissipation and Joule heating effects. *Journal of Nanofluids*, *11*(3), pp.401–409. https://doi.org/10.1166/jon.2022.1852
14. Thamizharasan, T. and Reddy, A.S., 2022. Pulsating hydromagnetic flow of au-blood Jeffrey nanofluid in a channel with joule heating and viscous dissipation. *Nanoscience and Technology: An International Journal*, *13*(2), pp. 1-13. 10.1615/NanoSciTechnolIntJ.2022039247
15. Zhu, A., Ali, H., Ishaq, M., Junaid, M.S., Raza, J. and Amjad, M., 2022. Numerical study of heat and mass transfer for williamson nanofluid over stretching/shrinking sheet along with Brownian and thermophoresis effects. *Energies*, *15*(16), p.5926. https://doi.org/10.3390/en15165926
16. Asjad, M.I., Zahid, M., Ali, B.M., Jarad, F., 2022. Unsteady MHD Williamson fluid flow with the effect of bioconvection over permeable stretching sheet. *Mathematical Problems in Engineering*, *2022*, p.7980267. https://doi.org/10.1155/2022/7980267.
17. Qayyum, M., Abbas, T., Afzal, S., Saeed, S.T., Akgül, A., Inc, M., Mahmoud, K., Alsubaie, A.S., 2022. Heat transfer analysis of unsteady MHD Carreau fluid flow over a stretching/shrinking sheet. *Coatings*, *12*, p.1661. https://doi.org/10.3390/coatings12111661
18. Islam, M.A., Ali, M.Y., , Gani, S.M., 2022. Unsteady heat and mass transfer slip flow over an exponentially permeable stretching sheet. *Journal of Scientific Research*, *14*, pp.443–457. https://doi.org/10.3329/jsr.v14i2.55577.
19. Jagadeesh, S., Chenna Krishna Reddy, M., Tarakaramu, N., Ahmad, H., Askar, S. and Shukhratovich Abdullaev, S., 2023. Convective heat and mass transfer rate on 3D Williamson nanofluid flow via linear stretching sheet with thermal radiation and heat absorption. *Scientific Reports*, *13*(1), p.9889. https://doi.org/10.1038/s41598-023-36836-4
20. Kune, R., Naik, S.H.S. and Nisar, K.S., 2023. Analysis of Williamson nanofluid MHD over a nonlinear stretchable sheet by velocity slip and Newtonian heating. *International Journal of Modern Physics B*, *38*, p.2450207. https://doi.org/10.1142/S0217979224502072
21. Saini, S.K., Agrawal, R. and Kaswan, P., 2023. Activation energy and convective heat transfer effects on the radiative Williamson nanofluid flow over a radially stretching surface containing Joule heating and viscous dissipation. *Numerical Heat Transfer, Part A: Applications*, pp.1–24. https://doi.org/10.1080/10407782.2023.2226815
22. Sarfraz, M., Khan, M. and Ahmed, A., 2023. Study of thermophoresis and Brownian motion phenomena in radial stagnation flow over a twisting cylinder. *Ain Shams Engineering Journal*, *14*(2), p.101869. https://doi.org/10.1016/j.asej.2022.101869
23. Ramanjini, V., Krishna, G.G., Mishra, S.R., Kumari, S.S. and Sree, H.K., 2022. An unsteady axisymmetric Williamson nanofluid flow over a radially stretching Riga plate for the inclusion of mixed convection and thermal radiation. *Partial*

Differential Equations in Applied Mathematics, *6*, p.100456. https://doi.org/10.1016/j.padiff.2022.100456

24. Shahzad, A., Ali, R., Kamran, M., Khan, S.U.D., Khan, S.U.D. and Farooq, A., 2020. Axisymmetric flow with heat transfer over exponentially stretching sheet: A computational approach. *Physica A: Statistical Mechanics and its Applications*, *554*, p.124242. https://doi.org/10.1016/j.physa.2020.124242
25. Ibrahim, W. and Gamachu, D., 2019. Nonlinear convection flow of Williamson nanofluid past a radially stretching surface. *AIP Advances*, *9*(8). https://doi.org/10.1063/1.5113688
26. Narender, G., Govardhan, K. and Sarma, G.S., 2020. Magnetohydrodynamic stagnation point on a Casson nanofluid flow over a radially stretching sheet. *Beilstein Journal of Nanotechnology*, *11*(1), pp.1303–1315. https://doi.org/10.3762/bjnano.11.114
27. Azam, M., Mabood, F., Xu, T., Waly, M. and Tlili, I., 2020. Entropy-optimized radiative heat transportation in the axisymmetric flow of Williamson nanofluid with activation energy. *Results in Physics*, *19*, p.103576. https://doi.org/10.1016/j.rinp.2020.103576
28. Azam, M., Khan, M. and Alshomrani, A.S., 2017. Effects of magnetic field and partial slip on unsteady axisymmetric flow of Carreau nanofluid over a radially stretching surface. *Results in Physics*, *7*, pp.2671–2682. https://doi.org/10.1016/j.rinp.2017.07.025
29. Neethu, T.S., Sabu, A.S., Mathew, A., Wakif, A. and Areekara, S., 2022. Multiple linear regression on bioconvective MHD hybrid nanofluid flow past an exponential stretching sheet with radiation and dissipation effects. *International Communications in Heat and Mass Transfer*, *135*, p.106115. https://doi.org/10.1016/j.icheatmasstransfer.2022.106115
30. Revathi, G., Sajja, V.S., Babu, M.J., Srinivasa Babu, K.S., Suneel Kumar, A., Raju, C.S.K. and Yook, S.J., 2023. Multiple linear regression analysis on the flow of ternary hybrid nanofluid by a quadratically radiated stretching surface with and second-order slip. *Waves in Random and Complex Media,* pp.1–18. https://doi.org/10.1080/17455030.2023.2181645
31. Bhargavi, N. and Poornima, T., 2024. Radiative impact on Jeffery trihybrid convective nanoflow over an extensible Riga plate: Multiple linear regression analysis. *Contemporary Mathematics*, 2024 Mar. 14 [cited 2025 Jul. 12];5(1) pp.1036–1053. https://ojs.wiserpub.com/index.php/CM/article/view/4058
32. Kumar, M.D., Ahammad, N.A., Raju, C.S.K., Yook, S.J., Shah, N.A. and Tag, S.M., 2023. Response surface methodology optimization of dynamical solutions of Lie group analysis for nonlinear radiated magnetized unsteady wedge: Machine learning approach (gradient descent). *Alexandria Engineering Journal*, *74*, pp.29–50. https://doi.org/10.1016/j.aej.2023.05.009

14 Maxwell Thin-Film Flow with Viscous Dissipation

M. Revathi Devi, Narsu Sivakumar, B. Madhusudhana Rao, R. Nanda Kumar, Dhanapal Prakash, and K. Prabakaran

14.1 INTRODUCTION

The Maxwell thin-film fluid flow is the study of how viscous and elastic fluids move through a thin sheet that is stretched and has surface tension acting on it. Maxwell fluids have been employed significantly in industrial and engineering fields such as coating technologies, lubrication, and polymer processing, because of their ability to display both elastic and viscous behavior. Several studies explored thin-film flows with various conditions, employing different numerical as well as analytical methods. Many focus on specific aspects, such as the influence of specific parameters or the use of particular solution techniques.

Traditional numerical methods, like the finite element method (FEM) and finite difference method (FDM), are commonly employed to solve thin-film flows (Reddy *et al.* [1] and Shilpa and Leela [2]). Even though these techniques are very accurate, they can be computationally expensive especially when dealing with complicated geometries and boundary conditions. For the study of magneto Casson nanofluid flow, for instance, Reddy *et al.* [1] used the implicit finite difference approach, taking chemical reactions, thermal radiation, and viscous dissipation into consideration. In a similar manner, Shilpa and Leela [2] addressed Maxwell, Oldroyd-B and Jeffrey nanofluids in a vertical annulus with a non-uniform heat source/sink and an induced magnetic field using the Galerkin finite element method (GFEM). The selection of numerical method is determined by the specific problems complexity and computational resources. Further investigations have employed alternative numerical techniques. The homotopy analysis method (HAM) for instance, has demonstrated efficiency in addressing non-linear differential equations that occur with the thin-film flow problem (Jawad *et al.* [3], Nasir *et al.* [4], Ullah *et al.* [5], Khan *et al.* [6] and Murugan *et al.* [7]). Jawad *et al.* [3] explored the MHD Darcy–Forchheimer thin-film nanofluid flow incorporating Navier partial slip and Joule dissipation through HAM. Nasir *et al.* [4] used HAM to examine the transmission of heat along with an unsteady stretched sheet and Darcy–Forchheimer thin-film nanofluid flow of SWCNTs. Then, Ullah *et al.* [5] utilized HAM to examine viscoelastic MHD thin-film nanofluid flow characteristics across an unsteady vertical stretching sheet, along with entropy analysis. Additionally, the combined effects of Hall current and thermophoresis on MHD mixed convective heat and mass transfer in a second-grade thin-film fluid were

DOI: 10.1201/9781003611820-17

investigated by Khan *et al.* [6] using HAM. Later, Murugan *et al.* [7] examined the Maxwell thin-film fluid properties with MHD as well as viscous dissipation effects over an inclined stretching sheet using HAM. These investigations demonstrate the ability of HAM is at addressing complicated nonlinear systems. The governing nonlinear equations for MHD thin-film flow of a third-grade fluid with slip boundary conditions have also been solved through the adomian decomposition method (ADM), as explained by Gul *et al.* [8]. The comparison of various methods emphasizes the significance of choosing the best approach depending on the specifics of the situation.

The process by which internal friction in a fluid turns mechanical energy into thermal energy is known as viscous dissipation. In high-speed or high-viscosity flows, when the heat produced influences the total temperature distribution, this effect is significant. The influence of viscous dissipation on heat transport is measured in the current study using the Eckert number (*Ec*). A change in thermal boundary layer and the surface temperature gradient can be affected by an increase in *Ec*, which indicates that the fluid is generating more heat inside. A more precise measurement of energy transfer in thin-film fluid flows is offered by integrated the concept of viscous dissipation within the governing equations. The strength of viscous dissipation is affected through variables such as fluid viscosity and flow velocity. Many of the studies observed above incorporated viscous dissipation into their mathematical models. The effects of viscous dissipation and nonlinear thermal radiation on ferrofluid flow along with heat transfer through a porous rotating disk have been investigated by Sharma [9]. Then Murugan *et al.* [10] examined the ohmic viscous dissipation effects with MHD and thermal radiation through the porous medium in a hybrid nanofluid flow. Additionally, Raju *et al.* [11] explored the energy transfer characteristics with the Falkner–Skan slip flow for the non-Newtonian flow via non-linearly heated moving wedge with the heat source/sink and viscous dissipation effects.

The connection between applied magnetic fields and electrically conducting fluids is addressed by magnetohydrodynamics (MHD). The existence of a magnetic field influence fluid motion and heat transfer in various kinds of industrial and technical applications, including nuclear reactor cooling, electromagnetic pumps, and plasma physics. The ratio of electromagnetic effects to viscous forces can be expressed by the magnetic parameter in the governing equations. Due to the Lorentz effect, that acts as a resistive force and influences the temperature and velocity boundary layers, a greater magnetic field may decrease the flow velocity. Applications employing electrically conducting fluids across porous or non-porous media need an understanding of MHD effects. Kumar and Kumar [12] analysed the significance of MHD through the Williamson fluid along with Blasius and Sakiadis unsteady flow along with a variable conductivity effect. The influence of a magnetic field on MHD dissipative Casson nanofluid liquid film flow was investigated by Alali and Megahed [13]. Later Devi et al. [14] and Murugan et al. [15] explored their research on influence of MHD on different physical phenomena such as different shape factor effects on nano fluids and rotating cone environment. Additionally, Tarakaramu et al. [16] presented the theoretical analysis of impacts of MHD through the 3D nanofluid flow with the convective boundary conditions. These investigations demonstrate how MHD effects have a significant influence on the flow properties of thin films and other flow geometry also.

A systematic review of these research studies is essential to establish a foundation for applying regression-based machine learning methods. In order to enhance the accuracy of prediction and parameter efficiency, a regression-based machine learning approach has been incorporated within the evaluation. Machine learning can efficiently simulate more complicated non-linear relationships between physical factors and corresponding transfer characteristics of heat and mass by using data-driven models. The numerical data from the numerical solvers are analysed using Regression algorithms, such as Linear Regression, Polynomial Regression, or Neural Networks. Despite the potential benefits, there are still significant research gaps. More investigation must be conducted to explore the feasibility of various machine learning models to particular thin film flow problems. Additionally, more advanced feature selection methods that are suited to this particular application must be developed. It is essential to research how data quantity and quality affect model accuracy. The addition of quantifying uncertainty methods is essential for accurate predictions. Furthermore, combining machine learning models with traditional numerical techniques may result in hybrid techniques that benefit on each technique's strengths. Complex thin film flow issues may be handled with greater accuracy and computationally efficient as outcome.

The study of characteristics of energy and mass transfer in thin film is important to several engineering specialties, such as material science, chemical processing, and microfluidics. A subclass of viscoelastic fluids known as Maxwell fluids display complicated rheological behaviour that has a significant impact on the transfer of both heat and mass. It is difficult to find analytical solutions to this problem because of the existence of MHD effects and viscous dissipation. Thus, strong numerical and machine learning approaches are needed to accurately predict and fully understand these systems. This analysis explores the application of regression-based machine learning techniques to examine Maxwell thin-film heat and mass transport considering MHD effects and viscous dissipation into account.

14.2 PHYSICAL DESCRIPTION

Here, the magnetic field is taken into account for mass and heat transfer in an unsteady two-dimensional Maxwell thin-film flow over a surface along with viscous dissipation. Figure 14.1 demonstrates that the fluid is placed on the stretching surface, in particular, which constructs a small slit in the origin. The flow commences at zero velocity and develops the form of a thin film,$U_w \left(= \frac{cx}{1-at}\right)$ can be taken as the stretching velocity, which is also referred to as an operative elasticity, where α is referred to as enhancement of timing $(0 \leq \alpha < 1)$, c is referred as elasticity, and the sheet motion is only one aspect affecting the movement. According to the stretched surface, the free surface moves more slowly. When the layer of a thin-film thickness grows the speed will actually get reduced. Based on the surface, the distribution for temperature can be referred to as $T_w(x,t) = T_0 - T_{ref}\left(\frac{cx^2}{2u_f}\frac{1}{(1-at)^{1.5}}\right)$ and volume fraction is referred to as $C_w(x,t) = C_0 - C_{ref}\left(\frac{cx^2}{2u_f}\frac{1}{(1-at)^{1.5}}\right)$, which are considered to modify the distance of x from the slit.

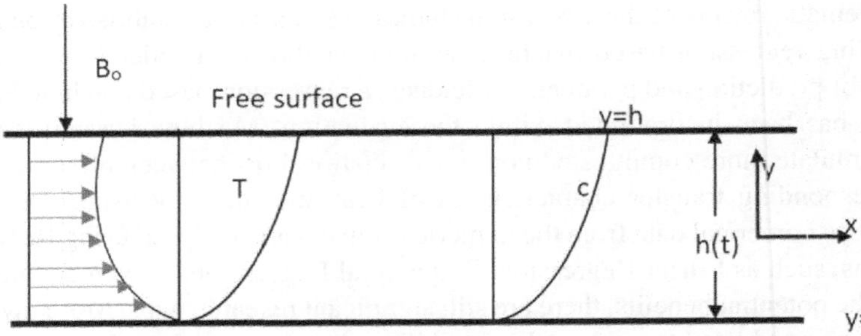

FIGURE 14.1 Physical geometry.

Here C_0, T_0 are the volume fraction and temperature toward the silt respectively, C_{ref} T_{ref} are the reference volume fraction nanoparticle (constant) and temperature reference (constant), respectively ($0< C_{ref}<C_0,\ 0< T_{ref}<T_0$). Presumed that the base fluid and the suspended nanoparticles are in thermal equilibrium.

14.3 MATHEMATICAL MODELING

From all those considerations, the governing equation for the boundary layer is listed below [7,17],

$$\frac{\partial u}{\partial x}+\frac{\partial v}{\partial y}=0 \tag{14.1}$$

$$u\frac{\partial u}{\partial x}+v\frac{\partial u}{\partial y}+\frac{\partial u}{\partial t}+\lambda_1\left(u^2\frac{\partial^2 u}{\partial x^2}+2uv\frac{\partial^2 u}{\partial x\partial y}+v^2\frac{\partial^2 u}{\partial y^2}\right)=\frac{\mu}{\rho}\frac{\partial^2 u}{\partial y^2}-\frac{\sigma B_0^2}{\rho}\left(\lambda_1 v\frac{\partial u}{\partial y}+u\right) \tag{14.2}$$

$$\begin{aligned}u\frac{\partial T}{\partial x}+v\frac{\partial T}{\partial y}+\frac{\partial T}{\partial t}&=\alpha\frac{\partial^2 T}{\partial y^2}+\frac{1}{\rho c_p}\left(\mu\left(\frac{\partial u}{\partial y}\right)^2-\rho\lambda_1\left(v^2\left(\frac{\partial u}{\partial y}\right)^2 2uv\frac{\partial u}{\partial x}\frac{\partial u}{\partial y}\right)\right)\\&+\tau\left[D_B\left(\frac{\partial T}{\partial y}\right)\left(\frac{\partial C}{\partial y}\right)+\frac{D_T}{T_0}\left(\frac{\partial T}{\partial y}\right)^2\right]+\frac{\sigma B_0^2}{\rho c_p}\left(\lambda_1 v\frac{\partial u}{\partial y}+u\right)^2\end{aligned} \tag{14.3}$$

$$u\frac{\partial C}{\partial x}+v\frac{\partial C}{\partial y}+\frac{\partial C}{\partial t}=\frac{D_T}{T_0}\frac{\partial^2 T}{\partial y^2}+D_B\frac{\partial^2 C}{\partial y^2} \tag{14.4}$$

Here, u and v are the velocity components in the x and y directions, respectively. The Brownian coefficient is represented by D_B, the thermophoresis coefficient is symbolized by D_T, the local temperature is T, and the respective boundary conditions are [7,17]

$$u=U_w, v=0, T=T_w, C=C_w \text{ at } y=0$$

$$\left(\mu_{nf}\frac{\partial u}{\partial y}\right)=\frac{\partial T}{\partial y}=\frac{\partial C}{\partial y}=0, v=\frac{\partial H}{\partial t}\text{ at } y=H(t) \tag{14.5}$$

The film thickness is represented by $H(t)$ and the similarity transformations are as follows [7,17]:

$$\psi(x,y,t)=x\left(\frac{c\upsilon_f}{1-\alpha t}\right)^{0.5} f(\zeta), T(x,y,t)=T_0-T_{ref}\left(\frac{cx^2}{2\upsilon_f}\right)(1-\alpha t)^{-1.5}\theta(\zeta)$$
$$C(x,y,t)=C_0-C_{ref}\left(\frac{cx^2}{2\upsilon_f}\right)(1-\alpha t)^{-1.5}\phi(\zeta), \zeta=\left(\frac{c}{\upsilon_f(1-\alpha t)}\right)^{0.5} y=\frac{\beta}{H(t)}y \tag{14.6}$$

The stream function $\psi(x,y)$ is defined as $u=\frac{\partial\psi}{\partial y}, v=-\frac{\partial\psi}{\partial x}$. The film thickness when dimensionless is assumed to be [7,17]

$$\beta=\left(\frac{c}{\upsilon_f(1-\alpha t)}\right)^{0.5} H(t) \tag{14.7}$$

Substituting Eq. (14.6) into Eq. (14.7) we get,

$$\frac{dH}{dt}=-\frac{\alpha\beta}{2}\left(\frac{\upsilon_f}{c}\right)^{0.5}\frac{1}{(1-\alpha t)^{0.5}} \tag{14.8}$$

Using Eqs. (14.6) in (14.1) to (14.4), we get the system of ordinary differential equations (ODEs) as follows:

$$f'''-M\left(f'-\lambda ff''\right)-\lambda f^2f'''+2\lambda ff'f''+ff''-f'^2-A\left(\frac{\zeta}{2}f''+f'\right)=0 \tag{14.9}$$

$$\frac{1}{Pr}\theta''+Ec\left(f''^2-2\lambda ff'^2f''-\lambda f^2f''^2\right)-N_b\theta'\phi'+N_t\theta'^2+MEc\left(f'-\lambda ff''\right)^2+f\theta'$$
$$-2f'\theta-\frac{A}{2}\left(3\theta+\zeta\theta'\right)=0 \tag{14.10}$$

$$\phi''+\frac{N_t}{N_b}\theta''+Le\left(f\phi''-2f'\phi-\frac{A}{2}\left(3\phi+\zeta\phi'\right)\right)=0 \tag{14.11}$$

where the prime is referred to as the differentiation with respect to the similarity variable ζ and M, Pr, N_b, N_t, Sc, A, and Ec are magnetic parameter, Prandtl number, Brownian motion, thermophoresis parameter, Schmidt number, dimensionless unsteady, and Eckert number, respectively.

$$M=\frac{\sigma}{\rho}B_0^2\frac{1-\alpha t}{c}, \lambda=\lambda_1\frac{c}{1-\alpha t}, Le=\frac{\upsilon}{D_b}, N_t=\frac{\tau D_T(T_w-T_0)}{\upsilon T_0},$$

$$A=\frac{\alpha}{c},\Pr=\frac{\upsilon}{\alpha},N_b=\frac{\tau D_B(C_w-C_0)}{\upsilon},Ec=\left(\frac{cx}{1-\alpha t}\right)^2\frac{1}{c_p\left(T_w-T_0\right)} \tag{14.12}$$

The boundary conditions in the reshaped form are taken as follows:

$$f(0)=0,f'(0)=1,\theta(0)=1,\phi(0)=1$$

$$f''(\beta)=0,f(\beta)=\frac{A\beta}{2},\theta'(\beta)=0,\phi'(\beta)=0 \tag{14.13}$$

The drag coefficient for surface (Cf), the Nusselt number (Nu), and the Sherwood number (Sh) are calculated as follows:

$$C_{fx}=\frac{\tau_w}{\rho U_w^2}\text{ where }\tau_w=\mu\left.\frac{\partial u}{\partial y}\right|_{y=0}$$

$$Nu_x=\frac{xq_w}{k(T_w-T_0)}\text{ where }q_w=-k\left.\frac{\partial T}{\partial y}\right|_{y=0}$$

$$Sh_x=\frac{xq_m}{D_B(C_w-C_0)}\text{ where }q_m=-D_B\left.\frac{\partial C}{\partial y}\right|_{y=0} \tag{14.14}$$

Using the similarity transformation Eq. (14.6), the equations above are reduced to the form as follows:

$$\begin{aligned}\mathrm{Nu}_x\mathrm{Re}^{-\frac{1}{2}}&=-\theta'(0)\\ C_f\mathrm{Re}^{\frac{1}{2}}&=f''(0)\\ \mathrm{Sh}_x\mathrm{Re}^{-\frac{1}{2}}&=-\phi'(0)\end{aligned} \tag{14.15}$$

where $\mathrm{Re}=\frac{xU_w}{\upsilon}$ refers to the Reynolds number.

14.4 METHODOLOGY

In order to resolve the coupled non-linear boundary value problem (BVP) governing fluid flow, heat transfer, and concentration, we incorporate the scipy.integrate.solve_bvp function. A computational domain can be discretized into 100 points. An initial guess is provided to aid solver convergence. solve_bvp solves the system iteratively using an adaptive collocation technique. The solver repeats until the error tolerance ($\mathrm{tol}=10^{-6}$) is encountered. In order to ensure accuracy, the maximum number of grid points is set at 5000. This methodology offering an efficient Python-based numerical solution for solving the non-linear boundary value problem governing the heat, fluid flow, and mass transfer phenomena. The adaptive mesh refinement and tolerance-based convergence criteria ensure that the results are both accurate and reliable.

14.5 MACHINE LEARNING APPROACH

Compared to conventional numerical approaches, the machine learning approach has a number of benefits. First, it can effectively manage challenging relationships among input and output variables as well as high-dimensional information. Second, especially for complexities, it involves less computational effort than traditional approaches. Third, when the model is trained, it provides efficient output variable prediction, which facilitates generic research and effective design optimization. Fourth, it makes it easier to determine essential features and how they communicate, providing valuable insight about the physics at play.

A regression-based machine learning model is created to forecast the skin friction coefficient (*Cf*) and Nusselt number (Nu) without the need for complete numerical simulations, improving computing efficiency. A supervised learning technique is employed and governing parameters are systematically changed to create a dataset. For successful prediction and prevent overfitting, the dataset has been split into training, validation, and testing sets. The trained model's ability to predict thermal and mass transport parameters effectively is illustrated by its strong correlation coefficients (*R*). This machine learning technique is a good choice for real-time applications in cooling systems, industrial coatings, and heat exchangers based on nanofluids as it significantly reduces computing costs while maintaining excellent accuracy. So, the dataset is generated (see Table 14.1 and 14.2) by systematically varying the governing parameters such as *Ec, M, A,* Le, *Nt* and *Nb*.

TABLE 14.1
Variation on Nusselt Number for Various Governing Parameters

Ec	M	A	Le	N_t	N_b	Nusselt Number (Nu_x)
0.1	0.1	0.1	2.0	0.4	0.2	0.4996
0.3	0.1	0.1	2.0	0.4	0.2	0.3712
0.5	0.1	0.1	2.0	0.4	0.2	0.2373
0.7	0.1	0.1	2.0	0.4	0.2	0.0974
0.1	0.2	0.1	2.0	0.4	0.2	0.5024
0.1	0.4	0.1	2.0	0.4	0.2	0.4967
0.1	0.6	0.1	2.0	0.4	0.2	0.4909
0.1	0.8	0.1	2.0	0.4	0.2	0.4852
0.1	1.0	0.1	2.0	0.4	0.2	0.4795
0.1	0.1	0.3	2.0	0.4	0.2	0.5764
0.1	0.1	0.1	2.0	0.2	0.2	0.4996
0.1	0.1	0.1	2.0	0.6	0.2	0.4122
0.1	0.1	0.1	2.0	0.8	0.2	0.3126
0.1	0.1	0.1	2.0	1.0	0.2	0.1982
0.1	0.1	0.1	1.0	0.4	0.2	0.4776
0.1	0.1	0.1	1.5	0.4	0.2	0.4904
0.1	0.1	0.1	2.5	0.4	0.2	0.4953
0.1	0.1	0.1	3.0	0.4	0.2	0.5032
0.1	0.1	0.1	2.0	0.4	0.0	0.5182

TABLE 14.1 (Continued)
Variation on Nusselt Number for Various Governing Parameters

Ec	*M*	*A*	Le	N_t	N_b	Nusselt Number (Nu_x)
0.1	0.1	0.1	2.0	0.4	0.1	0.4733
0.1	0.1	0.1	2.0	0.4	0.3	0.4666
0.1	0.1	0.1	2.0	0.4	0.4	0.4591
0.1	0.1	0.1	2.0	1.2	0.2	0.4519
0.1	0.1	0.1	2.0	1.6	0.2	0.4519

TABLE 14.2
Variation on Skin Friction for Various Governing Parameters

Ec	*M*	*A*	Le	N_t	N_b	Skin Friction (*Cf*)
0.2	0.00	0.1	2.0	0.4	0.2	–0.9868
0.2	0.10	0.1	2.0	0.4	0.2	–1.0212
0.2	0.20	0.1	2.0	0.4	0.2	–1.0551
0.2	0.30	0.1	2.0	0.4	0.2	–1.0885
0.2	0.40	0.1	2.0	0.4	0.2	–1.1215
0.2	0.50	0.1	2.0	0.4	0.2	–1.1541
0.2	0.60	0.1	2.0	0.4	0.2	–1.1862
0.2	0.70	0.1	2.0	0.4	0.2	–1.2180
0.2	0.80	0.1	2.0	0.4	0.2	–1.2493
0.2	0.90	0.1	2.0	0.4	0.2	–1.2803
0.2	1.00	0.1	2.0	0.4	0.2	–1.3109
0.2	1.10	0.1	2.0	0.4	0.2	–1.3411
0.2	1.20	0.1	2.0	0.4	0.2	–1.3710
0.2	1.30	0.1	2.0	0.4	0.2	–1.4005
0.2	1.40	0.1	2.0	0.4	0.2	–1.4297
0.2	1.50	0.1	2.0	0.4	0.2	–1.4586
0.2	1.60	0.1	2.0	0.4	0.2	–1.4872
0.2	1.70	0.1	2.0	0.4	0.2	–1.5154
0.2	1.80	0.1	2.0	0.4	0.2	–1.5433
0.2	1.90	0.1	2.0	0.4	0.2	–1.5710
0.2	2.00	0.1	2.0	0.4	0.2	–1.5983
0.2	2.10	0.1	2.0	0.4	0.2	–1.6254
0.2	2.20	0.1	2.0	0.4	0.2	–1.6522
0.2	2.30	0.1	2.0	0.4	0.2	–1.6787
0.2	2.40	0.1	2.0	0.4	0.2	–1.7050
0.2	2.50	0.1	2.0	0.4	0.2	–1.7310
0.2	2.60	0.1	2.0	0.4	0.2	–1.7568
0.2	2.70	0.1	2.0	0.4	0.2	–1.7823
0.2	2.80	0.1	2.0	0.4	0.2	–1.8075
0.2	2.90	0.1	2.0	0.4	0.2	–1.8326
0.2	3.00	0.1	2.0	0.4	0.2	–1.8574

14.6 RESULT AND DISCUSSION

The traditional Maxwell thin-film fluid flow through a stretched sheet is addressed in this work. The Python BVP solver, which effectively addresses highly non-linear ODEs, is used to solve the described flow model. An adaptive collocation approach is used for ensuring reliable numerical solutions. The system's behavior is also further analyzed and predicted via a regression-based machine learning technique, which improves comprehension of flow properties and parameter relationships.

Figure 14.2 describes the plots for three curves corresponding to velocity, temperature, and concentration profile. The blue line indicates the velocity profile, which starts at zero and increased, the orange line shows the temperature profile, which increases slowly, the green line represents the concentration profile, which starts at 1 and declines gradually. This gives the clear formation of stable boundary layer where the velocity increases gradually, temperature rises slowly due to the heat diffusion, and finally concentration declined due to mass transfer.

Figure 14.3 analyzes the effects in magnetic parameter M on velocity, temperature, and concentration profile. When a magnetic effect applied to the conducting fluid, it produces a resistive force known as magnetic drag. This drag opposes the fluid motion, which make the deceleration effects, which shows the velocity profile decreases as M increases. So, M enhances, the velocity declined at all points. Due to the Lorentz force, which generates magnetic drag in electrically conducting fluid. The enhancement of the magnetic effect strength declines the velocity, which is most common in MHD flow, where higher magnetic effect can pull down the fluid motion. The temperature profile rises with increasing M. This is because the drop in velocity affects convective heat transfer. The fluid retains heat longer when flowing

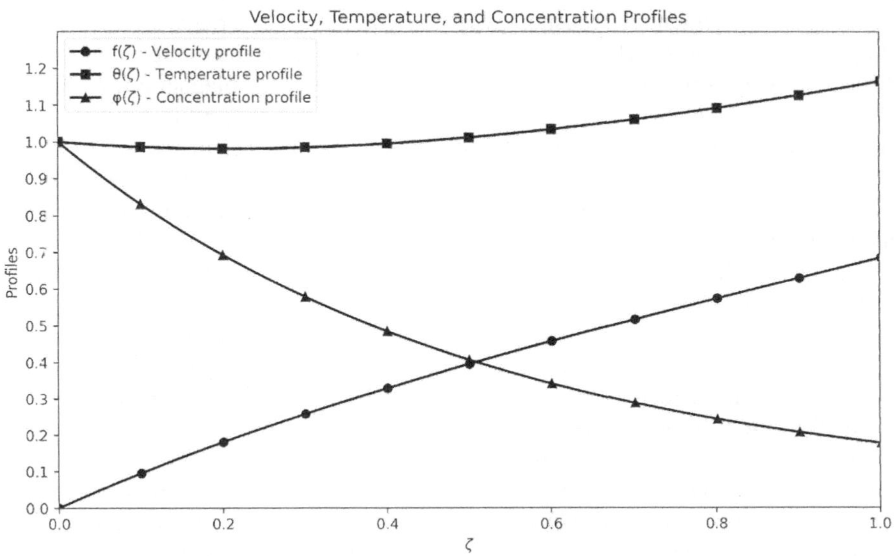

FIGURE 14.2 General behavior of the system.

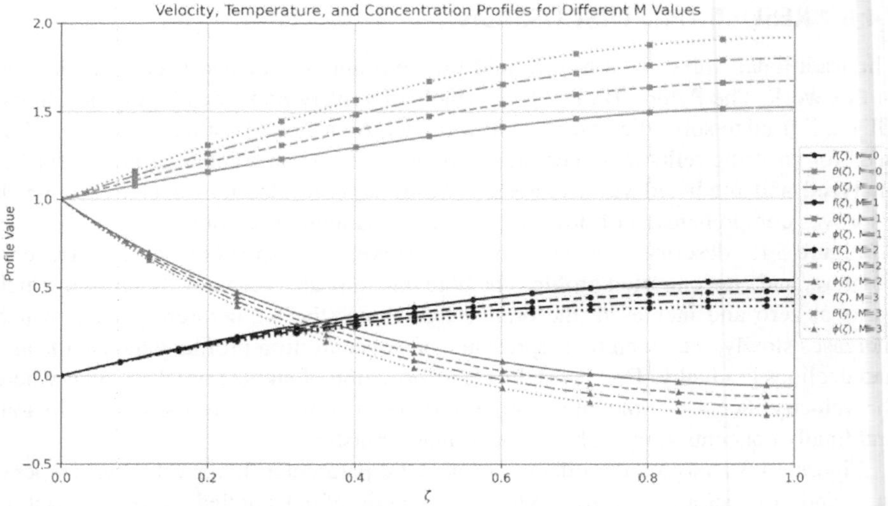

FIGURE 14.3 Impact of M on velocity, temperature, and concentration profiles.

more slowly, which raises the temperature. The concentration profile drops dramatically as M rises. Because motion is slowed by the magnetic field, mixing and diffusion are decreased. The fluid loses concentration more quickly as a result of the steeper concentration gradient.

Figure 14.4 represents the impact of the unsteady parameter on temperature, velocity, and concentration profiles. When A rises, the velocity falls. A higher A correlates to a stronger unsteadiness effect, which dampens the flow. It indicates that in an unsteady flow, momentum transport is weakened leading to a reduced velocity. The temperature drops as the A rises. Since unsteady effects enhance the speed of thermal diffusion, heat propagates faster, reducing temperature. The magnetic field has the opposite effect, raising the temperature. The concentration profile declines more gradually as A rises. This implies that unsteadiness stimulates diffusion, which makes it possible for concentration to disperse over the area more uniformly. Physically, as the system evolves more unsteady, heat and mass transfer become more important, and the boundary layer adjusts as well.

Figure 14.5 depicts that the influence of Ec on velocity, temperature, and concentration profiles. While higher Ec indicates increased conversion of kinetic energy into internal energy due to viscous dissipation, which improves the momentum boundary layer, the velocity rises at the surface as Ec rises, which suggests a stronger flow due to energy dissipation effects. The temperature profile grows drastically as Ec rises. This is why the fluid is heated by the viscous dissipation, which results in stronger thermal storage. With an increase in Ec, the concentration falls. This occurs because solutal boundary layers become thinner due to enhanced diffusion caused by the increased temperature. Hence, stronger velocity fields, greater temperatures, and thinner concentration layers are all a consequence of stronger Ec. This is essential for processes that involve significant amounts of viscous dissipation or high-speed flows.

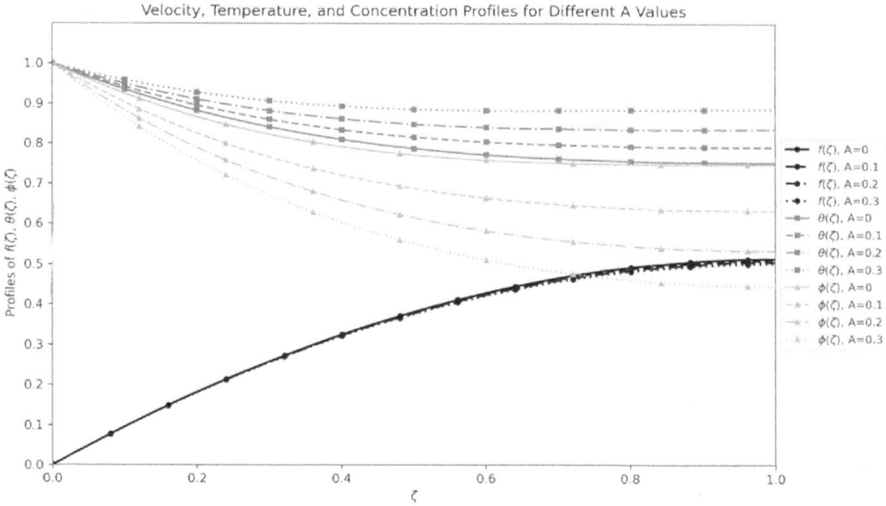

FIGURE 14.4 Impact of A on velocity, temperature, and concentration profiles.

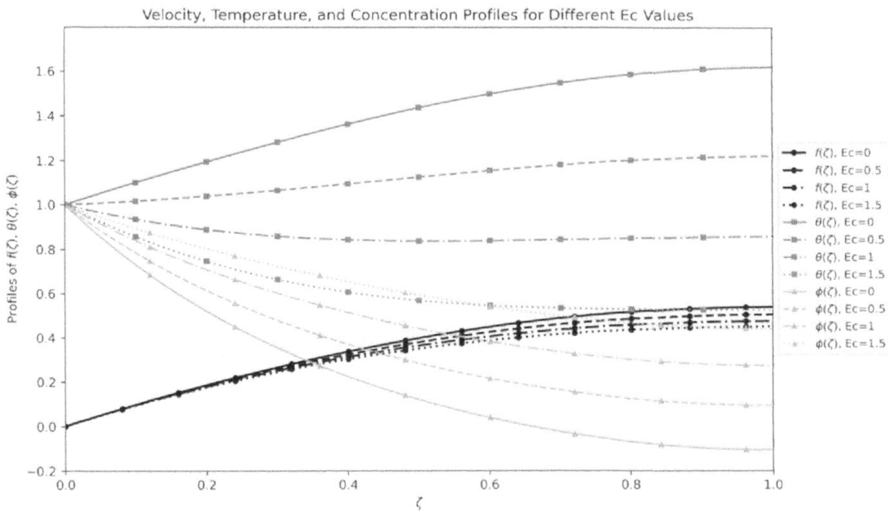

FIGURE 14.5 Impact of Ec on velocity, temperature, and concentration profiles.

Figure 14.6 demonstrate the impact of N_b on temperature, velocity, and concentration profiles. The velocity field gradually reduces as N_b rises. Greater Brownian motion raises momentum diffusion, leading to higher flow resistance along with decreased velocity. The temperature propels with higher N_b. Heat transmission in the fluid is increased by Brownian motion, which improves thermal diffusion. The concentration reduces as N_b increases. Physically, greater dispersion of nanoparticles due to stronger Brownian motion lowers local concentration. Brownian motion increases thermal diffusion but decreases solute concentration due to enhanced mixing.

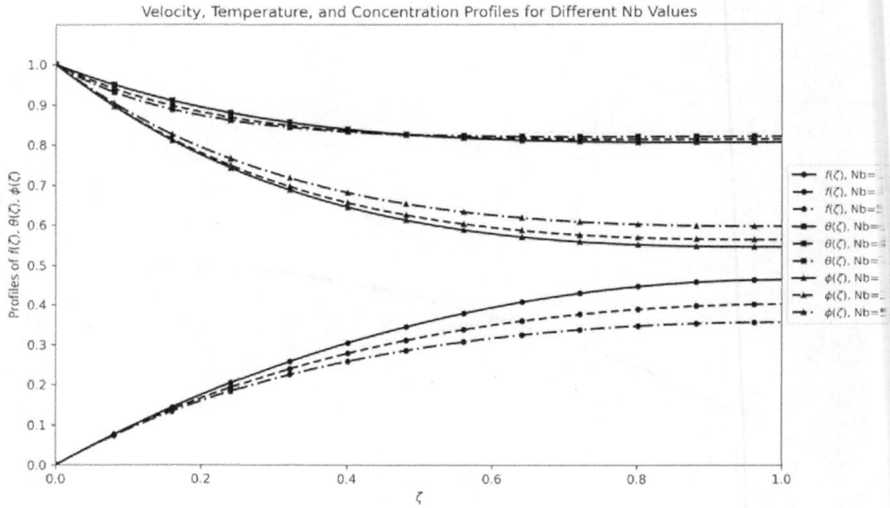

FIGURE 14.6 Impact of N_b on velocity, temperature, and concentration profiles.

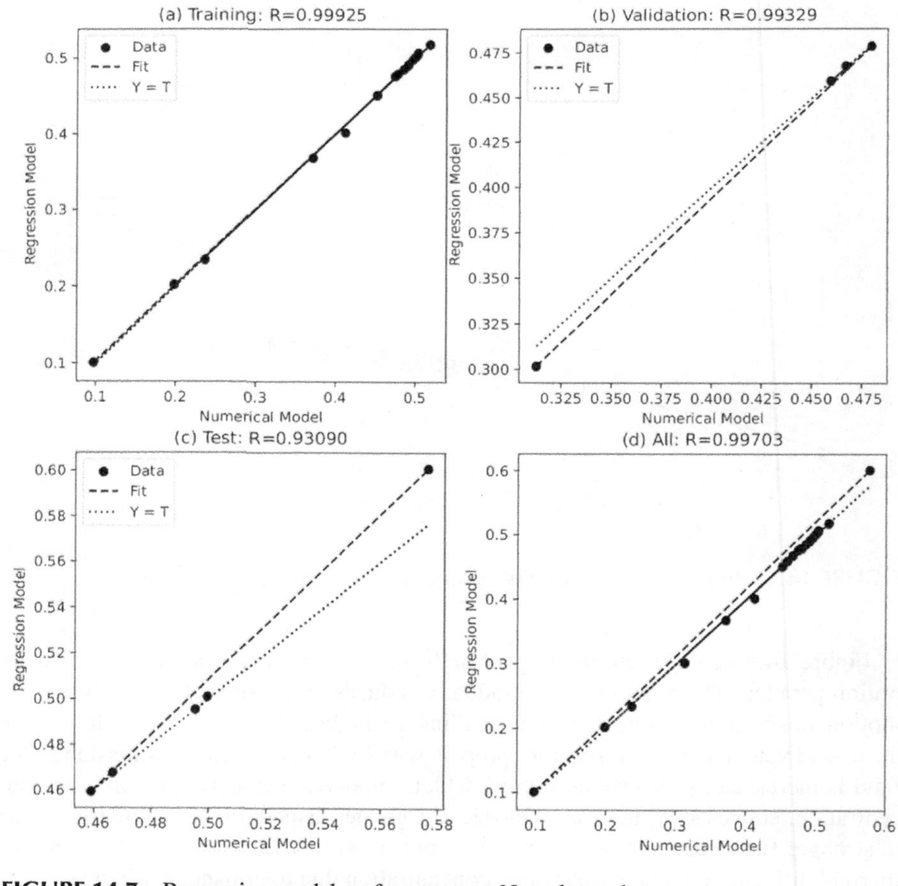

FIGURE 14.7 Regression model performance on Nusselt number.

Figure 14.7(a) represents regression model predictions and numerical findings for the Nusselt number in the training dataset are compared in a scatter plot. Excellent agreement between the regression model and numerical data is indicated by the strong correlation coefficient ($R = 0.9992$), which validates that the heat transport behavior was accurately learned. Numerical solutions and regression model predictions for the validation dataset's Nusselt number are compared in Figure 14.7(b). The model retains high prediction accuracy with a correlation value of $R = 0.99329$, demonstrating only small variations from the best match. The model's ability to predict the Nusselt number in the test dataset is shown in a scatter plot Figure 14.7(c). This model is still very accurate, even though the lower correlation coefficient ($R = 0.93090$) shows some small differences. Nusselt number predictions from regression models plotted versus numerical outcomes for all datasets (training, validation, and testing combined) in a scatter plot Figure 14.7(d). The overall strong correlation ($R = 0.99703$) indicates to the regression model's predictability in forecasting heat transfer properties.

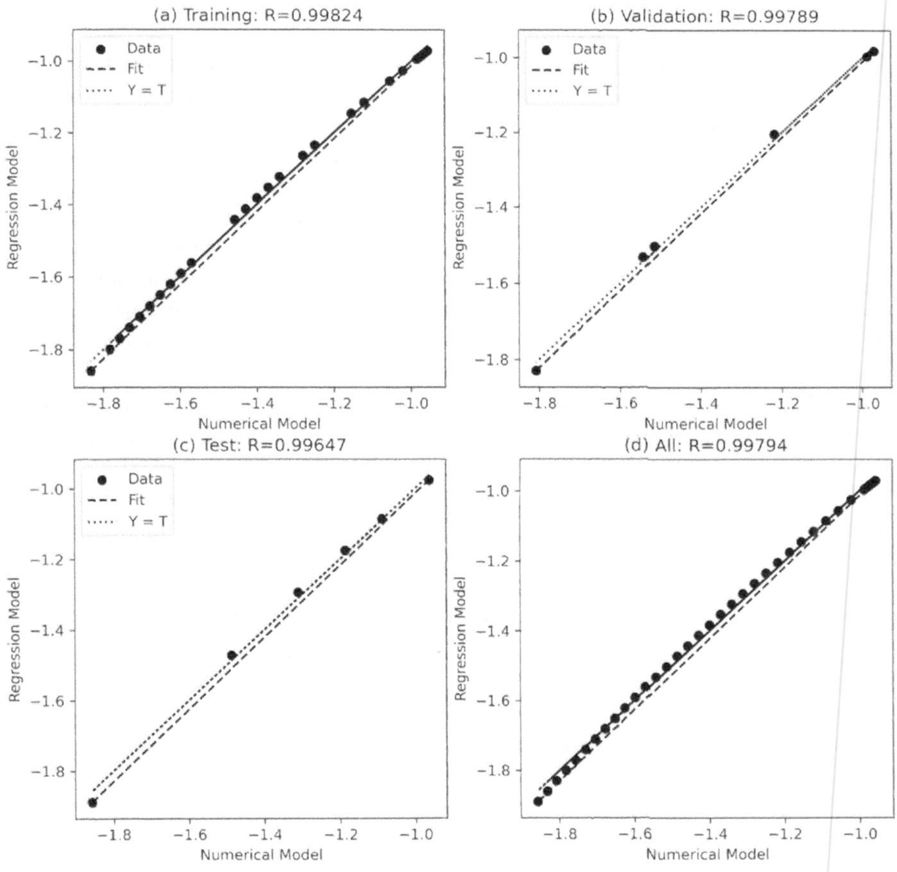

FIGURE 14.8 Regression model performance on skin friction.

Figure 14.8(a) is a scatter diagram showing the training dataset's skin friction coefficient numerical solutions with regression model predictions. Excellent agreement is indicated by the strong correlation coefficient ($R = 0.99824$), which provides accurate flow behavior learning. Numerical findings for the skin friction coefficient in the validation dataset are compared with the predictions of the regression model shown in Figure 14.8(b). Strong generalization skills can be seen by the model, with a correlation value of $R = 0.99789$. The capacity of the model to predict the skin friction coefficient in the test dataset is shown in a scatter plot in Figure 14.8(c). Even though there may be small variations, the model still has a great predictive ability overall, as seen by the somewhat lower correlation value ($R = 0.99647$). Regression model predictions towards skin friction coefficient numerical values for all datasets (training, validation, and testing combined) are plotted in a scatter plot in Figure 14.8(d). The accuracy of the regression model in predictions of shear stress behavior is shown by the strong correlation ($R = 0.99794$).

14.7 CONCLUSION

Regression-based machine learning approaches offer a way to predict and analyze Maxwell thin-film heat and mass transfer with viscous dissipation along with MHD effects. By incorporating the advantages of machine learning, researchers can effectively model complicated systems and acquire valuable insights into the underlying physics. However, additional study is necessary to address existing gaps to improve the accuracy and quality of these methods for this specific application. Future developments in this area might be greatly aided by combining sophisticated machine learning techniques with conventional numerical approaches. Without a doubt, the ongoing evolution and enhancement of these techniques will make a substantial contribution to the knowledge and design of thin-film systems in many different kinds of engineering applications. Here some of the important key points are listed as follows:

- The regression-based model offers a highly accurate and computationally effective alternative to direct numerical simulations for both shear stress (Cf) and heat transfer (Nu).
- The regression model's good predictions and stability for unknown data are supported by the strong correlation values found throughout the training, validation, and testing datasets.
- The results presented show that fluid flow and heat transfermay be effectively addressed using machine learning-based techniques, which can lower computational expenditures without compromising accuracy.
- The study suggests that higher Ec strengthens velocity and temperature but minimizes concentration because of viscous dissipation.
- Increasing N_b influences temperature but minimizes velocity and concentration because of the enhanced Brownian motion effects.
- Stronger magnetic fields (M) reduce velocity, strengthen temperature, and reduce concentration caused by Lorentz force effects.
- Increased unsteadiness (A) stimulates more uniform concentration diffusion whereas reducing temperature and velocity.

REFERENCES

1. Reddy, Y. D., Goud, B. S., Chamkha, A. J., & Kumar, M. A. (2022). Influence of radiation and viscous dissipation on mhd heat transfer Casson nanofluid flow along a nonlinear stretching surface with chemical reaction. *Heat Transfer*, 51(4), 3495–3511. https://doi.org/10.1002/htj.22460
2. Shilpa, B., & Leela, V. (2023). Galerkin finite element analysis of heat and mass transfer of Jeffrey, maxwell and Oldroyd-b nanofluids in a vertical annulus with an induced magnetic field and a nonuniform heat source/sink. *International Journal of Ambient Energy*, 44(1), 1887–1903. https://doi.org/10.1080/01430750.2023.2196988
3. Jawad, M., Shah, Z., Islam, S., Bonyah, E., & Khan, A. (2018). Darcy-forchheimer flow of mhd nanofluid thin film flow with joule dissipation and Naviers partial slip. *Journal of Physics Communications*, 2(11), 115014. https://doi.org/10.1088/2399-6528/aaeddf
4. Nasir, S., Shah, Z., Islam, S., Bonyah, E., & Gul, T. (2019). Darcy forchheimer nano fluid thin film flow of swcnts and heat transfer analysis over an unsteady stretching sheet. *AIP Advances*, 9(1). https://doi.org/10.1063/1.5083972
5. Ullah, A., Shah, Z., Kumam, P., Ayaz, M., Islam, S., & Jameel, M. (2019). Viscoelastic mhd nanofluid thin film flow over an unsteady vertical stretching sheet with entropy generation. *Processes*, 7(5), 262. https://doi.org/10.3390/PR7050262
6. Khan, N., Zuhra, S., Shah, Z., Bonyah, E., Khan, W., Islam, S., & Khan, A. (2019). Hall current and thermophoresis effects on magnetohydrodynamic mixed convective heat and mass transfer thin film flow. *Journal of Physics Communications*, 3(3), 035009. https://doi.org/ 10.1088/2399-6528/aaf830
7. Murugan, R. D., Sivakumar, N., Tarakaramu, N., Alhazmi, H., & Abdullaev, S. (2024). Entropy and energy transfer analysis of a Maxwell thin-film fluid over an inclined surface with viscous dissipation effect. *ZAMM – Journal of Applied Mathematics and Mechanics / Zeitschrift für Angewandte Mathematik und Mechanik*, 104(6), e202300381. https://doi.org/10.1002/zamm.202300381
8. Gul, T., Shah, R., Islam, S., & Arif, M. (2013). Mhd thin film flows of a third grade fluid on a vertical belt with slip boundary conditions. *Journal of Applied Mathematics*, 2013(1), 707286. https://doi.org/10.1155/2013/707286
9. Sharma, K. (2022). Fhd flow and heat transfer over a porous rotating disk accounting forcoriolis force along with viscous dissipation and thermal radiation. *Heat Transfer*, 51(5), 4377–4392. https://doi.org/10.1002/htj.22504
10. Murugan, R. D., Sivakumar, N., Tarakaramu, N., Ahmad, H., & Askar, S. (2024). Entropy generation on MHD motion of hybrid nanofluid with porous medium in presence of thermo-radiation and ohmic viscous dissipation. *Discover Applied Sciences*, 6(4), 199. https://doi.org/10.1007/s42452-024-05866-6
11. Raju, S. S. R., Renuka Devi, R. L. V., Asogwa, K. K., Raju, S. S. K., Raju, C. S. K., Kathyayani, G., & Siva Kumar, N. (2024). Falkner–Skan slip flow of non-Newtonian fluid over a moving and nonlinearly radiated wedge with variable heat source/sink and viscous dissipation. *International Journal of Modern Physics B,* 38(01), 2450004. https://doi.org/10.1142/S0217979224500048
12. Siva Kumar, N., & Rushi Kumar, B. (2021). Blasius and Sakiadis unsteady flow of chemically reacted MHD Williamson fluid with variable conductivity: A comparative study. In *Advances in Fluid Dynamics: Selected Proceedings of ICAFD 2018* (pp. 877–887). Springer Singapore. https://doi.org/10.1007/978-981-15-4308-1_67

13. Alali, E., & Megahed, A. (2022). Mhd dissipative casson nanofluid liquid film flow due to an unsteady stretching sheet with radiation influence and slip velocity phenomenon. *Nanotechnology Reviews*. 11(1), 463–472. https://doi.org/10.1515/ntrev-2022-0031
14. Devi, M. R., Sivakumar, N., Noeiaghdam, S., & Fernandez-Gamiz, U. (2024). Multiple shape factor effects of nanofluids on marangoni mixed convection flow through porous medium. *Results in Engineering,* 23, 102512. https://doi.org/10.1016/j.rineng.2024.102512
15. Murugan, R. D., Sivakumar, N., Tarakaramu, N., Sarhan, N., & Awwad, E. M. (2024). Mixed convection hybrid nanofluid flow over a rotating cone in a rotating fluid environment with interfacial nanolayer effect. *Numerical Heat Transfer, Part B: Fundamentals*, 1–20. https://doi.org/10.1080/10407790.2024.2364779
16. Tarakaramu, N., Sivakumar, N., Tamam, N., Satya Narayana, P. V., & Ramalingam, S. (2024). Theoretical analysis of Arrhenius activation energy on 3D MHD nanofluid flow with convective boundary condition. *Modern Physics Letters B*, 38(16), 2341009. https://doi.org/10.1142/S0217984923410099
17. Gul, T., & Pervez, R. (2022). Thin-film Maxwell hybrid nanofluid flow over an unsteady inclined stretching sheet in terms of nonlinear mixed convection. *Waves in Random and Complex Media,* 1–19. https://doi.org/10.1080/17455030.2022.2089372s

NOMENCLATURE

λ_1	Time Relaxation
k	Thermal Conductivity ($m^2\ s^{-1}$)
c_p	Specific Heat (J Kg $m^{-1}\ s^{-2}$)
ζ	Similarity Variable
ρ	Denstiy (Kg m^{-3})
υ	Kinematic Viscosity ($m^2\ s^{-1}$)
Pr	Prandtl Number
Θ	Dimesionless Temperature
μ	Dynamic Viscosity (Kg $m^{-1}\ s^{-1}$)
β	Non-Dimensionless Film Thickness
M	Dimensionless Magnetic Parameter
A	Unsteadiness Parameter
N_t	Thermophoresis Parameter
N_b	Brownian Parameter
T	Temperature (K)
Ec	Eckert Number
λ_2	Buoyancy Force Parameter
T_0	Fluid–Initial Temperature (K)
T_{ref}	Fluid – Reference Temperature (K)
C_0	Fluid–Initial Concentration
C_{ref}	Fluid–Reference Concentration
$h(t)$	Thickness of the Film
B_0	Magnetic Coefficient

Nu	Local Nusselt Number
Cf	Skin Friction
Sh	Sherwood number
Re	Local Reynolds Number
q_w	Local Heat Flux-Wall
x	Cartesian Coordinate for Horizontal-Axis (m)
y	Cartesian Coordinate for Vertical-Axis (m)
u	Velocity Component in Horizontal direction (m s^{-1})
v	Velocity Component in Vertical direction (m s^{-1})

Part IV

Bio-Convective and Reactive Flow Dynamics

15 Casson Hybrid Nanofluid Flow with Chemical Reactions

Sudha Mahanthesh Sachhin, Ulavathi Shettar Mahabaleshwar, Gadhigeppa Myacher Sachin, and Basma Souayeh

15.1 INTRODUCTION

Hybrid nanofluid flow combines multiple nanoparticles to enhance thermal conductivity and heat transfer efficiency, making it ideal for cooling systems, such as in electronics and nuclear reactors. Hybrid nanofluids improve the performance of solar collectors and heat exchangers, contributing to energy conservation. They are also used in medical fields, like targeted drug delivery and biomedical imaging. Additionally, hybrid nanofluids play a vital role in lubrication, reducing friction and wear in mechanical systems. Their versatile properties make them a critical component in advancing modern technology. Das *et al.* [1] studied the Darcy model on hybrid nanofluid movement across expanding cylinder with heat transfer. Lone *et al.* [2] explored the thermal transfer impact on ternary nanofluid movement across a stretchable surface in porous media. Mahabaleshwar *et al.* [3] examined the different objectives of heat transfer of hybrid nanofluid via stretching surface. Akshatha *et al.* [4] examined the thermal exchange effect on ternary nanofluid movement across stretching sheet with slip effects.

A chemical reaction in flow refers to the interaction of substances within a moving fluid, leading to the transformation of reactants into products. This phenomenon is essential in various fields, such as energy production, where combustion reactions drive engines and turbines. It plays a significant role in wastewater treatment by breaking down contaminants for cleaner water. In the pharmaceutical industry, precise chemical reactions in flow systems ensure efficient drug synthesis. Catalytic converters utilize these reactions to minimize vehicle emissions. Additionally, such reactions impact atmospheric chemistry, influencing climate and air quality, making them vital for environmental and industrial advancements. Mishra *et al.* [5] studied the chemical reaction influence on the nanofluid suspensions with bio mixers. Sachhin *et al.* [6] explored the chemical reaction and heat and mass exchange on

DOI: 10.1201/9781003611820-19

hybrid nanofluid movement across an expanding sheet. Santhi *et al.* [7] examined the impact of unsteady flow of hybrid nanofluid movement with magnetic field. Vijay and Sharma [8] explored the radiation effect on incompressible movement of nanoliquid flow over a rotating disk.

Thermal radiation influences heat transfer within the flow. It is crucial in high-temperature environments, such as in combustion chambers and space vehicle re-entry, where radiation significantly impacts heat distribution. In solar energy systems, thermal radiation aids in energy capture and utilization. It plays a vital role in cooling technologies, enhancing efficiency in electronics and machinery. Thermal radiation also affects atmospheric flows, influencing weather and climate patterns. Its study is essential for designing efficient systems in energy, aerospace, and environmental applications. Kumar *et al.* [9] studied the impact of heat exchange on nanoliquid movement over infinite vertical surface with magnetic effect. Turkyilmazoglu [10] examined the impact of convective MHD and radiation on electrically conducting liquid movement across porous disc. Misra and Sinha [11] examined the magnetic field and thermal radiation influence on blood movement in capillary with slip conditions. Li *et al.* [12] examined the mixed convection effect on radiative liquid movement across heated vertical annulus. The authors of [13–20] explored the heat transfer analysis on the different fluid flow across various boundaries using numerical approaches.

Inspired by the abovementioned studies, authors noted that there is a lack of investigation on hybrid nanofluids with the influence of heat transfer, authors utilized the investigation gap and studied the heat and mass transfer effect on Casson hybrid nanofluid flow over a porous stretching/shrinking sheet, the present work has many applications in industrial processes like polymer extrusion, metal casting, and manufacturing of plastic films, where precise material stretching enhances the quality and uniformity, thermal radiation plays a crucial role in high-temperature systems, such as furnaces and heat exchangers, to optimize heat transfer efficiency.

15.2 MATHEMATICAL FORMULATION

Considered two-dimensional hybrid nanofluid flow over a stretching/shrinking surface, the surface having the velocity $u = ax$, magnetic field and electric field applied to the flow, $u$and$v$ are the velocity components along x and y axis, $T$and$C$ represents the temperature and concentration of the fluid flow, considered non-Newtonian Casson fluid flow, heat source and thermal radiation effects included in the temperature equation and chemical reaction impact added in the concentration equation. Figure 15.1 represents the schematic diagram.

Rheological expression of Casson fluid [19, 20]

$$\tau_{ij} = \begin{cases} 2(\mu_B + p_y / \sqrt{2\pi})e_{ij}, & \pi > \pi_c, \\ 2(\mu_B + p_y / \sqrt{2\pi_c})e_{ij}, & \pi < \pi_c, \end{cases} \tag{15.1}$$

the yield stress P_y is calculated as

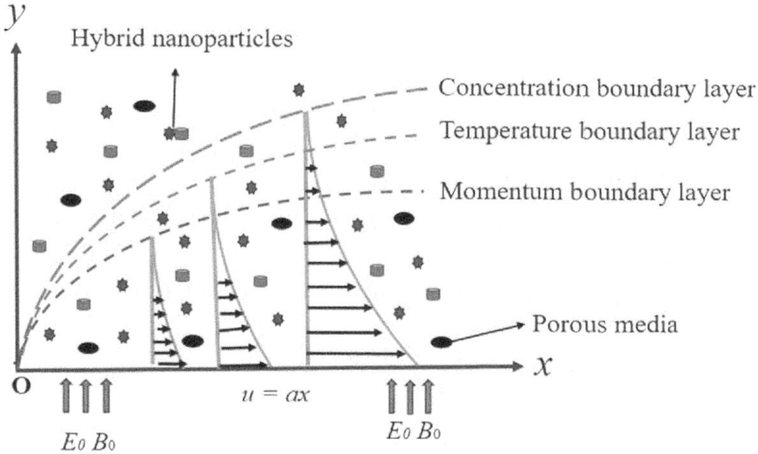

FIGURE 15.1 Schematic diagram.

$$P_y = \frac{\mu_b \sqrt{2\pi_c}}{\lambda} \tag{15.2}$$

15.2.1 Physical Description

Conservation of mass equation [3, 6].

$$\nabla . \vec{q} = 0, \tag{15.3}$$

the conservation of the velocity equation in vector form [3, 6].

$$\left[\vec{q_t}\left(\vec{q}\cdot\nabla\right)\vec{q}\right] = -\frac{\nabla p}{\rho_{hnf}} + \left(1+\frac{1}{\lambda}\right)\nu_{hnf}\nabla^2\vec{q} - \frac{\nu_{hnf}}{K*}\vec{q} + \frac{\sigma_{hnf}}{\rho_{hnf}}\left(E_0 B_0 - B_0^2\vec{q}\right). \tag{15.4}$$

Governing equations for the considered flow [3, 6, 17].

$$\frac{\partial u}{\partial x} + \frac{\partial v}{\partial y} = 0. \tag{15.5}$$

$$u\frac{\partial u}{\partial x} + v\frac{\partial u}{\partial y} = \left(1+\frac{1}{\lambda}\right)\frac{\mu_{hnf}}{\rho_{hnf}}\frac{\partial^2 u}{\partial y^2} + \frac{\sigma_{hnf}}{\rho_{hnf}}\left(E_0 B_0 - B_0^2 u\right) - \frac{\mu_{hnf}}{\rho_{hnf} K*}u. \tag{15.6}$$

$$u\frac{\partial T}{\partial x} + v\frac{\partial T}{\partial y} = \frac{\kappa_{hnf}}{(\rho Cp)_{hnf}}\frac{\partial^2 T}{\partial y^2} - \frac{1}{(\rho Cp)_{hnf}}\frac{\partial q_r}{\partial y} + \frac{Q_0\left(T-T_\infty\right)}{(\rho Cp)_{hnf}} + \frac{\sigma_{hnf}}{(\rho Cp)_{hnf}}\left(uB_0 - E_0\right)^2. \tag{15.7}$$

$$u\frac{\partial C}{\partial x}+v\frac{\partial C}{\partial y}=D_{hnf}\left(\frac{\partial^2 C}{\partial x^2}+\frac{\partial^2 C}{\partial y^2}\right)-Cr*(C-C_\infty) \tag{15.8}$$

15.2.2 Boundary Conditions [6, 15]

$$\begin{aligned} &u=u_w=ax \quad v=0, \quad T=T_w, \quad C=C_w \quad \text{at} \quad y=0. \\ &u=0, \quad v=0, \quad T\to T_\infty, \quad C\to C_\infty \quad \text{as} \quad y\to\infty. \end{aligned} \tag{15.9}$$

The heat flux of radiation by Rosseland approximation as [3, 6, 17].

$$q_r=-\frac{4\sigma*}{3k*}\frac{\partial T^4}{\partial y}, \tag{15.10}$$

$\Rightarrow \frac{\partial q_r}{\partial y}=-\frac{16\sigma*T_\infty^3}{3k*}\frac{\partial^2 T}{\partial y^2}$, where $T^4=(4T-3T\infty)\,T_\infty^3$.

15.2.3 Similarity Conversions [3, 6, 17]

$$\begin{aligned} &u=axf'(\eta), \qquad v=-f(\eta)\sqrt{a\nu_f}, \\ &\theta(\eta)=\frac{T-T_\infty}{T_w-T_\infty} \quad \varphi=\frac{C-C_\infty}{C_w-C_\infty}. \end{aligned} \tag{15.11}$$

Substituting Eq. (15.10) to Eq. (15.7)

$$\begin{aligned} u\frac{\partial T}{\partial x}+v\frac{\partial T}{\partial y}&=\left(\frac{\kappa_{hnf}}{(\rho C_p)_{hnf}}+\frac{1}{(\rho C_p)_{hnf}}\frac{16\sigma^*T_\infty^3}{3k^*}\right)\frac{\partial^2 T}{\partial y^2} \\ &+\frac{Q_0(T-T_\infty)}{(\rho C_p)_{hnf}}+\frac{\sigma_{hnf}}{(\rho C_p)_{hnf}}(uB_0-E_0)^2 \end{aligned} \tag{15.12}$$

the modified equations are as below:

$$\left(1+\frac{1}{\lambda}\right)A_1\frac{d^3 f}{d\eta^3}+A_2 f(\eta)\frac{d^2 f}{d\eta^2}-A_2\left(\frac{df}{d\eta}\right)^2+A_3 ME-A_3 M\frac{df}{d\eta}-A_1 L\frac{df}{d\eta}=0 \tag{15.13}$$

$$(A_4+Nr)\frac{d^2\theta}{d\eta^2}+A_5\Pr\left(f(\eta)\frac{d\theta}{d\eta}\right)+Ni\Pr\theta(\eta)+\Pr Ec\left[A_3 M\left(E-\frac{df}{d\eta}\right)^2\right]=0 \tag{15.14}$$

$$A_6\varphi_{\eta\eta}(\eta)-Scf(\eta)\varphi_\eta(\eta)-ScKc*\varphi(\eta)=0 \tag{15.15}$$

15.2.4 Modified Boundary Conditions [3, 6, 17]

$$f(0)=0, \quad f'(0)=1, \quad \theta(0)=1, \quad \varphi(0)=1, \quad \text{at} \quad \eta=0,$$
$$\theta(\infty)=0, \quad \varphi(\infty)=0, \quad f'(\infty)=0, \quad \text{as} \quad \eta\to\infty. \tag{15.16}$$

15.3 ENGINEERING QUANTITIES [3, 6, 17]

15.3.1 Skin Friction

$$C_f=\frac{2\tau_w}{\rho_{hnf}u_w^2}\Rightarrow\frac{(\mathrm{Re}_x)^{0.5}C_f}{2}=\frac{A_1}{A_2}f''(0),\text{ where }\tau_w=\mu_{hnf}\left(\frac{\partial u}{\partial y}\right)_{y=0}. \tag{15.17}$$

15.3.2 Nusselt Number

$$\mathrm{Nu}=\frac{xq_w}{\kappa_f(T_w-T_\infty)}\Rightarrow(\mathrm{Re}_x)^{-0.5}\mathrm{Nu}=-A_4\theta'(0),\text{ where }q_w=-\kappa_{hnf}\left(\frac{\partial T}{\partial y}\right)_{y=0}, \tag{15.18}$$

15.3.3 Sherwood Number

$$\mathrm{Sh}=\frac{xF_w}{D_f(C_w-C_\infty)}\Rightarrow(\mathrm{Re}_x)^{-0.5}\mathrm{Sh}=-A_6\varphi'(0),\text{ where }F_w=-D_{hnf}\left(\frac{\partial T}{\partial y}\right)_{y=0}. \tag{15.19}$$

15.4 NUMERICAL METHOD WITH SOLUTION

To solve higher order differential equations, we introduce new variables as follows:

$$y_1=f,\ y_2=f',\ y_3=f'',\ y_4=\theta,\ y_5=\theta',\ y_6=\varphi,\ y_7=\varphi'. \tag{15.20}$$

The governing Eqs. (15.13)–(15.15) are as follows:

$$\begin{aligned}
&y_2^1=y_3,\ y_3^1=-\left(\left(\frac{1}{1+\lambda^{-1}}\right)\left(A_2y_1y_3-A_2y_2^2+A_3ME-A_3My_2-A_1Ly_2\right)\right),\\
&y_4^1=y_5,\ y_5^1=-\left(\left(\frac{1}{(A_4+Nr)}\right)\left(\mathrm{Pr}A_5y_1y_5+Ni\mathrm{Pr}y_4+\mathrm{Pr}Ec\left[A_3M(E-y_2)^2\right]\right)\right),\\
&y_6^1=y_7,\ y_7'=-\left(\frac{Sc}{A_6}y_1y_7-\frac{Sc}{A_6}Kc*y_6\right),
\end{aligned} \tag{15.21}$$

$$\left.\begin{aligned}
&\frac{\rho_{hnf}}{\rho_f}=(1-\phi_2)\left(1-\phi_1+\phi_1\left(\frac{\rho_{s_1}}{\rho_f}\right)\right)+\phi_2\left(\frac{\rho_{s_2}}{\rho_f}\right),\\
&\frac{\sigma_{hnf}}{\sigma_f}=\frac{\sigma_{s_2}+2\sigma_{bf}+2\phi_2(\sigma_{s_2}-\sigma_f)}{\sigma_{s_2}+2\sigma_{bf}-\phi_2(\sigma_{s_2}-\sigma_f)}.\\
&\text{where}\\
&\frac{\sigma_{bf}}{\sigma_f}=\frac{\sigma_{s_1}+2\sigma_f+2\phi_1(\sigma_{s_1}-\sigma_f)}{\sigma_{s_1}+2\sigma_f-\phi_1(\sigma_{s_1}-\sigma_f)}.\\
&\frac{\mu_{hnf}}{\mu_f}=\frac{1}{(1-\phi_1)^{2.5}(1-\phi_2)^{2.5}}.\\
&\frac{D_{hnf}}{D_f}=\frac{1}{(1-\phi_1)(1-\phi_2)}.\\
&\frac{(\rho C_p)_{hnf}}{(\rho C_p)_f}=(1-\phi_2)\left(1-\phi_1+\phi_1\left(\frac{(\rho C_P)_{s_1}}{(\rho C_P)_f}\right)\right)+\phi_2\left(\frac{(\rho C_P)_{s_2}}{(\rho C_P)_f}\right).\\
&\frac{\kappa_{hnf}}{\kappa_f}=\frac{\kappa_{s_2}+2\kappa_{bf}+2\phi_2(\kappa_{s_2}-\kappa_f)}{\kappa_{s_2}+2\kappa_{bf}-\phi_2(\kappa_{s_2}-\kappa_f)},\\
&\text{where}\\
&\frac{\kappa_{bf}}{\kappa_f}=\frac{\kappa_{s_1}+2\kappa_f+2\phi_1(\kappa_{s_1}-\kappa_f)}{\kappa_{s_1}+2\kappa_f-\phi_1(\kappa_{s_1}-\kappa_f)}.
\end{aligned}\right\}\quad(15.22)$$

15.5 RESULTS AND DISCUSSION

The current chapter explores the heat and mass transfer effect on Casson sodium alginate hybrid nanofluid movement over permeable stretching/shrinking surface, numerical procedure utilized to calculate the ODEs, in this section studied the several physical parameters effect on fluid flow through graphs as follows, Table 15.1 represents non-dimensional parameters, Table 15.2 represents the thermophysical properties. Table 15.3 represents the comparison values with previous publications.

Figures 15.2(a) and (b) portrays the parameter L effect on velocity, enhancing the porosity reduces the momentum of the fluid movement, physically, porous media reduce the momentum of fluid flow due to the presence of numerous interconnected voids and solid particles that create resistance. This resistance, often referred to as drag or friction, dissipates the fluid's energy and slows its movement through the medium.

Figures 15.3(a) and (b) portrays the parameter E effect on velocity, enhancing the electric field rises the momentum. Physically, E increases the momentum of fluid flow by exerting a force on the charged particles within the fluid. This force accelerates the particles, reducing resistance and enhancing the overall flow rate, especially in electrically conductive or dielectric fluids.

Figure 15.4(a) and (b) represents the magnetic field effect on momentum, enhancing the magnetic field reduces the momentum, physically, a magnetic field can

TABLE 15.1
Non-Dimensional Parameters

Constants	$A_1 = \frac{\mu_{hnf}}{\mu_f}, A_2 = \frac{\rho_{hnf}}{\rho_f}, A_3 = \frac{\sigma_{hnf}}{\sigma_f}, A_4 = \frac{\kappa_{hnf}}{\kappa_f}, A_5 = \frac{(\rho Cp)_{hnf}}{(\rho Cp)_f}, A_6 = \frac{D_{hnf}}{D_f}$
Thermal radiation	$Nr = \frac{16\sigma * T_\infty^3}{3\kappa_f k*}$
Heat source	$Ni = \frac{Q_0}{a(\rho Cp)_f}$
Prandtl number	$\Pr\left(= \frac{\mu_f (C_p)_f}{\kappa_f}\right)$
Magnetic field	$M = \frac{\sigma_f B_0^2}{a\rho_f}$
Porosity	$L = \frac{\nu_f}{K*a}$
Eckert number	$Ec = \frac{(u_w)^2}{Cp(T_w - T_\infty)}$

TABLE 15.2
Nanoparticle's Thermophysical Properties [6]

Properties	SA	Al_2O_3	SiO_2
$\sigma\,(S/m)$	2.6×10^{-4}	3.69×10^{-12}	1×10^{-18}
$\rho(\text{kgm}^{-3})$	989	3970	2650
$C_p(\text{JkgK}^{-1})$	4175	765	730
$\kappa(\text{WmK}^{-1})$	0.6376	40	1.5

TABLE 15.3
Comparison $-\theta'(0)$ for Various Numbers of Pr, while Taking $\phi_1, \phi_2 = 0$, $Ec = 0$, $Ni = 0$

Pr	Waini *et al.* [13]	Sreedevi *et al.* [14]	Present results
2.0	0.911353	0.911341	0.911113
6.13	1.759682	1.759676	1.759610
7.0	1.895400	1.895397	1.895458
20.0	3.353902	3.353915	3.353991

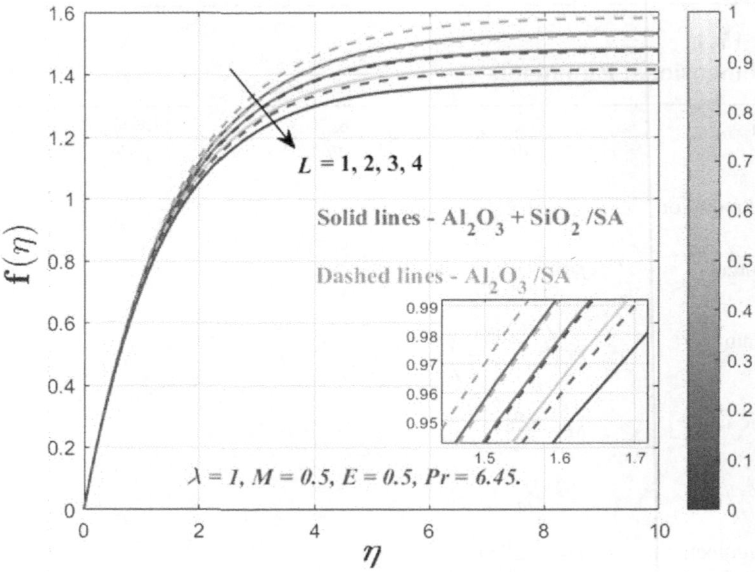

(a)

FIGURE 15.2(a) Porosity parameter effect on transverse velocity.

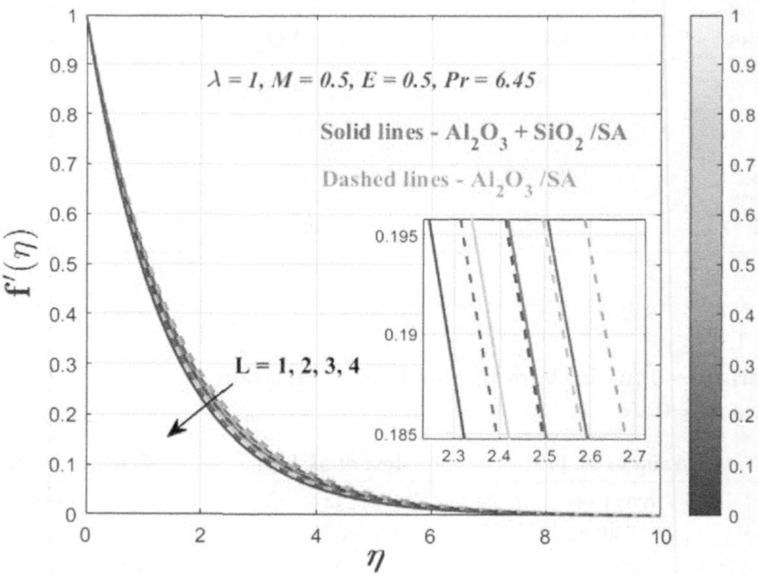

(b)

FIGURE 15.2(b) Porosity parameter effect on axial velocity.

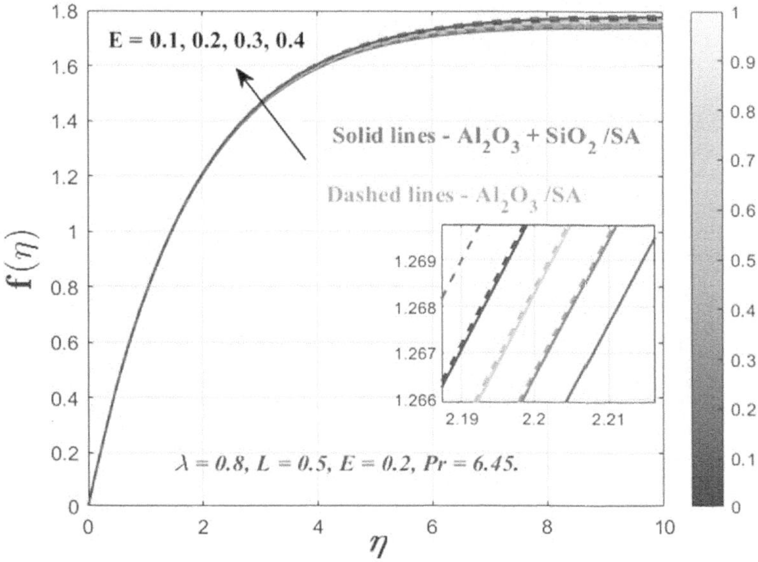

(a)

FIGURE 15.3(a) Electric field effect on transverse velocity.

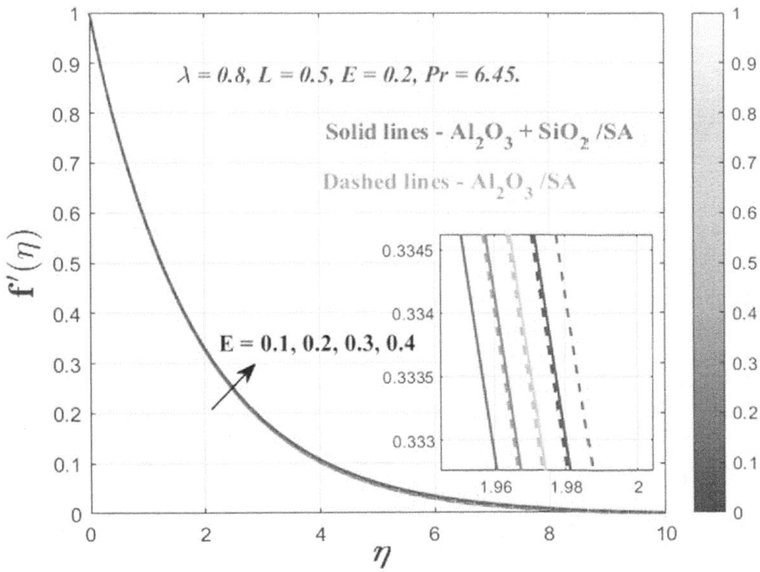

(b)

FIGURE 15.3(b) Electric field effect on axial velocity.

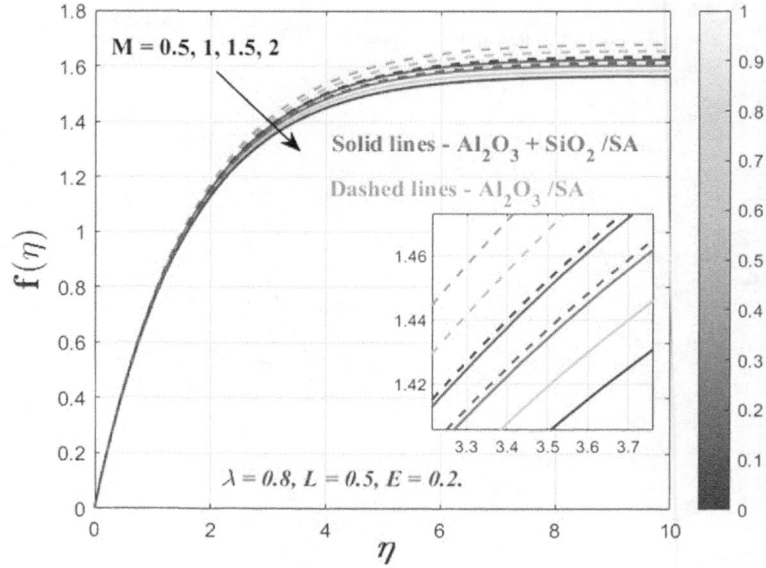

(a)

FIGURE 15.4(a) Magnetic parameter effect on transverse velocity.

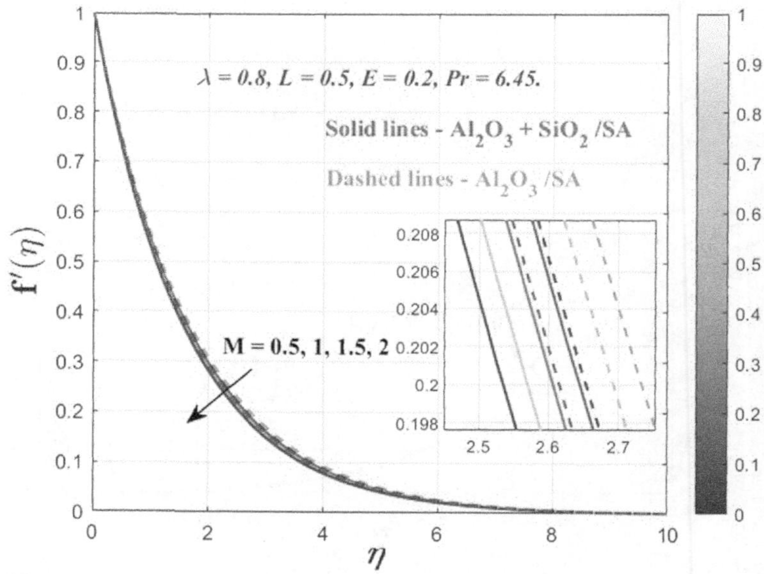

(b)

FIGURE 15.4(b) Magnetic field effect on axial velocity.

increase the momentum flow when the fluid is electrically conductive. Through the magnetohydrodynamic (MHD) effect, the collision between the magnetic field and the flow induced currents generates a Lorentz force, which accelerates the liquid and enhances its momentum.

Figure 15.5 represents the heat source parameter effect on temperature of the fluid, increasing the heat source rises the temperature of the flow, physically. A heat source increases the temperature of fluid flow by transferring thermal energy to the fluid. This energy raises the kinetic activity of the fluid molecules, resulting in a higher energy as the heat spreads through conduction.

Figure 15.6 represents the *Nr* parameter effect on energy of the fluid movement, rising the thermal radiation rises the temperature of the fluid, physically, *Nr* increases the energy of fluid flow by emitting electromagnetic waves that transfer energy to the fluid. When the fluid absorbs this energy, it causes an increase in the molecular motion, leading to a rise in its temperature.

Figure 15.7 represents the Schmidt number effect on $\varphi(\eta)$ of the fluid movement, enhancing the Sc number decays the $\varphi(\eta)$ of the fluid flow, physically, the Schmidt number influences the concentration distribution in fluid flow by characterizing the relative effectiveness of momentum diffusion to mass diffusion. A higher Sc number indicates slower mass diffusion compared to velocity, leading to steeper concentration gradients.

Figure 15.8 represents the Eckert parameter effect on temperature of the flow, rising the Eckert number rises the temperature, physically, the *Ec* number impacts the energy of fluid flow by representing the ratio of kinetic energy to thermal energy.

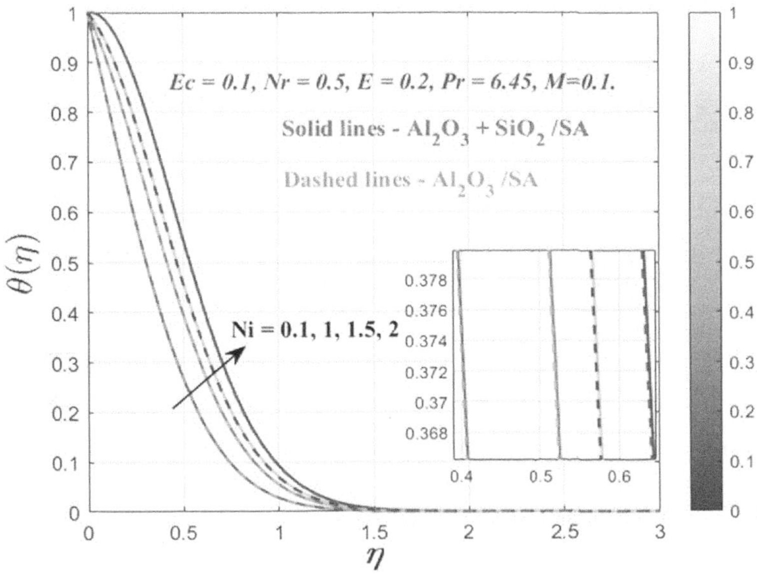

FIGURE 15.5 Heat source parameter effect on temperature profile.

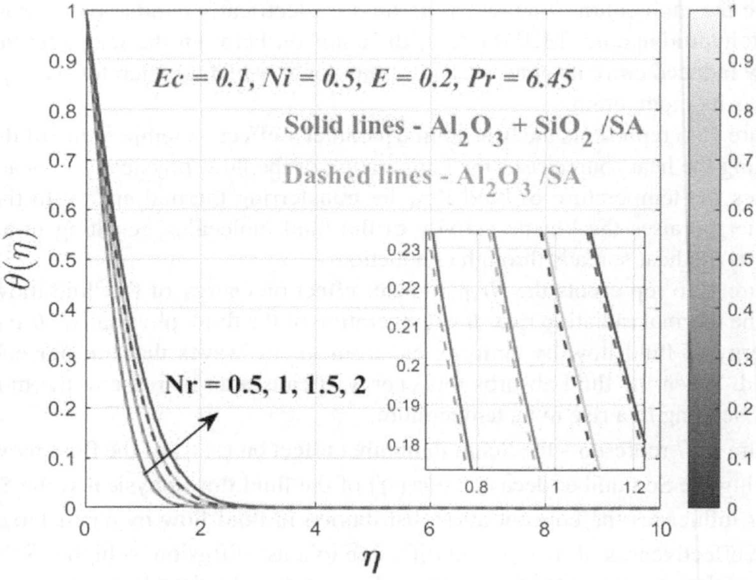

FIGURE 15.6 Thermal radiation parameter effect on temperature profile.

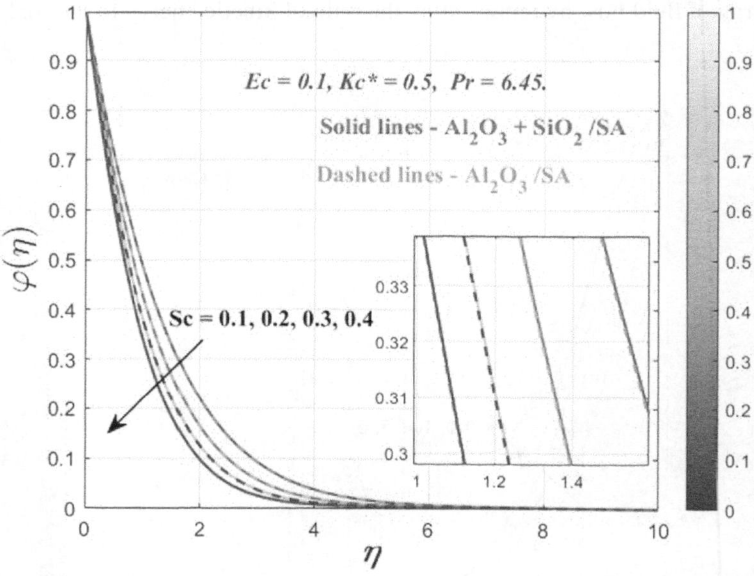

FIGURE 15.7 Schmidt number effect on concentration profile.

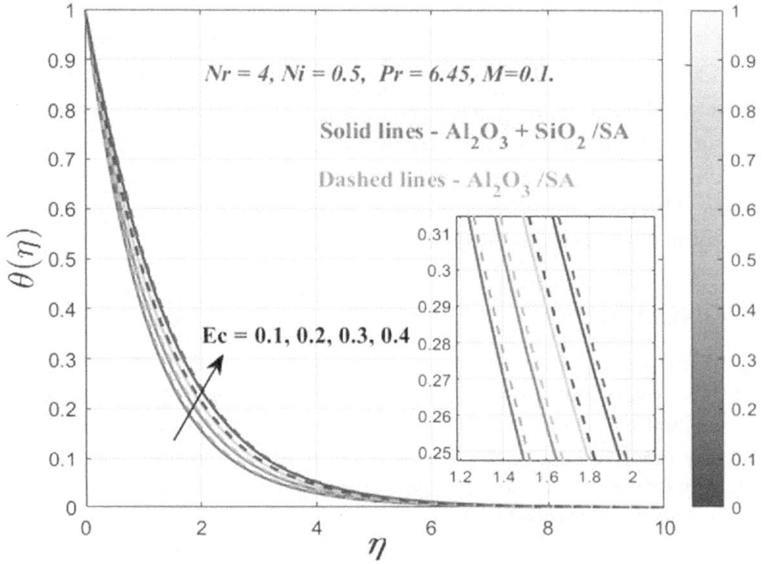

FIGURE 15.8 Eckert parameter effect on temperature profile.

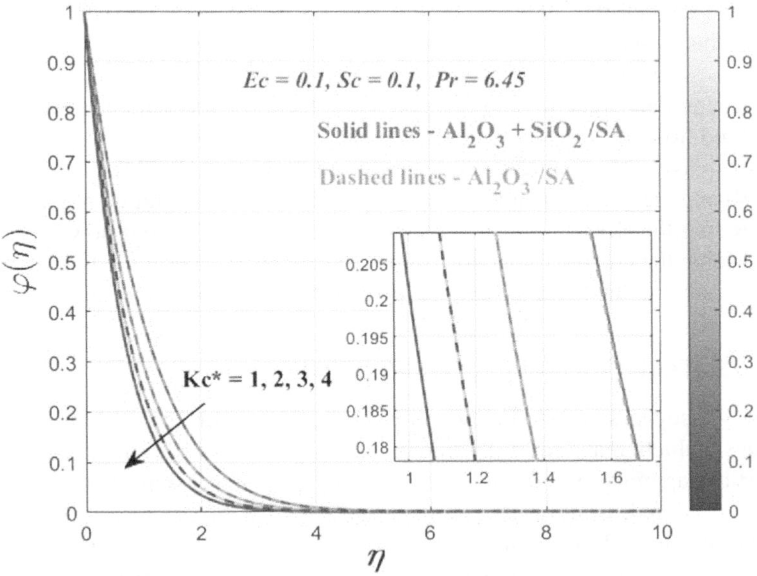

FIGURE 15.9 Chemical reaction parameter effect on concentration profile.

A higher Eckert number signifies increased conversion of kinetic energy into internal energy due to viscous dissipation, leading to a rise in fluid temperature.

Figure 15.9 represents the chemical reaction parameter effect on $\varphi(\eta)$, enhancing the chemical reaction decays the $\varphi(\eta)$, physically, chemical reactions affect the $\varphi(\eta)$ of fluid flow by altering the distribution of reactive species. Reactive processes can either consume or produce certain components, leading to changes in concentration profiles. The rate and type of reaction significantly influence how quickly and uniformly these changes occur within the flow.

15.6 CONCLUSION

The current chapter explores the heat and mass transfer effect on Casson hybrid nanofluid flow over a porous stretching/shrinking surface, and a numerical procedure is utilized to solve the ODEs. The present analysis of heat and mass effect on chemically reactive Casson sodium alginate hybrid nanofluid flow over a permeable stretching or shrinking sheet has promising future applications. It can contribute to advancements in biomedical engineering, particularly in drug delivery systems using biocompatible materials. The findings are valuable for optimizing industrial processes like polymer extrusion and chemical coatings. This research also holds potential for enhancing energy systems, such as solar collectors and heat exchangers, by improving thermal performance. Additionally, it can aid in environmental engineering, including wastewater treatment and pollutant control technologies. In this section, we studied the effect of several physical parameters on fluid flow and the results of the present study reveal the following:

- Rising the magnetic field and porosity effects decreases the momentum of the fluid flow.
- Increasing electric field enhances the velocity of the fluid flow.
- Rising the radiation and heat source increases the temperature of the fluid flow.
- Rising the Eckert number rises the temperature of the fluid flow.
- Rising the chemical reaction decays the concentration of the fluid flow.
- Enhancing the Schmidt number decays the concentration of the fluid flow.

15.7 LIMITATIONS AND FUTURE STUDY

This study assumes steady-state, two-dimensional flow with constant thermophysical properties, which may not fully represent real-world conditions. The effects of nanoparticle aggregation, variable permeability, and multi-step chemical reactions are not considered, limiting its practical applicability. Future research can explore unsteady and three-dimensional effects, temperature-dependent properties, and the role of external forces like magnetic and electric fields. Experimental validation and machine learning-based optimization can further enhance the model's accuracy. Additionally, extending the study to biomedical and industrial applications will broaden its relevance in engineering and technology.

REFERENCES

1) T. K. Das, A. Paul, and J. M. Nath, "Thermo-convection driven Sodium Alginate-based Darcy–Forchheimer EMHD Williamson hybrid nanofluid flow with varying thermal distribution," *Mod. Phys. Lett. B*, vol. 39, no. 05, p. 2450405, Feb. 2025, doi: 10.1142/S0217984924504050.
2) S. A. Lone, L. A. AL-Essa, F. S. Alduais, A. Al-Bossly, A. Dawar, and A. Saeed, "Optimization of heat transfer in bi-directional flow of sodium alginate-based ternary hybrid nanofluid over an extending heated surface with velocity slip conditions," *J. Therm. Anal. Calorim.*, vol. 150, pp. 1545–1556, Jan. 2025, doi: 10.1007/s10973-024-13872-4.
3) U. S. Mahabaleshwar, S. M. Sachhin, L. M. Pérez, and H. F. Oztop, "An impact of inclined MHD on biviscosity Bingham hybrid nanofluid flow over porous stretching/shrinking sheet with heat transfer," *J. Mol. Liq.*, vol. 398, p. 124244, Mar. 2024, doi: 10.1016/j.molliq.2024.124244.
4) H. D. Akshatha, S. M. Sachhin, U. S. Mahabaleshwar, R. K. Lodhi, and K. Ramesh, "Dynamics of sodium alginate-based ternary nanofluid flow over a stretching sheet with Al2O3, SiO2, and TiO2 nanoparticles," *Multiscale Multidiscip. Model. Exp. Des.*, vol. 8, no. 1, p. 32, Nov. 2024, doi: 10.1007/s41939-024-00591-4.
5) N. K. Mishra, S. Anwar, P. Kumam, T. Seangwattana, M. Bilal, and A. Saeed, "Numerical investigation of chemically reacting jet flow of hybrid nanofluid under the significances of bio-active mixers and chemical reaction," *Heliyon*, vol. 9, no. 7, Jul. 2023, doi: 10.1016/j.heliyon.2023.e17678.
6) S. M. Sachhin, U. S. Mahabaleshwar, N. Swaminathan, D. Laroze, and L. Pedraja-Rejas, "An effect of waste discharge concentration and electromagnetic field on chemically reactive Bingham fluid flow with the analysis entropy generation and mass transfer," *Hybrid Advances*, vol. 8, p. 100344, Mar. 2025, doi: 10.1016/j.hybadv.2024.100344.
7) M. Santhi, K. V. Suryanarayana Rao, P. Sudarsana Reddy, and P. Sreedevi, "Heat and mass transfer characteristics of radiative hybrid nanofluid flow over a stretching sheet with chemical reaction," *Heat Transf.*, vol. 50, no. 3, pp. 2929–2949, 2021, doi: 10.1002/htj.22012.
8) N. Vijay and K. Sharma, "Entropy generation analysis in MHD hybrid nanofluid flow: Effect of thermal radiation and chemical reaction," *Numer. Heat Transf. Part B Fundam.*, vol. 84, no. 1, pp. 66–82, Jul. 2023, doi: 10.1080/10407790.2023.2186989.
9) M. A. Kumar, Y. D. Reddy, V. S. Rao, and B. S. Goud, "Thermal radiation impact on MHD heat transfer natural convective nano fluid flow over an impulsively started vertical plate," *Case Stud. Therm. Eng.*, vol. 24, p. 100826, Apr. 2021, doi: 10.1016/j.csite.2020.100826.
10) M. Turkyilmazoglu, "Thermal radiation effects on the time-dependent MHD permeable flow having variable viscosity," *Int. J. Therm. Sci.*, vol. 50, no. 1, pp. 88–96, Jan. 2011, doi: 10.1016/j.ijthermalsci.2010.08.016.
11) J. C. Misra and A. Sinha, "Effect of thermal radiation on MHD flow of blood and heat transfer in a permeable capillary in stretching motion," *Heat Mass Transf.*, vol. 49, no. 5, pp. 617–628, May 2013, doi: 10.1007/s00231-012-1107-6.
12) B.-W. Li, W. Wang, and J.-K. Zhang, "Combined effects of magnetic field and thermal radiation on fluid flow and heat transfer of mixed convection in a vertical cylindrical annulus," *J. Heat Transf.*, vol. 138, no. 062501, Mar. 2016, doi: 10.1115/1.4032609.
13) I. Waini, A. Ishak, and I. Pop, "Unsteady flow and heat transfer past a stretching/shrinking sheet in a hybrid nanofluid," *Int. J. Heat Mass Transf.*, vol. 136, pp. 288–297, Jun. 2019, doi: 10.1016/j.ijheatmasstransfer.2019.02.101.

14) P. Sreedevi, P. Sudarsana Reddy, and A. Chamkha, “Heat and mass transfer analysis of unsteady hybrid nanofluid flow over a stretching sheet with thermal radiation,” *SN Appl. Sci.*, vol. 2, no. 7, p. 1222, Jun. 2020, doi: 10.1007/s42452-020-3011-x.
15) B. Venkateswarlu, S. Chavan, S. W. Joo, and S. C. Kim, “Irreversibility analysis of radiative TiO2+Al2O3+Cu/H2O ternary nanofluid flow over a bidirectional stretching sheet with cattaneo-christov heat flux: nanotechnology applications,” *Phys. Scr.*, vol. 99, no. 11, p. 115264, Oct. 2024, doi: 10.1088/1402-4896/ad7d46.
16) B. Venkateswarlu, S. Chavan, S. Joo, and S. Kim, “Impact of hybrid nanofluids on thermal management of cylindrical battery modules: A numerical study,” *J. Energy Storage*, vol. 99, p. 113266, Oct. 2024, doi: 10.1016/j.est.2024.113266.
17) B. Venkateswarlu, S. Chavan, S. W. Joo, and S. C. Kim, “Entropy analysis of electromagnetic trihybrid nanofluid flow with temperature-dependent viscosity in a Darcy-Forchheimer porous medium over a stretching sheet under convective conditions,” *J. Mol. Liq.*, vol. 393, p. 123660, Jan. 2024, doi: 10.1016/j.molliq.2023.123660.
18) B. Venkateswarlu, S. Chavan, S. W. Joo, S. C. Kim, and K. S. Nisar, “A numerical investigation of heat transfer performance in a prismatic battery cooling system using hybrid nanofluids,” *Case Stud. Therm. Eng.*, vol. 66, p. 105719, Feb. 2025, doi: 10.1016/j.csite.2024.105719.
19) K. Bhattacharyya, “MHD Stagnation-point flow of Casson fluid and heat transfer over a stretching sheet with thermal radiation,” *J. Thermodyn.*, vol. 2013, no. 1, p. 169674, 2013, doi: 10.1155/2013/169674.
20) S. A. Lone, S. Anwar, A. Saeed, and G. Bognár, “A stratified flow of a non-Newtonian Casson fluid comprising microorganisms on a stretching sheet with activation energy,” *Sci. Rep.*, vol. 13, no. 1, p. 11240, Jul. 2023, doi: 10.1038/s41598-023-38260-0.

16 MHD Heat and Mass Transfer in Non-Newtonian Fluids

Varshegaa Sivanarayana, Ragulkumar Elumalai, Sambath Paulsamy, Francis Peter, and Mir Waqas Alam

16.1 INTRODUCTION

In many industrial and engineering systems, the study of heat and mass transfer in free convective flows over a vertical cone is crucial for optimizing processes, such as geothermal energy extraction, cooling technologies, and chemical reactors. While Newtonian fluids have been traditionally studied, non-Newtonian fluids like Casson, Maxwell, and Williamson are more suitable for such scenarios due to their unique rheological properties shear-thinning, viscoelasticity, and yield stress behaviors that enhance heat and mass transfer. The Cattaneo–Christov heat and mass flux model is employed to account for finite thermal propagation speed, improving accuracy in dynamic thermal systems. This work explores the combined effects of MHD, non-uniform heat source/sink, chemical reactions, and thermal relaxation in non-Newtonian fluid flow over a vertical cone, offering valuable insights for optimizing real-world industrial processes.

Heat transfer over a vertical cone has been a focus of research due to its practical relevance in engineering systems, such as geothermal energy, cooling technologies, and chemical reactors. The foundational work by Hering and Grosh [1] on laminar free convection from a non-isothermal cone laid the groundwork for understanding thermal behavior in conical geometries. This was further extended by Weaver and Viskanta [2], who investigated natural convection caused by horizontal temperature and concentration gradients with variable thermophysical properties. Pop and Watanabe [3] added to this by studying free convection under uniform suction or injection for constant wall heat flux, providing insight into how boundary conditions affect the heat transfer process. The study of both heat and mass transfer over cones became important as industries sought to optimize these processes. Hasan and Mujumdar [4] explored coupled heat and mass transfer in natural convection, showing how temperature and concentration fields interact along a vertical cone. Chamkha and Quadri [5] expanded this by incorporating the effects of magnetohydrodynamics (MHD) and heat generation/absorption, a critical factor in many industrial systems. The classification of thermal patterns, as explored by Luhmann *et al.* [6], further demonstrated the real-world application of these heat transfer principles in natural settings like karst

DOI: 10.1201/9781003611820-20

springs. In recent years, non-Newtonian fluids have garnered significant attention because their complex flow behavior more accurately reflects the properties of many industrial fluids. Barth and Carey [7] introduced non-Newtonian fluids to natural convection benchmark problems, providing valuable insights into their distinct thermal properties. Kumar and Sivaraj [8] expanded this understanding by examining the heat and mass transfer in MHD viscoelastic fluid flows over cones, highlighting how variable viscosity impacts transfer rates. Peter *et al.* [9] recently conducted numerical studies on radiative hybrid nanofluids, advancing computational approaches to analyze such complex systems. The influence of MHD on fluid flow over cones has also been well studied. Dinarvand [10] explored free-convection boundary-layer flow around a heated, rotating cone under a transverse magnetic field, illustrating how magnetic fields affect both the velocity and temperature profiles. Palani and Ragavan [11] incorporated chemical reactions into MHD flows, while Sambath *et al.* [12] examined radiative chemical reaction impacts on MHD mass transfer over cones. El-Kabeir *et al.* [13] investigated MHD effects in micropolar fluids with heat generation and absorption, providing comprehensive insights into how chemical reactions and magnetic fields interact with thermal systems.

The study of non-Newtonian fluids, particularly Casson fluids, has gained momentum due to their ability to model real-world materials such as blood and slurries. Casson [14] introduced a model that describes the flow of fluids with yield stress. This was applied by Mythili *et al.* [15,16] to study Casson fluid flow over vertical cones under the influence of non-uniform heat sources, higher-order chemical reactions, and MHD effects, providing a more accurate representation of industrial fluid systems. Benazir *et al.* [17] extended this work to unsteady MHD flows, while Durairaj *et al.* [18] further explored heat generation and absorption in chemically reacting Casson fluids over cones. Singh *et al.* [19] examined the effect of non-uniform heating and viscous dissipation on the natural convection of Casson fluids over cones, addressing how spatial variations in thermal properties influence heat transfer. Other non-Newtonian fluids, such as Williamson and Maxwell fluids, have also been studied. Khan *et al.* [20] investigated Williamson nanofluid flows in the presence of MHD and chemical reactions, while Sathyanarayana and Goud [21] conducted a numerical study on MHD Williamson-nanofluid flows over cones under suction/injection effects. Kodi *et al.* [22] analyzed MHD mixed convection flows with Maxwell nanofluids, exploring the impact of variable heat conductivity and diffusion. Maxwell fluids, known for their viscoelastic properties, have been widely studied in the context of MHD flows. Sathyanarayana and Goud [23] examined the behavior of Maxwell fluids in free convective non-Newtonian nanofluid flows over cones, particularly under the influence of magnetic fields. The Cattaneo–Christov heat flux model, which accounts for finite thermal wave propagation, has been adopted in these studies to improve the accuracy of heat transfer predictions in dynamic systems. Ramandevi *et al.* [24] applied this model to non-Newtonian fluid flows with viscous dissipation and non-uniform heat sources, demonstrating its relevance for MHD flows. Raju *et al.* [25] similarly studied mass and thermal transfer in MHD non-Newtonian nanofluids in the presence of non-uniform heat sources. In addition, Koteswara Reddy *et al.* [26] used the

Buongiornos model to investigate mixed convection in MHD Maxwell, Jeffrey, and Oldroyd-B nanofluids, focusing on how thermophoresis and Brownian motion affect heat and mass transfer. This work contributes to the broader understanding of heat transfer in complex nanofluid systems, which are essential in optimizing industrial processes.

The reviewed literature highlights the complex interplay between these factors (MHD, chemical reactions, and non-uniform heat sources and sinks) and their significant impact on thermal and concentration profiles. This research aims to improve the understanding of heat and mass transfer mechanisms in the flow of non-Newtonian fluids over a vertical cone. It employs advanced computational tools to consider the impacts of MHDs, chemical reactions, and non-uniform heat sources and sinks. By examining these variables, the study aims to deliver valuable insights for optimizing heat and mass transfer processes in industrial settings. These settings often involve fluid flow systems impacted by electromagnetic forces, heat-generating or absorbing chemical reactions, and localized temperature fluctuations. Continued research in this area is essential for developing more efficient heat and mass transfer systems, ultimately contributing to the optimization of industrial processes.

16.2 FORMULATION OF THE PROBLEM

16.2.1 Problem Statement

This paper presents a comprehensive study of the MHD flow of three distinct non-Newtonian fluids—Williamson, Casson, and Maxwell fluids—over a vertical cone characterized by its radius r and half-angle ω. The flow is driven in an upward direction, influenced by a uniform temperature T_∞, uniform concentration C_∞, and free-stream velocity U_∞, both aligned parallel to the surface of the cone. To facilitate a thorough analysis, a coordinate system is employed where the y-axis is oriented perpendicularly to the cone's surface while the x-axis runs parallel along it, as depicted in the flow configuration diagram in Figure 16.1. The investigation further considers the effects of chemical reactions and non-uniform heat sources/sinks to control heat and mass flux in the system. Assuming that the ambient concentration and temperature are always below the wall concentration C_W and temperature T_W (i.e., $C_W > C_\infty$ and $T_W > T_\infty$), the constant ambient conditions are denoted by T_∞ and C_∞. The governing equations are derived using the Boussinesq approximation, accounting for small density variations.

16.2.2 Dimensional Governing Equations and Boundary Conditions

$$\frac{\partial (ru)}{\partial x} + \frac{\partial (rv)}{\partial y} = 0 \tag{16.1}$$

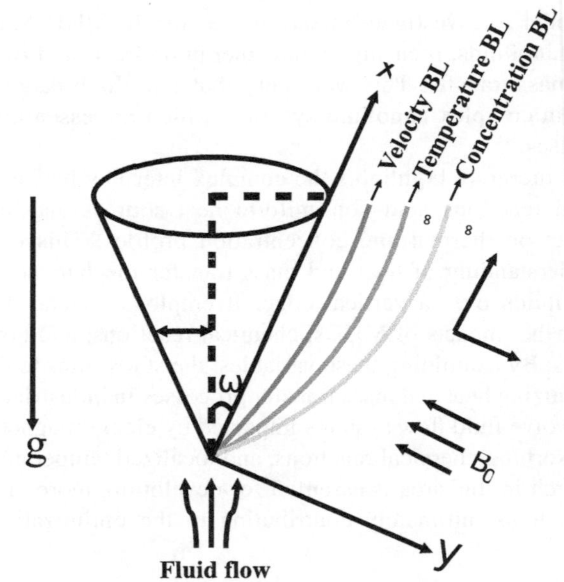

FIGURE 16.1 Schematic representation of the physical model.

$$u\frac{\partial u}{\partial x}+v\frac{\partial v}{\partial y}=\Lambda_1\left[\nu\left(1+\frac{1}{\beta}\right)\frac{\partial^2 u}{\partial y^2}\right]$$

$$+\Lambda_2\left[\nu\frac{\partial^2 u}{\partial y^2}-m_1\left(u^2\frac{\partial^2 u}{\partial x^2}+v^2\frac{\partial^2 u}{\partial y^2}+2uv\right)\frac{\partial^2 u}{\partial x\partial y}\right]$$

$$+\Lambda_3\left[\nu\frac{\partial^2 u}{\partial y^2}+\sqrt{2}\gamma\Gamma\frac{\partial u}{\partial y}\frac{\partial^2 u}{\partial y^2}\right]+g\beta_T\left(T-T_\infty\right)\cos\omega-\frac{\sigma B_0^2 u}{\rho}+g\beta_C\left(C-C_\infty\right)\cos\omega \tag{16.2}$$

$$u\frac{\partial T}{\partial x}+v\frac{\partial T}{\partial y}+\tau_T\left[u\frac{\partial u}{\partial x}\frac{\partial T}{\partial x}+v\frac{\partial v}{\partial y}\frac{\partial T}{\partial y}+u\frac{\partial v}{\partial x}\frac{\partial T}{\partial y}+v\frac{\partial u}{\partial y}\frac{\partial T}{\partial x}\right.$$

$$\left.+2uv\frac{\partial^2 u}{\partial x\partial y}+u^2\frac{\partial^2 T}{\partial x^2}+v^2\frac{\partial^2 T}{\partial y^2}\right]$$

$$=\alpha\frac{\partial^2 T}{\partial y^2}+\frac{q'''}{\rho C_p} \tag{16.3}$$

$$u\frac{\partial C}{\partial x}+v\frac{\partial C}{\partial y}=D\frac{\partial^2 C}{\partial y^2}-K_r\left(C-C_\infty\right)$$

$$-\tau_C\left[u\frac{\partial u}{\partial x}\frac{\partial C}{\partial x}+v\frac{\partial v}{\partial y}\frac{\partial C}{\partial y}+u\frac{\partial v}{\partial x}\frac{\partial C}{\partial y}+v\frac{\partial u}{\partial y}\frac{\partial C}{\partial x}+2uv\frac{\partial^2 u}{\partial x\partial y}+u^2\frac{\partial^2 C}{\partial x^2}+v^2\frac{\partial^2 C}{\partial y^2}\right] \tag{16.4}$$

with the following boundary conditions:

$$u=v=0,\ T=T_w,\ C=C_w \text{ at } y=0$$
$$u\to u_\infty,\ T\to T_\infty,\ C\to C_\infty \text{ as } y\to\infty$$

16.2.3 Fluid Models and Corresponding Stress Tensors

The mathematical model that has been presented requires changes in order to accommodate the particular circumstances that are described in Eq. (16.2).

1. When $\Lambda_1 = 1$, $\Lambda_2 = 0$, $\Lambda_3 = 0$ represents Casson fluid model.
2. When $\Lambda_2 = 1$, $\Lambda_1 = 0$, $\Lambda_3 = 0$ represents Maxwell fluid model.
3. When $\Lambda_3 = 1$, $\Lambda_1 = 0$, $\Lambda_2 = 0$ represents Williamson fluid model.

The parameter for the non-uniform heat source or sink, represented by q''', is defined by the following equation:

$$q''' = \frac{ku}{xv}\left[A^*\left(T_w-T_\infty\right)f'+B^*\left(T-T_\infty\right)\right]$$

In this context, positive values of A^* and B^* correspond to internal heat generation, while negative values indicate heat absorption coefficients.

16.2.3.1 Casson Fluid

The rheological behavior of an incompressible, isotropic Casson fluid [17,27] is described by the stress tensor τ_{ij}

$$\tau_{ij} = \begin{cases} 2\left(\mu_\beta + \dfrac{p_y}{\sqrt{2\pi}}\right)e_{ij}, & \pi > \pi_c \\ 2\left(\mu_\beta + \dfrac{p_y}{\sqrt{2\pi_c}}\right)e_{ij}, & \pi < \pi_c \end{cases}$$

16.2.3.2 Maxwell Fluid

The Maxwell fluid [28,29] model is governed by the constitutive equation:

$$T = -pI + S$$

16.2.3.3 Williamson Fluid

The Williamson fluid [30,31] model is defined by

$$\tau = \frac{\mu_0}{1-\Gamma\dot{\gamma}}A_1$$

16.2.4 Similarity Transformations and Stream Function Representation

The stream function $\psi(x,y)$ is a fundamental concept in fluid dynamics used to describe fluid flow. It allows for the velocity components u and v in the flow field to be expressed in terms of their partial derivatives. Specifically, the velocity in the x-direction, u, is given by $u = \frac{1}{r}\frac{\partial \psi}{\partial y}$, while the velocity in the y-direction, v, is given by $v = -\frac{1}{r}\frac{\partial \psi}{\partial x}$. These expressions help define the flow characteristics using the stream function $\psi(x,y)$. By applying similarity transformations, the governing partial differential equations, which are non-linear, can be transformed into a set of non-linear ordinary differential equations. The similarity variables are defined as follows:

$$r = x\sin\omega, u = \frac{\nu}{x}(Gr)^{1/2}.f'(\eta), \eta = \frac{y}{x}(Gr)^{1/4}, Gr = \frac{g\beta_T(T_w - T_\infty)\cos\omega.x^3}{\nu^2},$$
$$\theta(\eta) = \frac{T - T_\infty}{T_w - T_\infty}, v = \frac{\nu}{4x}(Gr)^{1/4}\left[\eta.f'(\eta) - 7f(\eta)\right], \phi(\eta) = \frac{C - C_\infty}{C_w - C_\infty} \tag{16.5}$$

16.2.5 Dimensionless Governing Equations and Boundary Conditions

The dimensionless forms of the momentum and energy can be derived by similarity transformations. Using Eq. (16.5), Eqs. (16.1–16.4) becomes

$$\left[\Lambda_1\left(1+\frac{1}{\beta}\right) + \Lambda_2\left(1 - G_2\left(\frac{49}{16}\right)f^2\right) + \Lambda_3\left(1 + G_3 f''\right)\right]f''' \tag{16.6}$$
$$-\frac{1}{2}f'^2 + \frac{7}{4}ff'' - Mf' + \theta + Nr\phi = 0$$

$$\theta'' + Pr\left(\frac{7}{4}\right)f\theta' + A^*f' + B^*\theta - C_1\left(ff'\theta' + f^2\theta''\right) = 0 \tag{16.7}$$

$$\phi'' + Sc\left(\frac{7}{4}\right)f\phi' - C_2\left(ff'\phi' + f^2\phi''\right) - K_0 Sc\phi = 0 \tag{16.8}$$

The following are the changed boundary conditions:

$$\theta = 1 = \phi, f = 0 = f' \text{ at } \eta = 0 \tag{16.9}$$
$$f' = 0 = \theta = \phi \text{ as } \eta \to \infty$$

where,

$$G_2 = \frac{m_1\nu(Gr)^{1/2}}{x^2}, G_3 = \frac{\sqrt{2}\Gamma\nu(Gr)^{3/4}}{x^2}, Sc = \frac{\nu}{D}, M = \frac{\sigma B_0^2(Gr)^{-0.5}x^2}{\mu},$$
$$Nr = \frac{\beta_C(C_w - C_\infty)}{\beta_T(T_w - T_\infty)}, Pr = \frac{\nu}{\alpha}, K_0 = \frac{k_c x^2}{\nu(Gr)^{1/2}}$$

16.2.6 Dimensional and Dimensionless Forms of Key Physical Parameters

Important physical parameters include the local Sherwood number Sh, local skin-friction coefficient C_f, and local Nusselt number Nu. These may be presented in both their dimensional [22,32] and dimensionless forms as follows:
Dimensional forms:

$$C_f = \frac{\tau_w}{\rho U^2}; Nu = \frac{q_w x}{k\left(T_w - T_\infty\right)}; Sh = \frac{j_w x}{D_B\left(C_w - C_\infty\right)}$$

where,

$$\tau_w = \mu\left(\frac{\partial u}{\partial y}\right)_{y=0}; q_w = -k\left(\frac{\partial T}{\partial y}\right)_{y=0}; j_w = -D_B\left(\frac{\partial C}{\partial y}\right)_{y=0}$$

are respectively, the shear stress and rate of heat and mass flux at the surface and $U\left(=\frac{\nu Gr_x^{1/2}}{x}\right)$ being the reference velocity.
Dimensionless forms:

$$(Gr)^{1/4} C_f = f''(0)\left[\left(1+\frac{1}{\beta}\right)\Lambda_1 + \Lambda_2 + \Lambda_3\left(1+\frac{G_2}{2}\left(f''(0)\right)\right)\right];$$
$$(Gr)^{1/4} Nu = -\left(\theta'(0)\right); (Gr)^{1/4} Sh = -\left(\phi'(0)\right)$$

16.3 NUMERICAL INVESTIGATION

To get the solution numerically for Eqs. (16.6–16.8), with the conditions on the boundary (16.9), Eqs. (16.6–16.8) are simplified into a system of differential equations with first-order under the presumption that $f = f_1, f' = f_2, f'' = f_3 \theta = f_4, \theta' = f_5, \phi = f_6, \phi' = f_7$

$$f''' = \frac{\frac{1}{2}f_2^2 - \frac{7}{4}f_1 f_3 + Mf_2 - f_6}{\left[\Lambda_1\left(1+\frac{1}{\beta}\right) + \Lambda_2\left(1 - G_2\left(\frac{49}{16}\right)f_1^2\right) + \Lambda_3\left(1 + G_3 f_3\right)\right]}$$

$$\theta'' = -Pr\frac{7}{4}f_1 f_5 - A^* f_2 - B^* f_4 + C_1\left[f_1 f_2 f_5 + \left(f_1\right)^2 \theta''\right]$$

$$\phi'' = -Sc\frac{7}{4}f_1 f_7 + K_0 Sc f_6 + C_2\left[f_1 f_2 f_7 + \left(f_1\right)^2 \phi''\right]$$

with boundary conditions $f_1 = 0, f_2 = 0, f_3 = S_1, f_4 = 1, f_5 = S_2, f_6 = 1, f_7 = S_3$. These S_1 and S_2, and S_3 represent the unknown boundary conditions. In order to arrive at a certain end point, the first order differential equations that have been modified are numerically integrated as an initial-valued problem. Because the initial conditions for the variables that are unknown are considered to be unknown, this assumption is made. To determine

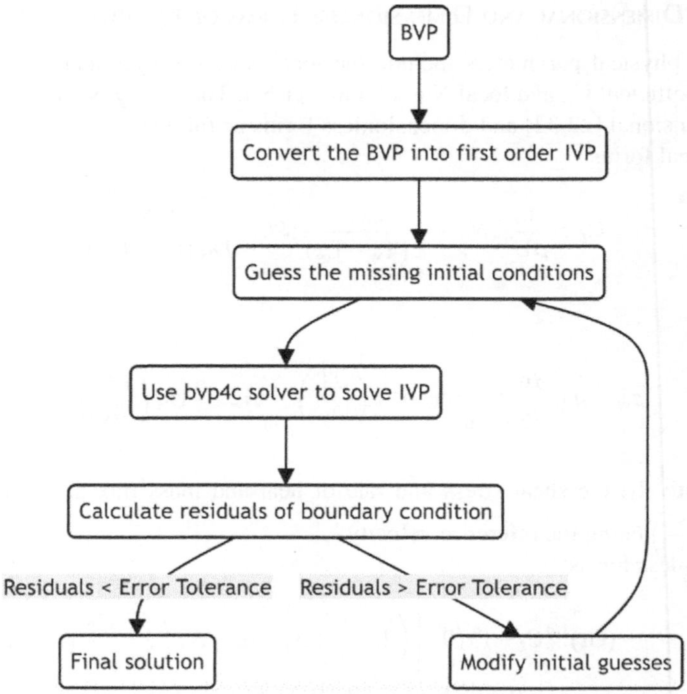

FIGURE 16.2 Flowchart of shooting technique.

whether or not the missing initial condition was correct, we can compare the values that were computed with those that were given for each of the individual variables at the end of the process. We are able to accomplish this by searching for variations in improved values that indicate that we need to locate the starting conditions that are missing. MATLAB® is used to perform the computations, and the program also incorporates the flowchart of the numerical technique, which is referred to as Figure 16.2.

16.4 RESULTS AND DISCUSSIONS

An illustration of the heat and mass transfer that occurs between three non-Newtonian fluid flows (Casson, Maxwell, and Williamson) is provided by this model. $M = 1$, $Nr = 0.5, K_0 = 0.3$, $A^* = 0.2$, $B^* = -0.5$, $Sc = 1$, $Pr = 6.8$, and $G_2 = G_3 = \beta = 0.3$ were the values assigned to each parameter in this model. All of the values were constant, unless it was specifically stated differently. The Casson fluid is represented by the solid line in each and every illustration, whereas the Maxwell fluid is represented by the dashed line, and the Williamson fluid is represented by the dotted line.

16.4.1 MAGNETIC PARAMETER

The magnetic parameter M represents the influence of an external magnetic field on the fluid flow. It is defined as the ratio of the magnetic force to the viscous force. When a magnetic field is applied, it generates a Lorentz force, which acts perpendicular to both the direction of the fluid flow and the magnetic field. This force opposes the fluid motion, acting as a resistive force.

Velocity Profile

As M increases, the Lorentz force becomes stronger, leading to a reduction in fluid velocity. This is evident in Figure 16.3, where the velocity profiles for all three fluids (Casson, Maxwell, and Williamson) decrease with increasing M. Casson fluids, which initially have higher velocities, show a more significant reduction compared to Maxwell and Williamson fluids. This is because Casson fluids have a higher initial momentum, making them more sensitive to the resistive effects of the magnetic field. Maxwell fluids, on the other hand, exhibit the lowest velocities, indicating greater resistance to flow under magnetic influence due to their viscoelastic nature.

Temperature Profile

The magnetic field also affects the temperature distribution within the fluid. As shown in Figure 16.4, increasing M leads to a rise in temperature. This is because the interaction between the magnetic field and the fluid's electrical conductivity generates additional heat, enhancing thermal diffusion. Casson fluids retain the highest temperatures, while Maxwell fluids dissipate heat more efficiently due to their lower thermal conductivity.

Concentration Profile

The concentration profile is similarly influenced by the magnetic field. Figure 16.5 illustrates that increasing M results in higher concentration levels. The magnetic field accelerates mass diffusion, leading to increased concentration. Casson fluids maintain

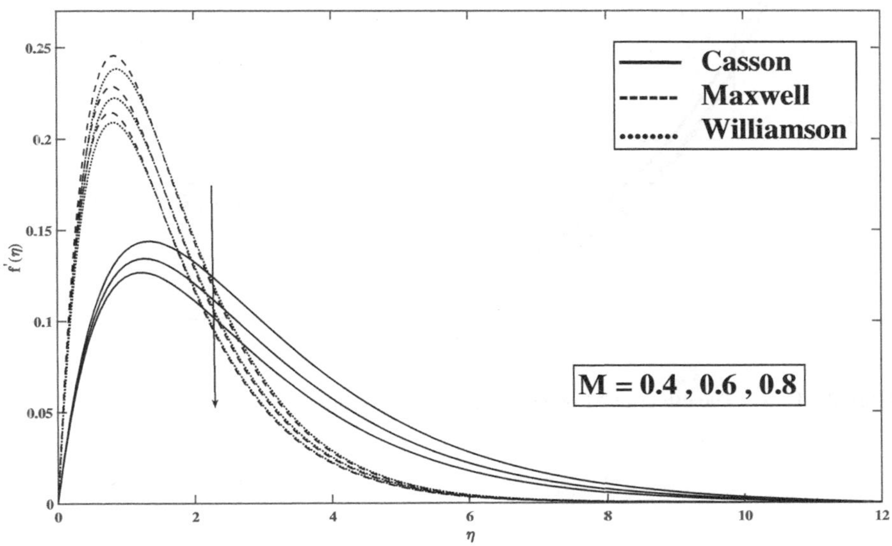

FIGURE 16.3 The influence of M on velocity distribution.

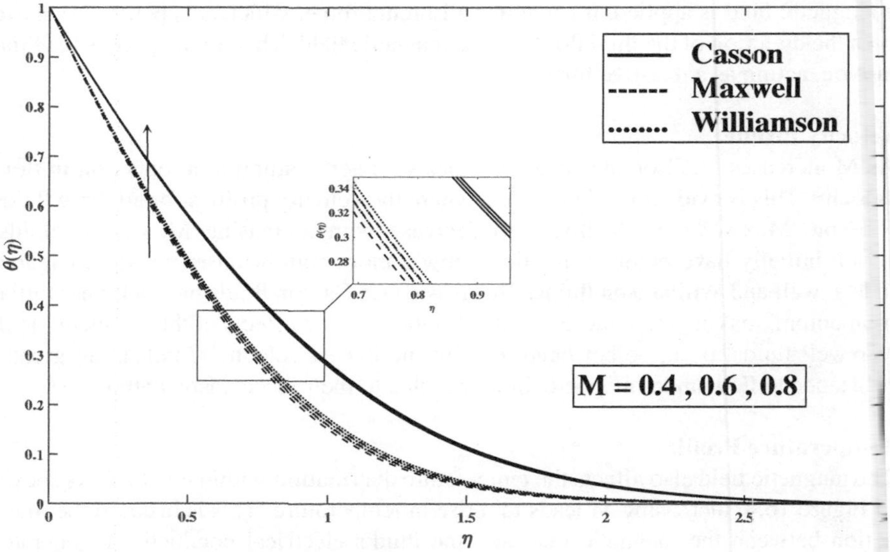

FIGURE 16.4 The impact of M on temperature distribution.

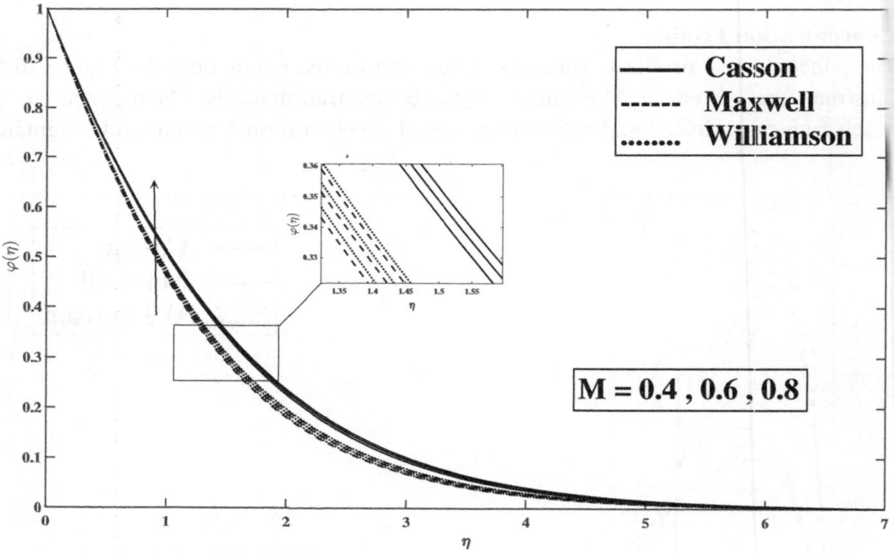

FIGURE 16.5 The effect of M on concentration distribution.

higher concentrations, while Maxwell fluids exhibit faster diffusion rates, resulting in lower concentration levels.

16.4.2 Prandtl Number

Increasing Pr results in a decrease in fluid velocity. This is because higher Pr values indicate lower thermal diffusivity, which reduces the fluid's ability to transfer heat. Consequently, the buoyancy-driven flow is weakened, leading to slower fluid motion. Casson fluids exhibit higher velocities compared to Maxwell and Williamson fluids, reflecting their lower resistance to flow. The temperature profile is also affected by the Prandtl number. Higher Pr values lead to a decrease in temperature. This is because the reduced thermal diffusivity slows down the heat transfer process, leading to faster temperature decay. Casson fluids retain higher temperatures, while Maxwell fluids dissipate heat more effectively due to their lower thermal conductivity.

16.4.3 Buoyancy Ratio Parameter

The buoyancy ratio parameter Nr represents the ratio of buoyancy forces due to concentration differences to those due to temperature differences. It is a measure of the relative strength of concentration-driven buoyancy compared to temperature-driven buoyancy.

Velocity Profile

As shown in Figure 16.6, increasing Nr enhances fluid velocity. This is because a higher buoyancy ratio strengthens the buoyancy forces driving the flow, leading to

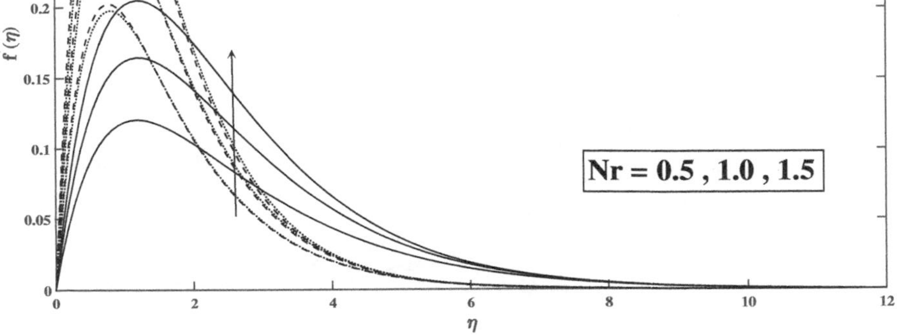

FIGURE 16.6 The impact of Nr on velocity distribution.

increased fluid motion. Casson fluids exhibit higher velocities, while Maxwell fluids show the lowest, indicating greater resistance to buoyancy-driven flow.

16.4.4 Effect of Heat Source A* and Sink B* Parameter

The heat source parameter A* represents space-dependent heat generation, while B* represents temperature-dependent heat absorption. These parameters are crucial in understanding how localized heat generation or absorption affects the fluid flow and temperature distribution.

Temperature Profile

The temperature profile is significantly influenced by the heat source and sink parameters. As seen in Figure 16.7, A* increases temperature, while B* decreases it (Figure 16.8). Heat sources elevate thermal energy, raising temperatures, while heat sinks absorb heat, reducing temperature levels. Casson fluids retain more heat, while Maxwell fluids dissipate it more efficiently.

Concentration Profile

The concentration profile is also affected by the heat source and sink parameters. As shown in Figure 16.9, A* decreases concentration, while B* increases it (Figure 16.10). Heat sources accelerate mass diffusion, reducing concentration levels, while heat sinks slow diffusion, increasing concentration. Casson fluids resist mass diffusion more effectively, while Maxwell fluids exhibit faster diffusion.

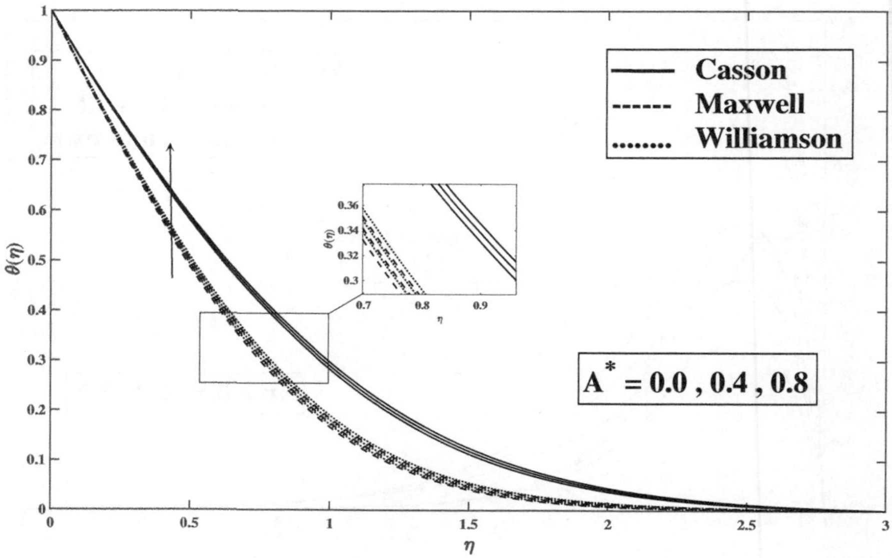

FIGURE 16.7 The effect of A* on the profile of temperature.

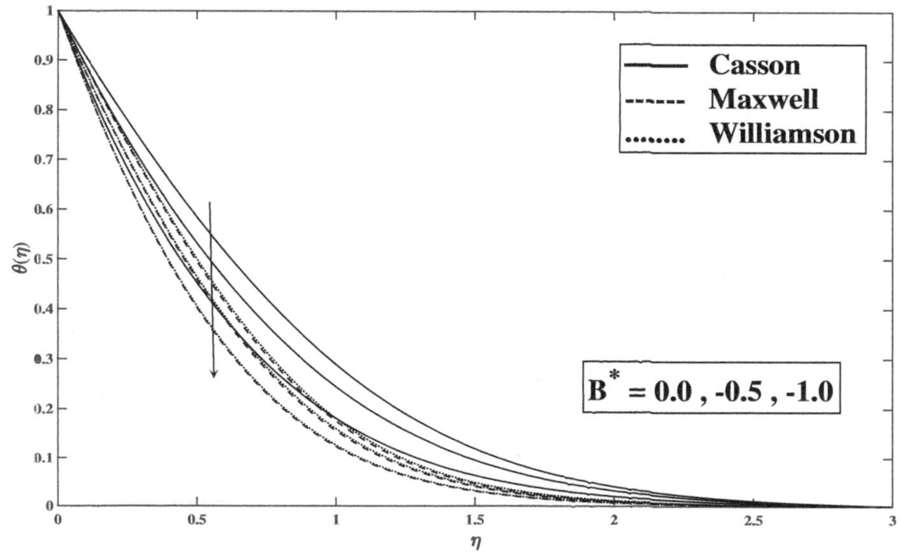

FIGURE 16.8 The impact of B* on the profile of temperature.

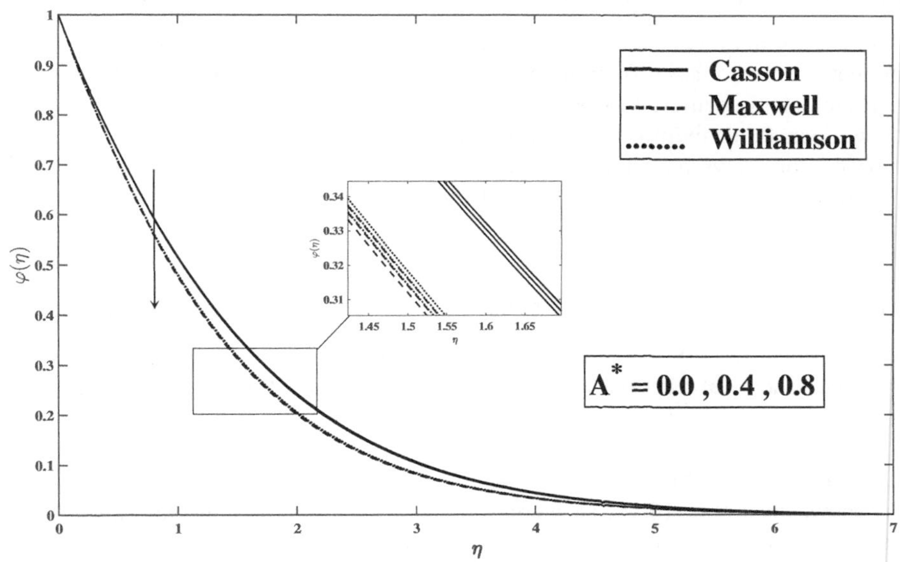

FIGURE 16.9 The effect of A* on the profile of concentration.

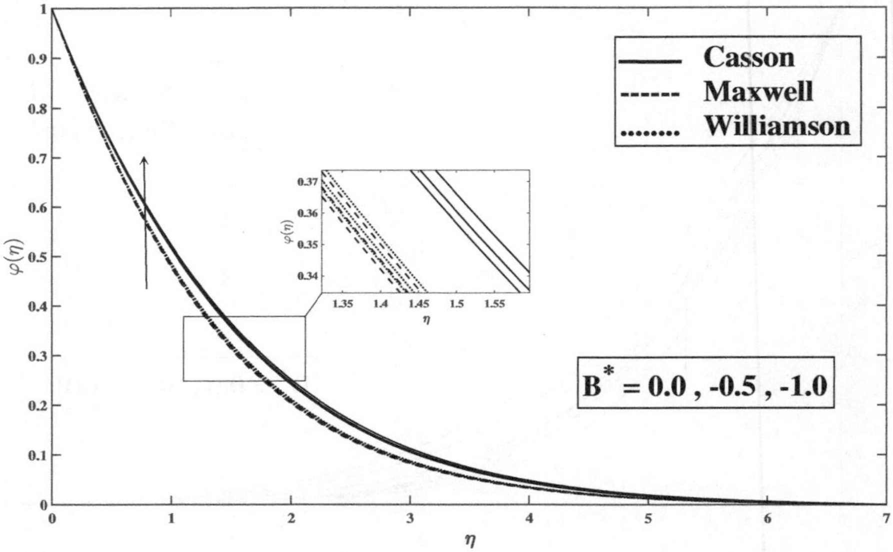

FIGURE 16.10 The impact of B* on the profile of concentration.

16.4.5 Schmidt Number (Sc)

An increase in Sc leads to a reduction in concentration. This is because a higher Sc indicates slower mass diffusion, meaning that the fluid is less effective at spreading particles or molecules. In this context, Maxwell fluids (which are a type of non-Newtonian fluid) exhibit the lowest concentrations, while Casson fluids (another type of non-Newtonian fluid) retain higher concentrations. This is likely due to the different rheological properties of these fluids, which affect how they transport mass.

16.4.6 Chemical Reaction Parameter

The chemical reaction parameter K_0 represents the rate of chemical reaction within the fluid. Chemical reactions can generate or absorb heat, affecting both the temperature and concentration profiles.

Concentration Profile

As shown in Figure 16.11, an increase in K_0 decreases the concentration. This is because chemical reactions can accelerate mass diffusion. As reactions occur, they may break down or transform the molecules, leading to a faster spread of particles and thus a reduction in concentration. Maxwell fluids exhibit the fastest diffusion, likely due to their lower viscosity, while Casson fluids show slower diffusion, possibly due to their higher viscosity or yield stress.

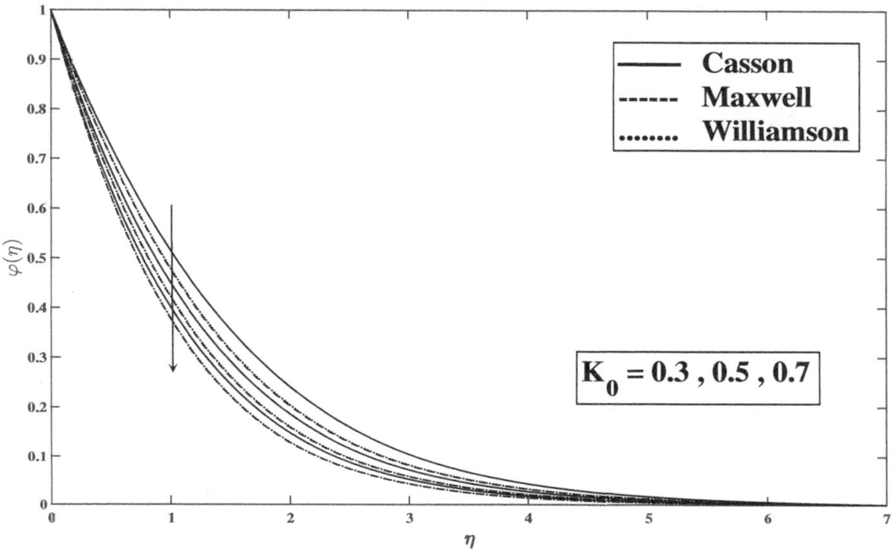

FIGURE 16.11 The impact of K0 on the profile of concentration.

16.4.7 Effect of Cattaneo–Christov Heat C_1 and Mass C_2 Flux

The Cattaneo–Christov heat flux (C_1) and mass flux (C_2) parameters are used to account for the finite time required for heat and mass to propagate through a fluid. This is important in dynamic systems where the traditional Fourier's law (for heat) and Fick's law (for mass) may not fully capture the behavior due to relaxation effects.

Temperature Profile

As seen in Figure 16.12, C_1 decreases the temperature. This is because the relaxation effects reduce thermal diffusion. The fluid does not immediately respond to changes in temperature, leading to a slower spread of heat and thus lower temperature levels. Casson fluids are less sensitive to these effects, possibly due to their higher viscosity, while Maxwell fluids show more pronounced reductions in temperature.

Concentration Profile

As shown in Figure 16.13, increasing C_2 decreases the concentration. Similar to the temperature profile, the relaxation effects slow down mass diffusion. The fluid takes time to respond to changes in concentration, leading to a slower spread of particles and thus lower concentration levels. Casson fluids retain higher concentrations, likely due to their higher viscosity or yield stress, while Maxwell fluids exhibit the lowest concentrations due to their lower viscosity and faster response to changes.

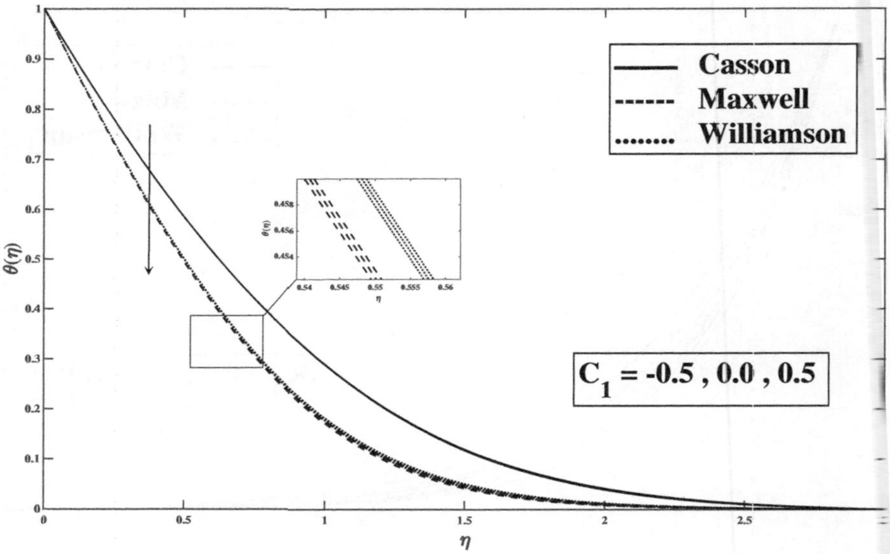

FIGURE 16.12 The effect of C1 on the profile of temperature.

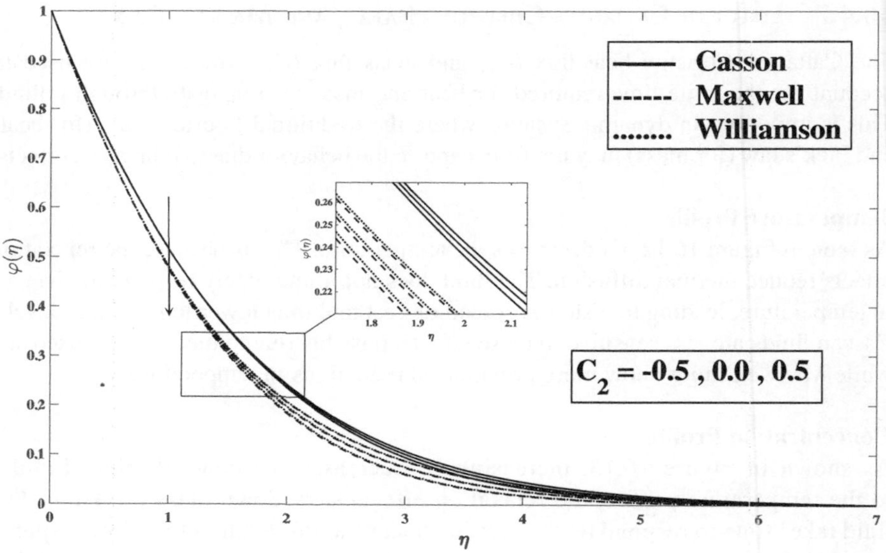

FIGURE 16.13 The effect of C2 on the profile of concentration.

16.4.8 Physical Quantities of Interest

The analysis of the skin friction coefficient (C_f), Nusselt number (Nu), and Sherwood number (Sh) across Casson, Maxwell, and Williamson fluid models provides valuable insights into how different physical parameters influence fluid flow, heat transfer, and mass transfer.

- The magnetic parameter (M) uniformly reduces both C_f and Nu across all fluid models. Physically, this reflects the damping effect of the magnetic field on fluid motion. The Lorentz force generated by the magnetic field opposes the flow, reducing frictional forces (lower C_f) and suppressing convective heat transfer (lower Nu).

TABLE 16.1
Local Skin Friction

M	Nr	A^*	B^*	K_0	Pr	Sc	C_1	C_2	Casson	Maxwell	Williamson
1	0.5	0.2	-0.5	0.3	6.8	1	-0.2	-0.3	0.889	0.6686	0.6756
0.4									0.9354	0.7431	0.7476
0.6									0.919	0.7146	0.7202
0.8									0.9036	0.69	0.6965
	0.5								0.889	0.6686	0.6756
	1								1.1931	0.9049	0.9125
	1.5								1.4921	1.1394	1.1435
		0							0.8836	0.6624	0.6694
		0.4							0.8946	0.675	0.682
		0.8							0.9061	0.6882	0.6952
			0						0.9382	0.6877	0.6954
			-0.5						0.889	0.6686	0.6756
			-1						0.8497	0.6519	0.6584
				0.3					0.889	0.6686	0.6756
				0.5					0.8698	0.6539	0.6607
				0.7					0.8541	0.6424	0.6491
					6.8				0.889	0.6686	0.6756
					7.8				0.8763	0.6584	0.6654
					8.8				0.8649	0.6494	0.6565
						0.5			0.915	0.701	0.7079
						1			0.889	0.6686	0.6756
						1.5			0.8689	0.649	0.656
							-0.5		0.8893	0.6693	0.6763
							0		0.8889	0.6682	0.6752
							0.5		0.8884	0.6671	0.6741
								-0.5	0.8889	0.6691	0.6761
								0	0.8893	0.6679	0.6749
								0.5	0.8898	0.6666	0.6737

TABLE 16.2
Local Nusselt Number

M	Nr	A^*	B^*	K_0	Pr	Sc	C_1	C_2	Casson	Maxwell	Williamson
1	0.5	0.2	-0.5	0.3	6.8	1	-0.2	-0.3	0.947	1.1511	1.1374
0.4									0.9618	1.2023	1.1832
0.6									0.9566	1.1832	1.1663
0.8									0.9517	1.1663	1.1511
	0.5								0.947	1.1511	1.1374
	1								1.0106	1.2575	1.2335
	1.5								1.0662	1.3473	1.3109
		0							0.959	1.1735	1.1595
		0.4							0.9348	1.128	1.1145
		0.8							0.9095	1.0799	1.0669
			0						0.7096	0.9693	0.9523
			-0.5						0.947	1.1511	1.1374
			-1						1.153	1.318	1.3066
				0.3					0.947	1.1511	1.1374
				0.5					0.9407	1.1397	1.1266
				0.7					0.9355	1.1307	1.1181
					6.8				0.947	1.1511	1.1374
					7.8				0.9722	1.1928	1.1782
					8.8				0.9959	1.2313	1.2161
						0.5			0.9557	1.1758	1.1602
						1			0.947	1.1511	1.1374
						1.5			0.9402	1.1354	1.1227
							-0.5		0.9471	1.1509	1.1372
							0		0.9469	1.1512	1.1375
							0.5		0.9468	1.1516	1.1379
								-0.5	0.9469	1.1515	1.1377
								0	0.9471	1.1505	1.1368
								0.5	0.9473	1.1494	1.1359

- The non-uniform heat source parameter (A^*) increases C_f and Sh while decreasing Nu. Physically, this means that localized heat sources enhance fluid motion by creating buoyancy-driven flows, which increases friction (higher C_f) and mass transfer (higher Sh). However, the heat sources also flatten temperature gradients, reducing the efficiency of heat transfer (lower Nu).
- Conversely, the non-uniform heat sink parameter (B^*) decreases C_f and Sh while increasing Nu. Physically, heat sinks absorb thermal energy, slowing down fluid motion and reducing friction (lower C_f) and mass transfer (lower Sh). At the same time, heat sinks create steeper temperature gradients, enhancing heat transfer efficiency (higher Nu).
- The chemical reaction parameter (K_0) consistently lowers C_f and Nu but raises Sh. Physically, chemical reactions introduce additional resistance to fluid flow, reducing frictional forces (lower C_f) and heat transfer (lower Nu).

TABLE 16.3
Local Sherwood Number

M	Nr	A*	B*	K_0	Pr	Sc	C_1	C_2	Casson	Maxwell	Williamson
1	0.5	0.2	-0.5	0.3	6.8	1	-0.2	-0.3	0.6351	0.6601	0.657
0.4									0.639	0.6822	0.6775
0.6									0.6376	0.6738	0.6697
0.8									0.6363	0.6665	0.6629
	0.5								0.6351	0.6601	0.657
	1								0.6511	0.698	0.692
	1.5								0.6665	0.7311	0.7215
		0							0.6346	0.6583	0.6552
		0.4							0.6356	0.6621	0.6589
		0.8							0.6365	0.666	0.6626
			0						0.6386	0.6639	0.6606
			-0.5						0.6351	0.6601	0.657
			-1						0.6324	0.6569	0.6539
				0.3					0.6351	0.6601	0.657
				0.5					0.7635	0.7901	0.7875
				0.7					0.8774	0.9033	0.9011
					6.8				0.6351	0.6601	0.657
					7.8				0.634	0.6576	0.6546
					8.8				0.6331	0.6554	0.6525
						0.5			0.5005	0.4781	0.4759
						1			0.6351	0.6601	0.657
						1.5			0.7506	0.7985	0.7947
							-0.5		0.6351	0.6604	0.6573
							0		0.6351	0.66	0.6568
							0.5		0.635	0.6595	0.6564
								-0.5	0.6359	0.6606	0.6574
								0	0.6339	0.6596	0.6565
								0.5	0.6319	0.6588	0.6557

However, chemical reactions accelerate mass diffusion by breaking down or transforming molecules, leading to enhanced mass transfer (higher Sh).

- The Cattaneo–Christov heat flux (C_1) and mass flux (C_2) parameters exhibit varying effects across fluid models. Physically, these parameters account for the finite time required for heat and mass to propagate through the fluid. Thermal and mass relaxation effects dampen fluid motion, reduce heat transfer (lower Nu), and slow down mass diffusion (lower Sh). However, the extent of these effects depends on the specific rheological properties of the fluid, with Casson fluids being less sensitive and Maxwell fluids showing more pronounced reductions.

These findings highlight the complex interplay between physical parameters and fluid behavior. Understanding these relationships is crucial for applications in industries such as chemical engineering, energy systems, and materials processing, where precise control of fluid dynamics, heat transfer, and mass transfer is essential.

16.5 CONCLUSIONS

This study explored the behavior of Casson, Maxwell, and Williamson non-Newtonian fluids in the context of heat and mass transfer processes influenced by various physical and thermophysical parameters. The research aimed to provide insights into how parameters such as the magnetic field strength (M), Prandtl number (Pr), buoyancy ratio (Nr), heat source and sink (A^*, B^*), Schmidt number (Sc), chemical reaction (K_0), and Cattaneo–Christov heat and mass flux (C_1, C_2) impact velocity, temperature, and concentration profiles in boundary layer flows.

KEY RESULTS

1. Increasing M reduced fluid velocity due to the resistive Lorentz force but enhanced temperature and concentration diffusion.
2. Heat sources (A^*) elevated temperature levels while reducing concentration, whereas heat sinks (B^*) reduced temperature but increased concentration.
3. Both C_1 and C_2 caused a decline in temperature and concentration profiles, reflecting reduced thermal and mass transfer rates due to relaxation effects.
4. Higher K_0 increased temperature profiles by enhancing heat generation but decreased concentration profiles due to accelerated mass diffusion.
5. Among the three fluid models, Casson fluids exhibited the highest temperature and concentration retention across all parameter variations, making them more resistant to heat and mass diffusion. Maxwell fluids displayed the lowest velocity, temperature, and concentration profiles, indicating greater resistance to flow and more efficient dissipation. Williamson fluids showed intermediate behavior, with moderate variations in response to changes in M, Pr, A^*, B^*, C_1, C_2, and K_0, suggesting a more balanced response between heat and mass transfer effects.

The study offers critical insights into the interplay of physical parameters in non-Newtonian fluids, which is essential for applications such as heat exchangers, cooling systems, and chemical process equipment. The results highlight the unique strengths of Casson fluids in retaining heat and mass and of Maxwell fluids in facilitating faster dissipation and diffusion. This study assumed steady, two-dimensional flow with constant fluid properties, which limits its applicability to more dynamic or complex real-world systems. Effects such as unsteady flows, variable properties, and surface roughness were not considered, which could alter the outcomes. Future research could expand on these findings by exploring a broader range of parameters and validating the results through experimental studies.

Abbreviations

A^*	Coefficient for space-dependent heat source and sink
B^*	Coefficient for temperature-dependent heat source and sink
B_2^0	Magnetic parameter
BL	Boundary layer
C	Concentration
C1	Dimensionless thermal relaxation time parameter
C2	Dimensionless concentration relaxation time parameter
D	Concentration diffusivity
g	Gravity
G2	Maxwell fluid parameter
G3	Weissenberg (viscoelasticity) number
Kr	Dimensional chemical reaction parameter
K0	Dimensionless chemical reaction parameter
m1	Deborah number in relation to relaxing time
M	Dimensionless magnetic parameter
Nr	Buoyancy ratio parameter
Pr	Prandtl number
Q'''	Parameter for heat source and sink
Sc	Schmidt number
T	Temperature
u,v	Velocity components
x,y	Coordinates

Greek Symbols

α	Thermal diffusivity
β	Casson fluid parameter
β_T	Volumetric expansion of thermal
Γ	Positive time constant
τ_T	Heat flux relaxation time
τ_C	Mass flux relaxation time
θ	Dimensionless function of temperature
$\Lambda_1, \Lambda_2, \Lambda_3$	Non-Newtonian controlling parameter
μ	Dynamic viscosity
ν	Kinematic viscosity

ϕ	Dimensionless function of concentration
ψ	Stream function
ρ	Density

Subscripts

w	Condition of wall
∞	Condition of ambient

REFERENCES

[1] Hering, R., & Grosh, R. (1962). Laminar free convection from a non-isothermal cone. *International Journal of Heat and Mass Transfer*, 5(11), 1059–1068.

[2] Weaver, J., & Viskanta, R. (1991). Natural convection due to horizontal temperature and concentration gradients—1. Variable thermophysical property effects. *International Journal of Heat and Mass Transfer*, 34(12), 3107–3120.

[3] Pop, I., & Watanabe, T. (1992). Free convection with uniform suction or injection from a vertical cone for constant wall heat flux. *International Communications in Heat and Mass Transfer*, 19(2), 275–283.

[4] Hasan, M., & Mujumdar, A. (1984). Coupled heat and mass transfer in natural convection under flux condition along a vertical cone. *International Communications in Heat and Mass Transfer*, 11(2), 157–172.

[5] Chamkha, A. J., & Quadri, M. M. A. (2002). Combined heat and mass transfer by hydromagnetic natural convection over a cone embedded in a non-Darcian porous medium with heat generation/absorption effects. *Heat and Mass Transfer*, 38(6), 487–495.

[6] Luhmann, A. J., Covington, M. D., Peters, A. J., Alexander, S. C., Anger, C. T., Green, J. A., Runkel, A. C., & Alexander Jr, E. C. (2011). Classification of thermal patterns at karst springs and cave streams. *Groundwater*, 49(3), 324–335.

[7] Barth, W. L., & Carey, G. F. (2006). On a natural-convection benchmark problem in non-Newtonian fluids. *Numerical Heat Transfer, Part B: Fundamentals*, 50(3), 193–216.

[8] Kumar, B. R., & Sivaraj, R. (2013). Heat and mass transfer in MHD viscoelastic fluid flow over a vertical cone and flat plate with variable viscosity. *International Journal of Heat and Mass Transfer*, 56(1–2), 370–379.

[9] Peter, F., Sambath, P., & Dhanasekaran, S. (2023). Numerical investigation of radiative hybrid nanofluid flows over a plumb cone/plate. *Mathematics*, 11(20), 4331.

[10] Dinarvand, S. (2011). The laminar free-convection boundary-layer flow about a heated and rotating down-pointing vertical cone in the presence of a transverse magnetic field. *International Journal for Numerical Methods in Fluids*, 67(12), 2141–2156.

[11] Palani, G., & Ragavan, A. (2015). Free convection flow past a vertical cone with a chemical reaction in the presence of transverse magnetic field. *Journal of Engineering Physics and Thermophysics*, 88, 1256–1263.

[12] Sambath, P., Pullepu, B., Hussain, T., & Shehzad, S. A. (2018). Radiated chemical reaction impacts on natural convective MHD mass transfer flow induced by a vertical cone. *Results in Physics*, 8, 304–315.

[13] El-Kabeir, S., Modather, M., & Abdou, M. (2007). Chemical reaction, heat and mass transfer on MHD flow over a vertical isothermal cone surface in micropolar fluids with heat generation/absorption. *Applied Mathematical Sciences*, 1(34), 1663--1674.

[14] Casson, N. (1959). Flow equation for pigment-oil suspensions of the printing ink-type. *Rheology of Disperse Systems*, 1, 84–104.

[15] Mythili, D., Sivaraj, R., Rashidi, M., & Yang, Z. (2015). Casson fluid flow over a vertical cone with non-uniform heat source/sink and high order chemical reaction. *Journal of Naval Architecture and Marine Engineering*, 12(2), 125–136.

[16] Mythili, D., & Sivaraj, R. (2016). Influence of higher order chemical reaction and non-uniform heat source/sink on Casson fluid flow over a vertical cone and flat plate. *Journal of Molecular Liquids*, 216, 466–475.

[17] Jasmine Benazir, A., Sivaraj, R., & Makinde, O. D. (2016). Unsteady magnetohydrodynamic Casson fluid flow over a vertical cone and flat plate with non-uniform heat source/sink. *International Journal of Engineering Research in Africa*, 21, 69–83.

[18] Durairaj, M., Ramachandran, S., & Mehdi, R. M. (2017). Heat generating/absorbing and chemically reacting Casson fluid flow over a vertical cone and flat plate saturated with non-Darcy porous medium. *International Journal of Numerical Methods for Heat & Fluid Flow*, 27(1), 156–173.

[19] Singh, A. K., Kumar, A., & Singh, A. K. (2022). Effect of non-uniform heating and viscous dissipation on natural convective flow of Casson fluid over a vertical cone. *Indian Journal of Physics*, 96(2), 481–489.

[20] Khan, M., Malik, M., Salahuddin, T., Rehman, K., Naseer, M., & others. (2017). MHD flow of Williamson nanofluid over a cone and plate with chemically reactive species. *Journal of Molecular Liquids,* 231, 580–588.

[21] Sathyanarayana, M., & Goud, T. R. (2023). Numerical study of MHD Williamson-nano fluid flow past a vertical cone in the presence of suction/injection and convective boundary conditions. *Archives of Thermodynamics*, 44, 115–138.

[22] Kodi, R., Ganteda, C., Dasore, A., Kumar, M. L., Laxmaiah, G., Hasan, M. A., Islam, S., & Razak, A. (2023). Influence of MHD mixed convection flow for maxwell nanofluid through a vertical cone with porous material in the existence of variable heat conductivity and diffusion. *Case Studies in Thermal Engineering*, 44, 102875.

[23] Sathyanarayana, M, & Goud, T. R. (2023). Maxwell fluid performance on free convective non-newtonian nanofluid flow over a cone in presence of magnetic field, *Heat and Mass Transfer. Communications in Mathematics and Applications*, 14(2), 1019.

[24] Ramandevi, B., Reddy, J. R., Sugunamma, V., & Sandeep, N. (2018). Combined influence of viscous dissipation and non-uniform heat source/sink on MHD non-Newtonian fluid flow with Cattaneo-Christov heat flux. *Alexandria Engineering Journal*, 57(2), 1009–1018.

[25] Raju, C., Sandeep, N., & Malvandi, A. (2016). Free convective heat and mass transfer of MHD non-Newtonian nanofluids over a cone in the presence of non-uniform heat source/sink. *Journal of Molecular Liquids*, 221, 108–115.

[26] Koteswara Reddy, G., Yarrakula, K., Raju, C., & Rahbari, A. (2018). Mixed convection analysis of variable heat source/sink on MHD Maxwell, Jeffrey, and Oldroyd-B nanofluids over a cone with convective conditions using Buongiorno's model. *Journal of Thermal Analysis and Calorimetry*, 132, 1995–2002.

[27] Khan, M. N., Ahmad, S., Wang, Z., Ahammad, N. A., & Elkotb, M. A. (2023). Bioconvective surface-catalyzed Casson hybrid nanofluid flow analysis by using thermodynamics heat transfer law on a vertical cone. *Tribology International*, 188, 108859.

[28] Hayat, T., Abbas, Z., & Sajid, M. (2006). Series solution for the upper-convected Maxwell fluid over a porous stretching plate. *Physics Letters A*, 358(5–6), 396–403.

[29] Hayat, T., Muhammad, T., Shehzad, S., & Alsaedi, A. (2015). Three-dimensional boundary layer flow of Maxwell nanofluid: Mathematical model. *Applied Mathematics and Mechanics*, 36, 747–762.

[30] Amanulla, C., Nagendra, N., & Suryanarayana Reddy, M. (2017a). MHD flow and heat transfer in a williamson fluid from a vertical permeable cone with thermal and momentum slip effects: A mathematical study. *Frontiers in Heat and Mass Transfer (FHMT)*, 8 , 1–11.

[31] Amanulla, C., Nagendra, N., & Suryanarayana Reddy, M. (2017b). Thermal and momentum slip effects on hydromagnetic convection flow of a Williamson fluid past a vertical truncated cone. *Frontiers in Heat and Mass Transfer (FHMT)*, 9(1), 10–5098

[32] Hossain, M., & Paul, S. (2001). Free convection from a vertical permeable circular cone with non-uniform surface temperature. *Acta Mechanica*, 151(1), 103–114.

Part V

Microchannels and Solar Applications

17 Cooling Solutions Using Microchannel Systems

Muhammad Anas Wazir and Kareem Akhtar

17.1 INTRODUCTION

Thermal management has become a crucial aspect for ensuring the reliability and efficiency of advanced electronic systems. Due to the miniaturization of electronic components, there has been an increase in the energy density, which leads to high heat flux generation. Thermal management is particularly important in high-performance computing and data centers where managing heat is necessary for obtaining optimum performance with reduced energy consumption. Fluid dynamics provides the foundation of advanced thermal management solutions. For designing efficient thermal management systems understanding the behavior of fluid flow in laminar and turbulent flow regimes is also important. Due to the compactness of high-power density devices advanced cooling methods such as microchannel heat sinks and phase change mechanisms are used.

17.2 ADVANCEMENT IN MICROCHIPS TECHNOLOGY

Over the last few decades, there has been significant advancement in integrated circuit (IC) technology. IC technology has developed very fast over the years with its applications in aerospace, power electronics, radars and smart grids as shown in Figure 17.1. The first microchip with about 100 square millimeters was introduced in 1970, which consists of 2000 transistors [1]. With the miniaturization of these microchips' transistors count increased during 1990 with even a million transistors in one microchip. Transistor scaling and architecture advancements were made in 2000, which led to approximately 100 million transistors in 2010s in a single microchip. Today single digit square millimeter range microchips are available with billions of transistors.

Intel Corporation developed their 10 nm processor with 100 million transistors in 2018 but with surface area of 100 square millimeters [2]. The latest advancements in microprocessors are their compact size along with an increase in their computing power. This compactness and high computing power led to a significant rise in heat generated by these electronic systems [1]. This high heat flux could lead to the generation of hot spots in microchips without effective heat dissipation, which is a great threat to the integrity of microchips. Thus, thermal management is an important factor for efficient operation and reliability of these microchips [3]. With the miniaturization

DOI: 10.1201/9781003611820-22

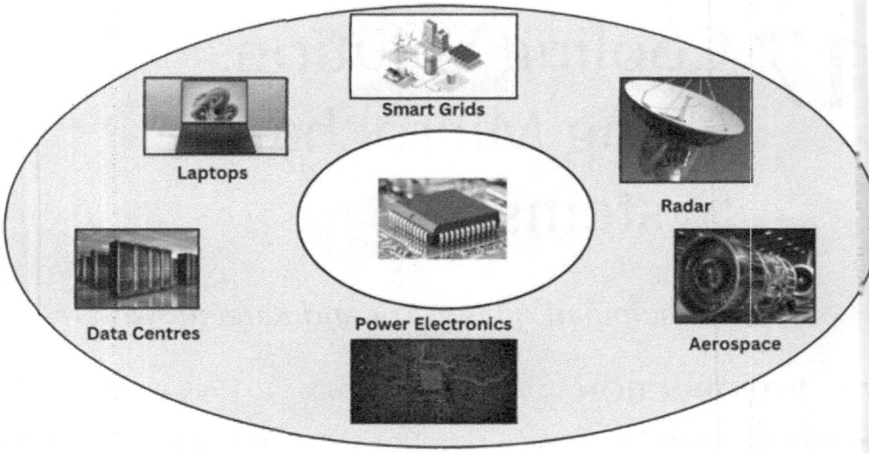

FIGURE 17.1 Integrated circuit technology applications.

of mechanical and electrical components efficient heat dissipation from these systems is required for the safe operation of these microelectromechanical systems (MEMS) (Figure 17.1).

17.3 DEVELOPMENT OF COOLING SYSTEMS AND THEIR LIMITATIONS

The evolution of thermal management systems has been related to advancements in fluid flow methods to address the increasing thermal management demands of modern electronic systems. Among several factors involved in the failure of electronic components, temperature instability is the most contributing factor as shown in Figure 17.2.

The earliest thermal management systems used natural convection and buoyancy-driven fluid motion principles to dissipate heat through the passive movement of air [4]. However, due to low heat transfer efficiency natural convection methods proved inadequate for high computing applications. Thermoelectric cooling uses the Peltier effect to dissipate heat from the system with precise temperature control without moving parts. This method requires high power consumption and has low efficiency, which led to further exploration into cooling systems [5].

Forced convection using fans and blowers enhances the airflow in high heat generating components. Despite its effectiveness, air cooling began to encounter limitations as microdevices required high heat dissipation capabilities [5]. Due to the limitations of air-cooling methods, liquids with high thermal conductivity and heat capacity compared to air were used for superior heat removal capabilities. Phase change cooling utilizes latent heat transfer for heat transfer for heat dissipation, but they are also not effective in extreme heat flux conditions [6]. This led to the development of microchannel based cooling systems, which use water or other specific coolants circulated through the channels and heat exchangers for effective heat dissipation [7].

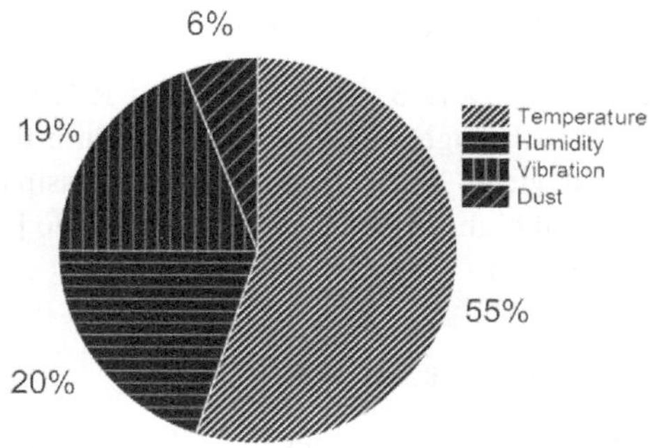

FIGURE 17.2 Major causes of electronic failure [8].

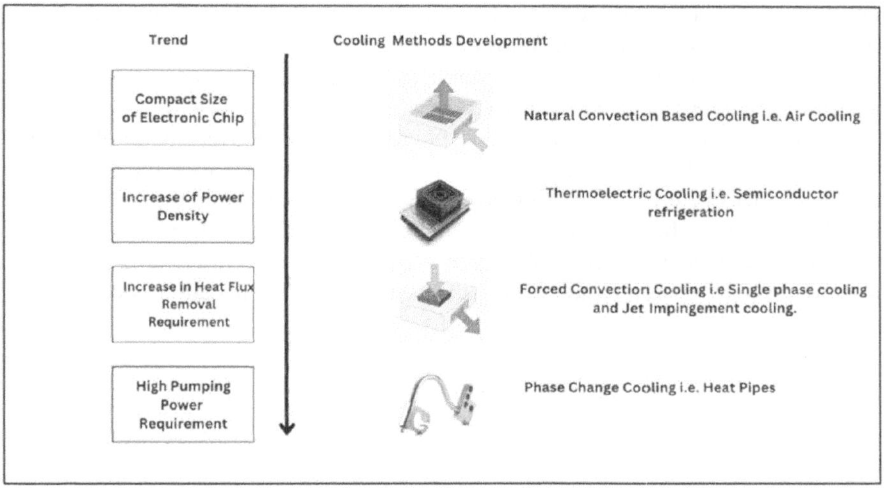

FIGURE 17.3 Trends in cooling methods development for microchips [9].

Figure 17.3 demonstrates the trend of advanced cooling systems development along with their characteristics (Figures 17.2 and 17.3).

17.4 DEVELOPMENT OF MICROCHANNEL BASED COOLING SYSTEMS

The number of transistors in a microchip doubles every two years according to Moore's Law [10]. Heat flux removal requirements also increase with an increase in the number of transistors. Traditional cooling systems fail to meet the demands of MEMS. Hence, microchannel based cooling systems, microchannel heat sinks (MCHS) have been developed to effectively remove heat from these microsystems. Among all the cooling methods, MCHS are the most effective to remove heat flux from microchips [11].

17.4.1 Microchannel Heat Sinks (MCHS)

MCHS were first developed by D.B. Tuckerman and R.F.W. Pease in 1981 [11]. MCHS uses microchannels through which the working fluid flows for effective heat dissipation in electronic devices. MCHS can effectively dissipate high heat flux due to its small channel size and high surface area to volume ratio [12]. The schematic of MCHS is shown in Figure 17.4.

17.4.2 Design Parameters for MCHS

An optimal MCHS gives enhanced heat transfer efficiency with less pressure drop. Few design parameters such as shape and material of MCHS, number, shape and arrangement pattern of pin-fin must be considered to design an optimal MCHS [13]. Material of the MCHS heat sink has a significant role in heat transfer. Materials with high thermal conductivity, such as copper and aluminum, are used to design MCHS. Copper costs more than aluminum but its thermal conductivity is about twice compared to aluminum. Zinc is also used to fabricate MCHS as zinc degrades its porosity during casting process, but its thermal conductivity is less than copper and aluminum. Pin-fin parameters like shape, height, alignment have a major effect on the efficiency of MCHS. The inclusion of pin-fins in MCHS increases the surface area, which enhances heat transfer efficiency. However, the increase in the number of pin-fins also increases the pressure drop generated in MCHS, which increases pumping power requirements [14]. The volumetric flow rate, cross-sectional area of flow, and pressure drop are design parameters in the case of forced convection systems [15].

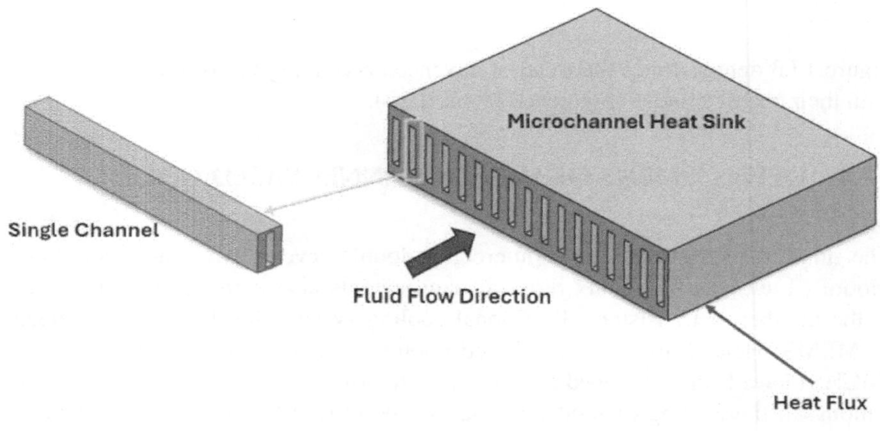

FIGURE 17.4 Schematic of microchannel heat sinks.

17.5 INNOVATION IN MICROCHANNEL BASED THERMAL MANAGEMENT TECHNIQUES

Due to the compactness of MEMS heat flux removal requirements from these systems have also increased. Hence, several methods are developed to enhance the MCHS efficiency to meet demands for high heat dissipation. These methods are broadly divided into active and passive cooling methods [9]. Active and passive cooling methods are further divided into several types as shown in Figure 17.5.

17.5.1 Active Methods

Active techniques use external power sources to enhance heat transfer in MCHS. The objective is to create mixing in the fluid flow by generating turbulence by using both or either active or passive techniques. Active techniques mainly include jet impingement, pulsating flow, vibration and electro-osmotic flow technique. Active techniques are less often used in MCHS due to the small size of MCHS.

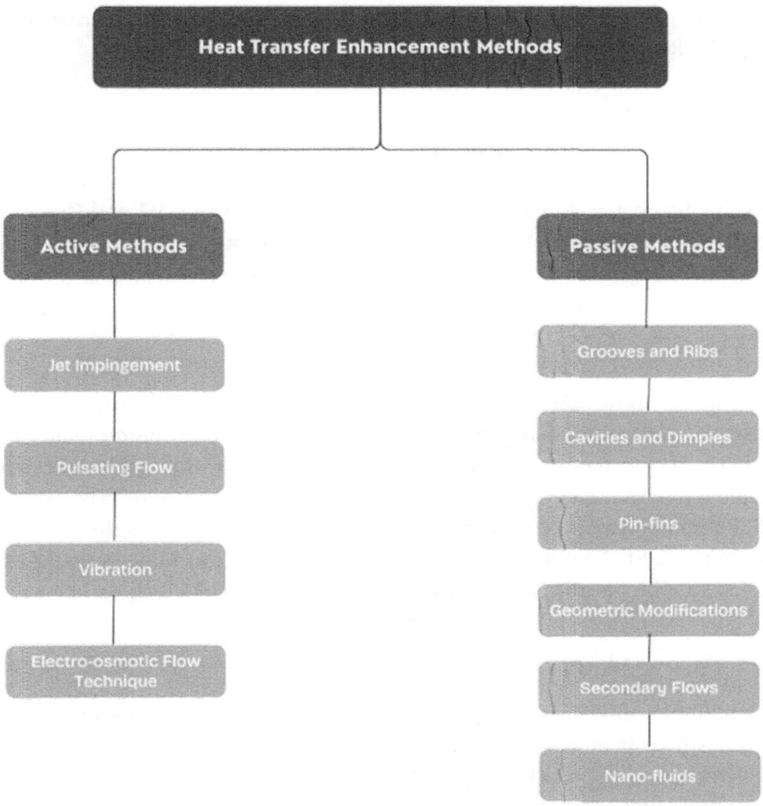

FIGURE 17.5 Thermal enhancement techniques.

17.5.1.1 Jet Impingement

In the jet impingement method, a rapidly flowing fluid coolant strikes the heated surface directly with small flow distance. This method enhances the heat transfer efficiency by convective heat transfer by forced convection. The best synergy between temperature and velocity gradient is obtained when flow strikes the heated surface vertically according to field synergy principle [16]. When the coolant strikes the heated surface, it has several effects on droplets depending upon their size and mass. These droplets form vortices when they leave the nozzle, which generates turbulence and disrupts the thermal boundary layer leading to better heat dissipation.

17.5.1.2 Pulsating Flow

Pulsating flow refers to flow within small channels where the velocity of flow increases and decreases over time causing wake formation. Pulsating flow is generally evaluated in terms of parameters like frequency and amplitude. This wake formation enhances the fluid mixing, which disrupts the boundary layer and improves heat transfer. The size, shape of channel, working fluid viscosity, density and pulsation frequency and amplitude are the different factors that influence heat transfer efficiency in microchannels by pulsating flow. Enhanced heat removal and less pressure drop is obtained by decreasing the intensity of flow and increasing the frequency [17, 18].

17.5.1.3 Vibration

Vibration based cooling methods in microchannels heat sinks enhance the heat transfer by inducing mixing and disrupting the thermal boundary layer. Vibration induced chaotic convection increase heat transfer rate up to 11.5% at different air flow velocities [19]. Although vibration cooling techniques are an effective approach for high power density electronics, optimization of vibrational structural parameters is necessary to minimize structural wear, which reduces reliability.

17.5.1.4 Electro-Osmotic Flow Technique

Electro-osmotic flow (EOF) cooling utilizes the movement of fluid flow in microchannel heat sinks using applied electric field. Electric field creates pumping action without mechanical components, which makes this technique ideal for microscale temperature control applications in compact electronics. The effectiveness of EOF technique depends on the surface charge of the channel walls and properties of the cooling fluid. A significant disruption in fluid flow is obtained using EOF particularly at small hydraulic diameter microchannel heat sinks [20].

17.5.2 Passive Methods

Passive methods utilize natural phenomena like conduction, convection, and radiation to dissipate heat without any external source such as fan. These techniques include grooves, ribs, cavities, pin-fins, and secondary channels. Compared to active methods these methods require low power consumption and increase system reliability. Few passive techniques for thermal enhancement in MCHS are described in detail below.

17.5.2.1 Grooves and Ribs for Enhanced Surface Area

Grooves and ribs increase the surface area of heat transfer in MCHS, which promotes high heat flux removal. Due to the increase in surface area ribs and grooves generate turbulence, which reduces the thickness of thermal boundary layer and increases heat transfer coefficient. Ribs and grooves in several shapes like triangular, rectangular, trapezoidal helical etc., as described in Figure 17.6, can be placed in several arrangements such as staggered or aligned arrangement in fluid flow path. Microscale ribs and grooves increase the Nusselt number by 1.55 times when both grooves and ribs are added on walls of MCHS [21]. Cylindrical ribs and cavities give 71% higher Nusselt number compared to conventional channel, which shows greater heat dissipation [22]. Trapezoidal grooves lead to a 2.35% increase in friction factor and 52% improvement in Nusselt number [23]. Semi-circular grooves give 16% improvement in heat transfer with high pressure drop compared to a conventional channel [24]. Among the different studies discussed, cylindrical ribs and cavities gave the best performance of heat transfer enhancement leading to enhanced heat dissipation and reduced pressure drop as shown in Figures 17.7 and 17.8. Rectangular and semi-circular ribs give less pressure drop compared to circular shaped ribs and cavities in MCHS.

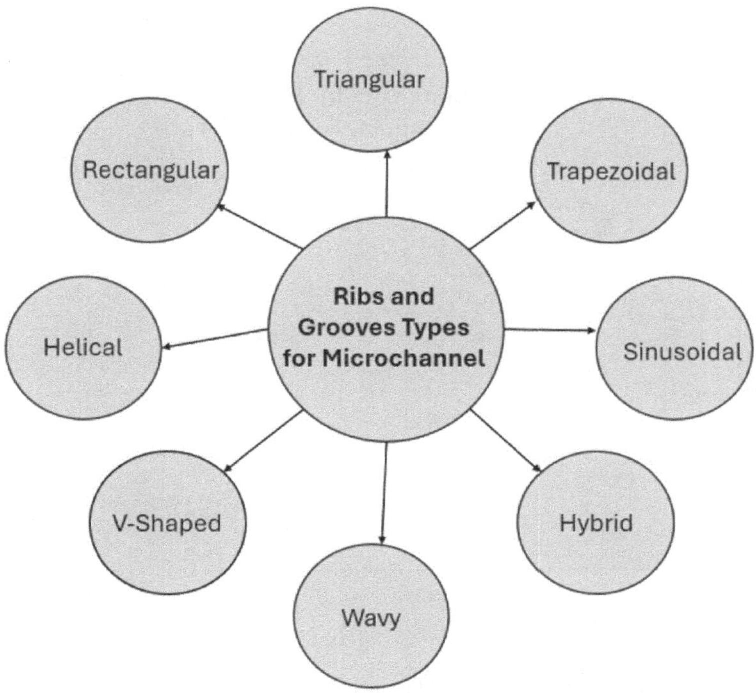

FIGURE 17.6 Types of ribs and grooves.

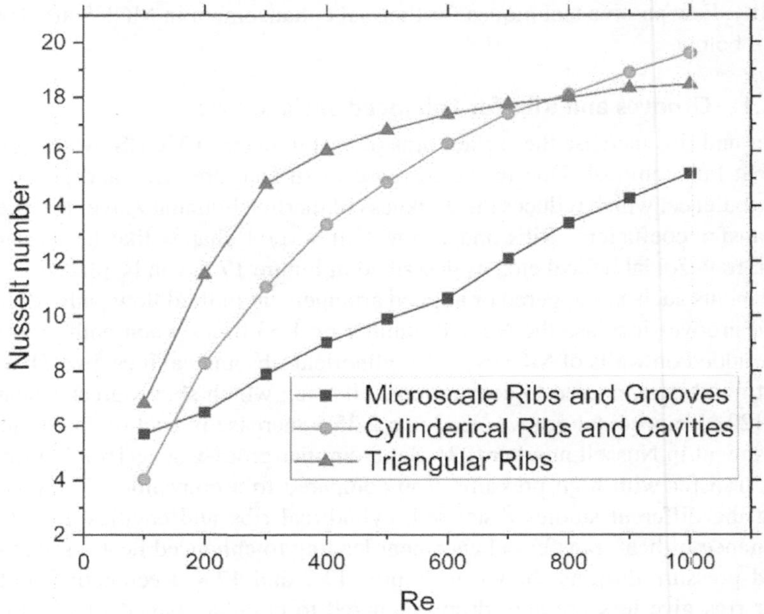

FIGURE 17.7 Nusselt number vs Reynold number for ribs and grooves [20, 21, 22].

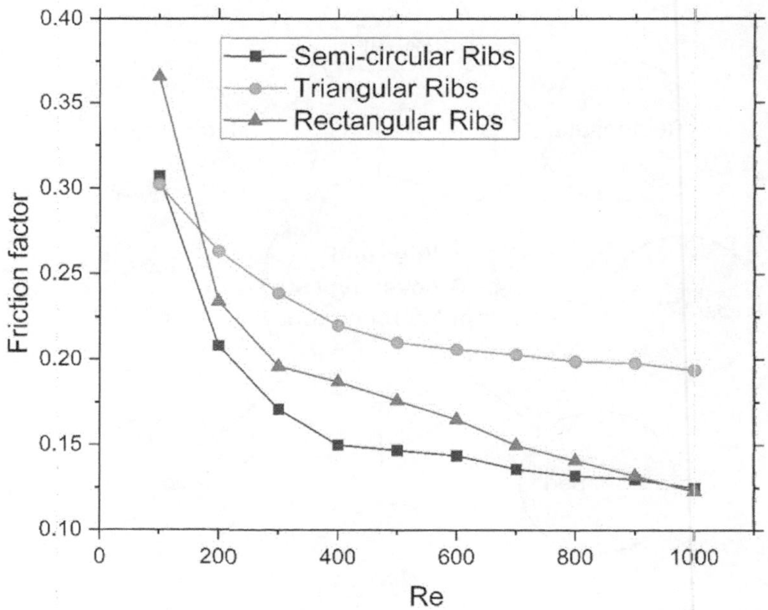

FIGURE 17.8 Friction factor vs Reynold number for ribs and grooves [20, 21, 22].

17.5.2.2 Cavities and Dimples

Cavities and dimples are an effective passive cooling method to improve heat transfer in MCHS. Different dimple and cavities shapes as described in Figure 17.9 are included in fluid flow path in MCHS. These cavities and dimples create vortices and swirls in the flow, which causes mixing and improves heat dissipation [25, 26]. Dimpled surface reduces the local fluid flow resistance and improves the thermal performance of MCHS in comparison to conventional MCHS. Studies show that the dimpled channel shows 15% improvement in Nusselt number and 2% decrease in pressure drop compared to smooth channel [27]. The shape and arrangement of cavities and dimples have a significant role in improving the performance of MCHS. Tear drop dimpled channel give the highest value of heat transfer coefficient, which is 18% more than spherical and elliptical dimple have the lowest heat transfer, which is 10% lower than spherical dimples [28]. The average heat transfer enhancement is 1.6–1.7 times and friction factor 2 times more than smooth MCHS using spherical dimples. Teardrop dimples increase the pressure drop by 35% compared to spherical dimples [29]. Among different shapes of dimples and cavities tear drop dimples gave the highest heat transfer and pressure drop described in terms of Nusselt number and friction compared to spherical and elliptical dimple as shown in Figures 17.10 and 17.11.

17.5.2.3 Pin-Fins

Pin-fins are included in the fluid flow path for disturbing the fluid flow for enhanced mixing. These pin-fins create turbulence in the flow, which improves thermal performance by increasing the heat transfer surface area in MCHS. The geometry, shape, diameter, height, and spacing between pin-fins are the different parameters that influence the thermal performance of MCHS. Different pin-fin shapes like triangular, rectangular, elliptical, square, circular etc. described in Figure 17.12 are used in MCHS to enhance thermal and hydraulic characteristics of MCHS with each pin-fin with its advantages in terms of heat transfer augmentation and limitations of pressure drop. The enhanced heat transfer using pin-fins is also due to the wake formation at the downstream side of pin-fins and the flow separation effects [30, 31]. The staggered and inline arrangement of pin-fins results in fluid flow path optimization leading to improved performance. The staggered arrangement of pin-fins give high heat dissipation than the inline arrangement [32]. Studies indicate the pin-fin MCHS shows 51.8–94.8% increase in Nusselt number compared to conventional channel, which indicates superior heat dissipation performance [33]. Although pin-fin MCHS results in high thermal performance, it increases the friction factor in MCHS which results in high pumping power requirement [33]. Another benefit of micro pin-fins is that they can effectively dissipate heat at low coolant flow rate. This is beneficial in compact cooling applications like microprocessors, lasers and power electronics. Pin-fins also results in uniform temperature distribution, which gives reduced hotspots leading to systems integrity [34]. With the progression of advanced manufacturing methods like 3D printing and laser micromachining the manufacturing of micro pin-fins has become more feasible [35, 36]. Among different pin-fin shapes circular

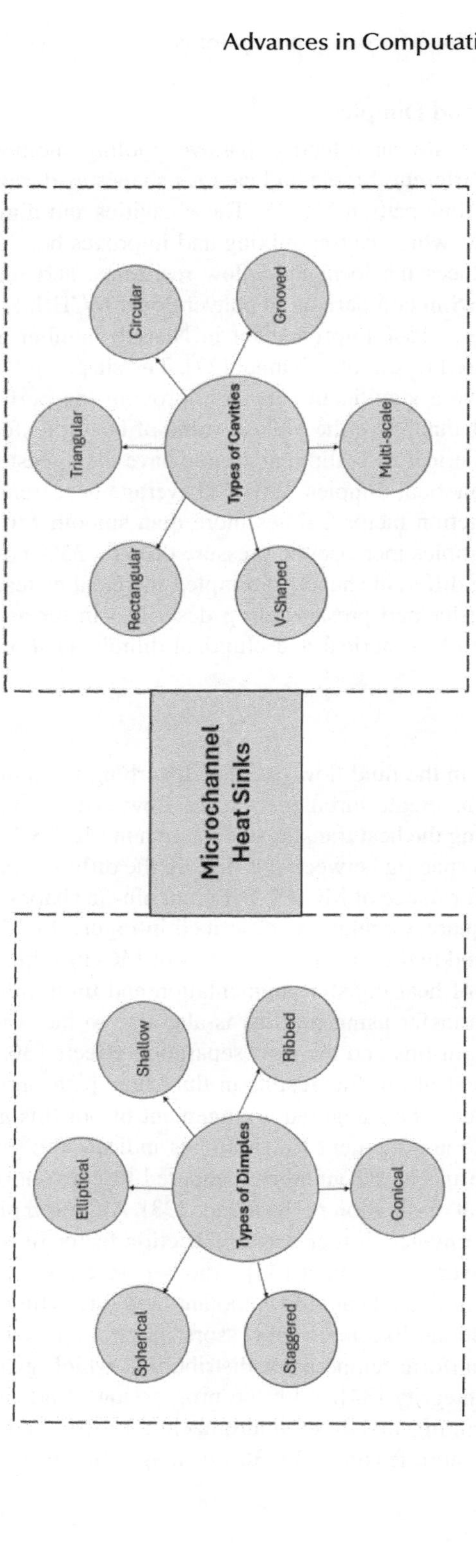

FIGURE 17.9 Types of cavities and dimples.

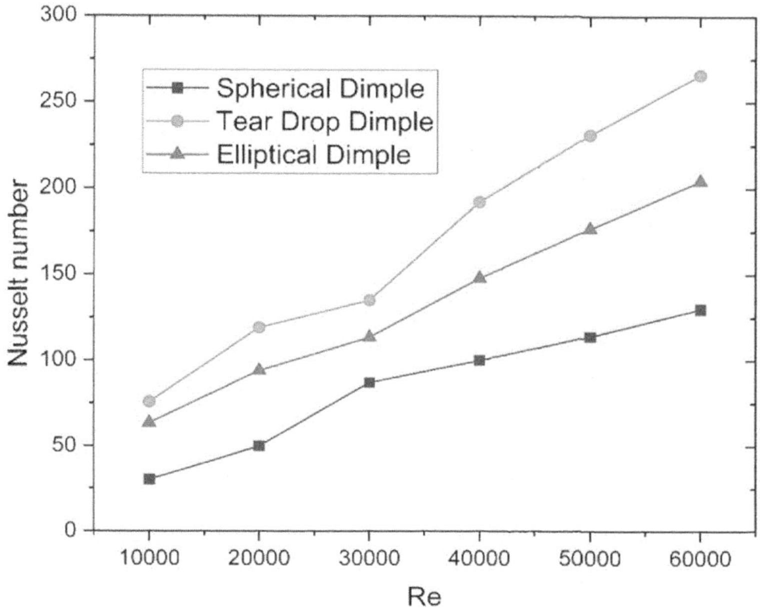

FIGURE 17.10 Nusselt number vs Reynold number for dimples [29].

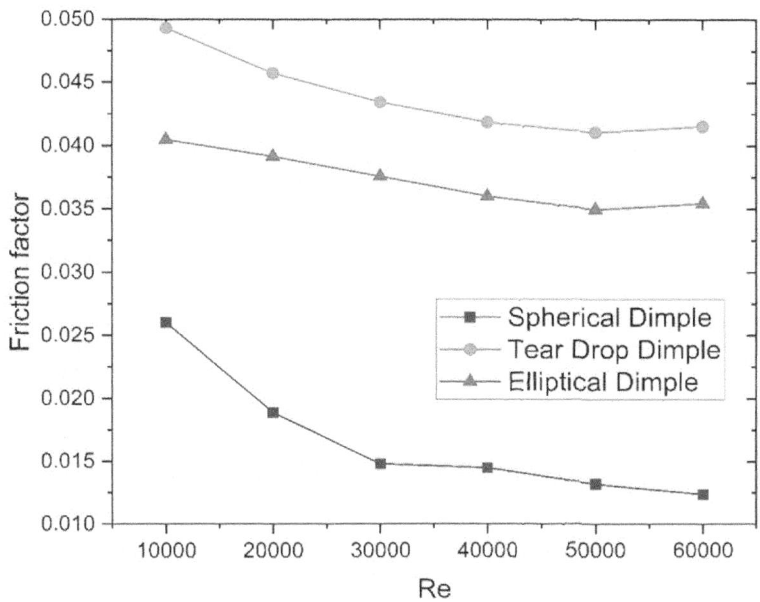

FIGURE 17.11 Friction factor vs Reynold number for dimples [29].

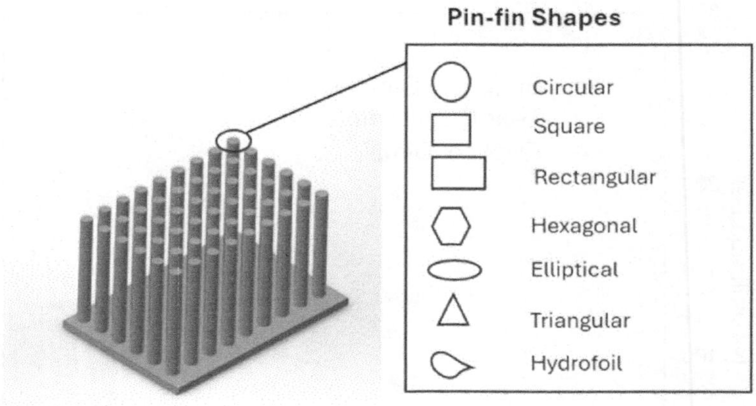

FIGURE 17.12 Pin-fin shapes.

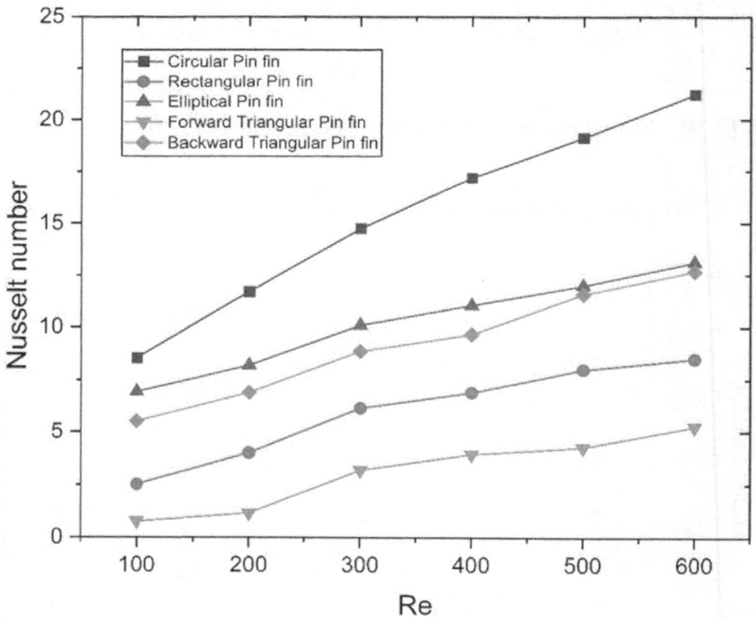

FIGURE 17.13 Nusselt number vs Reynold number for pin-fins [31, 33].

pin-fins gave the maximum Nusselt number and friction factor compared to other pin-fin configurations as shown in Figure 17.13 and 17.14.

17.5.2.4 Geometric Modifications

Geometric modifications in MCHS have emerged as an efficient passive method of heat transfer improvement. These geometric modifications include varying the

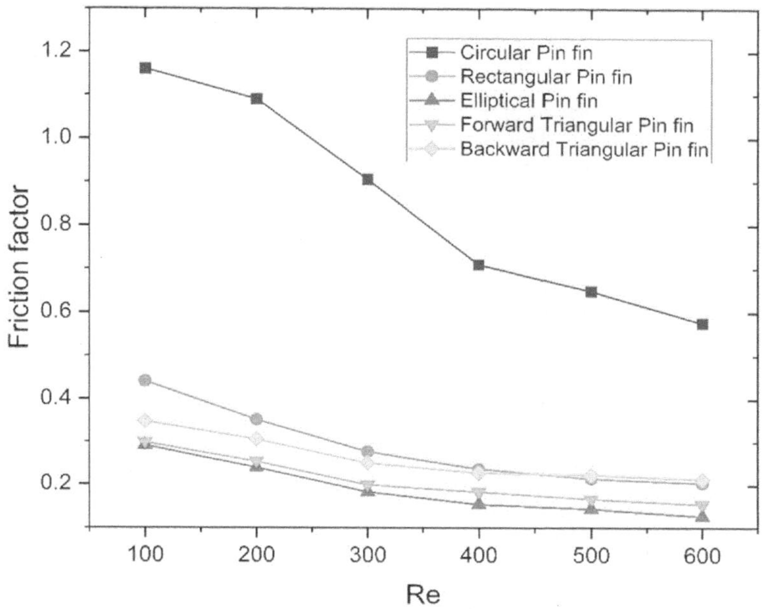

FIGURE 17.14 Friction factor vs Reynold number for pin-fins [31, 33].

size and shape of MCHS. In terms of the size of MCHS, the hydraulic diameter and the aspect ratio of MCHS are the most significant parameters. High heat transfer efficiency in single phase flows is obtained by reducing the hydraulic diameter. However, in flow boiling small channel results in restriction of bubbles, which lead to reduced heat transfer. Narrow and deep MCHS have high heat transfer efficiency. The geometric shape of the MCHS has a significant effect on heat transfer. Rectangular, triangular, circular, elliptical and trapezoidal cross section as shown in Figure 17.15 are employed on the walls of MCHS [37, 38]. The heat transfer coefficient of circular and square shaped MCHS was 30% more than conventional MCHS but it is still lower than triangular channel [39]. The rectangular shape gives the highest heat transfer performance and high pressure drop, and the triangular shape has the least hydraulic resistance and convective heat transfer coefficient. Trapezoidal shape MCHS gives the least value of total convection thermal resistance and shape gives the highest thermal resistance. Channel curvature also affects the thermal and hydraulic performance of microchannel heat sinks. Fluid moving through curved channels experience centrifugal forces, which increase fluid velocity along outside walls. This high velocity and increased surface area by curvature promotes efficient heat transfer between channel walls and fluid. Trapezoidal channel gave the highest value of heat transfer enhancement in terms of Nusselt number compared to square, semi-circular and trapezoidal channel shapes as described in Figure 17.16.

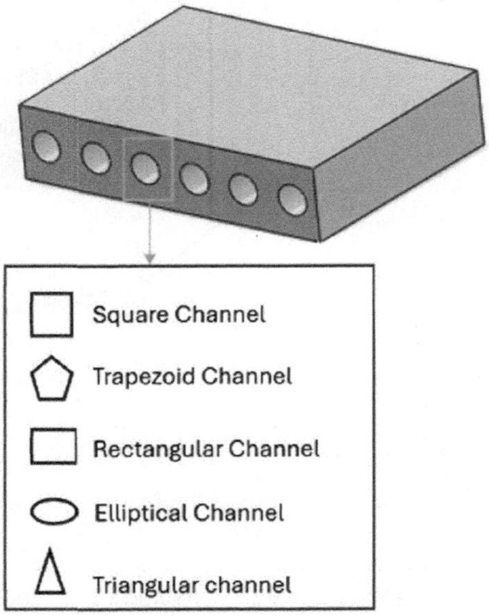

FIGURE 17.15 Heat sink geometries.

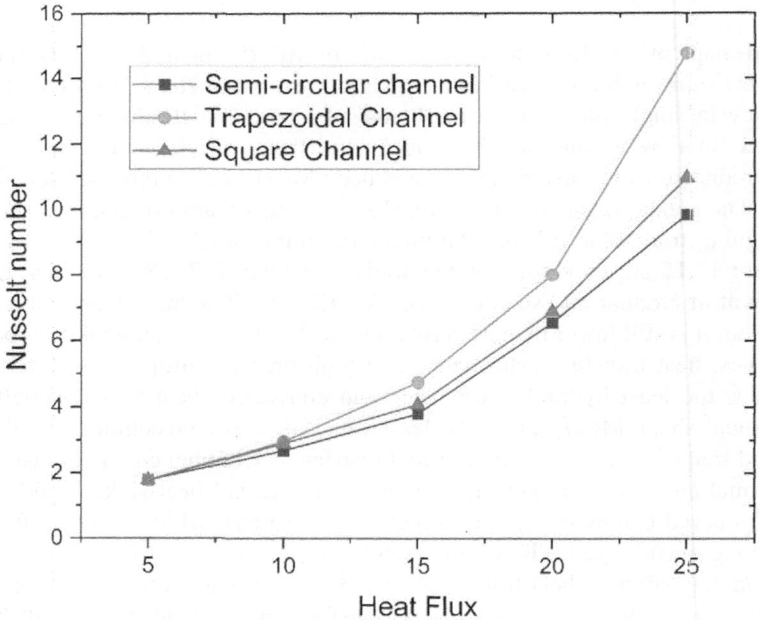

FIGURE 17.16 Nusselt number vs Reynold number for channel cross-sections [40].

17.5.2.5 Secondary Flow Channel for Enhanced Mixing

Secondary flow channels are channels inside the main coolant channels in MCHS. These secondary flow channels increase the fluid mixing and reduce the thermal boundary layer thickness promoting more effective heat transfer. Vortices are also generated in the fluid flow due to these secondary passages, which lead to thermal enhancement. Secondary channels can be added in several patterns such as serpentine or meandering pattern in MCHS. The size, spacing, and curvature of the secondary channel can be adjusted for optimal heat transfer efficiency. In the case of multilayer arrangement, MCHS secondary flow channels are placed between piled up layers of microchannels.

Studies indicate that secondary flows gave exceptional thermal performance compared to rectangular ribs, triangular cavities and MCHS with both rectangular ribs and triangular cavities [41]. Trapezoidal MCHS with secondary flow passages improved the heat transfer coefficient by 60% compared to the conventional channel [42]. This is due to the combined effect of redevelopment of thermal boundary layer and fluid flow mixing in microchannels. The fabrication of secondary flow channels with high precision and high-pressure loss are the limitations associated with secondary flow channels. Secondary flow also gives rise to random flow distribution, which results in flow maldistribution and thermal non-uniformity. Therefore, these issues must be considered for designing optimized MCHS with secondary flows.

17.5.2.6 Nanofluids

Common working fluids like water and ethylene glycol have low thermal conductivity, which results in reduced heat transfer efficiency. To overcome this low thermal conductivity issue, small nanoscale particles are suspended into the main working fluid thus making nanofluids. Commonly used nanofluids include CuO-water, Al_2O_3-water and SiO_2-water. These nanofluids increase the thermal conductivity of working fluid, which significantly enhances the thermal properties of MCHS. High pressure drop is also generated in MCHS with nanofluids that require more pumping power requirement. Although nanofluids offer high heat transfer enhancement yet nanoparticle agglomeration and sedimentation reduce their thermal conductivity, which affects the long-term stability of nanofluids. Due to agglomeration CuO-water based nanofluids lost up to 30% of their thermal conductivity in 30 days [43]. Al_2O_3-water based nanofluids give high heat transfer performance compared to SiO_2-water based nanofluids because of high thermal conductivity of Al_2O_3.

17.6 MULTI-OBJECTIVE OPTIMIZATION OF MCHS COOLING SYSTEM

Microchannel heat sinks are the most effective method for managing high heat fluxes in compact electronics and high-performance systems. However, designing an efficient microchannel heat sink involves multiple objectives, such as maximizing heat transfer, minimizing pressure drop, and optimizing the overall size and cost of the system. Several optimization techniques have been used to develop efficient MCHS.

17.6.1 Particle Swarm Optimization

Particle swarm optimization (PSO) is inspired by the behavior of the population of moving individuals like bird's flocks and fish schools. PSO starts with initializing swarm of particles in which each particle of swarm represents a prospective solution. In each iteration, particles have their velocity and position of particles, and they are evaluated by fitness function. There is personal best value and global best value of each particle in space and after *n* iterations all particles converge to the best solution.

17.6.2 Genetic Algorithms Based Optimization

Genetic algorithm-based optimization method is one of the simplest random based evolutionary algorithms based on Darwin theory of evolution. It uses an initial set of solutions and evaluates them based on fitness function. Solutions with high fitness value are selected for reproduction and new solutions are generated based on the parent's genes. The old generation parents are then mutated to create new generations of solutions to find the optimal solution. Genetic algorithms are used to solve complex optimization problems like smart energy management and bus network optimization.

17.6.3 Pareto Optimization

Pareto optimization is a commonly used optimization technique in which several sets of independent values are generated for a design variable, which leads to an optimum value of objective function. In pareto optimization, a set of pareto frontiers is generated for finding pareto optimal solutions. Different pareto solutions are analyzed to find the optimal solution for the conflicting input variables.

17.6.4 Response Surface Methodology

Response surface methodology (RSM) is the combination of several mathematical and statistical methods used to perform design parameter optimization using some objective function. RSM is used to model and analyze the output variables that are influenced by several input variables. RSM methodology is often utilized with other optimization techniques to reduce computational costs and give cost-effective thermal management solutions.

17.7 FUTURE DIRECTIONS

Thermal management has always been a major challenge because of the growing evolution of electronic devices. The use of eco-friendly and biodegradable coolants in advanced thermal management systems has become crucial to minimize environmental impact. The development of hybrid cooling solutions utilizing both active and passive methods will contribute to reduced energy consumption in compact electronics. The integration of nanotechnology, artificial intelligence, and sustainability principles will have a significant impact on advancements in microchannel based cooling systems.

- The manufacturing of microchannel based cooling systems using advanced additive manufacturing methods should be explored. The manufacturing of microchannel based cooling systems with unconventional fin shapes such as conical, elliptical and triangular could be done using 3D printing methods.
- Artificial intelligence and machine learning algorithms' optimization has revolutionized microchannel cooling. These methods will lead to the design of optimized systems using predictive models to simulate fluid flow behavior and heat transfer for improving energy efficiency and reliability.
- Future work should focus on developing nanofluids for enhanced heat transfer efficiency in microchannel heat sink applications. In the future, the use of developing nanomaterials besides traditional metallic or oxide nanoparticles for enhanced surface properties and thermal stability should be explored.
- The study of multiple layered structure and patterns with various channel width, height, and shapes of microchannel can be investigated for improving heat transfer uniformity, flow mixing and fluid flow distribution.

17.8 APPLICATIONS

Electronics components are used in a variety of applications like computers, fusion reactors, rocket nozzles, automobiles, avionics, and space systems. These applications require effective dissipation of heat and better temperature uniformity. Microchannel cooling can maintain the turbine blade surface temperature below 540°C and prevent it from hot corrosion effects.

- Cooling of modern military and commercial aircraft is done inside avionics enclosures. The heat is rejected to air, which is removed from the compressor of aircraft. The heat dissipation capacity per module increased by about 200W utilizing microchannel cooling systems.
- Thermal management plays a crucial role in the integrity of modern power electronics and electric machines in hybrid vehicles. Microchannel based cooling systems can be utilized to tackle high power density and reduced volume of power electronics.
- In refrigeration cooling microchannel heat sinks can be implemented in two configurations. In direct refrigeration cooling configuration microchannel heat sink can act as an evaporator and in-direct refrigeration cooling configuration heat from the device is rejected by microchannel heat sink to main coolant, which reject heat by heat exchanger.
- Microchannel based cooling systems are expected to play a significant role in aerospace applications due to their compact size, reduced weight, and minimal coolant inventory requirement.

REFERENCES

[1] S. Furber, "Microprocessors: The engines of the digital age," *Proc. R. Soc. Math. Phys. Eng. Sci.*, vol. 473, no. 2199, p. 20160893, Mar. 2017, doi: 10.1098/rspa.2016.0893.

[2] T. Krazit, "Intel unveils its first 10-nanometer ice lake processors," Aug. 01, 2019.

[3] J. R. Black, "Electromigration—A brief survey and some recent results," *IEEE Trans. Electron Devices*, vol. 16, no. 4, pp. 338–347, Apr. 1969, doi: 10.1109/T-ED.1969.16754.

[4] L. A. Florio and A. Harnoy, "Combination technique for improving natural convection cooling in electronics," *Int. J. Therm. Sci.*, vol. 46, no. 1, pp. 76–92, Jan. 2007, doi: 10.1016/j.ijthermalsci.2006.03.007.

[5] M. Di Capua H and W. Jahn, "Performance assessment of thermoelectric self-cooling systems for electronic devices," *Appl. Therm. Eng.*, vol. 193, p. 117020, Jul. 2021, doi: 10.1016/j.applthermaleng.2021.117020.

[6] G. Righetti, C. Zilio, L. Doretti, G. A. Longo, and S. Mancin, "On the design of phase change materials based thermal management systems for electronics cooling," *Appl. Therm. Eng.*, vol. 196, p. 117276, Sep. 2021, doi: 10.1016/j.applthermaleng.2021.117276.

[7] I. A. Ghani, N. A. C. Sidik, and N. Kamaruzaman, "Hydrothermal performance of microchannel heat sink: The effect of channel design," *Int. J. Heat Mass Transf.*, vol. 107, pp. 21–44, Apr. 2017, doi: 10.1016/j.ijheatmasstransfer.2016.11.031.

[8] U. Ghani, M. A. Wazir, K. Akhtar, M. Wajib, and S. Shaukat, "Microchannel heat sinks—A comprehensive review," *Electron. Mater.*, vol. 5, no. 4, pp. 249–292, 2024, doi: 10.3390/electronicmat5040017.

[9] Z.-Q. Yu, M.-T. Li, and B.-Y. Cao, "A comprehensive review on microchannel heat sinks for electronics cooling," *Int. J. Extreme Manuf.*, vol. 6, no. 2, p. 022005, Jan. 2024, doi: 10.1088/2631-7990/ad12d4.

[10] I. Tuomi, "The lives and death of Moore's Law," *First Monday*, vol. 7, Apr. 2002, doi: 10.5210/fm.v7i11.1000.

[11] D. Tuckerman and R. F. W. Pease, "High-performance heat sinking for VLSI," *Electron Device Lett. IEEE*, vol. 2, pp. 126–129, Jun. 1981, doi: 10.1109/EDL.1981.25367.

[12] "Hydrothermal investigation of the performance of microchannel heat sink with Ribs employed on side walls." Accessed: Feb. 06, 2025. [Online]. Available: www.degruyter.com/document/doi/10.1515/jnet-2020-0104/html

[13] P. Bhandari, K. S. Rawat, Y. K. Prajapati, D. Padalia, L. Ranakoti, and T. Singh, "A review on design alteration in microchannel heat sink for augmented thermohydraulic performance," *Ain Shams Eng. J.*, vol. 15, no. 2, p. 102417, Feb. 2024, doi: 10.1016/j.asej.2023.102417.

[14] M. Dogan and M. Sivrioglu, "Experimental investigation of mixed convection heat transfer from longitudinal fins in a horizontal rectangular channel," *Int. J. Heat Mass Transf.*, vol. 53, no. 9, pp. 2149–2158, Apr. 2010, doi: 10.1016/j.ijheatmasstransfer.2009.12.031.

[15] W. Qu and I. Mudawar, "Experimental and numerical study of pressure drop and heat transfer in a single-phase micro-channel heat sink," *Int. J. Heat Mass Transf.*, vol. 45, no. 12, pp. 2549–2565, Jun. 2002, doi: 10.1016/S0017-9310(01)00337-4.

[16] H. C. Cui, C. Y. Shi, M. J. Yu, Z. K. Zhang, Z. C. Liu, and W. Liu, "Optimal parameter design of a slot jet impingement/microchannel heat sink base on multi-objective optimization algorithm," *Appl. Therm. Eng.*, vol. 227, p. 120452, Jun. 2023, doi: 10.1016/j.applthermaleng.2023.120452.

[17] "(PDF) Numerical investigation of two-phase laminar pulsating nanofluid flow in an interrupted microchannel heat sink." Accessed: Feb. 06, 2025. GSJ: Volume 8, Issue 7, July 2020, Online 2320-9186. Available: www.researchgate.net/publication/383414446_Numerical_investigation_of_two-phase_laminar_pulsating_nanofluid_flow_in_an_interrupted_microchannel_heat_sink

[18] M.-A. Hessami, A. Berryman, and P. Bandopdhayay, "Heat Transfer Enhancement in an Electrically Heated Horizontal Pipe Due to Flow Pulsation," *Proceedings of the ASME 2003 International Mechanical Engineering Congress and Exposition. Heat Transfer,* Washington, DC, USA. November 15–21, 2003. vol. 374. pp. 49-56. ASME. doi: 10.1115/IMECE2003-55146.

[19] J. S. Go, "Design of a microfin array heat sink using flow-induced vibration to enhance the heat transfer in the laminar flow regime," *Sens. Actuators Phys.*, vol. 105, no. 2, pp. 201–210, Jul. 2003, doi: 10.1016/S0924-4247(03)00101-8.

[20] "Trapezoidal microchannel heat sink with pressure-driven and electro-osmotic flows for microelectronic cooling | Request PDF." Accessed: Feb. 06, 2025. [Online]. doi: 10.1109/TCPMT.2013.2272478. Available: www.researchgate.net/publication/260583714_Trapezoidal_Microchannel_Heat_Sink_With_Pressure-Driven_and_Electro-Osmotic_Flows_for_Microelectronic_Cooling

[21] G. Wang, D. Niu, F. Xie, Z. Yang, X. Zhao, and G. Ding, "Experimental and numerical investigation of a microchannel heat sink (MCHS) with micro-scale ribs and grooves for chip cooling," *Appl. Therm. Eng.*, vol. 85, Jun. 2015, doi: 10.1016/j.applthermaleng.2015.04.009.

[22] F. Ahmad, T. Cheema, M. Mohib, and M. M. Ur Rehman, *Flow and Heat Transfer Characteristics of Microchannel Heat Sink with Cylindrical Ribs and Cavities*. 2019. 9th International Mechanical Engineering Conference (IMEC 2019).

[23] H. E. Ahmed and M. I. Ahmed, "Optimum thermal design of triangular, trapezoidal and rectangular grooved microchannel heat sinks," *Int. Commun. Heat Mass Transf.*, vol. 66, pp. 47–57, Aug. 2015, doi: 10.1016/j.icheatmasstransfer.2015.05.009.

[24] P. Kumar, "Numerical investigation of fluid flow and heat transfer in trapezoidal microchannel with groove structure," *Int. J. Therm. Sci.*, vol. 136, pp. 33–43, Feb. 2019, doi: 10.1016/j.ijthermalsci.2018.10.006.

[25] "Parametric numerical study of the flow and heat transfer in microchannel with dimples – University of Huddersfield Research Portal." Accessed: Feb. 06, 2025. [Online]. https://doi.org/10.1016/j.icheatmasstransfer.2016.06.002. Available: https://pure.hud.ac.uk/en/publications/parametric-numerical-study-of-the-flow-and-heat-transfer-in-micro

[26] M. M. Ur Rehman, T. Cheema, F. Ahmad, A. Abbas, and M. sohail Malik, "Numerical investigation of heat transfer enhancement and fluid flow characteristics in a microchannel heat sink with different wall/design configurations of protrusions/dimples," *Heat Mass Transf.*, vol. 56, Jan. 2020, doi: 10.1007/s00231-019-02697-9.

[27] M. Xu, H. Lu, L. Gong, J. C. Chai, and X. Duan, "Parametric numerical study of the flow and heat transfer in microchannel with dimples," *Int. Commun. Heat Mass Transf.*, vol. 76, pp. 348–357, Aug. 2016, doi: 10.1016/j.icheatmasstransfer.2016.06.002.

[28] "Experimental and numerical study of heat transfer and flow friction in channels with dimples of different shapes," *J. Heat Transfer. ASME Digital Collection.* Accessed: Feb. 06, 2025. [Online]. https://doi.org/10.1115/1.4029036. Available: https://asmedigitalcollection.asme.org/heattransfer/article-abstract/137/3/031901/374912/Experimental-and-Numerical-Study-of-Heat-Transfer?redirectedFrom=fulltext

[29] Y. Rao, B. Li, and Y. Feng, "Heat transfer of turbulent flow over surfaces with spherical dimples and teardrop dimples," *Exp. Therm. Fluid Sci.*, vol. 61, pp. 201–209, Feb. 2015, doi: 10.1016/j.expthermflusci.2014.10.030.

[30] "Thermo-hydraulic investigation of open micro prism pin fin heat sink having varying prism sides – ScienceDirect." Accessed: Feb. 06, 2025. [Online]. https://doi.org/10.1016/j.aej.2023.02.016. Available: www.sciencedirect.com/science/article/pii/S111001682300114X

[31] "Thermal enhancement of microchannel heat sink using pin-fin configurations and geometric optimization – IOPscience." Accessed: Feb. 06, 2025. [Online]. doi: 10.1088/2631-8695/ad3400. Available: https://iopscience.iop.org/article/10.1088/2631-8695/ad3400

[32] "Comparison of the cooling performance of staggered and in-line arrays of electronic packages," *J. Electron. Packag. ASME Digital Collection.* Accessed: Feb. 06, 2025. [Online]. https://doi.org/10.1115/1.2792123. Available: https://asmedigitalcollection.asme.org/electronicpackaging/article-abstract/118/1/27/404573/Comparison-of-the-Cooling-Performance-of-Staggered?redirectedFrom=fulltext

[33] G. Wang *et al.*, "Experimental and numerical investigation of hydrothermal performance of a microchannel heat sink with pin fins," *Case Stud. Therm. Eng.*, vol. 60, p. 104631, Aug. 2024, doi: 10.1016/j.csite.2024.104631.

[34] "Design modifications in micro pin fin configuration of microchannel heat sink for single phase liquid flow: A review – ScienceDirect." Accessed: Feb. 06, 2025. [Online]. https://doi.org/10.1016/j.est.2023.107548. Available: www.sciencedirect.com/science/article/pii/S2352152X23009453

[35] M. M. Sarafraz and F. C. Christo, "Thermal and flow characteristics of liquid flow in a 3D-printed micro-reactor: A numerical and experimental study," *Appl. Therm. Eng.*, vol. 199, p. 117531, Nov. 2021, doi: 10.1016/j.applthermaleng.2021.117531.

[36] "Fabrication of micro pin fins on inclined V-shaped microchannel walls via laser micromilling | Advances in Manufacturing." Accessed: Feb. 06, 2025. [Online]. https://doi.org/10.1007/s40436-021-00382-x. Available: https://link.springer.com/article/10.1007/s40436-021-00382-x

[37] H. Wang, Z. Chen, and J. Gao, "Influence of geometric parameters on flow and heat transfer performance of micro-channel heat sinks," *Appl. Therm. Eng.*, vol. 107, pp. 870–879, Aug. 2016, doi: 10.1016/j.applthermaleng.2016.07.039.

[38] M. Goodarzi, I. Tlili, Z. Tian, and M. R. Safaei, "Efficiency assessment of using graphene nanoplatelets-silver/water nanofluids in microchannel heat sinks with different cross-sections for electronics cooling," *Int. J. Numer. Methods Heat Amp Fluid Flow*, vol. 30, no. 1, pp. 347–372, Jun. 2019, doi: 10.1108/HFF-12-2018-0730.

[39] M. Magnini and O. K. Matar, "Numerical study of the impact of the channel shape on microchannel boiling heat transfer," *Int. J. Heat Mass Transf.*, vol. 150, p. 119322, Apr. 2020, https://doi.org/10.1016/j.ijheatmasstransfer.2020.119322.

[40] A. A. Alfaryjat, H. A. Mohammed, N. M. Adam, M. K. A. Ariffin, and M. I. Najafabadi, "Influence of geometrical parameters of hexagonal, circular, and rhombus microchannel heat sinks on the thermohydraulic characteristics," *Int. Commun. Heat Mass Transf.*, vol. 52, pp. 121–131, Mar. 2014, doi: 10.1016/j.icheatmasstransfer.2014.01.015.

[41] "A review of passive methods in microchannel heat sink application through advanced geometric structure and nanofluids: Current advancements and challenges." Accessed: Feb. 06, 2025. [Online]. https://doi.org/10.1515/ntrev-2020-0094. Available: www.degruyter.com/document/doi/10.1515/ntrev-2020-0094/html?lang=en&srsltid=AfmBOorvkbdNEJDY5gP73VPE-6miGYLjirxSfoN3D8EBkwCl9dShYhJE

[42] "Hydrothermal investigation of a microchannel heat sink using secondary flows in trapezoidal and parallel orientations." Accessed: Feb. 06, 2025. [Online]. https://doi.org/10.3390/en13215616. Available: www.mdpi.com/1996-1073/13/21/5616
[43] "On the thermal performance of a fractal microchannel subjected to water and kerosene carbon nanotube nanofluid." *Sci. Rep*. Accessed: Feb. 06, 2025. [Online]. https://doi.org/10.1038/s41598-020-64142-w. Available: www.nature.com/articles/s41598-020-64142-w

18 Solar Energy and Thermal Management Systems

Ali F. Ali Fadiel, Monaem Elmnifi, and Hasan Shakir Majdi

18.1 INTRODUCTION TO SOLAR ENERGY IN COOLING SYSTEMS

18.1.1 Overview of Renewable Energy Sources

Concerning global renewable resources, solar energy possesses a unique value due to its ability to provide clean and sustainable energy amid rising climatic complications and the reliance on fossil fuels. Solar thermal energy has recently been utilized in cooling systems, especially within regions that receive intense solar irradiance. Technologies that harness solar power include photovoltaic cells for electricity and solar thermal collectors for heating.

When harnessed through photovoltaic cells, the solar system's electricity generation aligns perfectly with cooling requirements, as peak cooling coincides with the hottest times when solar radiation is strongest. Such qualities make solar energy the most feasible resource for air conditioning and refrigeration, which minimizes greenhouse gas emissions and promotes environmental sustainability by reducing fossil fuel electricity dependency.

Investing in solar cooling technologies enhances energy security and promotes economic growth by fostering job creation within green technology. Most importantly, countries can opt for cutting-edge solar technology for their cooling needs, allowing them to offset their operational electricity costs. The major hurdles, however, are the enormous initial capital investments and maintenance required. Furthermore, advancements in technology are improving the efficiency of systems and lowering the cost simultaneously, allowing for wider applications of solar water heaters [2], [11], [7], and [5].

Renewable energy sources can improve thermal comfort in buildings by providing heating and cooling, Figure 18.1. Examples of renewable energy applications in buildings include solar thermal water heaters, biomass boilers, heat pumps, and natural cooling systems. Reducing energy use in buildings and industries is crucial for transitioning to a renewable energy system [23].

Renewable energy may also facilitate heating and cooling operations in industrial and agricultural sectors, encompassing food processing and pulp and paper manufacturing. Hydrogen from renewable energy can fulfill the energy demands of high-intensity industrial processes in the iron, steel, and chemical sectors. Renewable

 DOI: 10.1201/9781003611820-23

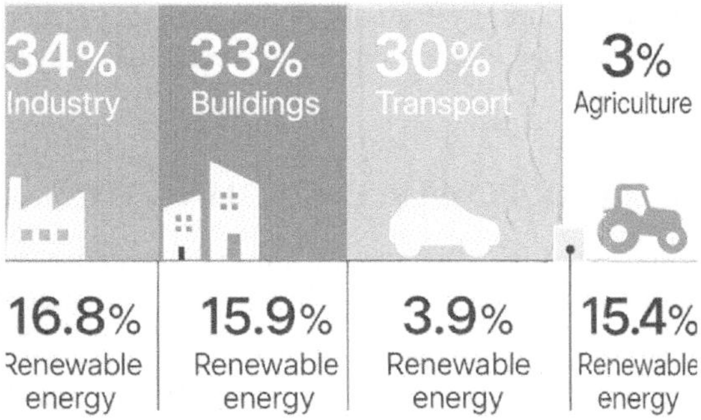

FIGURE 18.1 Proportion of renewable energy in total final energy consumption by source in 2021. (Source: www.ren21.net/why-is-renewable-energy-important/.)

energy can be utilized in transportation through sustainable biofuels, high-percentage mixes, and drop-in biofuels, powering electric vehicles and producing electro-fuels like hydrogen. Enhancing energy efficiency and conservation is crucial for reducing fuel consumption in the transportation industry.

Globally, renewable sources accounted for around 30% of electricity in 2023. However, apart from power, positive developments are more elusive. Using electricity, such as for lights and appliances, constitutes merely 23% of global energy requirements. Immediate measures are required to increase the significantly lower proportions of renewables in heating and fuels, which account for three-quarters of the energy supply.

18.1.2 Importance of Solar Thermal Energy

Solar energy is essential for developing sustainable cooling strategies in regions with exceptional direct solar irradiation. This energy source is beneficial because the maximum cooling needs and solar radiation levels are simultaneously high, making running air conditioning units easy and cost-effective. Compared to air conditioning systems that utilize fossil fuels as a primary energy source and increase greenhouse gas emissions, solar thermal systems use the heat from the sun to provide more eco-friendly cooling.

His ability to provide cooling or refrigeration via absorption and adsorption chillers proves the effectiveness of solar thermal energy. These modern technologies utilize the waste heat from solar collectors to achieve the desired cooling effect while replacing the untreated refrigerant used in other systems. In addition, the surge in thermal storage devices has made it possible for these chillers to counter the intermittent supply problem associated with solar energy by storing the excess heat for later use.

Adopting solar thermal systems improves energy self-sufficiency while significantly reducing dependence on fossil fuels. These systems are beneficial economically over time, unlike the initial investment required to incorporate them within commercial or residential structures. Moreover, governments across the globe are adopting solar thermal technology-friendly policies as they recognize its potential and are willing to stimulate its integration through established systems [7], [3], [10] [12], and [23].

18.2 TYPES OF SOLAR-POWERED COOLING TECHNOLOGIES

Solar cooling system technologies are classified into three main categories: solar electrical, thermal, and combined power and cooling cycles, as depicted in Figure 18.2.

18.2.1 Elements and Temperature Operational Ranges of Solar Cooling Systems

The solar cooling system comprises three fundamental components: the solar energy collection unit, refrigeration cycles, and applications across various temperature ranges. The appropriate cycle for each application can primarily be determined based on specific applications that necessitate a diverse range of chilling, which cannot be accomplished by a singular refrigeration cycle. The multi-effect system is the most efficient method for achieving various refrigeration effects and temperature ranges, utilizing solar energy, and mitigating environmental issues.

18.2.2 Photovoltaic Thermal Cooling

A solar-powered cooling system comprises a photovoltaic panel and an electric refrigeration unit. Completed converting light into electricity using the photoelectric effect.

A variety of solar-powered cooling systems are designed to achieve independent functions. Photovoltaic cells are composed of semiconductor materials,

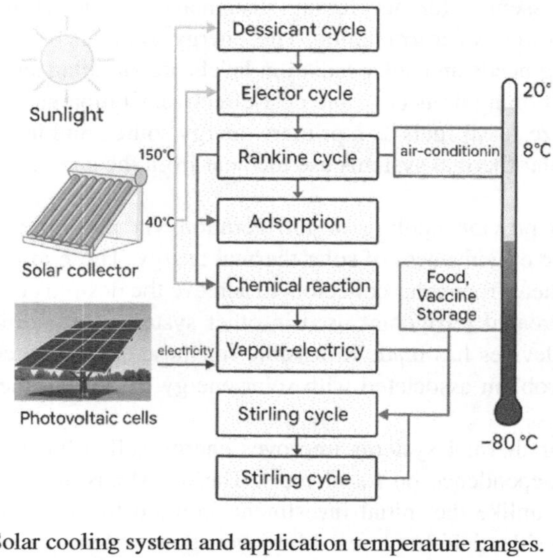

FIGURE 18.2 Solar cooling system and application temperature ranges.

monocrystalline thin films, polycrystalline structures, and silicon wafers, with silicon being the dominant component on the market.

Polycrystalline thin films exhibit more efficiency than silicon wafers, with efficiencies ranging from 10 to 17%. In contrast, single-crystalline thin films can achieve 15 to 20% efficiencies when employing a multi-junction cell structure. Conversely, the performance of silicon wafers is inferior, and their costs are elevated compared to thin-film technologies. The energy produced by solar photovoltaic cells is allocated to either the thermoelectric system, Stirling cycle, or traditional vapor compression systems [8], [2].

18.2.3 Thermoelectric Cooling (Peltier Cooling System)

The Peltier effect allows thermoelectric devices to establish temperature differences between two semiconductor materials with unique properties. When an electric current passes through two dissimilar conductors, their junctions experience heating or cooling based on the Peltier effect. Single or multiple pairs of *n*-type and *p*-type semiconductor materials receive direct current (DC), making one conductor end cool by extracting heat from the environment. Heat absorption happens when electrons transfer between *p*-type and *n*-type materials, changing their energy levels from low to high, Figure 18.3. The establishment of temperature differences between conductor ends leads to voltage production. Heat moves from the cold to the hot area using the *n*-type and *p*-type semiconductors before dissipating into the semiconductor material. Changing the current direction allows both heat flow directions to be altered, enabling the system to perform heating and cooling functions. The use of thermoelectric cooling technology in industry remains limited relative to vapor compression cycles, but Peltier coolers excel by involving no moving components that traditional refrigeration systems depend on. The Peltier cooler possesses distinctive characteristics, including the absence of circulating fluid, an almost perpetual lifespan, minimal leakage risk, and compact dimensions. Thermoelectric devices lack chlorofluorocarbons, making them environmentally

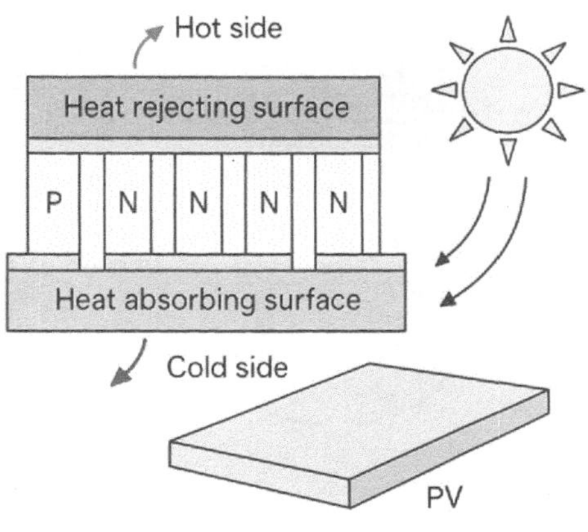

FIGURE 18.3 Pettier cooling system.

sustainable. They encompass fully reversible cycles, precise temperature regulation, and function efficiently in delicate applications. The main disadvantages are increased expenses and reduced efficiency. Numerous researchers are currently engaged in the development of cost-effective and highly efficient Peltier coolers. Thermoelectric can cool electronic devices, refrigerators, and air conditioners. Thermoelectric devices are utilized in military, aerospace, instrumentation, medical, and industrial industries. Describe the thermoelectric operating principle, the materials utilized in thermoelectric applications, and their uses. They also investigated the applications of thermoelectric devices in refrigeration and energy production. They discussed the development of innovative materials that could improve thermoelectric devices for diverse applications, including sensors in thermal energy. The principal disadvantage of thermoelectric systems is their low coefficient of performance (COP). Nevertheless, it holds considerable promise in specific applications, such as cooling electronic devices, where thermoelectric systems are preferred for their compactness and low energy usage [5],[8], [25].

18.2.4 Absorption Cooling Systems

Systems driven by absorption cooling harness solar thermal energy to provide cooling using a heat-driven refrigeration cycle, Figure 18.4. These systems are particularly effective for regions with warmer climates as they help track the peak cooling times,

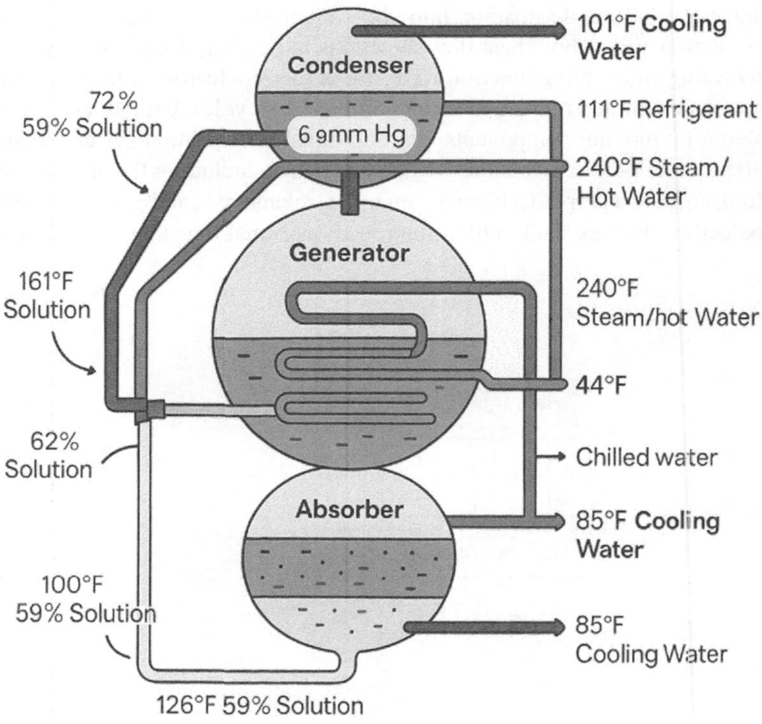

FIGURE 18.4 Schematic of a solar absorption system.

which usually overlap with the hours of high solar insolation. The absorption process usually utilizes a refrigerant that evaporates to absorb heat from the surrounding space and a liquid absorbent that vaporizes. This mixture releases refrigerant vapor while being heated, which is then condensed into a liquid using solar thermal energy.

Absorption chillers have recently become more popular than other thermally-driven technologies due to their improved efficiency and demonstrated reliability. They can be categorized into two types: single-effect and multi-effect systems. Examples of chillers that are prone to common failure are single-effect chillers. They are, however, widely used as they operate successfully under moderate temperatures. On the other hand, multi-effect chillers provide improved performance as efficiency is increased at the cost of more stringent operational parameters, which limits the regions that can use them to those with high levels of direct solar radiation.

Local climatic conditions and the availability of solar energy greatly determine the economic feasibility of absorption cooling systems. These systems can affordably substitute traditional cooling methods by providing a sustainable alternative to fossil fuels, but the initial costs remain challenging. Furthermore, adequate insulation is essential to reduce heat leaks from the storage tank and the piping systems. For this reason, these systems frequently depend on government grants or subsidies to be economically competitive in the market, even if such systems are highly efficient [15], [8], [24], and [27].

18.2.5 Adsorption Cooling Systems

Adsorption cooling systems run on thermodynamic cycles and rely on thermal energy from the sun's heat. They have two primary constituents: an adsorbent and a refrigerant. They start with the refrigerant vapor that condenses onto the adsorbent. When exposed to solar heat, the adsorbent removes the refrigerant vapor, which causes the refrigerant to evaporate and draw further heat away from the environment, Figure 18.5.

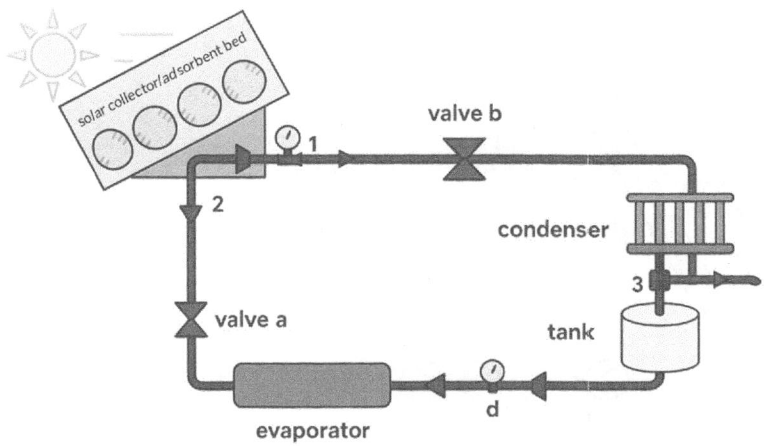

FIGURE 18.5 Schematic representation of the solar adsorption refrigeration system.

Chillers powered by adsorption can operate with low temperature sources, unlike conventional vapor compression systems, which can only work under high-temperature conditions. For that reason, they are compatible with solar thermal collectors that collect energy at peak sunlight. They can also use harmless refrigerants like water or other organic fluids, further reducing environmental pollution [27].

While researchers strive to design more favorable systems, adsorption systems remarkably assume lower COP than absorption systems. Improving adsorbents can foster lower costs and reduced complexity, posing significant challenges for data science. Regardless, the promising opportunities of operations cost and greenhouse gas emission reductions in highly sunny regions showcase adsorption cooling as a sustainable option for climate control [1], [15], [9], and [5].

18.2.6 Ejector Cooling Systems

Ejector cooling systems utilize the Venturi effect by passing the fluid through a nicely designed nozzle. Ejector systems effectively amalgamate solar energy or waste heat above 80 degrees Celsius, while ejectors have lower COP than absorption or adsorption systems, Figure 18.6. Moreover, maintenance efforts and mechanical failures are minimized compared to conventional vapor-compression systems.

These adaptable systems have become omnipresent in residential and industrial air conditioning in places with high solar radiation and few conventional energy sources. Ejector systems are more commonly regarded as decentralized engines for areas with abundant solar power and renewable energy fuels. Recently, shifts in approaches to solar cooler systems have enabled researchers to enhance their modeling perfection. Advanced experimental studies and modeling endeavors are being undertaken to

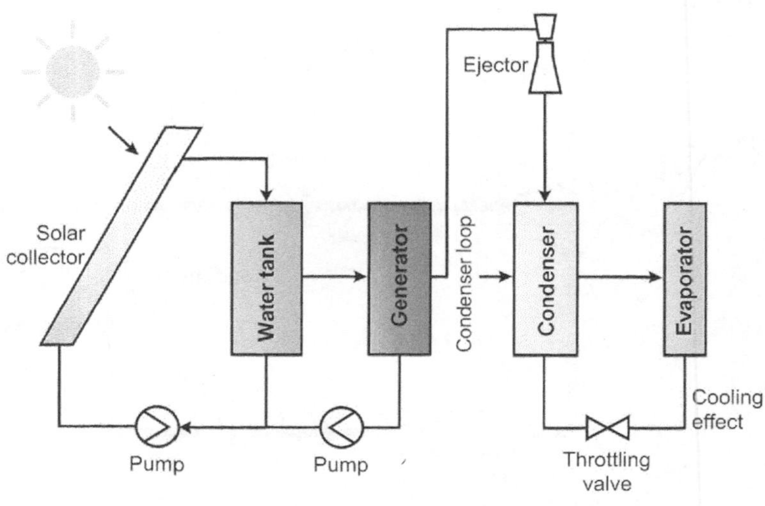

FIGURE 18.6 The diagram of a traditional solar-powered ejector refrigeration system.

optimize the cooler's efficiency and level up its competition against robust refrigeration technology.

The increased emphasis on conserving fuels makes solar ejector cooling systems an easy-to-implement solution for combating global warming and mitigating dependence on fossil fuels [1], [22], and [28].

18.2.7 Photovoltaic-Compression Cooling Systems

Photovoltaic technology and compression cooling systems employ photovoltaic (PV) panels that transform sunlight into electric energy to power standard vapor compression refrigeration units. This cooling technology is based on electricity generated by solar panels and may be used for various purposes, such as residential cooling or even large-scale commercial activities, Figure 18.7. The most notable advantage is that solar energy can be harnessed and used directly, alleviating the need for fossil fuels and reducing emissions of greenhouse gases.

Solar energy converted into cooling output is assessed using the solar COP (SCOP), which determines the system's efficiency. Progress in PV systems has increased electricity production efficiency in areas with high solar radiation. These improvements significantly decreased life cycle costs and enhanced economic benefits.

Overall system efficiency improves in hybrid systems that integrate thermal and electric power from PV panels. These systems cover changes in daily energy requirements and integrate optimal resource usage for cooling and heating systems. Adding battery storage significantly increases reliability since it allows a power supply during periods without sun.

While working on urban sustainable alternatives to air conditioning, efforts are directed towards developing photovoltaic-powered compression cooling systems, which would serve this purpose along with other global renewable energy and climate change initiatives [25].

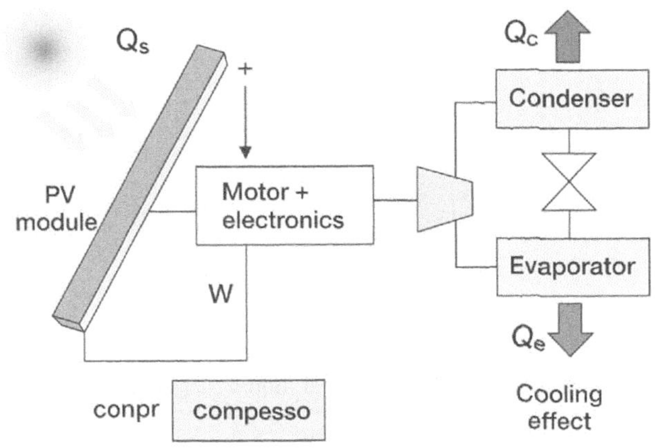

FIGURE 18.7 Solar powered vapor compression cycle.

18.2.8 Liquid and Solid Desiccant Systems

Utilize hygroscopic substances to regulate the humidity and temperature of the enclosed areas. Solid desiccants such as silica gel absorb moisture from the surrounding air, and liquid options like lithium bromide perform the same function, Figure 18.8 [26]. The system's most important characteristic is the regeneration done with solar thermal energy, which removes moisture from the desiccant to reactivate its functionality, Figure 18.9.

The air is passed through a desiccant-filled rotating disk for solid desiccant systems, where it gets dehumidified. To reactivate the material's ability to dry, it is saturated and then heated using solar energy. This enables an uninterrupted cooling cycle. Liquid desiccant systems can work with photovoltaic sources to operate fans and pumps that move air and the desiccant.

Because of using solar energy during peak demand, these technologies increase operational efficiency while reducing dependence on traditional electricity-based cooling. They also provide a significant environmental benefit. This makes them particularly advantageous in high and dry regions where effective cooling and dehumidification are essential. It is expected that investing in desiccant technology will improve performance and decrease costs, making it suitable not only for large commercial buildings but also for homes [1], [15], [2], [16], [9], and [19].

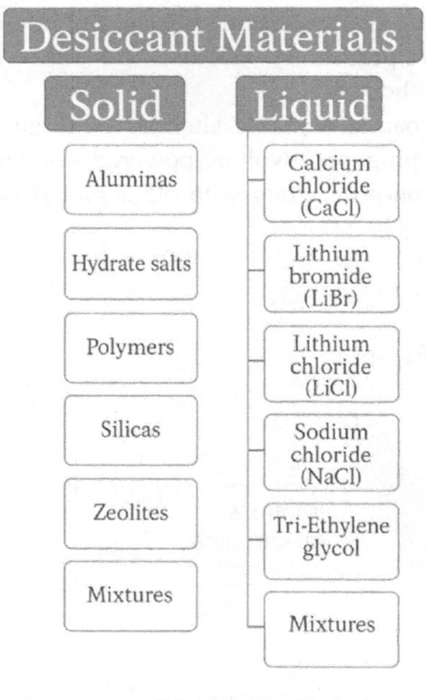

FIGURE 18.8 Difference between solid and liquid desiccant materials in a desiccant cooling cycle.

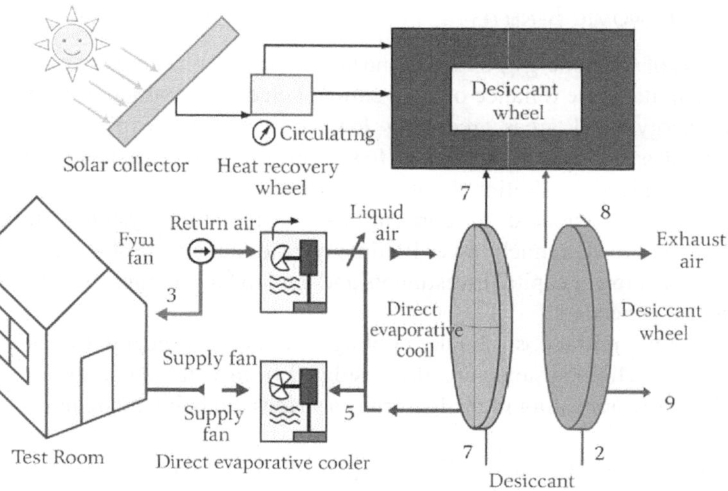

FIGURE 18.9 Illustration of a solid desiccant cooling system.

18.3 BENEFITS OF SOLAR-POWERED COOLING SYSTEMS

18.3.1 Environmental Advantages

Cooling systems powered by solar energy are environmentally beneficial as they lessen the use of fossil fuels and greenhouse gas emissions. Conventional cooling systems that rely on electricity produced by fossil fuels are significant sources of carbon pollution. Instead, solar-powered systems utilize energy directly from the sun, significantly reducing their carbon footprints. Research further indicates that adopting solar-assisted technologies can help reduce CO_2 emissions by tens of thousands of metric tons every year, making them an excellent option to enhance global emission reduction strategies and support sustainable development.

These systems work best in sunny areas, which conveniently have the most significant demand for energy during the summer months when solar radiation is most abundant. In addition, during the hottest summer months, their peak energy demand periods perfectly coincide with solar radiation. These factors economically enhance the feasibility of solar cooling solutions and aid in combating climate change by improving energy efficiency and reducing energy use. Furthermore, solar cooling technologies are essential for areas with insufficient or no access to grid electricity because they provide reliable cooling to preserve vital food and medical supplies.

Additionally, incorporating these systems within urban areas supports the construction of eco-friendly infrastructure. Utilizing solar cooling technologies is one of the more substantial developments in meeting inclusive climate change mitigation efforts, while ensuring ecological well-being, social comfort, and safety [10], [9], and [29].

18.3.2 Economic Benefits

The benefits of solar energy go far beyond simply saving on energy expenditure. These systems help decrease reliance on conventional electricity, which is highly important in high-energy-rated areas, reducing long-term electric expenses. Additionally, grid demand is decreased, leading to fewer power outages that may interfere with operations and lead to additional unforeseen costs.

Furthermore, compared to conventional refrigeration systems, solar cooling systems tend to have a much lower lifetime operational cost. Solar thermal or photovoltaic systems lower capital investments and tend to have longer lifespans since little maintenance is needed.

Although the initial cost of solar cooling is higher, tax incentives and rebates can help reduce it. These systems are also insulated from volatile energy prices, making them financially appealing in the long run as strides are being made in energy storage technology.

The application of solar cooling has presented favorable returns in commercial spaces. Evaluative studies show that systems like solar absorption chillers provide improved net present value and favorable payback periods against conventional air conditioners. With the increasing concern for climate change, businesses that adopt renewable technologies stand to gain a significant competitive advantage and consumer goodwill [7], [11], [5], and [21].

18.4 COMPARATIVE ANALYSIS OF TECHNOLOGIES

18.4.1 Solar Coefficient of Performance Evaluation

The COP is highly relevant to assessing the efficiency of solar cooling systems. It considers the energy input for achieving a sound cooling output. The values of COP differ tremendously with technology and circumstances; single-effect absorption chillers typically attain roughly 0.7, while double-effect systems can reach anywhere from 1.3 to 1.8, depending on the design and other environmental factors. This shows that solar thermal technologies' efficiency depends on the surrounding temperature and radiation.

Chillier performance characteristics vary significantly regarding PV (photovoltaic) systems that supply power to vapor compression chillers. Both chiller efficiency and PV module performance under sunlight impacted their COP. Research shows that the PV-powered cooling can have several levels of effectiveness, mainly when sophisticated control instruments and ideal conditions are utilized.

Also, it has been established that an increase in the COP of absorption chillers generates enormous savings in the energy expenses incurred for every square meter cooled. A 0.1 increase in COP, for instance, could result in savings of over $50 per square meter in certain climates. This information underscores the investment opportunities offered by solar cooling technologies, which are extremely appealing economically [1], [10], [2], [7], and [5].

18.4.2 Capital Cost Comparison

The cost of solar cooling systems, also called active solar cooling, is highly reliant on particular technologies applied, the size of the systems, and other environmental factors. Solar-assisted cooling (SAC) systems are generally well known for their higher initial investment than standard cooling systems, often costing anywhere between two thousand and five thousand euros for every kilowatt of cooling. Literature suggests that these costs can surpass two to five times those of traditional cooling systems.

Nonetheless, we have seen a reduction in the price of pre-engineered solar cooling kits. In 2007, they sold for around six thousand euros per kilowatt, with only six kits launching at roughly 4500 euros per kW by 2009. Despite these downward price movements, the costs accompanying fully integrated solar-powered cooling solutions are still unbelievably high.

Absorption systems are shown to have a lower capital cost than adsorption systems, ranging from investment cost savings to as high as forty to fifty percent. Furthermore, these systems are likely to become economically feasible as the cost of operations decreases and technology improves for SAC systems. While recognition of the traditional cost-intensive nature of SAC systems is widespread, life cycle cost analyses justify integrating these systems over the more conventional systems in several areas.

Once again, high energy prices and visible costs are not the only factors affecting the overall effectiveness of retrievable energy resources. Incorporating thermal energy storage increases the economic attractiveness of solar-powered cooling by utilizing energy efficiently during off-peak demand times [7], [6], and [13].

18.4.3 Life Cycle Cost Assessment

Estimating solar energy cooling systems' life cycle costs requires defining all costs from inception to retirement, including capital expenditures, operational expenses, maintenance, and disposal costs. This assessment is essential to stakeholders because it illuminates the economic feasibility of different solar cooling technologies, such as photovoltaic compression and absorption systems.

The adoption of solar cooling solutions often demands significant upfront capital investment. Still, this disadvantage is mitigated by lower facility operational expenditures due to increased energy efficiency, especially in comparison to traditional systems. Implementing energy storage also improves overall energy efficiency by enabling solar energy utilization during off-peak hours.

Some studies comparing two technologies of vapor compression and vapor absorption refrigeration systems to determine their life cycle cost and economic feasibility found that both are financially self-sustaining. This trend is also observable in vapor compression systems, which attained a superior benefit-cost ratio stemming from their low maintenance requirements and extensibility.

The introduction of government aid has a profound effect on the payback period (PBP) of these pieces of new technology, which, without such assistance, may reach two or more decades, but with appropriately allocated monetary support, may decrease drastically.

In conclusion, different solar cooling technologies have considerably varying life cycle costs, enabling stakeholders to make better investment and policy decisions regarding the use of renewable energy [7], [6], [5], and [30].

18.5 APPLICATIONS OF SOLAR-POWERED COOLING WORLDWIDE

18.5.1 Residential Air Conditioning Applications

Due to their efficiency and environmental friendliness, the adoption of solar-powered cooling systems for homes is rising. Traditional air conditioning in extremely hot areas causes high energy consumption and dependence on grid electricity. Unlike conventional units, solar thermal cooling systems utilize sunlight for cooling and operate with thermally driven chillers.

Studies demonstrate that solar-assisted absorption chillers can provide sufficient cooling for single-family homes in hotter regions while reducing the need for reliance on electrical grids. New technology, such as photovoltaic-driven DC air conditioning units, is a promising off-grid option that has the potential to fully fulfill cooling requirements by making maximum use of solar power during peak hours. The addition of thermal storage systems also allows for cooling without sunlight.

The economic factors resulting from government policies and the benefits of technological improvements offset the reduction of system costs, making residential solar cooling systems more attractive. Solar cooling offers operational efficiency, which can result in reduced carbon emissions and lower domestic electricity bills, which makes households more motivated to adopt these systems. Schools have already started adopting solar cooling for long-term cost savings and operational efficiency, which strongly advocates using renewable energy for residential use [1], [10], [2], [14], and [9].

18.5.2 Commercial and Industrial Applications

Toward the end of the 20th century, a need arose to focus on trigeneration systems which can perform three distinct functions under one magnificently advanced system. Implementing solar energy in cooling systems for the commercial and industrial sectors is gaining importance as a novel approach to solving substantial cooling demand with little reliance on traditional energy sources. Commercial buildings, particularly in hot regions, can significantly benefit from using thermal solar chillers, as this will save a considerable amount of energy. For example, hotels and shopping malls utilize absorption chillers that use solar collectors for cooling. These systems operate at much lower noise levels and vibrations compared to other systems, such as conventional vapor-compression systems.

Furthermore, it is worth mentioning more advanced configurations, such as hybrid trigeneration plants with cooling, which enable heating and cooling simultaneously and make them suitable for a wide range of facilities with varying energy needs. One good example is the primary focus in industrial refrigeration, using solar-assisted power cooling technologies to displace conventional refrigeration during peak usage periods. These developments are both workable and cost-effective when used in existing frameworks.

Also, the agricultural sector increasingly uses small mobile photovoltaic-powered cold storage and other devices required to protect fruits and other perishables. These units are self-contained and serve as mobile cold storage to preserve crops right after harvest. This advancement improves food security while substantially reducing post-harvest waste.

While efforts to promote sustainability exist, companies and industries embrace renewable energy sources. The constant advancement of custom-built improvement measures assures increasing profitability and effectiveness [7], [9], [12], [16], and [28].

18.6 CHALLENGES IN IMPLEMENTING SOLAR COOLING TECHNOLOGIES

18.6.1 Technical Barriers to Adoption

Difficulties in implementing cooling technologies that harness solar energy are numerous and diverse. For one, solar-based systems have intricate processes regarding their deployment and integration, which have to be designed and built by specialists with a background in solar energy, hydraulics, and air conditioning. Additional installation circuits make the installation even more challenging. In many cases, the contemporary cooling systems available in the market still do not measure up to the efficiency offered by modern cooling systems, which results in low COP. This inefficiency is exacerbated by these systems' high initial costs from all the solar components they comprise, such as the collectors and storage tanks.

Dependably providing energy creates another challenge because it requires additional storage. Dependence on backup units further drives maintenance and operational costs. Additionally, given the small number of manufacturers producing these advanced technologies compared to traditional cooling systems, there is a lack of competition among vendors. These stakeholders are usually reluctant to switch to solar-powered systems because they doubt their longevity and operational dependability.

Incorporating new technology is the optimal way to tackle these challenges. Focusing on innovations that enhance collector efficiency and decrease overall system costs is equally important. Furthermore, supportive policies that promote research and development can help make solar cooling a more viable option for residential and commercial buildings [2], [13], and [14].

18.6.2 Economic and Financial Constraints

Their economic and financial aspects may hinder the adoption of these systems. The highly anticipated investment is a deterring factor compared to conventional cooling systems. The funding covers costs for solar collectors, storage tanks, and other components vital for the system. This, combined with the purchase of a cooling system, may further curb transitioning users. For example, depending on local conditions and system design, the price of solar-assisted cooling systems can be two to five times that of the standard system.

Furthermore, even though the operational costs associated with these are usually lower than those of their counterparts, the prolonged period for return on investment maintains a point of contention. Studies show that without financial aid, the return on investment period exceeds 11 years. This is greatly impacted by the area's energy prices and financing resources. Moreover, subsidization helps significantly reduce these timelines since, for example, government aid is shown to decrease the return on investment period by almost forty percent in some cases.

In addition, there are continuous maintenance expenses and the requirement for sophisticated energy storage systems that add to the financial costs of these systems. Although new technologies, such as smart grids, could improve efficiency, they come with added expenditure, which may further complicate the issue. Therefore, economic risks and mysteriously low long-term profits present a significant barrier for most companies seeking solar cooling technologies [7], [2], [13], and [18].

18.7 FUTURE TRENDS IN SOLAR-POWERED COOLING SYSTEMS

18.7.1 Innovations in Technology Development

Furthermore, considerable efforts are underway to develop innovative ideas for the performance and sustainability of solar-powered cooling technologies. New photovoltaic-thermal (PVT) collector designs that integrate heating and electricity generation improve overall system integration and give solid assurance for the energy supply to cooling during all seasons. Hybrid systems with thermal and electrical components were developed to improve energy performance and economical standby energy backup.

Researchers look at new materials and nano-structured elements to enhance the cooling system's heat transfer. Including intelligent control systems with artificial intelligence increases efficiency, as real-time information from users is utilized for instantaneous changes based on the current conditions. Studies of cryogenic devices and classical solar-powered absorption chillers show potential in the reduction of carbon emissions and the cost of operation.

A primary dynamic model using, for example, TRNSYS allows for configuring a particular system in question and evaluating its thermodynamic performance with thermodynamic and criterion-based selection systems. It is also possible to analyze the impact of various parameters on system performance, for instance, how the size and angle of collectors and seasons influence system efficiency. Innovations focus on multifunctional aspects to improve operational characteristics and economic efficacy based on the life cycle assessments of diverse solar cooling technologies [2] and [17].

18.7.2 Policy and Regulatory Impacts

Formulating policies and their regulatory frameworks affects the scope and intensity of applying solar-powered cooling systems. Governments play an essential role in utilizing renewable energy resources by offering financial support, such as tax credits, which solves the initial economic barriers related to solar devices. Supportive policies provide systems like absorption chillers with lower capital costs than is the norm to enhance competition with standard cooling methods.

In addition, fostering cooperation between public and private entities can enhance resource sharing among governmental agencies, utility companies, and private-sector innovative firms. These partnerships help expedite the accumulation of novel solar cooling technologies and open new innovative funding opportunities that stimulate market growth. Such new financing models include energy performance contracts.

Generating demand for solar cooling systems requires public education on their benefits and applicability. Outreach campaigns can help various stakeholders appreciate these technologies' energy efficiency and ecological value for broader application in homes and business premises. In addition, modernizing these systems by adding the smart grid can enhance the energy control approaches, whereby planning is geared towards achieving international sustainability objectives.

Last but not least, persistent alterations in policies related to construction and energy efficiency standards may enhance the penetration of solar cooling systems. For instance, laws stipulating a minimum share of renewable energy consumption in new constructions would alter market conditions [15], [17], [10], and [14]

18.8 CONCLUSION AND RECOMMENDATIONS FOR STAKEHOLDERS

18.8.1 Strategic Approaches for Adoption

Based on the individual's requirements, residential, industrial, or commercial structures built at the site must be assessed for their preexisting infrastructures. The local climate factors must be considered, as they would hinder the system's effectiveness while evaluating the site's solar potential. The most appropriate technology must also be selected with the prior points in mind. Small furnishings can be acquired readily, while those required for larger installations can be customized. This is effective in cost reduction and simplification of the installation process.

Investing in training programs for engineers and technicians specializing in solar cooling technologies is crucial to improving energy usage efficiency and installation quality. Setting new benchmarks and procedural guidelines will enable better streamlining. Cross-sector partnerships in regions like private corporations, government departments, and research bodies will help drive innovation while promoting best practice sharing.

Promotion of financial options like tax incentives can help lower the steep upfront costs of solar cooling systems. Exploring investments can also help acquire wider acceptance. Further marketing campaigns for people in the community to highlight the benefits of solar cooling will further help promote the need for investment.

Investigations into technology, including novel materials, sophisticated systems, predictive control mechanisms, and hybrid systems that amalgamate various renewable energy sources, can enhance system efficacy and adaptation to climate change. Working with legislators to create the necessary sustainable regulatory conditions will ensure that cooling initiatives are feasible and prioritized in development agendas [3], [17], [13], [10], [2], [4], and [20].

18.8.2 Investment Opportunities in Solar Cooling Technologies

Funding for solar cooling technologies is on the rise thanks to the economic and ecological advantages they provide. By using solar energy for cooling, companies in hotter areas can reduce their use of traditional electricity. Take, for example, businesses and some commercial buildings in hot regions; they can drastically lower energy expenditures by using solar-powered air conditioning systems. Political economy studies reveal that solar-powered absorption chillers can repay their initial capital within a window of six to twelve periods, depending on the energy conditions of the region and the harvester design offered.

For more effective project financing, investors ought to analyze government support options like tax reliefs, subsidies, and feed-in tariffs, which increase the profitability of solar cooling installations by reducing the investment payback time and increasing the NPV. In addition, investment in innovation increases the case for new systems that are cheaper to run and more productive.

Moreover, the combination of thermal energy storage (TES) and solar cooling is advantageous because it permits high reliability and cost efficiency for residential and commercial buildings by storing excess thermal energy collected during peak sunlight. With the increasing concern around climate change and higher issues for energy, there is a definite increase in interest surrounding solar-powered cooling systems, making them a sound investment in any sustainability strategy [7], [4], [17], and [20].

18.9 SUMMARY OF KEY FINDINGS

Alternatives to conventional cooling systems have progressively shifted toward solar energy due to its low environmental impact and minimal operational costs. Solar energy cooling solutions, such as absorption, adsorption, and photovoltaic systems, are incredibly advanced. Each category responds differently to climate conditions, making it challenging to determine their effectiveness for particular geographies.

Ample research has established the importance of thermal energy storage in providing a constant cooling output performance for solar energy systems. Although the high initial capital investment may limit their adoption, solar cooling systems can be practical tools for improving sustainability. The long-term operational cost benefits combined with environmental effectiveness make them attractive on the global stage. Moreover, hybrid systems are a more complex but effective way to integrate advanced cooling solutions. Solar cooling technologies can be applied worldwide in residential, commercial, and industrial sectors. Sustainability efforts are constantly on the rise, and the growing concern surrounding investment in renewable energy makes this technology promising. Understanding these systems' economic impacts and technical challenges will be vital for investors and stakeholders [2], [4], and [5].

REFERENCES

[1] K.R. Ullah, R. Saidur, H.W. Ping, R.K. Akikur and N.H. Shuvo . "A review of solar thermal refrigeration and cooling methods". *Renewable and Sustainable Energy Reviews, 24*, 499–513. 2013. [Online]. Available www.sciencedirect.com/science/article/abs/pii/S1364032113001834

[2] F.L. Rashid, M.A. Eleiwi, H.I. Mohammed, A. Ameen and S. Ahmad. "A review of using solar energy for cooling systems: Applications, challenges, and effects". Jan 2023. [Online]. Available: www.mdpi.com/1996-1073/16/24/8075

[3] G.G. Maidment and A. Paurine . "Solar cooling and refrigeration systems". 2012. [Online]. Available www.sciencedirect.com/science/article/pii/B9780080878720003140?via%3Dihub

[4] A. VR. "The benefits and challenges of solar-powered refrigeration and air conditioning". Mar 2023. [Online]. Available: https://arka360.com/ros/solar-powered-cooling-benefits-challenges/

[5] M. Mortadi and A. El Fadar. "Performance, economic and environmental assessment of solar cooling systems under various climates". Jan 2022. [Online]. Available: www.sciencedirect.com/science/article/abs/pii/S0196890421011699

[6] S.. Alrwashdeh and H. Ammari. "Life cycle cost analysis of two different refrigeration systems powered by solar energy". Jan 2019. [Online]. Available: www.sciencedirect.com/science/article/pii/S2214157X19303545

[7] A.A. Al-Ugla, M.A.I. El-Shaarawi, S.A.M. Said and A.M. Al-Qutub. "Techno-economic analysis of solar-assisted air-conditioning systems for commercial buildings in Saudi Arabia". Jan 2016. [Online]. Available: www.sciencedirect.com/science/article/abs/pii/S1364032115011260

[8] A.J. Alazazmeh and E.M. Mokheimer. "Review of solar cooling technologies". *Journal of Applied Mechnical Engineering, 4*(5), 180–194. 2015. [Online]. Available: www.researchgate.net/publication/285604262_Review_of_Solar_Cooling_Technologies

[9] "Solar thermal cooling". Sep 2015. [Online]. Available: www.solarheateurope.eu/wp-content/uploads/2017/06/Solar-Thermal-Cooling.pdf

[10] M.A. Ebaid, T.A. Mohamed and H. Safwat. "A case study on optimizing industrial air conditioning with thermal solar energy in Egypt". Jul 2024. [Online]. Available: https://pmc.ncbi.nlm.nih.gov/articles/PMC11336367/

[11] J. Mahan. "What is solar cooling? Solar cooling systems explained". Jan 2023. [Online]. Available: https://cc-techgroup.com/solar-cooling/

[12] U. Desideri, S. Proietti and P. Sdringola. "Solar-powered cooling systems: Technical and economic analysis on industrial refrigeration and air-conditioning applications". Jan 2009. [Online]. Available: https://www.sciencedirect.com/science/article/pii/S0306261909000087

[13] H. Henning, D. Jähnig, A. Thür, P. Olofsson, C. Balaras, D. Mugnier, C. Holter, M. Motta, M. Calderoni, U. Jakob, E. Wiemken and P. Schossig. "Solar Cooling Position Paper". Jan 2011. [Online]. Available: www.iea-shc.org/data/sites/1/publications/IEA-SHC-Solar-Cooling-Position-Paper.pdf

[14] Q. Al-Yasiri, M. Szabó and M. Arıcı. "A review on solar-powered cooling and air-conditioning systems for building applications". Jan 2022. [Online]. Available: www.sciencedirect.com/science/article/pii/S2352484722001731

[15] A. Al-Alili, M.D. Islam, I. Kubo, Y. Hwang and R. Radermacher. "Modeling of a solar powered absorption cycle for Abu Dhabi". Jan 2012. [Online]. Available: www.sciencedirect.com/science/article/abs/pii/S0306261910004976

[16] "Solar air conditioning". Sep 2024. [Online]. Available: https://en.wikipedia.org/wiki/Solar_air_conditioning
[17] A. Al-Nini, H.H. Ya, N. Al-Mahbashi and H. Hussin. "A review on green cooling: Exploring the benefits of sustainable energy-powered district cooling with thermal energy storage". Jan 2023. [Online]. Available: www.mdpi.com/2071-1050/15/6/5433
[18] M. Asif. Solar thermal technologies for energy transition. 2024. [Online]. www.sciencedirect.com/science/article/abs/pii/B9780323903868000784
[19] S.K.F. Fong, T.T. Chow, J.Z. Lin and A.L.S. Chan. "Application potential of solar thermal technologies for buildings in Hong Kong". Jun 2007. [Online]. Available: https://citeseerx.ist.psu.edu/document?repid=rep1&type=pdf&doi=bdb4327860ccdcc9ce231bb3908f9f20a893873d
[20] Deye. "A guide to solar powered heating and cooling systems". Dec 2024. [Online]. Available: https://deye.com/a-guide-to-solar-powered-heating-and-cooling-systems/
[21] Sayadi, S., Akander, J., Hayati, A., & Cehlin, M. (2021). Review on District Cooling and Its Application in Energy Systems. In *Urban Transition-Perspectives on Urban Systems and Environments*. IntechOpen. www.intechopen.com/chapters/78943
[22] W. Amjad and A. Munir. "Decentralized solar-powered cooling systems for fresh fruit and vegetables to reduce post-harvest losses in developing regions: a review". Jan 2023. [Online]. Available: https://academic.oup.com/ce/article/7/3/635/7174941
[23] K.R. Ullah, R. Saidur, H.W. Ping, R.K. Akikur and N.H. Shuvo. A review of solar thermal refrigeration and cooling methods. *Renewable and Sustainable Energy Reviews*, *24*, 499–513. 2013. [Online]. Available: www.sciencedirect.com/science/article/abs/pii/S1364032113001834
[24] F. Ali Fadil and A.M. Esham. "Absorption refrigeration cycle technology". *Intech Open*. 2023. doi: 10.5772/intechopen.1001585. [Online]. Available: www.intechopen.com/chapters/1132388
[25] A. Allouhi, T. Kousksou, A. Jamil, P. Bruel, Y. Mourad and Y. Zeraouli. "Solar driven cooling systems: An updated review". *Renewable and Sustainable Energy Reviews*, *44*, 159–181. 2015.
[26] D.B. Jani. "A review on progressive development in desiccant materials". *International Journal of Material Science Innovations*, *2*, 73–82. 2020. [Online]. Available: www.researchgate.net/publication/340279951_A_review_on_progressive_development_in_desiccant_materials
[27] G. Adenane, B. Mohammed and G. Mea. "Effect of operating temperatures and working pairs on performance of solar adsorption cooling system". *International Journal of Heat & Technology*, *39*(4). 2021. [Online]. Available: www.iieta.org/journals/ijht/paper/10.18280/ijht.390426
[28] R. Wang and T. Ge (Eds.). *Advances in solar heating and cooling*. Woodhead Publishing. 2016.
[29] D.B. Jani, M. Mishra and P.K. Sahoo. "Solid desiccant air conditioning–A state-of-the-art review". *Renewable and Sustainable Energy Reviews*, *60*, 1451–1469. 2016. [Online]. Available: www.sciencedirect.com/science/article/pii/S1364032116002665
[30] Y. Elaouzy and A. El Fadar. "Solar and photovoltaics energy utilization and sustainability". 2023. [Online]. Available: www.sciencedirect.com/science/article/pii/B9780323939409000025

Part VI

Specialized Flow Configurations

19 Immiscible Micropolar Flow in Vertical Channels

Jayaprada Evani, Suresh Babu Baluguri, Vasavi Cheruku, Srinivas Gosukonda, and Bumireddy Neelakanta Reddy

19.1 INTRODUCTION

The study of fluid dynamics, which is essential to many branches of science and engineering, includes the motion and behavior of liquids, gases, blood, and plasmas. Astronomy, mechanical and chemical engineering, and biological systems are just a few of the fields in which it finds extensive use as referred by Chamka *et al.* [1]. The study of two-phase flow, involving the interaction between two distinct phases, plays a crucial role in various fields of science, engineering, and technology. This phenomenon is fundamental to processes such as oil and gas production, gas-liquid flows in boilers, and aerosol deposition in spray treatments. Additionally, two-phase flows are observed in diverse natural and industrial scenarios, including inkjets, fog, clouds, groundwater movement, ocean waves, and pest control. Madaliev *et al.* [2] focuses on the numerical simulation of the layer mixing problem using a new two-fluid turbulence model. It aims to simulate the mixing of two flows with different velocities and analyze the resulting longitudinal velocity and turbulent stress profiles. The results are compared with experimental data to validate the accuracy of the new model. Zhang *et al.* [3] examines the flow characteristics and interfacial interactions of two immiscible liquid phases on a vertical wall. It explores how different flow patterns and geometries affect the stability and behavior of the interface. The findings provide insights into optimizing industrial processes involving immiscible fluids. Kumar *et al.* [4] investigate the magneto-convective transport of immiscible binary fluids in an inclined channel. They examine how magnetic fields and fluid properties affect flow behavior and heat transfer. The results provide insights into optimizing fluid dynamics and thermal management in engineering applications. Nagabhushana *et al.* [5] study heat transfer in a nanofluid-filled vertical composite porous channel positioned between two fluid layers. Employing the Darcy–Forchheimer equation and the Tiwari–Das model, they analyze the influence of parameters such as the solid volume fraction, Brinkman number, Grashof number and on heat transfer and flow. Their findings emphasize the potential for improving heat transfer by utilizing various nanoparticles. Bayareh *et al.* [6] provide an overview of non-Newtonian nanofluid flow in both macro- and micro-channels using two-phase schemes. They examine how different flow behaviors and thermal properties affect heat transfer and

DOI: 10.1201/9781003611820-25

fluid dynamics. The findings aim to enhance the understanding and optimization of nanofluid applications in various engineering systems.

Magnetohydrodynamics (MHD) research has gained significance due to its diverse applications, such as in accelerators, liquid metal cooling systems, geothermal energy extraction, compact chemical industry beds, MHD power generators, plasma research, and aerodynamic boundary layer control. Asadulla *et al.* [7] investigated hydromagnetic flow in channels with uneven walls. Hosseini *et al.* [8] used the method of differential-transform to analyze flow in non-uniform channels under the influence of a magnetic field. Similarly, Hatami and Ganji [9] applied the weighted residual method to explore the effect of a magnetic field on nanofluid flow in tubes with permeable walls. Erdinc and Yõlmaz [10] examined heat transfer in communicating channels with non-parallel walls, finding that the heat transfer rate in convergent and divergent channels was significantly higher compared to parallel-walled channels.

The theory of slip effects explains phenomena where fluids do not adhere to boundaries, leading to various practical applications such as cleaning heart valves and internal cavities, reducing friction, and conserving energy. Padma Devi *et al.* [11] examines the entropy generation in two-immiscible MHD flow of pulsating Casson fluid in a vertical porous space with slip effects. Using numerical methods, it analyzes how parameters like Hartmann number, Darcy number, and radiation parameter influence flow behavior and entropy generation. The results highlight the interplay between magnetic fields, fluid properties, and thermal effects on entropy production. Geraeilinezhad *et al.* [12] investigate the pseudoplastic fluid flow and heat transfer in a microchannel under the influence of velocity slip effects. They examine how slip conditions affect flow behavior and heat transfer rates. The results provide insights into optimizing thermal management in microscale applications. Sharma *et al.* [13] examine the unsteady flow of MHD Williamson fluid within a vertical porous channel, considering the effects of porous media and thermal radiation. They examine how the applied magnetic field, velocity slip, and temperature jump conditions affect the fluid's velocity and temperature profiles. The results provide insights into optimizing flow behavior and heat transfer in such systems. Das *et al.* [14] explore the combined effects of viscous dissipation, gravity modulation, and Joule heating on MHD Casson fluid flow over a vertically moving plate with second-order slip velocity. It examines how these factors influence the fluid's velocity and temperature profiles, providing insights into optimizing flow behavior and heat transfer in engineering applications. Kaladhar *et al.* [15] analyze the entropy generation in natural convection Jeffrey fluid flow through a vertical channel with an inclined magnetic field under Navier-slip conditions. It examines how parameters like magnetic field angle, Soret parameter, and thermal radiation affect entropy production. Vyas *et al.* [16] investigate the Darcy–Forchheimer thermos fluidics of Micropolar–Casson fluid adjacent to a non-isothermal vertical plate with velocity slip, using the homotopy analysis method and Cattaneo–Christov flux. It examines how these factors influence velocity, temperature distribution, and heat transfer rates. Hamza *et al.* [17] explore the time-dependent MHD flow of an exothermic Arrhenius fluid in a vertical channel with convective boundary conditions. It investigates how the fluid's temperature and reaction rates affect flow behavior and heat transfer. The findings aim to enhance understanding of MHD flows in chemical engineering applications. Andersson [18] and Wang [19] investigated viscous fluid flow with slip velocity effects over a stretched surface. Mahabaleshwar *et al.* [20,21] studied boundary layer flow with slip effects in Newtonian fluids. Ebaid *et al.* [22] reported that the slip parameter increases the temperatures of Cu-water and TiO_2-water nanofluids, while Bhattacharyya *et al.* [23] found that the velocity profile improves with an increase in the velocity slip parameter. Jusoh *et al.* [24], Jamaludin *et al.* [25], and Khashi'ie *et al.* [26] investigated the effect of velocity slip on three-dimensional stretching and shrinking flows of nanofluids. Numerical solutions in heat and mass transfer are crucial as they provide diverse values for different combinations of relevant parameters, helping predict conditions for specific outcomes.

This study investigates the numerical solution for the flow of non-mixing viscous fluids and micropolar within a vertical channel, emphasizing the impact of velocity slip. Micropolar fluids, characterized by their microstructure, display distinct flow behaviors compared to conventional viscous fluids. Velocity slip at the boundaries plays a crucial role in altering flow dynamics, influencing velocity profiles, pressure distribution, and energy efficiency. The governing equations, along with boundary conditions and fluid properties, are solved using the sixth-order Runge–Kutta method with Mathematica's ND Solve. The findings offer valuable insights into optimizing multiphase flow systems, particularly in industrial and medical applications where precise flow prediction is essential for effective design and performance.

19.1.1 Mathematical Formulation

Consider two isothermal parallel plates as shown in Figure 19.1, divided into two regions, positioned at $y = -h_1$ and $y = h_2$.

- **Region-1**: Extending from $-h_1 \le y \le 0$ and is filled up with micropolar fluid.
- **Region-2**: Spanning from $0 \le y \le h_2$ and is filled with viscous fluid.
- The two regions are maintained at temperatures T_1 and T_2, respectively, with the condition $T_1 > T_2$.
- The flow is assumed to be the following:
  - One-dimensional, steady, and fully developed.
  - Laminar, immiscible, incompressible, with constant transport properties.
  - Influenced by buoyancy effects.
- The isothermal walls have constant concentrations and satisfy the relation $C_1 > C_{2.}$

The governing equations are derived based on the above-outlined assumptions:

Region-1:

$$\frac{\partial U_1}{\partial Y} = 0 \tag{19.1}$$

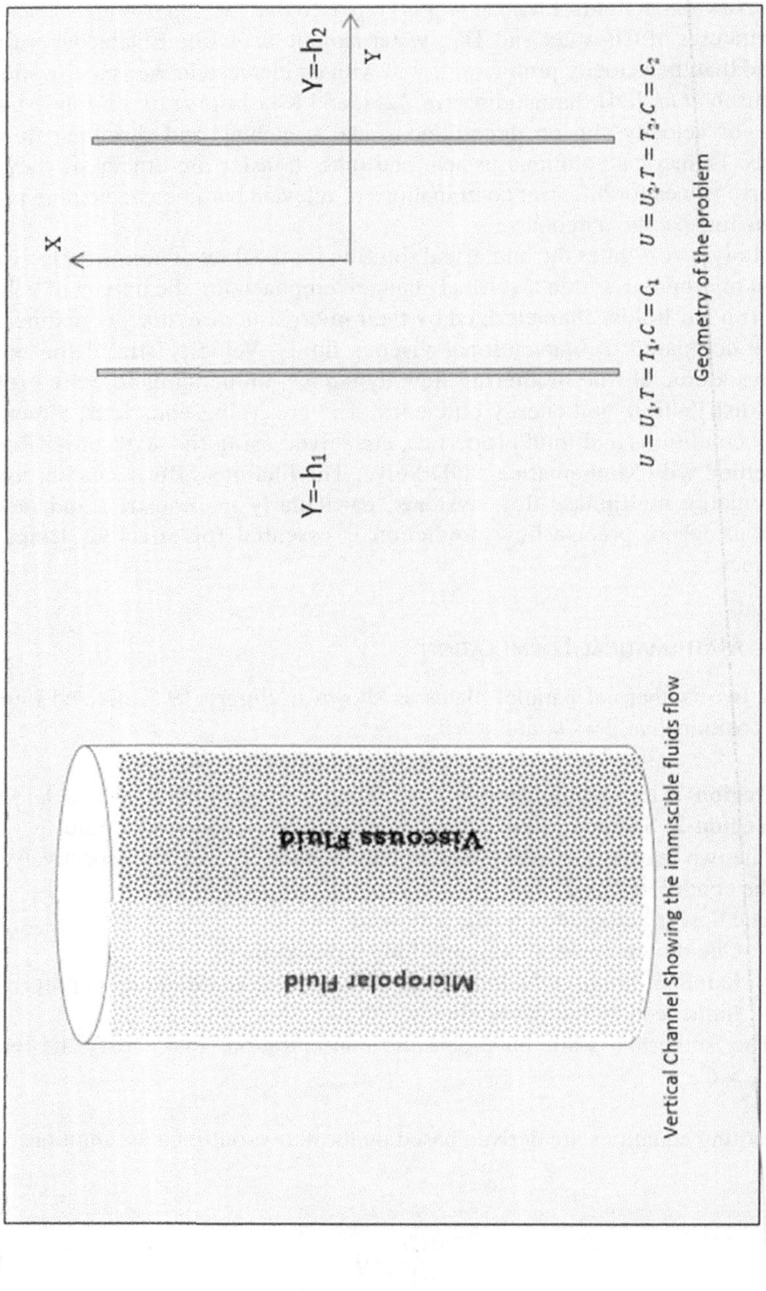

FIGURE 19.1 Schematic diagram.

$$\rho_1 = \rho_0\left[1-\beta_{1T}(T_1-T_0)-\beta_{1C}(C_1-C_0)\right] \tag{19.2}$$

$$\frac{\mu_1+K}{\rho_1}\frac{\partial^2 U_1}{\partial Y^2}+\frac{K}{\rho_1}\frac{\partial n}{\partial Y}+g\beta_{1T}(T_1-T_0)+g\beta_{1C}(C_1-C_0)-\frac{\sigma B_0{}^2 U_1}{\rho_1}=0 \tag{19.3}$$

$$\gamma\frac{\partial^2 n}{\partial Y^2}-K\left[2n+\frac{\partial U_1}{\partial Y}\right]=0 \tag{19.4}$$

where $\gamma = \left(\mu_1+\frac{K}{2}\right)j$

$$\frac{k_1}{\rho_1 C_p}\frac{\partial^2 T_1}{\partial Y^2}+\frac{1}{\rho_1 C_p}\left[\mu_1\left(\frac{\partial U_1}{\partial Y}\right)^2+\frac{\rho_1 D_1 K_{T1}}{C_{S1}}\frac{\partial^2 C_1}{\partial Y^2}\right]=0 \tag{19.5}$$

$$D_1\frac{\partial^2 C_1}{\partial Y^2}+\frac{D_1 K_{T1}}{T_M}\frac{\partial^2 T_1}{\partial Y^2}=0 \tag{19.6}$$

Region-2:

$$\frac{\partial U_2}{\partial Y}=0 \tag{19.7}$$

$$\rho_2 = \rho_0\left[1-\beta_{2T}(T_2-T_0)-\beta_{2C}(C_2-C_0)\right] \tag{19.8}$$

$$\frac{\mu_2}{\rho_2}\frac{\partial^2 U_2}{\partial Y^2}+g\beta_{2T}(T_2-T_0)+g\beta_{2C}(C_2-C_0)-\frac{\sigma B_0{}^2 U_2}{\rho_2}=0 \tag{19.9}$$

$$\frac{k_2}{\rho_2 C_p}\frac{\partial^2 T_2}{\partial Y^2}+\frac{1}{\rho_2 C_p}\left[\mu_2\left(\frac{\partial U_2}{\partial Y}\right)^2+\frac{\rho_2 D_2 K_{T2}}{C_{S2}}\frac{\partial^2 C_2}{\partial Y^2}\right]=0 \tag{19.10}$$

$$D_2\frac{\partial^2 C_2}{\partial Y^2}+\frac{D_2 K_{T2}}{T_M}\frac{\partial^2 T_2}{\partial Y^2}=0 \tag{19.11}$$

To address the aforementioned system of Eqs. (19.1) through (19.11), we applied the following boundary and interface conditions:

$U_1 = U_0 + S'\frac{\partial U_1}{\partial Y}$ at $Y=-h_1$, $U_2 = 0$ at $Y = h_2, U_1(0) = U_2(0)$

$T = T_1$ at $Y=-h_1, T = T_2,$ at $Y = h_2, T_1(0) = T_2(0),$

$C = C_1$ at $Y=-h_1 C = C_2$ at $Y = h_2, C_1(0) = C_2(0),$

$n = 0$ at $Y=-h_1, (\mu_1+K)\frac{\partial U_1}{\partial Y}+Kn = \mu_2\frac{\partial U_2}{\partial Y}\, at Y = 0,$

$\frac{\partial n}{\partial Y} = 0$ at $Y = 0, k_1\frac{\partial T_1}{\partial Y} = k_2\frac{\partial T_2}{\partial Y}\, at Y = 0, D_1\frac{\partial C_1}{\partial Y} = D_2\frac{\partial C_2}{\partial Y}\, at Y = 0.$

Equations (19.1) to (19.11) are transformed into a dimensionless form using the following non-dimensional variables:

$$y=\frac{Y}{h_1}\text{ (region-1)},y=\frac{Y}{h_2}\text{ (region-2)},u_1=\frac{U_1}{U_0},u_2=\frac{U_2}{U_0},\theta_1=\frac{T_1-T_0}{\Delta T},s=\frac{S'}{h_1}$$

$$C_S=\frac{C_{S1}}{C_{S2}},K_T=\frac{K_{T1}}{K_{T2}},D=\frac{D_1}{D_2},h=\frac{h_1}{h_2},m=\frac{\mu_1}{\mu_2},\alpha=\frac{k_1}{k_2},\ \rho=\frac{\rho_1}{\rho_2},b_1=\frac{\beta_{1T}}{\beta_{2T}},b_2=\frac{\beta_{1C}}{\beta_{2C}},\nu=\frac{\nu_1}{\nu_2}.$$

The governing equations are expressed in their dimensionless form as follows:

For Region-1:

$$\frac{\partial^2 N}{\partial y^2}-\frac{2K'}{2+K'}\left(2N+\frac{\partial u_1}{\partial y}\right)=0 \tag{19.12}$$

$$\left(1+K'\right)\frac{\partial^2 u_1}{\partial y^2}+K'\frac{\partial N}{\partial y}+\frac{Gr}{R}\theta_1+\frac{Gc}{R}c_1-Mu_1=0 \tag{19.13}$$

$$\frac{1}{\Pr R}\frac{\partial^2\theta_1}{\partial y^2}+\frac{Ec}{R}\left(\frac{\partial u_1}{\partial y}\right)^2+\frac{Du}{R}\frac{\partial^2 c_1}{\partial y^2}=0 \tag{19.14}$$

$$\frac{1}{ScR}\frac{\partial^2 c_1}{\partial y^2}+Sr\frac{\partial^2\theta_1}{\partial y^2}=0 \tag{19.15}$$

For Region-2:

$$\frac{\partial^2 u_2}{\partial y^2}+\frac{m}{b_1\rho h^2}\frac{Gr}{R}\theta_2+\frac{m}{b_2\rho h^2}\frac{Gc}{R}c_2-\frac{mM}{h^2}u_2=0 \tag{19.16}$$

$$\frac{\rho h}{\alpha}\frac{1}{\Pr R}\frac{\partial^2\theta_2}{\partial y^2}+\frac{\rho h}{m}\frac{Ec}{R}\left(\frac{\partial u_2}{\partial y}\right)^2+\frac{c_s h}{DK_T}\frac{D_u}{R}\frac{\partial^2 c_2}{\partial y^2}=0 \tag{19.17}$$

$$\frac{h}{D}\left(\frac{1}{ScR}\right)\frac{\partial^2 c_2}{\partial y^2}+\frac{h}{K_T D}Sr\frac{\partial^2\theta_2}{\partial y^2}=0 \tag{19.18}$$

The resultant dimensionless boundary and interface conditions are

$$u_1=1+s\frac{\partial u}{\partial y}\text{ at }y=-1\ u_2=0\text{ at }y=1,\ u_1\left(0\right)=u_2\left(0\right),$$

$$\theta_1=1\text{ at }y=-1,\theta_2=0\text{ at }y=1,\ \theta_1\left(0\right)=\theta_2\left(0\right),$$

$$c_1=1\text{ at }y=-1,c_2=0\text{ at }y=1,\ c_1\left(0\right)=c_2\left(0\right),$$

$$N=0\text{ at }y=-1,$$

$$\frac{\partial u_1}{\partial y} + \frac{K'}{1+K'}N = \frac{1}{mh(1+K')}\frac{\partial u_2}{\partial y} \text{ at } y = 0,$$

$$\frac{\partial N}{\partial y} = 0 aty = 0, \frac{\partial \theta_1}{\partial y} = \frac{1}{h\alpha}\frac{\partial \theta_2}{\partial y} \text{ at } y = 0,$$

$$\frac{\partial c_1}{\partial y} = \frac{1}{hD}\frac{\partial c_2}{\partial y} \text{ at } y = 0. \tag{19.19}$$

19.2 SOLUTION OF THE PROBLEM

The nonlinear system of equations is numerically solved using the sixth-order Runge–Kutta method. The computations are performed for various values of key governing parameters, including the thermal Grashof number (Gr), molecular Grashof number (Gc), magnetic field parameter (M), Reynolds number (R), material parameter (K'), Dufour number (Du), Schmidt number (Sc), Eckert number (Ec), and Soret number (Sr).The influence of these dimensionless parameters on velocity, temperature, and concentration is analyzed. The results are graphically represented to show the effects of each parameter.

19.3 RESULTS AND DISCUSSIONS

Figures 19.2 to 19.7 exhibit the effect of Grashof numbers (Gr) and molecular Grashoff number (Gc) on velocity and angular momentum profiles. It is observed that as Gr increases velocity and angular momentum increases substantially. The fluid velocity increases due to the enhancement of thermal and species buoyancy forces. The

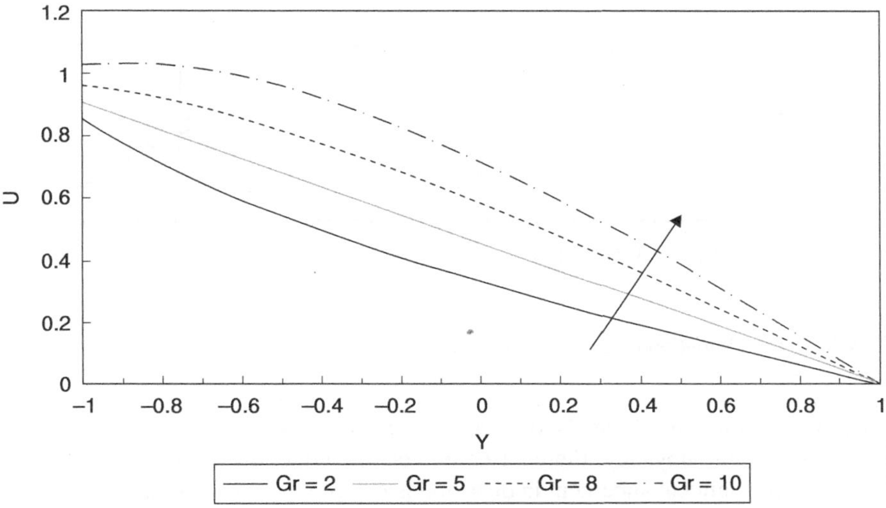

FIGURE 19.2 Variation of velocity with Gr.

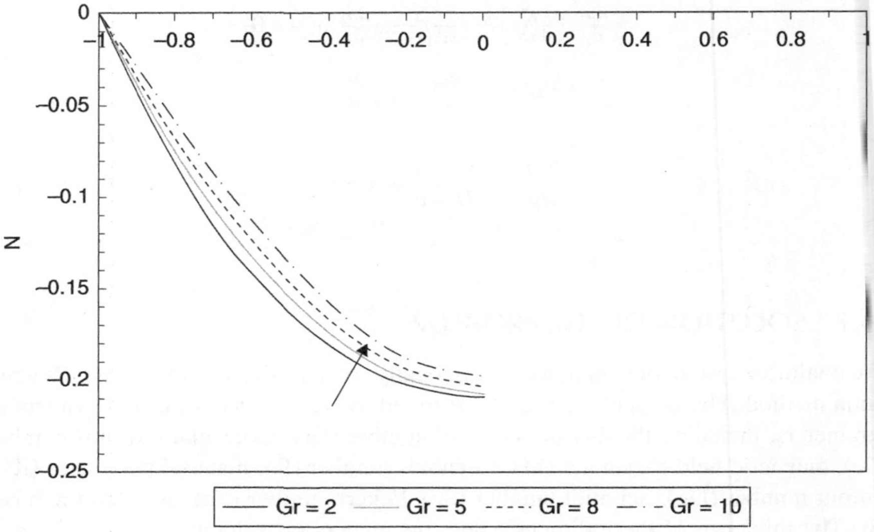

FIGURE 19.3 Variation of angular momentum with Gr.

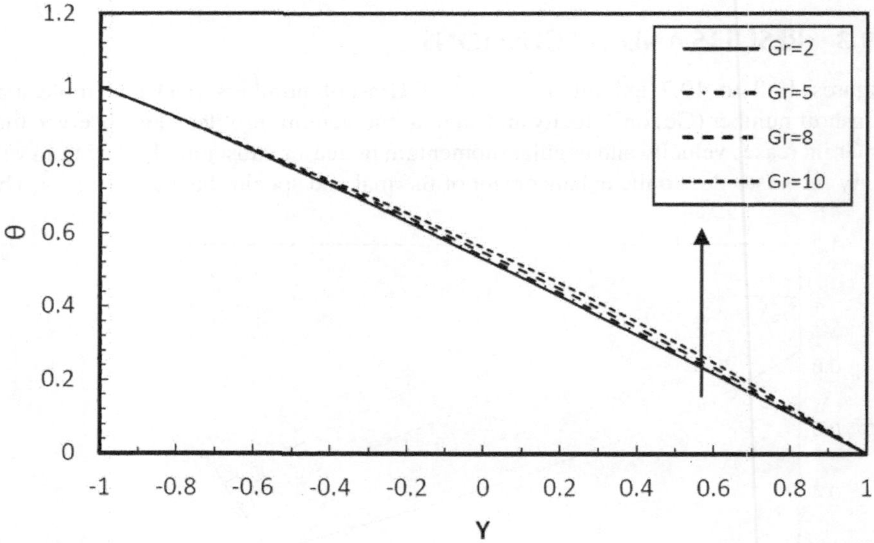

FIGURE 19.4 Variation of temperature with Gr.

effect of Grashof numbers leads to a rise in temperature as there is an increment in Gr duo to change in buoyancy. Figure 19.5 depicts the effect of Grashof numbers (Gr) on diffusion, slight enhancement in diffusion is recorded as Gr increases. Figure 19.6 describes the effect of Gc on velocity, due to the enhancement of bouncy forces,

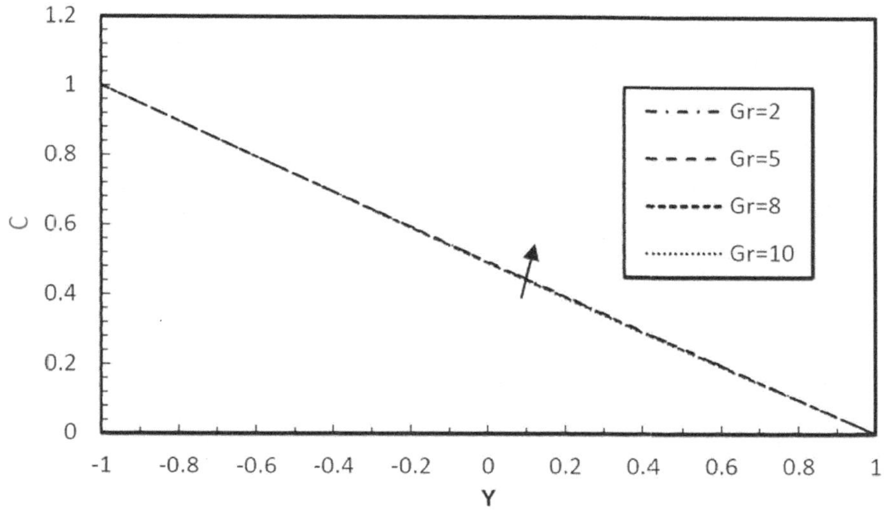

FIGURE 19.5 Variation of diffusion with Gr.

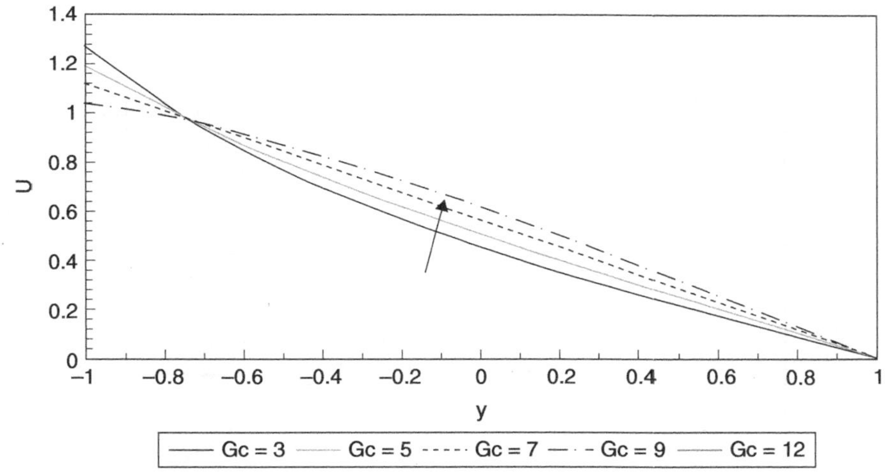

FIGURE 19.6 Variation of velocity with Gc.

velocity increases. Figure 19.7 reports the effect of Gc on diffusion. There is a slight increment in diffusion parameter with respect to molecular Grashof number (Gc).

Figures 19.8 and 19.9 illustrate the impact of the micropolar parameter K' on velocity and diffusion. It is observed that velocity decreases with increasing K' values. Notably, the application of a transverse magnetic field introduces a resistive Lorentz force, opposing the fluid motion and further reducing the velocity. As K' increases, the velocity profile continues to decline, while an opposite effect is observed in

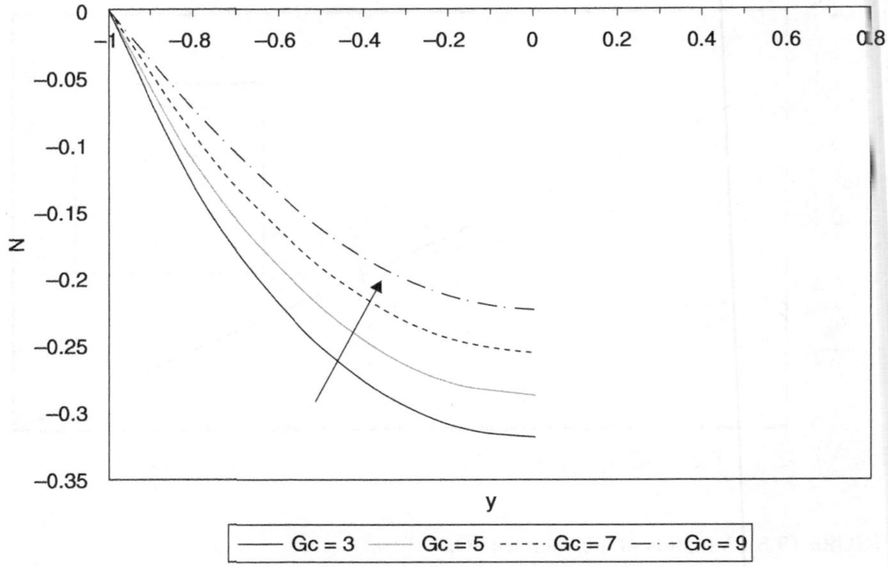

FIGURE 19.7 Variation of angular momentum with Gc.

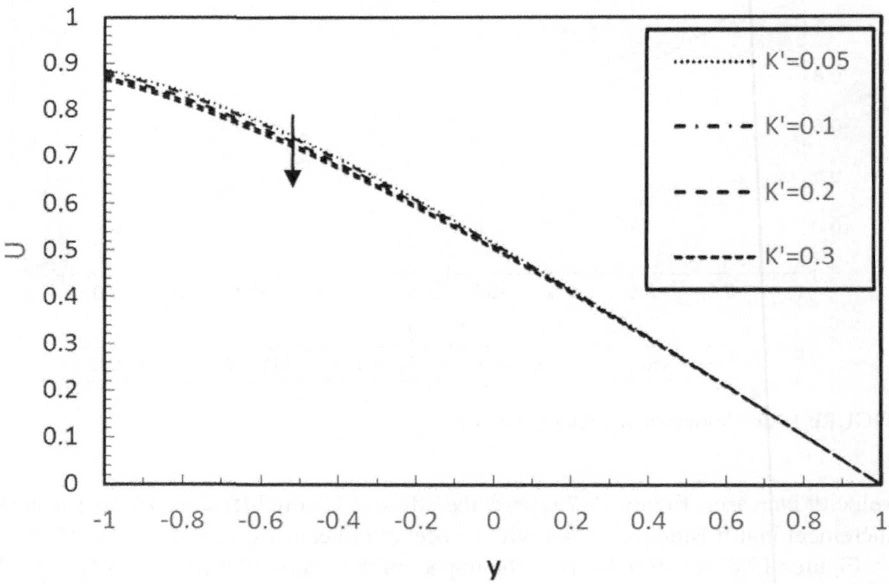

FIGURE 19.8 Variation of velocity with K'.

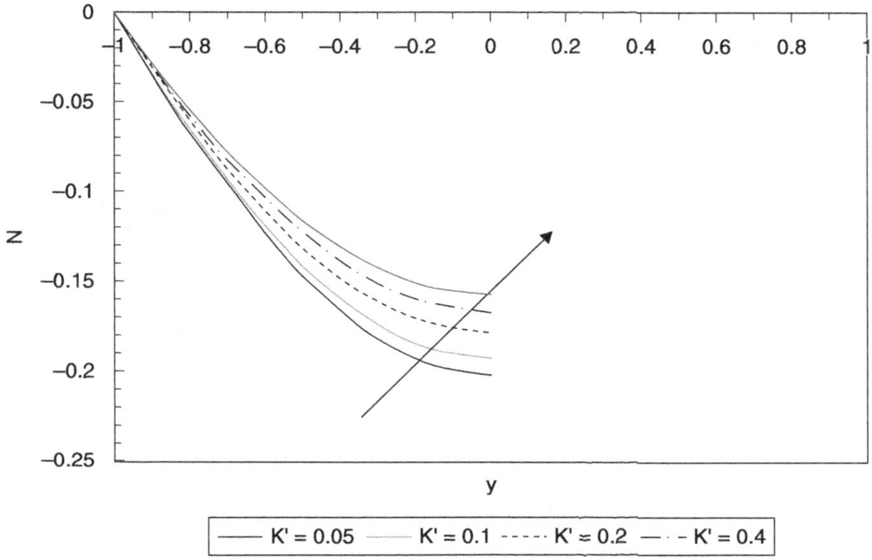

FIGURE 19.9 Variation of diffusion with K'.

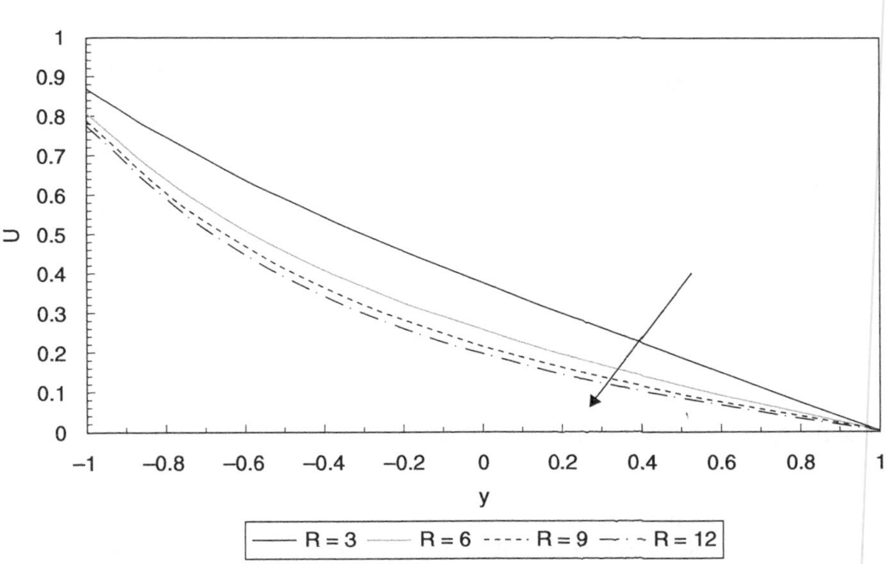

FIGURE 19.10 Variation of velocity with R.

diffusion. Figures 19.10 and 19.12 depict the influence of the radiation parameter RR on velocity and angular velocity. A growth in R results in a reduction in both parameters due to the dominance of conduction over radiation, reducing the momentum boundary layer's buoyancy thickness. Figure 19.11 illustrates the impact

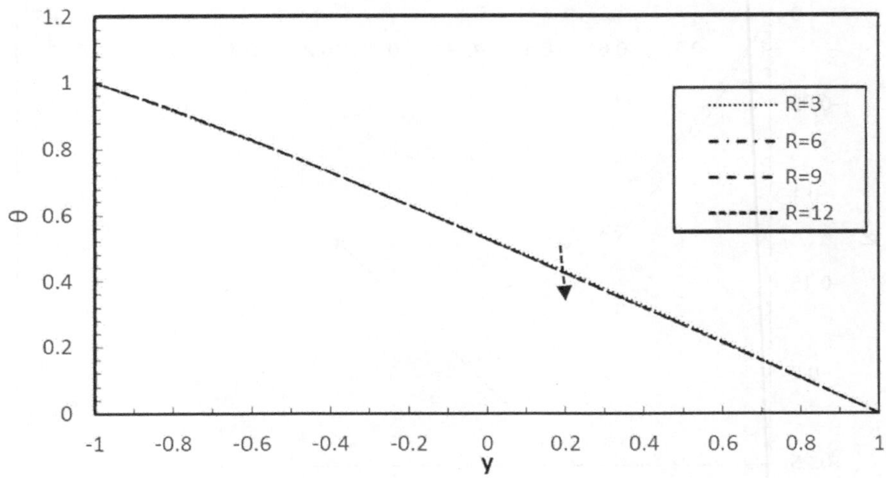

FIGURE 19.11 Variation of temperature with R.

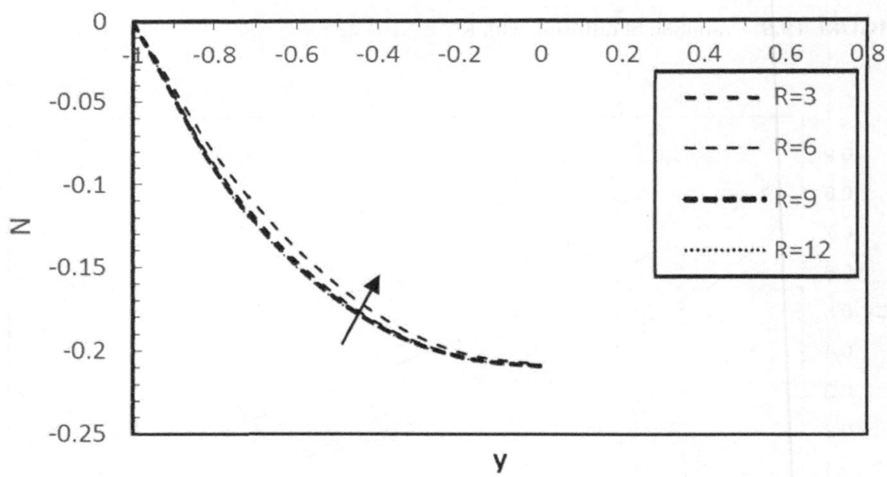

FIGURE 19.12 Variation of angular momentum with R.

of the Reynolds number (R) on temperature, showing an increase in temperature as (R) increases. Lastly, Figure 19.13 illustrates the effect of R on diffusion, showing a decrease in diffusion with increasing R.

The crux of the work is studying the impact of the slip parameter. Many possible values for this are considered when s (velocity slip) is varying as $s = -1.2, -0.9, -0.6, 0, 0.3, 0.6, 0.9, 1.2$ there is gradual decay in the velocity which is displayed in Figure 19.14. Figures 19.15 and 19.17 report the effect of slip on temperature and diffusion profiles at $s = -1.2, -0.9, -0.6, -0.3, 0, 0.3, 0.6, 0.9, 1.2$. As the slip

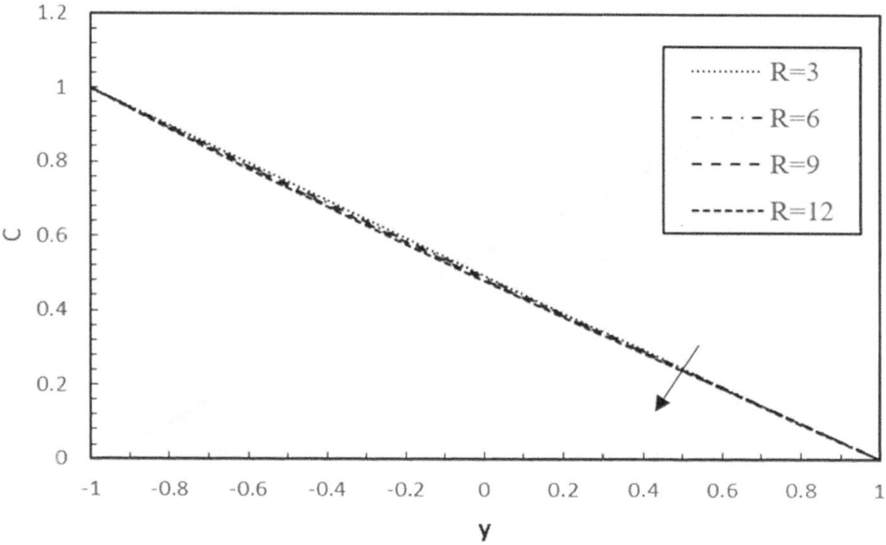

FIGURE 19.13 Variation of diffusion with R.

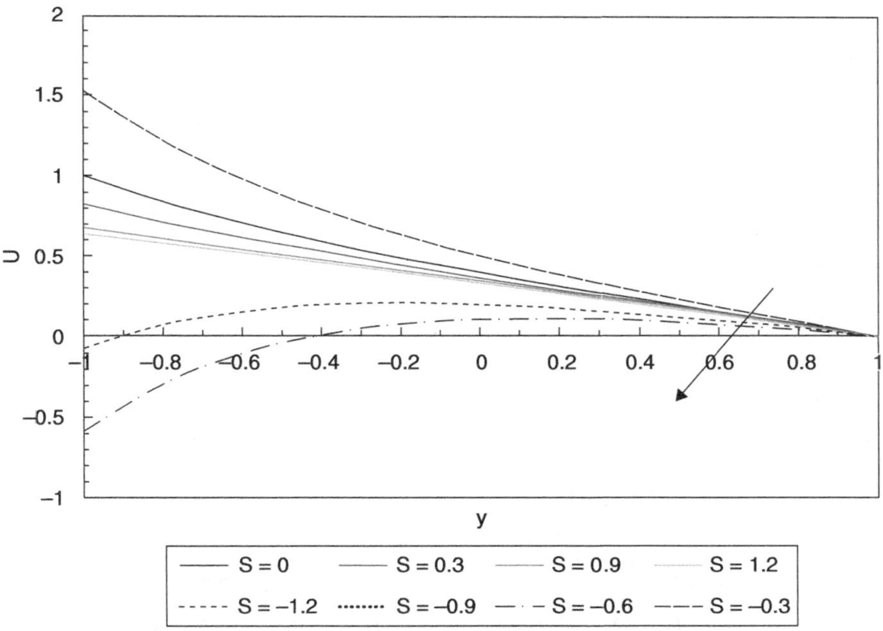

FIGURE 19.14 Variation of velocity with slip.

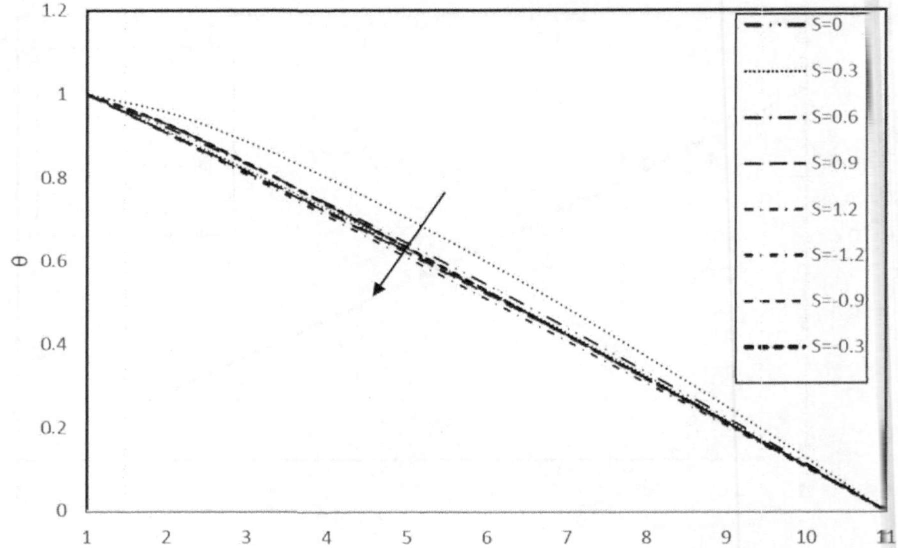

FIGURE 19.15 Variation of temperature with slip.

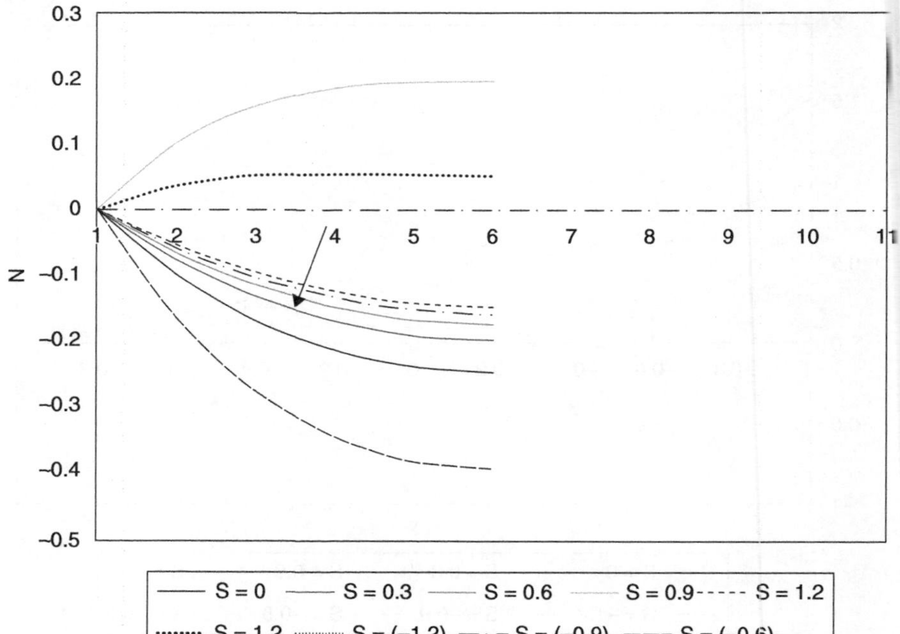

FIGURE 19.16 Variation of angular momentum with slip.

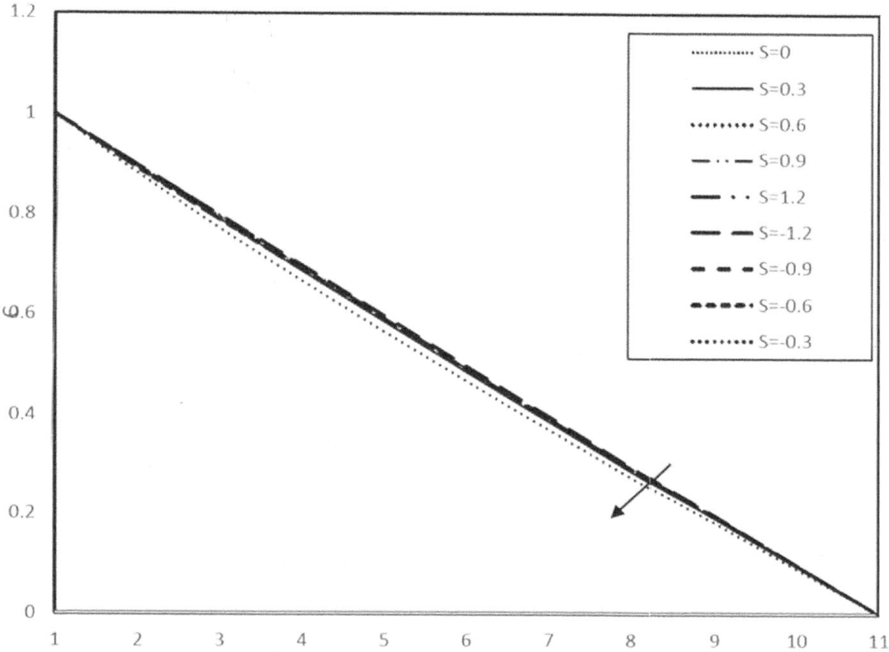

FIGURE 19.17 Variation of diffusion with slip.

parameter s is increasing, temperature and diffusion is decreasing. Figure 19.16 depicts the effect of slip parameter angular velocity, the effect is observed as similar velocity.

Figure 19.18 shows the calculated shear stress along with their responses to changes in governing parameters. The findings indicate that higher Gr and Gc values enhance buoyancy forces, leading to stronger shearing effects, especially at the left boundary. Increasing R reduces shear stress due to stronger inertial forces, which dominate over viscous forces, leading to lower shearing effects at both boundaries, with a more significant reduction at the left boundary. Higher M values dampen fluid motion due to magnetic damping, lowering shear stress, particularly at the left boundary where the effect is more pronounced. Increasing S reflects more compliant materials, reducing resistance to flow and thereby decreasing shear stress at both boundaries. Du and Sr show varied impacts on shear stress due to the coupling between thermal and mass diffusion processes, often resulting in higher shear stress at the right boundary. Changes in Sc have minor effects on shear stress, reflecting the balance between momentum diffusivity and mass diffusivity. Higher Ec values indicate more significant viscous dissipation, reducing shear stress at the left boundary while increasing it at the right boundary due to enhanced energy dissipation effects.

Figure 19.19 shows the calculated Nusselt number along with their responses to changes in governing parameters. Gr, Gc both show slight decreases in heat transfer at the left and right boundaries, indicating stable conditions. Increased R leads to

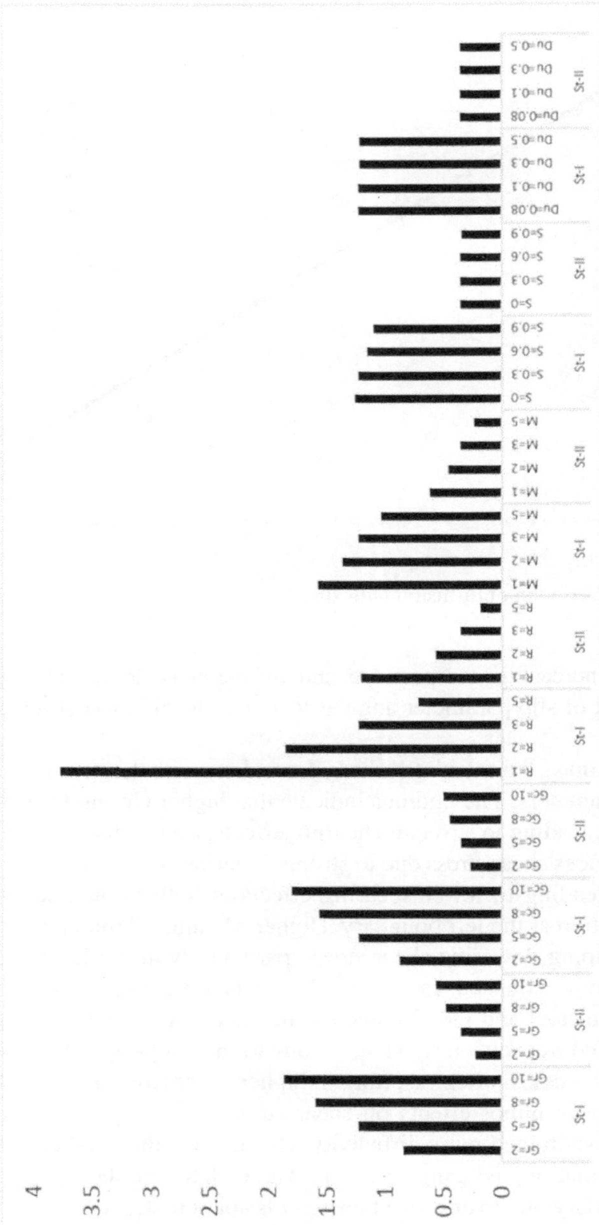

FIGURE 19.18 Shear stress values.

FIGURE 19.19 Nusselt number variations.

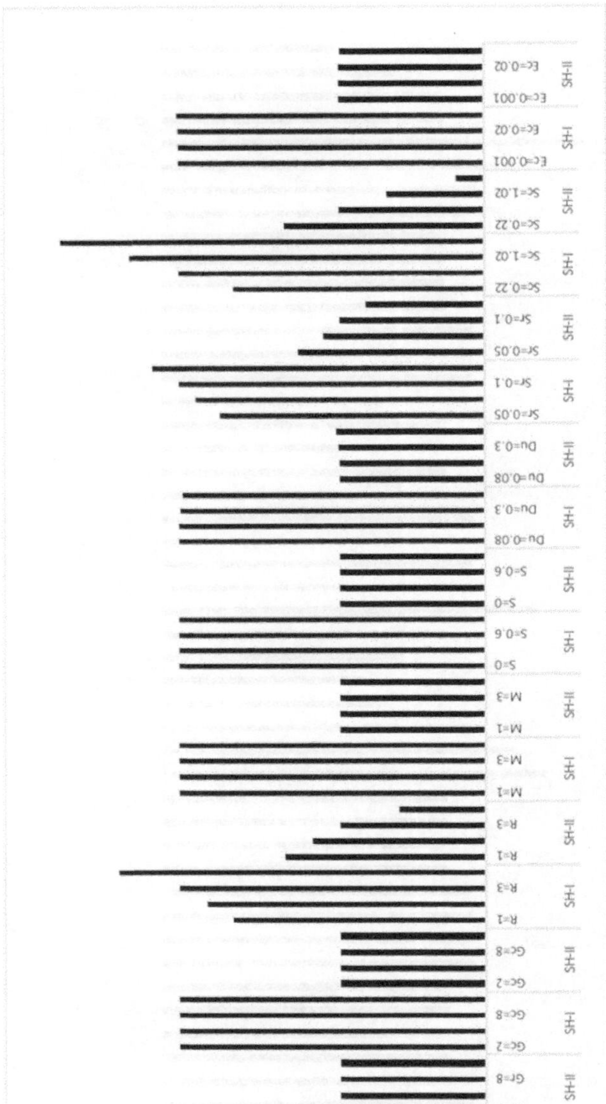

FIGURE 19.20 Sherwood variations.

significant reductions in heat transfer at both boundaries, with a more pronounced drop at the left boundary. Magnetic field (M) and slip parameter (S) exhibit minimal variations in Nusselt numbers, suggesting uniform effects on heat transfer. Dufour (Du) and Soret (Sr) numbers show varied impacts, enhancing heat transfer at one boundary while reducing it at the other due to thermal and mass diffusion. Schmidt number (Sc) shows constant Nusselt numbers, indicating negligible effects on heat transfer. Eckert number (Ec) higher values reduce heat transfer at the left boundary and enhance it at the right due to viscous dissipation effects.

Figure 19.20 shows the calculated Sherwood number, along with their responses to changes in governing parameters. The analysis of the Sherwood number reveals that at the left boundary, values are generally higher and stable compared to the right boundary, which shows lower and nearly constant values. The Reynolds number significantly impacts mass transfer, with SH-I increasing and SH-II decreasing, indicating stronger mass transfer at the left boundary. Parameters like the magnetic field and material compliance show minimal variations in Sherwood numbers at both boundaries, suggesting uniform mass transfer effects. The Dufour and Soret numbers influence the left and right boundaries differently, with SH-I decreasing slightly while SH-II increases. The Schmidt number displays constant values for both boundaries, reflecting negligible effects on mass transfer. Lastly, the Eckert number results in constant SH-I values and a slight decrease in SH-II, indicating minimal viscous dissipation effects. This summary highlights how different parameters uniquely affect mass transfer at the left and right boundaries.

19.4 CONCLUSIONS

The impact of all governing parameters was investigated and displayed graphically. Some of the salient findings are as follows:

- The thermal Grashof number (Gr) significantly increases velocity and angular momentum due to enhanced thermal buoyancy forces.
- An increase in Gr leads to a rise in temperature, reflecting its impact on buoyancy-driven thermal effects.
- Diffusion shows a slight enhancement with increasing Gr, indicating its influence on concentration gradients.
- The molecular Grashof number (Gc) enhances velocity by strengthening buoyancy forces acting on the fluid.
- A minor increase in diffusion is observed with higher Gc, highlighting its role in mass transport.
- The micropolar parameter (K′) reduces velocity due to the resistive Lorentz force introduced by the transverse magnetic field, while diffusion increases with higher K′.
- The Reynolds number (R) decreases velocity and angular velocity.
- Increasing the Reynolds number (R) results in a rise in temperature, reflecting the influence of inertial forces on thermal behavior.

- Higher Reynolds number (R) decreases diffusion, showing its inverse relationship with mass transport.
- Increased buoyancy forces (Gr and Gc) near the boundaries enhance shear stress, but it decreases with higher R, M, K′, Du, Sr, and Sc, while Ec shows the opposite effect.
- Heat transfer (Nusselt number) varies with parameters, increasing on one boundary and decreasing on the other, influenced significantly by R, Gr, Gc, Du, and Sr.
- Mass transfer (Sherwood number) generally increases near the boundary for Gr, Gc, R, Sr, Sc, and Ec but decreases for M, K', and Du at the right boundary.

REFERENCES

[1] A.J. Chamkha, S.S. Al-Rashidi, Analytical solutions for hydromagnetic natural convection flow of a particulate suspension through isoflux-isothermal channels in the presence of a heat source or sink, *Energy Convers. Manag.* 51 (4) (2010) 851–858.

[2] Z.M. Malikov, A.A. Mirzoev, M. Madaliev, Numerical simulation of the layer mixing problem based on a new two-fluid turbulence model, *AIP Conf. Proc.* 2612 (1) (2023) 030002.

[3] W. Zhang, Y. Chen, Y. Huang, Y. Ding, Q. Liao, M. Cheng, Study on flow characteristics and interfacial interaction of two immiscible liquid phases on vertical wall, *Chem. Eng. Sci.* 277 (2023) 118860.

[4] M. Kumar, P. K. Mondal, Magneto-convective transport of immiscible binary fluids in inclined channel, *J. Fluids Eng.* 145 (9) (2023) 091402.

[5] P. Nagabhushana, S. Ramprasad, C. Durga Prasad, H. Vasudev, C. Prakash, Numerical investigation on heat transfer of a nano-fluid saturated vertical composite porous channel packed between two fluid layers, *Int. J. Interact. Des. Manuf.* 18 (5) (2024) 2927–2941.

[6] M. Bayareh, An overview of non-Newtonian nanofluid flow in macro- and micro-channels using two-phase schemes, *Eng. Anal. Bound. Elem.* 148 (2023) 165–175.

[7] M. Asadullah, U. khan, N. Ahmed, R. Manzoor, S. Tauseef, MHD flow of a Jeffery fluid in converging and diverging channels, *Int. J. Mod. Math. Sci.* 6 (2) (2013) 92–106.

[8] R. Hosseini, S. Poozesh, S. Dinarvand, MHD flow of an incompressible viscous fluid through convergent or divergent channels in presence of a high magnetic field, *J. Appl. Math.* (2012) 157067.

[9] M. Hatami, D.D. Ganji, MHD nanofluid flow analysis in divergent and convergent Channels using WRM's and numerical method, *Int. J. Numer. Methods Heat Fluid Flow* 24 (5) (2014) 911–919.

[10] M.T. Erdinc, T. Yõlmaz, Numerical investigation of flow and heat transfer in communicating converging and diverging channels, *J. Therm. Eng.* 4 (5) (2018) 2318–2332.

[11] M. Padma Devi, S. Srinivas, K. Vajravelu, Entropy generation in two-immiscible MHD flow of pulsating Casson fluid in a vertical porous space with Slip effects, *J. Therm. Anal. Calorim.* 149 (14) (2024) 7449–7468.

[12] M. Geraeilinezhad, M. A. Akbari, M. R. Safaei, Numerical investigation of pseudoplastic fluid flow and heat transfer in a microchannel under velocity slip effect, *Eng. Anal. Bound. Elem.* 155 (2023) 501–510.

[13] K.R. Sharma, S. Jain, An unsteady MHD Williamson fluid flow in a vertical porous channel with porous media and thermal radiation, *Int. J. Adv. Eng. Sci. Appl. Math.* 16 (2024): 274–285.

[14] U.J. Das, I. Patgiri. Effects of viscous dissipation, gravity modulation and Joule heating on MHD Casson fluid flow over a vertically moving plate with second-order slip velocity, *Int. J. Ambient Energy* 45(1) (2024) 2331226.

[15] K. Kaladhar, R. Mahla, Entropy analysis of natural convection Jeffrey fluid flow through a vertical channel with an inclined magnetic field effect under Navier-slip conditions, *Eur. Phys. J. Plus* 138(8) (2023) 1–14.

[16] P. Vyas, Gajanand, Darcy-Forchhiemer thermofluidics of Micropolar-Casson fluid adjacent to a non-isothermal vertical plate with velocity slip using homotopy analysis method: Cattaneo-Christov flux, *Numer. Heat Transf. B: Fundam.* (2023) 1–24.

[17] M. M. Hamza, S. Abdulsalam, S. K. Ahmad, Time-dependent Magnetohydrodynamic (MHD) flow of an exothermic arrhenius fluid in a vertical channel with convective boundary condition, *Advances in Mathematical Physics* 2023(1) (2023) 7173925.

[18] H.I. Andersson, Slip flow past a stretching surface, *Acta Mech.* 158 (1–2) (2002) 121–125.

[19] C.Y. Wang, Flow due to a stretching boundary with partial slip—an exact solution of the Navier–Stokes equations, *Chem. Eng. Sci.* 57 (17) (2002) 3745–3747.

[20] U.S. Mahabaleshwar, K.R. Nagaraju, M.A. Sheremet, P.V. Kumar, G. Lorenzini, Effect of mass transfer and MHD induced Navier's slip flow due to a non-linear stretching sheet, *J. Eng. Thermophys.* 28 (4) (2019) 578–590.

[21] U.S. Mahabaleshwar, K.R. Nagaraju, M.A. Sheremet, D. Baleanu, E. Lorenzini, Mass transpiration on Newtonian flow over a porous stretching/shrinking sheet with slip, *Chinese J. Phys.* 63 (2020) 130–137.

[22] A. Ebaid, F. Al Mutairi, S.M. Khaled, Effect of ss slip boundary condition on the flow and heat transfer of Cu-water and TiO2-water nanofluids in the presence of a magnetic field, *Adv. Math. Phys.* 2014 (2014) 538950.

[23] K. Bhattacharyya, S. Mukhopadhyay, G.C. Layek GC, Slip effects on boundary layer stagnation-point flow and heat transfer towards a shrinking sheet, *Int. J. Heat Mass Transf.* 54 (1–3) (2011) 308–313.

[24] R. Jusoh, R. Nazar, I. Pop, Three-dimensional flow of a nanofluid over a permeable stretching/shrinking surface with velocity slip: A revised model, Phys. *Fluids* 30 (3) (2018) 033604.

[25] A. Jamaludin, R. Nazar, I. Pop, Three-dimensional magnetohydrodynamic mixed convection flow of nanofluids over a nonlinearly permeable stretching/ shrinking sheet with velocity and thermal slip, *Appl. Sci.* 8 (7) (2018) 1128.

[26] N.S. Khashi'ie, N.M. Arifin, R. Nazar, E.H. Hafidzuddin, N. Wahi, I. Pop, A stability analysis for magnetohydrodynamics stagnation point flow with zero nanoparticles flux condition and anisotropic slip, *Energies* 12 (7) (2019) 1268.

20 Micropolar Flow with Radiation and Magnetic Fields

Gadhigeppa Myacher Sachin,
Sudha Mahanthesh Sachhin,
Abhilash Mahantheshappa, Veeranna Bhoomika,
Ulavathi Shettar Mahabaleshwar, and
Basma Souayeh

20.1 INTRODUCTION

Micropolar fluids are a type of fluid that consists of microstructural elements capable of rotation, making them distinct from classical Newtonian fluids. These fluids are governed by the Eringen micropolar theory, which accounts for micro-rotational inertia and couple stresses. They have various applications in biomechanics (such as blood flow analysis), lubrication, polymer manufacturing, and chemical processing. Their importance lies in accurately modeling complex fluids that exhibit micro-rotational effects, which classical fluid mechanics cannot describe. Micropolar fluids help improve industrial processes, enhance biomedical research, and optimize engineering designs involving complex fluid interactions. Brewster [1] and Yücel [2] examined the radiation effect on fluid flow with its properties. Ishak *et al.* [3, 4] examined the stagnation point movement on an expanding sheet. Nazar *et al.* [5] studied the incompressible micropolar steady fluid flow over an expanded plane. Ahmadi [6] explored the couple stress and micro-rotation impact on micropolar fluid flow across a flat sheet. Cortell [7] examined the radiation impact on fluid flow over a flat sheet by using numerical approaches.

A stretching sheet plays a crucial role in industrial applications such as metal rolling, glass manufacturing, and aerodynamic studies. Its importance lies in enhancing product quality, improving efficiency in material processing, and understanding thermal behavior in engineering systems. Analyzing stretching sheet dynamics also aids in designing better manufacturing techniques and industrial equipment. Chen [8] examined the mixed convective fluid flow over a continuously expanding surface. The authors of [9, 10] studied the heat transfer impact on fluid flow across an expanding sheet. Sachhin *et al.* [11, 12] explored the velocity slip and heat transfer effect on nanofluid movement across an expanding/shrinking sheet.

A magnetic field influences fluid flow by exerting electromagnetic forces, altering velocity, pressure, and heat transfer properties. This interaction, known as

DOI: 10.1201/9781003611820-26

magnetohydrodynamics (MHD), is widely applied in plasma physics, cooling systems in nuclear reactors, electromagnetic pumps, and liquid metal processing. It is crucial in astrophysical phenomena, such as solar winds and planetary magnetospheres. The application of magnetic fields in fluid flow helps control turbulence, enhance heat dissipation, and improve industrial processes involving conductive fluids. Its importance lies in optimizing energy systems, improving material processing, and advancing research in space and fusion energy technologies. Understanding MHD enhances efficiency in various engineering and scientific applications. Maity *et al.* [13] explored the Casson fluid flow over a porous unsteady expanding sheet with the influence of magnetic field. Ilango and Lakshminarayana [14] examined the viscous dissipation and Soret impact on convective magnetic fluid flow across the stretching surface. Seddeek [15] examined the heat transfer effect on visco-elastic fluid movement through permeable media with magnetic field. Liu [16] examined the heat flux impact on second grade steady fluid flow over an electrically conducting surface. Subhas and Veena [17] examined the temperature boundary conditions' impact on visco-elastic fluid flow across impermeable expanding surface. The authors of [18–24] examined the thermal energy effect on fluid flow across porous media.

20.2 MATHEMATICAL ANALYSIS

2D micropolar fluid flow across stretching/shrinking sheet, radiation and heat source impacts are considered in the energy profile, u and v are the momentum terms along the x and y axis, magnetic field is applied to the fluid flow, Figure 20.1 represents the shimatic diagram, PDEs for the considered analysis are stated as below [3, 4, 11, 12]:

$$\frac{\partial u}{\partial x}+\frac{\partial v}{\partial y}=0, \tag{20.1}$$

$$u\frac{\partial u}{\partial x}+v\frac{\partial u}{\partial y}=\left(\nu+\frac{k}{\rho}\right)\frac{\partial^2 u}{\partial y^2}+\frac{k}{\rho}\frac{\partial N}{\partial y}-\frac{\sigma B_0^2}{\rho}u-\frac{\nu}{L_1}u, \tag{20.2}$$

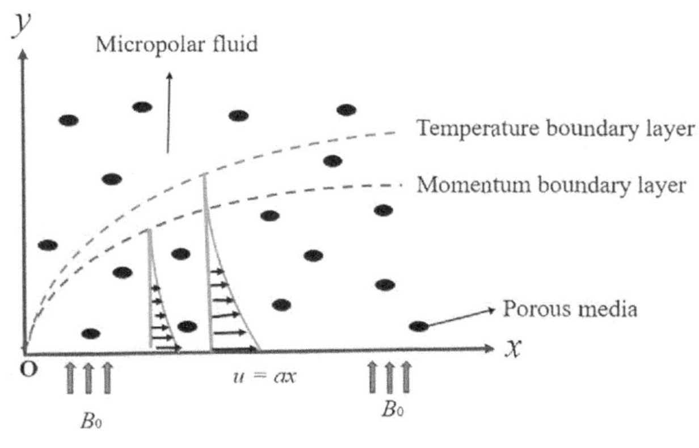

FIGURE 20.1 Schematic diagram.

$$u\frac{\partial N}{\partial x}+v\frac{\partial N}{\partial y}=\frac{\gamma}{\rho j}\frac{\partial^2 N}{\partial y^2}-\frac{k}{\rho j}\left(2N+\frac{\partial u}{\partial y}\right), \tag{20.3}$$

$$u\frac{\partial T}{\partial x}+v\frac{\partial T}{\partial y}=\frac{\kappa}{\rho Cp}\frac{\partial^2 T}{\partial y^2}-\frac{1}{\rho Cp}\frac{\partial q_r}{\partial y}+\frac{Q_0}{\rho Cp}\left(T-T_\infty\right), \tag{20.4}$$

Spin gradient viscosity γ can be written as [3, 4]

$$\gamma=\left(\mu+\frac{k}{2}\right)j=\mu\left(1+\frac{K}{2}\right)j, \tag{20.5}$$

here $K=\frac{k}{\mu}$ is material parameter, $j=\frac{\nu}{a}$ is reference length.
B. Cs for the fluid flow [3, 4]

$$\begin{aligned} &u=U_w=ax, \quad v=0, \quad T=T_w, \quad N=-n\frac{\partial u}{\partial y}, \quad \text{at} \quad y=0, \\ &u\to 0, \qquad\qquad N\to 0, \quad T\to T_\infty \qquad \text{as} \quad y\to\infty. \end{aligned} \tag{20.6}$$

T_∞ is ambient fluid temperature, n is a boundary layer term with $0\leq n\leq 1$, radiative heat flux defined as [11, 12]

$$q_r=-\frac{4\sigma*}{3k*}\frac{\partial T^4}{\partial y}, \tag{20.7}$$

here $\sigma*$ is the Stefan-Boltzmann term and $k*$ is the mean absorption term, expanding T^4 in Taylor's series and neglecting upper-order terms we obtain

$$T^4\approx 4T_\infty^3 T-3T_\infty^4. \tag{20.8}$$

Similarity transformations are as follows [3, 4, 11, 12]:

$$\begin{aligned} &\psi=\left(ax^2\nu\right)^{0.5}f(\eta), \qquad \eta=\left(\frac{a}{\nu}\right)^{0.5}y, \\ &N=ax\left(\frac{a}{\nu}\right)^{0.5}g(\eta), \qquad \theta(\eta)=\frac{T-T_\infty}{T_w-T_\infty}. \end{aligned} \tag{20.9}$$

By using similarity variables governing equations transformed to the following ODEs:

$$(1+K)f_{\eta\eta\eta}+ff_{\eta\eta}-f_\eta^2+Kg_\eta-Mf_\eta-Lf_\eta=0, \tag{20.10}$$

$$\left(1+\frac{K}{2}\right)g_{\eta\eta}+fg_\eta-f_\eta h-K\left(2g+f_{\eta\eta}\right)=0, \tag{20.11}$$

$$\frac{1}{\Pr}(1+Nr)\,\theta_{\eta\eta}+f\theta_{\eta}+Ni\theta=0, \tag{20.12}$$

here $Nr=\frac{16\sigma*T_{\infty}^{3}}{3k*\kappa}$ is thermal radiation, $M=\frac{\sigma B_0^2}{\rho a}$ is the magnetic field, $Ni=\frac{Q_0}{\rho Cpa}$ is heat source/sink parameter, $L=\frac{\nu}{L_1 a}$ porosity parameter,

dimensionless boundary conditions defined as

$$\begin{aligned} &f(0)=0, \quad f_{\eta}(0)=1, \quad g(0)=-nf_{\eta\eta}(0), \quad \theta(0)=1, \quad \text{at } \eta\to 0,\\ &f_{\eta}(\eta)\to 0, \quad g(\eta)\to 0, \quad \theta(\eta)\to 0, \quad \text{as } \eta\to\infty. \end{aligned} \tag{20.13}$$

20.3 NUMERICAL METHOD WITH SOLUTION

Converted higher order to a linear differential equation.

$$y_1=f,\ y_2=f',\ y_3=f'',\ y_4=g,\ y_5=g',\ y_6=\theta,\ y_7=\theta'. \tag{20.14}$$

By using (20.10)–(20.12) we get

$$\begin{aligned} &y_2^1=y_3,\ y_3^1=-\left(\left(\left(\frac{1}{1+K}\right)\left(y_1y_3-y_2^2+y_5K-My_2-Ly_2\right)\right)\right),\\ &y_4^1=--\left(\left(\left(\frac{1}{1+\frac{K}{2}}\right)\left(y_1y_5+y_2y_4-K\left(2y_4+y_3\right)\right)\right)\right),\\ &y_7^1=-\left(\left(\frac{1}{(1+Nr)\Pr}\right)\left(y_1y_7+Niy_6\right)\right), \end{aligned} \tag{20.15}$$

20.4 RESULTS AND DISCUSSIONS

The micropolar fluid movement across the stretching surface with the influence of MHD, heat source, and radiation effects was considered, the governing equations are solved to PDEs via similarity variables, Table 20.1 represents the comparision of Nusselt number values with the previous publications, and several physical parameters such as MHD, porosity term, heat source and radiation parameters were explained through graphs in the below section:

Figure 20.2 represents the micropolar parameter impact on axial and transverse velocity profile, increasing the micropolar parameter enhances the momentum, physically, the micropolar term significantly influences both the angular and transverse momentum. An increase in this parameter enhances the micro-rotation effects, leading to higher angular velocity while altering the transverse velocity distribution.

TABLE 20.1
Comparing the Choices of $-\theta'(0)$ for Different Choices of Nr,Pr,K While Taking $n = 0.5, Ni = 0, M = 0, L = 0$

Nr	Pr	K	Grubka [9]	Ali [10]	Chen [8]	Ishak [3]	Present Results
0	0	0.01	0.0099		0.00991	0.0099	0.0099
	0.72		0.4631	0.4617	0.46315	0.4631	0.4632
	1.0		0.5820	0.5801	0.58199	0.5820	0.5834
	3.0		1.1652	1.1599	1.16523	1.1652	1.1611
	10		2.3080	2.2960	2.30796	2.3080	2.3013
	100		7.7657		7.76536	7.7657	7.7611
1	1					0.3547	0.35473
		1				0.3893	0.3855
		2				0.4115	0.41000
		0				0.2588	0.25768
		1				0.2895	0.28999
		2				0.3099	0.31001

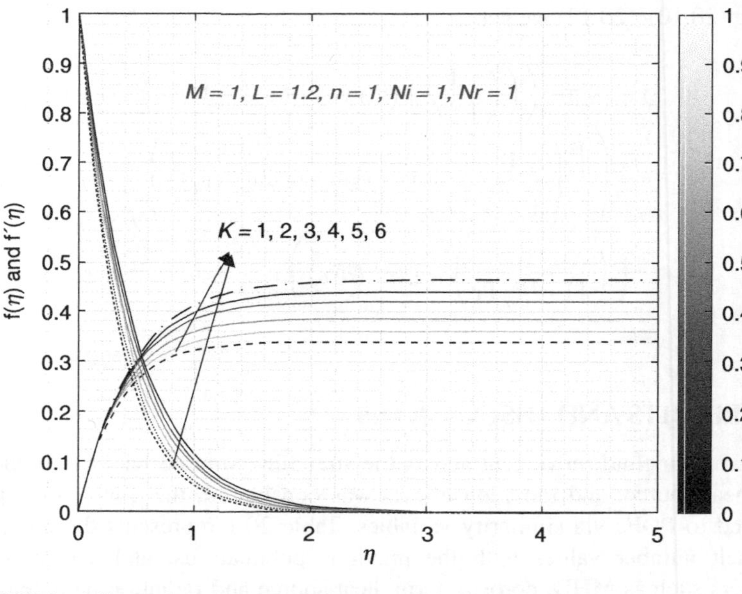

FIGURE 20.2 Micropolar parameter effect on velocity profile.

This impact is particularly noticeable in fluids with microstructural characteristics, where rotational motion plays a crucial role in momentum transfer. Consequently, the micropolar parameter must be carefully considered in flow analysis to accurately predict velocity behavior.

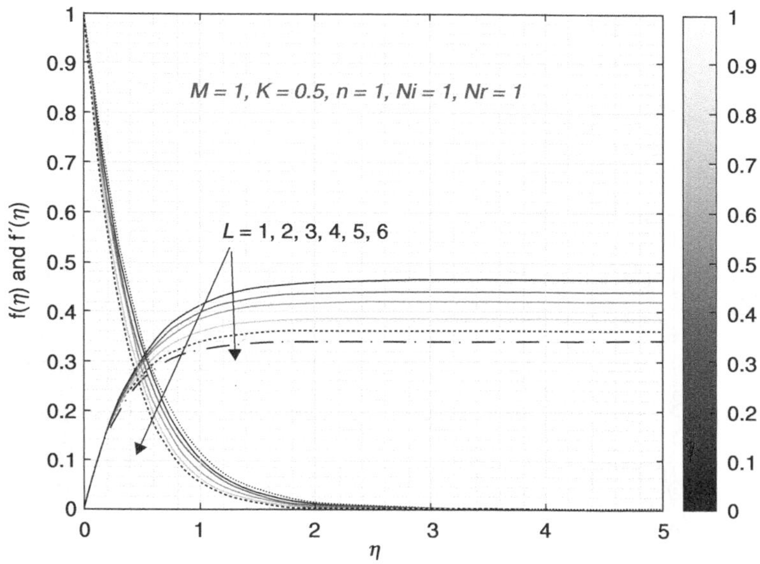

FIGURE 20.3 Porosity parameter effect on velocity profile.

Figure 20.3 represents the porosity term impact on axial and transverse momentum profile, increasing the porosity term decays the momentum, physically, the porosity term affects both the angular and transverse momentum by altering resistance within the medium. Higher porosity reduces flow resistance, leading to an increase in transverse velocity, while also influencing the rotational behavior of fluid particles, thereby impacting angular velocity. This parameter is particularly important in studying fluid dynamics in porous structures, where both linear and rotational motion are significantly modified.

Figure 20.4 represents the magnetic parameter impact on axial and transverse velocity profile, increasing the magnetic parameter decays momentum, physically, the magnetic term affects both angular and transverse velocity by introducing a resistive force in the momentum. As the magnetic field strength increases, the transverse velocity decreases due to the damping effect of the Lorentz force. Meanwhile, the angular velocity is also influenced, as the fluid's rotational motion responds to the magnetic interactions. This effect is particularly important in magnetohydrodynamic applications.

Figure 20.5 represents the boundary parameter impact on axial and transverse velocity profile, increasing the boundary term decays the momentum, physically, the boundary term influences fluid flow by modifying both angular and transverse velocity near the surface. An increase in this parameter alters velocity gradients, impacting the transverse velocity distribution. It also affects angular velocity by influencing the rotation of fluid particles close to the boundary. This plays a key role in flow stability and control in various applications.

Figure 20.6 portrays the porosity parameter impact on angular velocity profile, increasing the porosity term decays the momentum. Physically, the porosity term

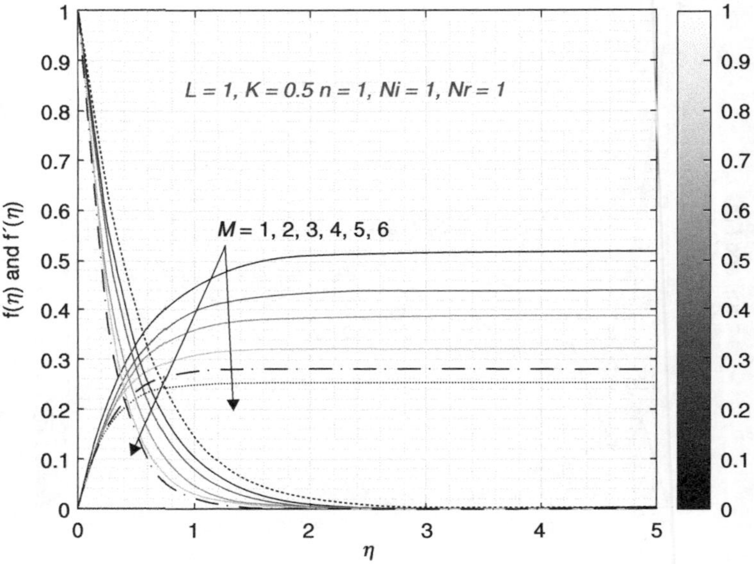

FIGURE 20.4 Magnetic term effect on momentum.

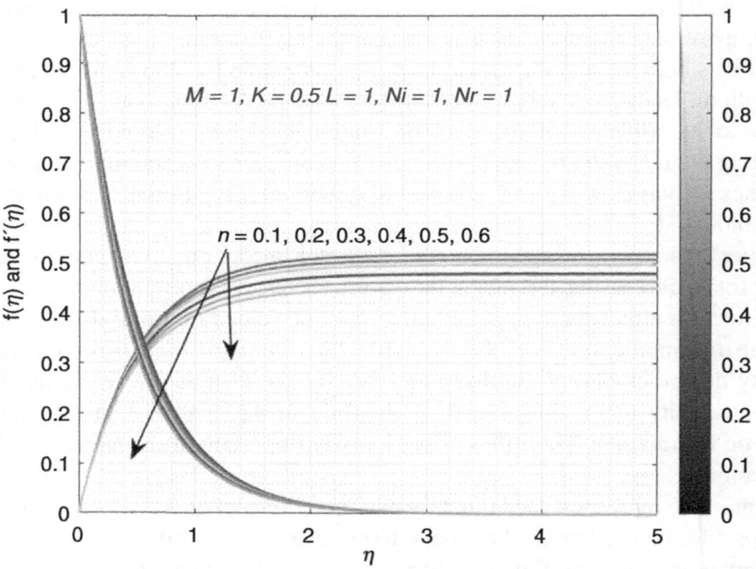

FIGURE 20.5 *n* parameter effect on velocity profile.

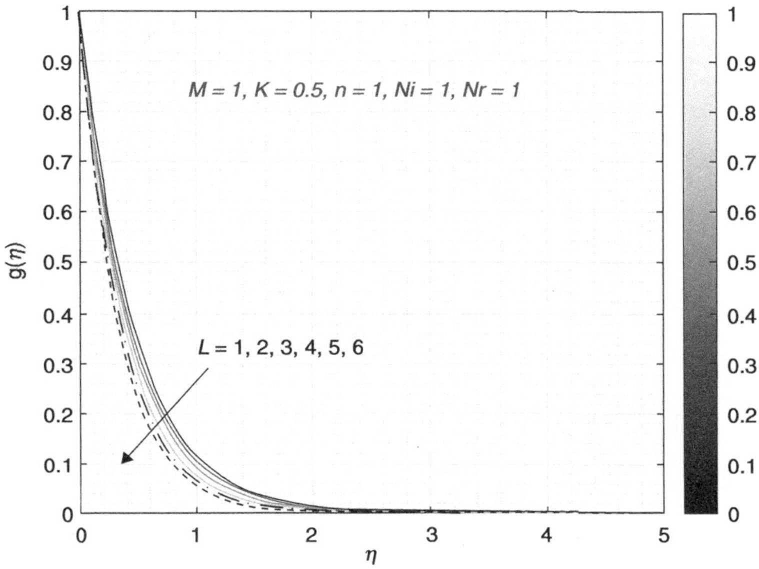

FIGURE 20.6 Porosity parameter effect on angular momentum profile.

plays a crucial role in determining the angular velocity of fluid flow by influencing resistance within the medium. As porosity increases, reduced obstruction allows fluid particles to exhibit greater rotational motion, leading to changes in angular velocity. This effect is significant in porous media flows, where both linear and rotational dynamics interact.

Figure 20.7 represents the micropolar term impact on angular momentum profile, rising the micropolar parameter rises the momentum. Physically, the micropolar term influences the angular momentum by enhancing the rotational behavior of fluid particles. As this parameter increases, it promotes stronger micro-rotations, which leads to a rise in angular velocity.

Figure 20.8 represents the magnetic term impact on angular momentum, increasing the magnetic parameter decays the momentum. Physically, the magnetic term affects the angular momentum by introducing a resistive force through the Lorentz effect. As the magnetic field strength increases, it dampens the rotational motion of fluid particles, reducing the angular velocity. This interaction is significant in magnetohydrodynamic flows, where the magnetic field influences both the linear and rotational components of the fluid motion.

Figure 20.9 represents the micropolar parameter impact on temperature profile, increasing the magnetic term decays the fluid temperature. Physically, the micropolar term influences the temperature distribution in fluid by affecting the fluid's microstructure and its thermal conductivity. As this parameter increases, it can lead to variations in heat transfer due to enhanced rotational motion of fluid particles. This change can either promote or hinder temperature diffusion, depending on the flow

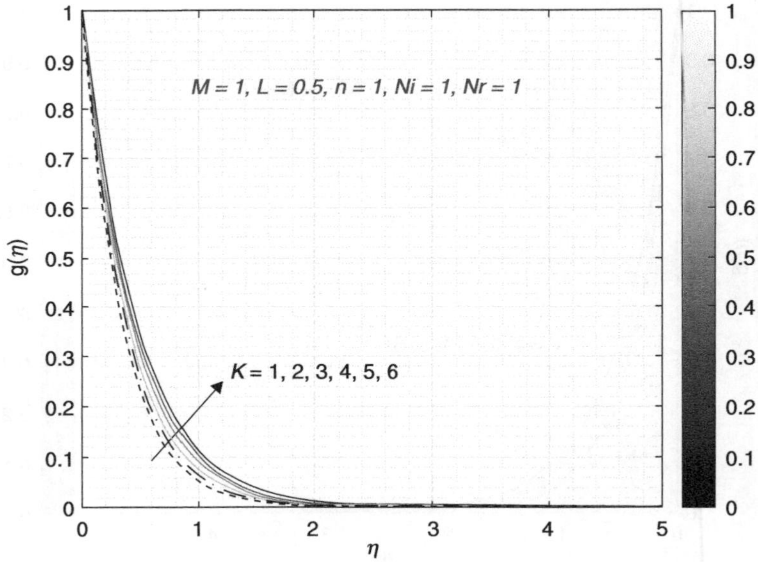

FIGURE 20.7 Micropolar parameter effect on angular velocity profile.

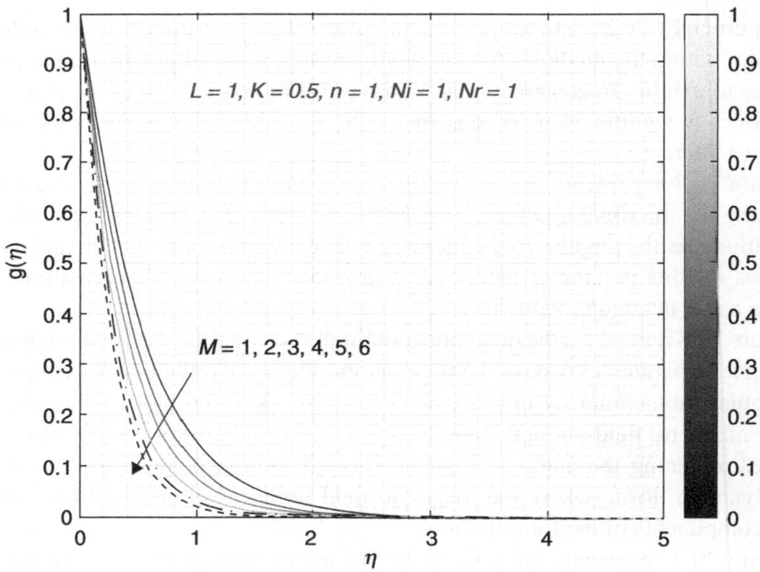

FIGURE 20.8 Magnetic parameter effect on angular momentum.

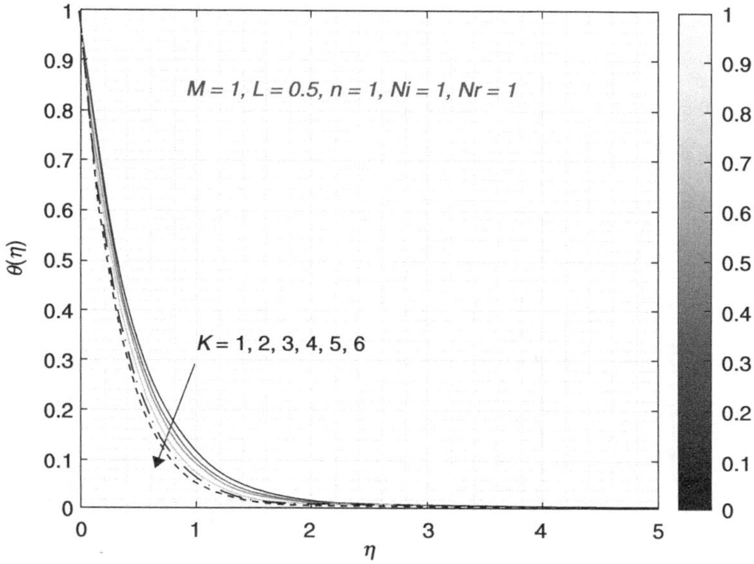

FIGURE 20.9 Micropolar parameter effect on temperature profile.

conditions and the interaction between thermal and rotational effects. Understanding this relationship is crucial in optimizing heat transfer in micropolar fluids.

Figure 20.10 represents the heat source term impact on temperature, rising the heat source term increases the fluid temperature. Physically, the heat source parameter plays a key role in determining the temperature profile of fluid by adding thermal energy into the system. As the intensity of the heat source increases, the fluid temperature rises, leading to a change in heat distribution and flow dynamics. This effect is particularly significant in systems where heat is generated internally, influencing both the temperature gradients and overall fluid behavior.

Figure 20.11 represents the radiation parameter impact on fluids temperature, increasing the thermal radiation parameter increases the temperature of the fluid. Physically, the thermal radiation term influences the temperature of fluid by affecting the heat exchange between the fluid and its surroundings through radiation. An increase in this parameter enhances the radiative heat transfer, which can raise or lower the fluid's temperature depending on the environment. This effect is particularly important in high-temperature flows, where radiation significantly contributes to overall heat transfer. Understanding this parameter helps in accurately modeling thermal behavior in various fluid systems.

Figure 20.12 represents the Prandtl number parameter impact on temperature, increasing the Prandtl number decays the fluids temperature. Physically, the Prandtl number affects the temperature distribution in fluid by balancing the rates of momentum and heat diffusion. A higher Prandtl number means that heat diffuses slower than momentum, causing sharper temperature gradients. On the other hand, a lower Prandtl number results in faster heat diffusion, leading to more uniform

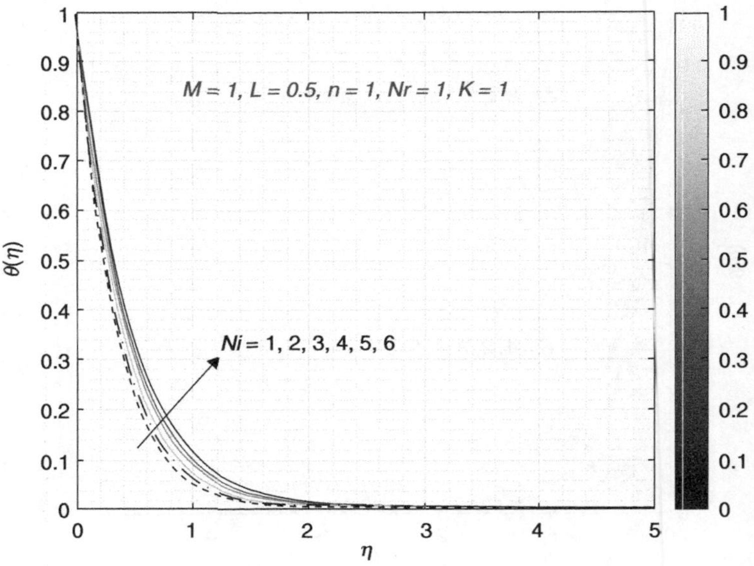

FIGURE 20.10 Heat source parameter effect on temperature.

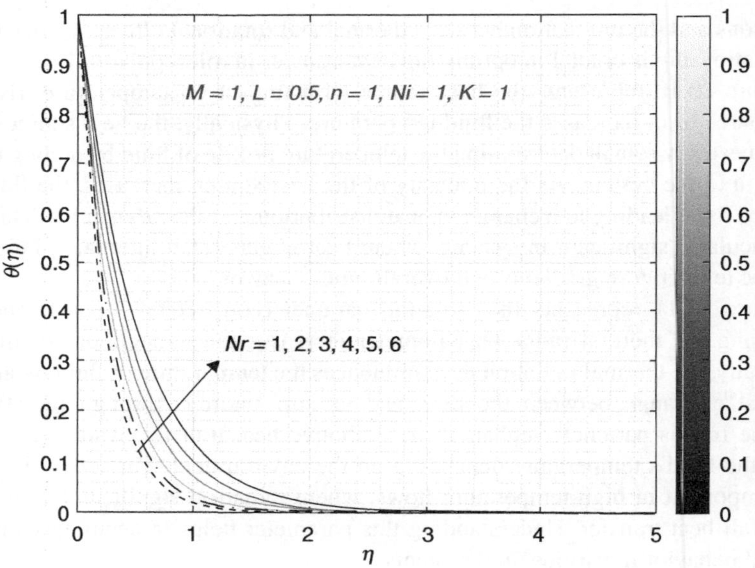

FIGURE 20.11 Radiation term effect on temperature.

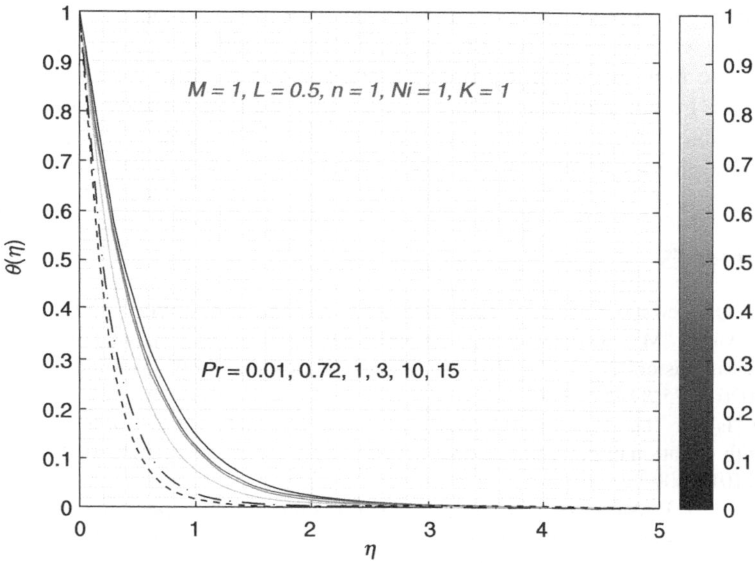

FIGURE 20.12 Prandtl number effect on temperature.

temperature distribution. This parameter is crucial for understanding thermal behavior in various fluid flow scenarios.

20.5 CONCLUSION

Micropolar fluid flow over an expanding surface with the influence of MHD, heat source, and radiation effects was considered, the governing equations are solved to PDEs via similarity variables, several physical terms such as magnetic field, porosity, heat source and radiation parameters were explained through graphs and the main results of the current study are as follows:

- Rising the M decays the angular and axial velocity of the flow.
- Rising the porosity term reduces the angular and axial momentum.
- Rising the micropolar parameter increases the angular and axial momentum.
- Enhancing the thermal radiation enhances the temperature.
- Rising the heat source parameter rises the temperature.
- Rising the Prandtl number decreases the temperature of the fluid flow.

The future scope of this work includes exploring the impact of varying MHD strengths and radiation effects on more complex fluid models, such as non-Newtonian or biofluids. Additionally, extending the analysis to three-dimensional flows and varying geometric conditions could offer more comprehensive insights into practical applications. The findings can be applied to industries like aerospace, materials processing, and microfluidics, where fluid behavior under magnetic fields and thermal radiation is critical. However, limitations include simplifying assumptions such as steady-state flow or constant properties, which may not fully represent real-world conditions. Future studies may focus on addressing these limitations and incorporating variable thermal properties or unsteady effects for a more accurate representation of complex systems.

REFERENCES

1. M. Q. Brewster, *Thermal Radiative Transfer and Properties*. Wiley, 1992.
2. A. Yücel, "Mixed convection in micropolar fluid flow over a horizontal plate with surface mass transfer," *Int. J. Eng. Sci.*, vol. 27, no. 12, pp. 1593–1602, Jan. 1989, doi: 10.1016/0020-7225(89)90153-5.
3. A. Ishak, "Thermal boundary layer flow over a stretching sheet in a micropolar fluid with radiation effect," *Meccanica*, vol. 45, no. 3, pp. 367–373, Jun. 2010, doi: 10.1007/s11012-009-9257-4.
4. A. Ishak, R. Nazar, and I. Pop, "Mixed convection stagnation point flow of a micropolar fluid towards a stretching sheet," *Meccanica*, vol. 43, no. 4, pp. 411–418, Aug. 2008, doi: 10.1007/s11012-007-9103-5.
5. R. Nazar, N. Amin, D. Filip, and I. Pop, "Stagnation point flow of a micropolar fluid towards a stretching sheet," *Int. J. Non-Linear Mech.*, vol. 39, no. 7, pp. 1227–1235, Sep. 2004, doi: 10.1016/j.ijnonlinmec.2003.08.007.
6. G. Ahmadi, "Self-similar solution of imcompressible micropolar boundary layer flow over a semi-infinite plate," *Int. J. Eng. Sci.*, vol. 14, no. 7, pp. 639–646, Jan. 1976, doi: 10.1016/0020-7225(76)90006-9.
7. R. Cortell Bataller, "Radiation effects in the Blasius flow," *Appl. Math. Comput.*, vol. 198, no. 1, pp. 333–338, Apr. 2008, doi: 10.1016/j.amc.2007.08.037.
8 C.-H. Chen, "Laminar mixed convection adjacent to vertical, continuously stretching sheets," *Heat Mass Transf.*, vol. 33, no. 5, pp. 471–476, Apr. 1998, doi: 10.1007/s002310050217.
9. L. J. Grubka and K. M. Bobba, "Heat transfer characteristics of a continuous, stretching surface with variable temperature," *J. Heat Transf.*, vol. 107, no. 1, pp. 248–250, Feb. 1985, doi: 10.1115/1.3247387.
10. M. Ali, "Heat transfer characteristics of a continuous stretching surface," *Wäarme-Stoffübertrag.*, vol. 29, pp. 227–234, Jan. 1994, doi: 10.1007/BF01539754.
11. S. M. Sachhin, U. S. Mahabaleshwar, and A. Chan, "Effect of slip and thermal gradient on micropolar nano suspension flow across a moving hydrogen fuel-cell membrane," *Int. J. Hydrog. Energy*, vol. 63, pp. 59–81, Apr. 2024, doi: 10.1016/j.ijhydene.2024.02.332.
12. S. M. Sachhin, U. S. Mahabaleshwar, S. N. R. Nayakar, H. S. R. Patil, and B. Souayeh, "Darcy–Brinkman model for Boussinesq–Stokes suspension tetra dusty nanofluid flow over stretching/shrinking sheet with suction/injection," *Numer. Heat Transf. Part Appl.*, vol. 0, no. 0, pp. 1–23, doi: 10.1080/10407782.2024.2359047.
13. S. Maity, S. K. Singh, and A. V. Kumar, "Unsteady three dimensional flow of Casson liquid film over a porous stretching sheet in the presence of uniform transverse magnetic field and suction/injection," *J. Magn. Magn. Mater.*, vol. 419, pp. 292–300, Dec. 2016, doi: 10.1016/j.jmmm.2016.06.004.
14. M. S. Ilango and P. Lakshminarayana, "Induced magnetic field and Soret–Dufour effects on viscous dissipative Casson fluid flow through porous medium over a stretching sheet,"

J. Therm. Anal. Calorim., vol. 149, no. 15, pp. 8713–8727, Aug. 2024, doi: 10.1007/s10973-024-13352-9.

15. M. A. Seddeek, "Heat and mass transfer on a stretching sheet with a magnetic field in a visco-elastic fluid flow through a porous medium with heat source or sink," *Comput. Mater. Sci.*, vol. 38, no. 4, pp. 781–787, Feb. 2007, doi: 10.1016/j.commatsci.2006.05.015.
16. I.-C. Liu, "Flow and heat transfer of an electrically conducting fluid of second grade in a porous medium over a stretching sheet subject to a transverse magnetic field," *Int. J. Non-Linear Mech.*, vol. 40, no. 4, pp. 465–474, May 2005, doi: 10.1016/j.ijnonlinmec.2004.07.008.
17. A. Subhas and P. Veena, "Visco-elastic fluid flow and heat transfer in a porous medium over a stretching sheet," *Int. J. Non-Linear Mech.*, vol. 33, no. 3, pp. 531–540, May 1998, doi: 10.1016/S0020-7462(97)00025-5.
18. R. Cortell, "Toward an understanding of the motion and mass transfer with chemically reactive species for two classes of viscoelastic fluid over a porous stretching sheet," *Chem. Eng. Process. – Process Intensif.*, vol. 46, no. 10, pp. 982–989, Oct. 2007, doi: 10.1016/j.cep.2007.05.022.
19. M. Turkyilmazoglu, "Flow of a micropolar fluid due to a porous stretching sheet and heat transfer," *Int. J. Non-Linear Mech.*, vol. 83, pp. 59–64, Jul. 2016, doi: 10.1016/j.ijnonlinmec.2016.04.004.
20. G. Bognár, "On similarity solutions of boundary layer problems with upstream moving wall in non-Newtonian power-law fluids," *IMA J. Appl. Math.*, vol. 77, no. 4, pp. 546–562, Jun. 2012, doi: 10.1093/imamat/hxr033.
21. S. A. Lone, F. Ali, A. Saeed, and G. Bognár, "Irreversibility analysis with hybrid cross nanofluid of stagnation point and radiative flow (TiO2+CuO) based on engine oil past a stretchable sheet," *Heliyon*, vol. 9, no. 4, p. e15056, Apr. 2023, doi: 10.1016/j.heliyon.2023.e15056.
22. I. F. Barna, G. Bognár, M. Guedda, L. Mátyás, and K. Hriczó, "Analytic self-similar solutions of the Kardar-Parisi-Zhang interface growing equation with various noise terms," *Math. Model. Anal.*, vol. 25, no. 2, pp. 241–256, Mar. 2020, doi: 10.3846/mma.2020.10459.
23. M. Klazly and G. Bognár, "A novel empirical equation for the effective viscosity of nanofluids based on theoretical and empirical results," *Int. Commun. Heat Mass Transf.*, vol. 135, p. 106054, Jun. 2022, doi: 10.1016/j.icheatmasstransfer.2022.106054.
24. G. Bognar, "Ferrofluid flow in magnetic field above stretching sheet with suction and injection," *Math. Model. Anal.*, Jul. 2020, Accessed: Feb. 20, 2025. [Online]. Available: www.academia.edu/121605678/Ferrofluid_Flow_in_Magnetic_Field_Above_Stretching_Sheet_with_Suction_and_Injection.

21 Magneto-Oberbeck Convection with Uniform Heat Flux

D.R. Sasi Rekha and R. Naveen Kumar

21.1 INTRODUCTION

Magneto-Oberbeck convection finds applications in various fields where the interaction of magnetic fields and thermal convection plays a crucial role. In nuclear reactors, particularly liquid-metal-cooled reactors, magnetic fields help regulate heat transfer and suppress turbulence in the coolant, ensuring stable operation. Callisto's plasma interaction simulations combined with an ensemble forward and inverse modeling approach to demonstrate the believable range of Callisto's internal features were discussed by Cochrane *et al.* [1]. By establishing a soft touch, rabbit hair improves the link with the low-grade energy source, drastically lowers frictional confrontation, and increases gadget strength and was studied by Deng *et al.* [2]. The persistent disruption of thermal, concentration equilibrium, and fluid is a prevalent aspect of many industrial operations, which introduces the notion of relaxation. This research investigates the relationship between sudden changes in temperature and relaxation phenomena. A collection of PDEs is expressed to depict the system's behavior. This collection is subsequently transformed into a boundary value problem and mathematically resolved using the spectral quasilinearization technique by Dhlamini *et al.* [3]. The experimental ability to measure changes in the convective heat transfer coefficient at the wall of the turbulent borderline. Foroozan *et al.* [4] done this by measuring two-dimensional fields of wall-temperature variations at two mild friction levels beneath a turbulent boundary layer with a zero-pressure gradient. The SIMPLE algorithm and the FVM approach are used to discretize and solve the associated equations. Two wires with electric currents put below and above the enclosure are used to apply the non-uniform magnetic field by Guo *et al.* [5].

The Reynolds–Colburn similarity is described as a fundamental method for determining convective heat transfer. Furthermore, the features of gases that engross, emit, and scatter are discussed, along with the radiation interactions between various surfaces. An overview of investigative solutions for heat transfer into tunnel walls is included, considering various types of simplified boundary conditions by Ingason *et al.* [6]. Ignoring this element could have serious repercussions for the development and real-world use of nanotechnology in heat transfer. To determine effective viscosity and thermal conductivity, the study also looks at the effects of three different models, thermal Rayleigh number, and nanoparticle concentration by Kaddiri [7]. The

DOI: 10.1201/9781003611820-27

cavity's walls are all isolated, and the governing equations are explained using the SIMPLER algorithm. The impact of several factors, including nanoparticle volume fraction, Hartmann number, and Reynolds number, was examined. The findings demonstrate that the isothermal lines become more compact, and the cold zone's dimensions close to the inlet grow as the Reynolds number rises by Kianpour *et al.* [8]. The natural convection regime of the boundary layer in a rectangular cavity with a constant side-to-side heat flux was discussed by Kimura *et al.* [9].

The variations impacting the NCBL dynamics during the transitional phase, linked to the layering of the adjacent liquid, are analyzed. In the earlier identified scaling for the unstable NCBL under consistent heat flux conditions continues to be mostly relevant to this situation, even though Ra, adds extra effects due to the time-varying characteristics of the applied temperature by Lin *et al.* [10]. The thermally isolated component method was utilized to implement a thermal confrontation model under uniform heat flux concentration boundary conditions, facilitating the calculation of the active thermal conductivity of the pebble bed by Luo *et al.* [11]. The research of Obalalu *et al.* [12] examines three different models of thermal radiation: nonlinear, quadratic, and linear, concentrating on the flow of a dusty nanofluid with a two-phase boundary layer over a horizontal plate that moves vertically. The nanofluid is self-possessed of aluminum oxide nanoparticles dispersed in a water-based medium. Furthermore, the analysis utilizes the quadratic Boussinesq approximation along with Prandtl's boundary layer approximation. The models provide improved accessibility, wider applicability, and better accurateness compared to conventional experiential correlations for CHF, establishing them as a significant focus in the field. This research intends to present the most recent advancements in applying machine learning methods for forecasting CHF and to improve these models for better analytical performance. It assesses the latest developments in the use of machine learning methods for CHF prediction, organizing the findings into two primary categories: predicting CHF values and identifying CHF events by Qi *et al.* [13]. According to the study of Saha *et al.* [14], vortex detaching at optimistic magnetic angles can be induced with a lower Richardson number and interface parameter than in the case of an associated magnetic field.

Among these techniques, the slope method stands out as it can be applied instantaneously, thus enabling time-resolved analysis; however, it does not yield additional insights when the Nusselt number is already determined. Consequently, while the average characteristics of the BL are extensively examined, particularly regarding heat transfer, its time-dependent automated dynamics, and surface unevenness remain significantly under-researched by Tegze *et al.* [15]. By combining the influence of buoyancy and Lorentz forces on the momentum equation and applying the single-phase nanofluid model in conjunction with the second-grade rheological model's constitutive equations and Corcione's correlations, we mathematically establish the governing equations along with their corresponding boundary conditions, drawing from the boundary layer system and other appropriate physical approximations by Wakif *et al.* [16]. The opening dimension, heat release rate, and air curtain jet velocity are all altered to reflect common fire situations. CCD cameras and a water-cooled heat flux gauge record the spill plume's heat flux profile and flame behavior for discussion. The findings indicate that the air curtain jet may significantly affect how plumes from a compartment-facade fire behave in terms of flames, Xu *et al.* [17]. This is inspired by the prominence of viscoelastic nanofluids in nanotechnology and biomedicine. A primary boundary layer formulation is effectively derived based on practical assumptions by applying Buongiorno's model alongside the conditions of convective heating and Stefan blowing by Zhang *et al.* [18]. The balance between film thickness and thermal conductivity can be linked to gases that are emitted during the assembly of the film, leading to significant structural damage. In this context, the development of the pore erection during the assembly of graphene films is analyzed, and the formation of gas leakage pathways is proposed as a strategy for generating thicker graphene films by Zheng *et al.* [19]. The heat transfer surface shifts from an inactive condition to an active condition, leading to the appearance of an active/inactive transition existing peak in the PDP curve, which grows larger as negative heat flux increases as discussed by Zhou *et al.* [20].

Examining the combined properties of magnetic fields and non-uniform thermal gradients on heat transfer and fluid flow is the primary innovative component of researching magneto-Oberbeck convection in a rectangular cavity with vertical uniform heat flux limits. This configuration simulates more realistic engineering and geophysical circumstances than standard investigations with defined temperature boundaries. It displays distinct flow patterns, heat transport behaviors, and stability traits driven by magnetohydrodynamics. The study aids in improving heat regulation in sophisticated industrial operations. It also improves knowledge of convection mechanisms in applications related to material processing and energy systems. In this chapter, the characteristics of heat transfer and flow in a vertically oriented rectangular cavity with uniform heat flux on both side walls are documented. The solution developed for the boundary layer scenario in a rectangular cavity with isothermal side walls is based on the principles outlined in the subsequent demonstration through an analytical result.

The objective of this research is to examine how a magnetic field affects natural convection within a rectangular cavity that has vertical walls and is exposed to a consistent heat flux. It seeks to explore the synergistic impacts of buoyancy and magnetohydrodynamic (MHD) forces on the behavior of fluid flow and the characteristics of heat transfer. A key focus is to examine how the Chandrasekhar number affects velocity distribution, temperature profiles, and overall heat transfer efficiency. Additionally, the study seeks to explore the stability and behavior of convection under varying Rayleigh number (Ra), considering different magnetic field strengths. The impact of cavity dimensions and aspect ratios on convective heat transfer patterns will also be assessed. To achieve these goals, analytical models will be developed to predict magneto-Oberbeck convection behavior in confined enclosures. Furthermore, the study aims to compare the results with classical Oberbeck–Boussinesq convection to highlight deviations caused by magnetic effects.

21.2 MATHEMATICAL FORMULATION

Figure 21.1 given below represents a two-dimensional rectangular cavity that is filled with a viscous; electrically conducting fluid with a horizontal uniform magnetic field H_o is present. The coordinate system consists of the vertical y axis, and the horizontal

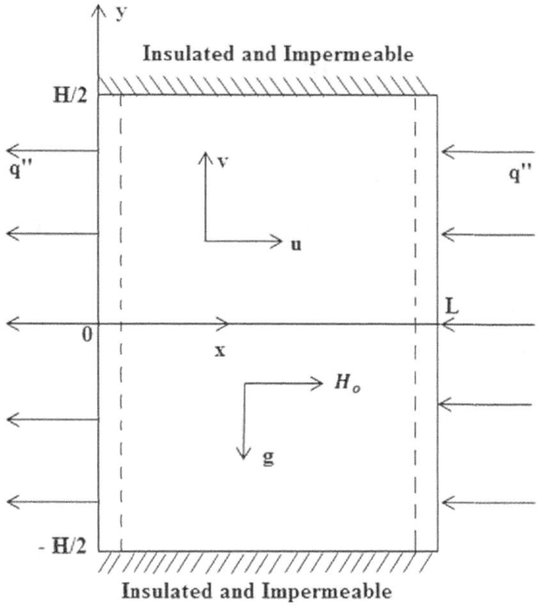

FIGURE 21.1 Schematic illustration.

x axis with velocity components u (along x) and v (along y). The domain is bounded by two vertical walls at $x=0$ and $x=L$, where a heat flux q'' is applied, indicating thermal energy transfer through these boundaries. The top and bottom walls, located at $y=H/2$ and $y=-H/2$, are insulated and impermeable, implying that no heat transfer occurs through these surfaces (adiabatic condition) and that there is no fluid penetration across them. Gravity g acts downward, which suggests that buoyancy effects may play a role in the fluid flow.

The following are a set of governing equations from Kimura *et al.* [9]:

$$\frac{\partial u}{\partial x}+\frac{\partial v}{\partial y}=0 \tag{21.1}$$

$$\frac{\partial^3 v}{\partial x^3}+g\beta\frac{\partial T}{\partial x}-Q^*\frac{\partial v}{\partial x}=0 \tag{21.2}$$

$$u\frac{\partial T}{\partial x}+v\frac{\partial T}{\partial y}=\frac{\partial^2 T}{\partial x^2} \tag{21.3}$$

where $Q^*=\frac{Q}{Ra^{2/5}}$, $Ra=\frac{g\beta q'' H^4}{\alpha v k}$, Rayleigh number and $Q=\frac{\sigma\,\mu^2\,H^2\,H_0^2}{\rho_0\,v}$, Chandrasekhar number.

For the boundary layer on the left, the dimensionless boundary conditions are

i. $$u = v = 0, \quad \frac{\partial T}{\partial x} = 1 \text{ at } x = 0\,,\ 1 \tag{21.4}$$

ii. $$\left.\begin{aligned} u &\rightarrow u_\infty(y) \\ v &\rightarrow 0 \\ T &\rightarrow T_\infty(y) \end{aligned}\right\} \text{as} \quad x \rightarrow \infty \tag{21.5}$$

The three important statements to determine $T_\infty\,(y)$, and the unknown functions.

(a) Mass conservation integral:

$$\psi_\infty = -\int_0^\infty v\,dx \tag{21.6}$$

(b) Energy conservation integral:

$$\left|\, u\,T\,\right|_0^\infty + \frac{d}{dy}\int_0^\infty v\,T\,dx = \left|\frac{\partial T}{\partial x}\right|_0^\infty \tag{21.7}$$

(c) Core Centro symmetry about $y = 0\,, x = L/2$ in Figure 21.1, which means that

i. $\psi_\infty\,(y) = \psi_\infty\,(-y)$

ii. $T_\infty{}'\,(y) = T_\infty{}'(-y)$ (21.8)

iii. $T_\infty\,(y) = -T_\infty\,(-y)$

21.2.1 Asymptotic Behavior of the Boundary-Layer Solutions

Due to the non-linear nature of Eqs. (21.1–21.3), an exact analytical treatment is not possible. The asymptotic behavior of the solution in the boundary-layer as $x \rightarrow \infty$ might be considered to yield a lot of valuable information. When the boundary conditions are relevant, this analysis will be applicable, thus let

$$T = T_\infty\,(y) + \theta\,(x,y) \tag{21.9}$$

$$\psi = u_\infty(y) + \varphi(x,y) \tag{21.10}$$

where T_∞ and u_∞ are core solution and θ and φ represents perturbation satisfying the condition $\theta, \varphi \rightarrow 0$ as $x \rightarrow \infty$. Substitute Eqs. (21.9) and (21.10), into Equations (21.1–21.3), and we get the following linear equation:

$$\frac{u_\infty\, v_x}{Pr} = \theta + v_{xx} - Q^*\, v \tag{21.11}$$

A solution that has been linearized as Kimura *et al.* [9]

$$v = \sum_{n=1}^{4} a_n(y)\ \exp[-\lambda_n(y)\ x] \tag{21.12}$$

where imaginary numbers with positive real components are denoted by $\lambda_n(y)$. Equation (21.12), when this solution is subjected to the current conditions, yields

$$\lambda_n^2\ (\lambda_n + u_\infty)\left(\lambda_n + \frac{u_\infty}{Pr}\right) - Q^*\ \lambda_n(\lambda_n + u_\infty) + T_\infty^{'(y)=0} \tag{21.13}$$

The behavior of the roots depends on the signs of $T_\infty^{'(y)}$.

21.2.2 An Approximate Solution Using the Oseen-Linearized Technique

Solving boundary layer equations with an approximation (21.1–21.3) is a prerequisite for determining an approximate solution to the convective problem. The technique used for this is a variation of the Oseen technique. If $\Pr \rightarrow \infty$ is selected in the following analysis, Eqs. (21.11) and (21.13), becomes

$$0 = v_{xx} - Q^*\ v + \theta \tag{21.14}$$

$$\lambda_n^3\ (\lambda_n + u_\infty) - Q^*\ \lambda_n(\lambda_n + u_\infty) + T_\infty^{'(y)=0} \tag{21.15}$$

The solution (21.12) close to the left vertical boundary that corresponds with the core flow is because these solutions are approximations for the entire boundary layer. Therefore we get,

$$v = a_1 e^{-\lambda_1 x} + a_2 e^{-\lambda_2 x} \tag{21.16}$$

The solution (21.16), satisfies the left vertical wall at $x = 0$. Hence we have $a_1 = -a_2$, where $a_1 = \frac{1}{(\lambda_1 - \lambda_2)\left[\lambda_1{}^2 + \lambda_2{}^2 + \lambda_1\ \lambda_2 - Q^*\right]}$

Equation (21.16), using (21.14) becomes

$$\theta + a_1\left(\lambda_1{}^2 e^{-\lambda_1 x} - \lambda_2{}^2 e^{-\lambda_2 x}\right) - a_1 Q^*\left(e^{-\lambda_1 x} - e^{-\lambda_2 x}\right) = 0 \tag{21.17}$$

Under the present circumstances, this solution yields according to the boundary conditions,

$$v = \frac{e^{-\lambda_1 x} - e^{-\lambda_2 x}}{(\lambda_1 - \lambda_2)\left[\lambda_1{}^2 + \lambda_2{}^2 + \lambda_1\lambda_2 - Q^*\right]} \tag{21.18}$$

Substitution of a_1 in (21.17), we get

$$\theta = -\frac{\lambda_1^2\, e^{-\lambda_1 x} - \lambda_2^2\, e^{-\lambda_2 x} - Q^*\,\left(e^{-\lambda_1 x} - e^{-\lambda_2 x}\right)}{(\lambda_1 - \lambda_2)\left[\lambda_1{}^2 + \lambda_2{}^2 + \lambda_1\lambda_2 - Q^*\right]} \tag{21.19}$$

From Eq. (21.9), we get

$$T = T_\infty\,(y) - \frac{\lambda_1^2\, e^{-\lambda_1 x} - \lambda_2^2\, e^{-\lambda_2 x} - Q^*\,\left(e^{-\lambda_1 x} - e^{-\lambda_2 x}\right)}{(\lambda_1 - \lambda_2)\left[\lambda_1{}^2 + \lambda_2{}^2 + \lambda_1\lambda_2 - Q^*\right]} \tag{21.20}$$

where the functions $\lambda_1\,(y)$, $\lambda_2\,(y)$ are unknown.

The important statement (a, b) provides

$$\psi_\infty = \frac{1}{\lambda_1\lambda_2\left[\lambda_1{}^2 + \lambda_2{}^2 + \lambda_1\lambda_2 - Q^*\right]} \tag{21.21}$$

$$\frac{d}{dy}\left[\frac{\lambda_1\lambda_2 + Q^*}{2\,\lambda_1\lambda_2(\lambda_1 + \lambda_2)\left[\lambda_1{}^2 + \lambda_2{}^2 + \lambda_1\lambda_2 - Q^*\right]^2}\right] - \psi_\infty\, T_\infty{}'\,(y) = -1 \tag{21.22}$$

Additionally, it is appropriate to formulate Eqs. (21.21, 21.3), in terms of the functions given the centrosymmetry conditions (c), Kimura *et al.* [9]:

$p\,(y) =$ even function
$q\,(y) =$ odd function

$$\text{where}\,\lambda_{1,\,2} = \frac{p}{4}(1-q) \pm \frac{1}{2}\sqrt{2\,(1+q)\,Q^* - \frac{p^2\,(1-q)^2\,(1+2q)}{4}} \tag{21.23}$$

Thus, Eqs. (21.21, 21.22), become

$$\psi_\infty = \frac{64}{\left(1-q^2\right)\left[p^2\left(1-q\right)^2-4Q^*\right]^2} \tag{21.24}$$

$$\frac{d}{dy}\left[\frac{64(1-q)\left[p^2(1-q^2)+4Q^*\right]}{p(1-q)^3(1+q)\left[p^2(1-q^2)-4Q^*\right]^3}\right] - \frac{64}{(1-q^2)\left[p^2(1-q)^2-4Q^*\right]^2}T'_\infty(y) = -1 \tag{21.25}$$

These two equations capture the idea of thorough analysis: it is clear that both ψ_∞ and the left-hand side of Eq. (25), must be even functions of y at the same time.

$$\begin{aligned} p &= \text{constant} \\ q &= 0 \\ T_\infty{}'(y) &= \text{constant} \end{aligned} \tag{21.26}$$

Furthermore, Eq. (21.23), as the λ's are constant.

$$\lambda_{1,2} = \frac{p}{4} \pm \frac{1}{2}\sqrt{2Q^* - \frac{p^2}{4}} = a \pm b \tag{21.27}$$

where $a = \frac{p}{4}$ and $b = \frac{1}{2}\sqrt{2Q^* - \frac{p^2}{4}}$

In summary, the solution is

$$v = \frac{8\,e^{-ax}\sinh bx}{b\left(4Q^* - p^2\right)}$$

$$T = T_\infty\left(y\right) + \frac{4\,e^{-ax}\left(b\,p\cosh bx + Q^*\sinh bx\right)}{b\left(4Q^* - p^2\right)}$$

if it is independent of y, then

$$T = \frac{4\,e^{-ax}\left(b\,p\cosh bx + Q^*\sinh bx\right)}{b\left(4Q^* - p^2\right)}$$

$$\psi_\infty = \frac{64}{\left(4Q^* - p^2\right)^2}, T_\infty\left(y\right) = \frac{y}{\psi_\infty} \tag{21.28}$$

where the constant p is unknown. Determining p will finish the solution. The downward heat conduction must precisely balance the vertical enthalpy flow through the flow.

$$\int_0^L \rho\, c_p\, v\, T\, dx = \int_0^L k \left(\frac{\partial T}{\partial y}\right) dx \tag{21.29}$$

and entering these into Eq. (21.29), together with Eq. (21.28), results

$$p^{11} - 20\, Q^* p^9 + 160\, (Q^*)^2 p^7 - 640\, (Q^*)^3 p^5 + 1280\, (Q^*)^4 p^3 - 1024\, (Q^*)^5 p$$
$$- 8192 \left(\frac{H}{L}\right) Ra^{\frac{1}{5}} \left(p^2 + 4Q^*\right) = 0 \tag{21.30}$$

Equations (21.28), reflect the final analytical solution for the boundary layer regime with Eq. (21.30), providing p.

The Nusselt number follows:

$$Nu = \frac{q''}{k\left(\frac{\Delta T}{H}\right)} = \frac{p^2 - 4Q^*}{8p}\, Ra^{\frac{1}{5}} \tag{21.31}$$

The following section discusses the findings of the computation of the average Nusselt number, temperature, and velocity.

21.3 RESULTS AND DISCUSSIONS

To comprehend the natural convection of an electrically conducting fluid in a rectangular enclosure with a (transverse) magnetic field perpendicular to gravity, an analytical analysis is conducted. Thermal insulation is used for the horizontal bottom and top walls, and the two side walls are kept at a constant heat flux state. The non-dimensional parameters are the Rayleigh number Ra, aspect ratio H/L, and the Chandrasekhar number Q. In this work, $H/L = 1$, 1.5, 2, 2.5, and 3 are chosen, and calculations are performed for Ra ranging from 10^1 to 10^6 and Q ranging from 0 to 200. A variety of analytical findings for temperature and flow fields are shown graphically. A detailed discussion is given of the changes in field characteristics brought about by the joint effects of buoyancy and the applied magnetic field. By calculating the average Nusselt number, the rate of heat transmission is determined.

In Figure 21.2, dimensional velocity charts against the stretched horizontal coordinate x are displayed for different Q values (0, 50, 100), when $H/L = 1$ and $Ra = 10^5$. The figure presents the velocity profile as a function of x for several values of the Chandrasekhar number Q, which likely represents a physical quantity influencing the flow, such as heat generation and magnetic field strength. The plotted curves show how the velocity evolves, reaching a peak before gradually decaying to zero. Therefore, the magnetic field has the effect of decreasing flow. As Q increases, the peak velocity shifts downward, indicating a suppression of flow dynamics, possibly

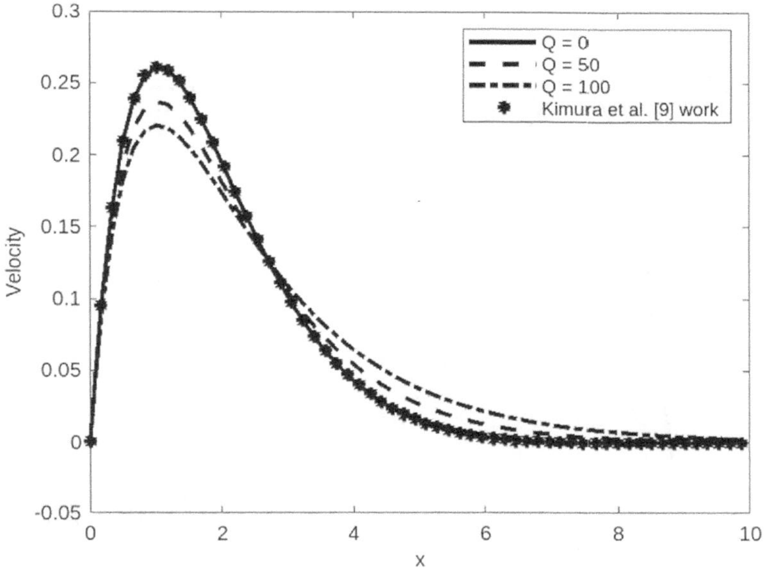

FIGURE 21.2 Velocity profile at $H/L = 1$ and $Ra = 10^5$.

due to an increasing resistive force such as viscous dissipation or magnetic drag. This behavior is characteristic of physical scenarios where an opposing force, such as Lorentz force in MHD flows or thermal dissipation in heat-driven flows, reduces the overall momentum transfer. The overall trend suggests that higher values of Q result in a more restrained velocity field, impacting the transport phenomena in the system. The comparison with Kimura *et al.* [9] reference data shows that the model accurately captures the velocity trends, validating its application in real-world fluid dynamics problems. Such insights are crucial in engineering applications, including heat exchangers, aerodynamics, and electromagnetic fluid control systems.

Figure 21.3 represents the velocity distribution along x for different values of the aspect ratio H/L. The curves exhibit a peak velocity followed by a gradual decline, indicating the development and dissipation of flow motion. As H/L increases, the peak velocity decreases, suggesting that increasing the height relative to the length results in a more constrained or suppressed velocity profile. This behavior is typical in fluid dynamics, where a higher aspect ratio may introduce greater resistance or reduce the influence of driving forces such as pressure gradients. The results suggest that the geometric configuration plays a crucial role in determining the velocity characteristics, which is important in applications such as heat transfer, aerodynamic design, and confined fluid flows in engineering systems.

Figure 21.4 presents the velocity distribution as a function of x for various values of the Rayleigh number (Ra), which characterizes the buoyancy-driven flow in natural convection. The velocity initially rises to a peak before gradually decaying to zero, reflecting the influence of thermal-driven motion. As Ra increases from 10^3 to 10^6, the peak velocity decreases, indicating that higher Rayleigh numbers result in a

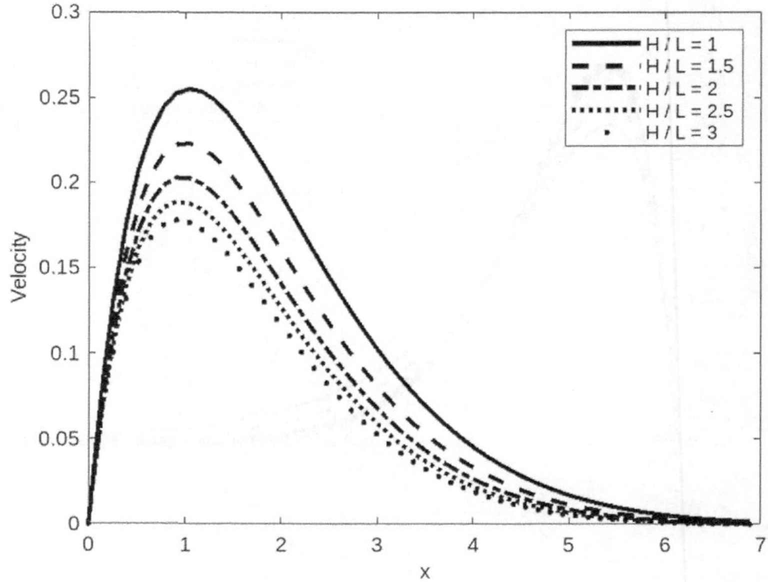

FIGURE 21.3 Velocity profile at $Q = 10$ and $Ra = 10^5$.

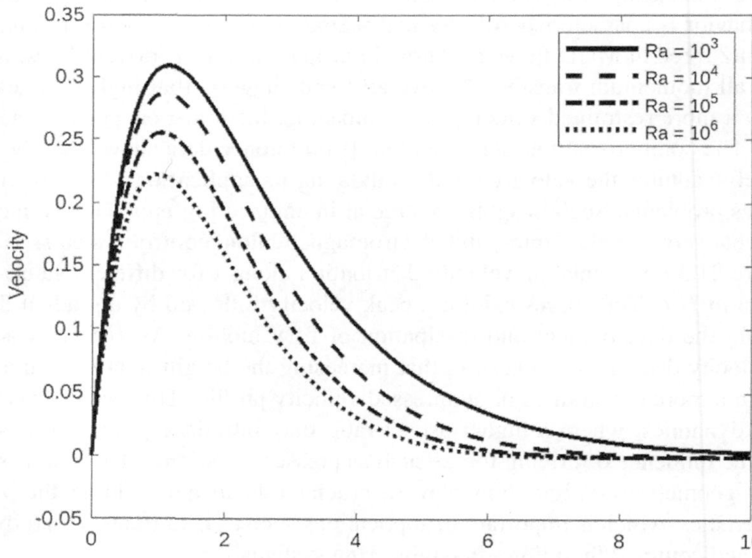

FIGURE 21.4 Velocity profile at $Q = 10$ and $H/L = 1$.

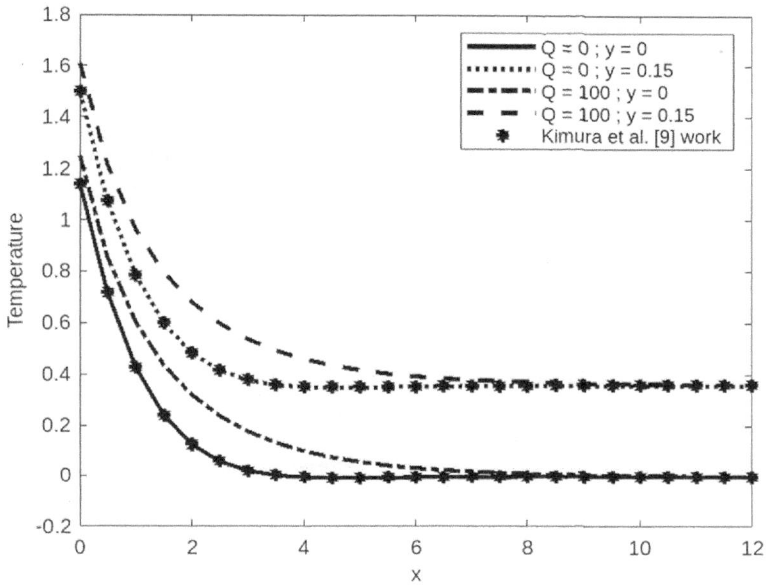

FIGURE 21.5 Temperature profile at $H/L = 1$ and $Ra = 10^5$.

more suppressed velocity profile. This behavior suggests that at larger *Ra*, the flow becomes more stratified, potentially due to increased thermal resistance or changes in the convection regime. Physically, the ratio of buoyancy forces to viscous forces is represented by the Rayleigh number, so a lower velocity at high *Ra* may imply that thermal energy is more efficiently dissipated, leading to less pronounced bulk motion. This understanding is essential in heat transfer applications, such as cooling systems, atmospheric convection, and geothermal energy extraction, where buoyancy-driven flow plays a crucial role in energy transport.

Figure 21.5 presents a temperature distribution as a function of x for different values of Q and transverse coordinate y. The solid lines represent analytical results, while the black star indicates data from Kimura *et al.* [9], serving as a reference for validation. The graph shows that temperature decreases along the x direction for all cases, but the rate of decrease depends on Q and y. Higher $Q = 100$ results in elevated temperature profiles compared to $Q = 0$, as expected due to additional thermal energy input. Additionally, increasing y (moving away from the boundary) leads to higher temperature values, likely due to reduced cooling effects. The close agreement with Kimura *et al.*'s data suggests that the model accurately captures the physical behavior of the system.

Figure 21.6 represents the temperature distribution along the x axis for different aspect ratios (H/L) and transverse positions (y). The curves indicate that temperature decreases as x increases, suggesting heat dissipation due to conduction. A higher aspect ratio (H/L) results in lower temperature profiles, implying that as the height-to-length ratio increases, the system exhibits enhanced heat dissipation, likely due

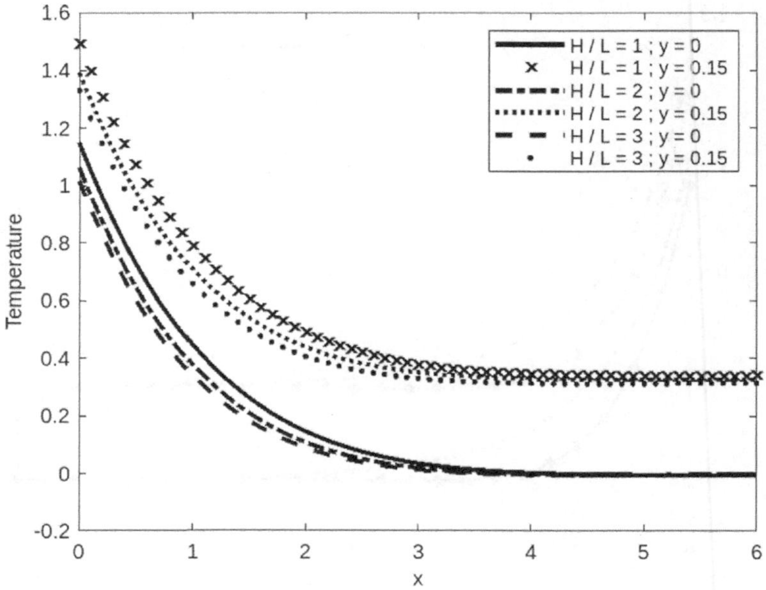

FIGURE 21.6 Temperature profile at $Q = 10$ and $Ra = 10^5$.

to increased surface area for conduction. Additionally, the temperature at $y = 0.15$ is consistently higher than at $y = 0$, reflecting reduced heat loss further from the boundary. This trend demonstrates how geometric factors influence thermal behavior, making aspect ratio a crucial parameter in controlling heat transfer efficiency.

Figure 21.7 presents the temperature distribution along the *x* axis for different Rayleigh numbers (*Ra*) and transverse positions (*y*). In a fluid system, the Rayleigh number is the proportion of buoyancy-driven convection to thermal diffusion. As *Ra* increases from 10^4 to 10^6, the temperature decreases more rapidly along *x*, indicating enhanced heat transfer due to stronger convective effects. Higher *Ra* values correspond to increased fluid motion, leading to more efficient heat transport and lower temperature retention. The curves also show that temperature is higher at $y = 0.2$ than at $y = 0$, suggesting that the thermal boundary layer near the surface is thinner, and convective heat transfer is more dominant at greater distances from the boundary. This behavior highlights the influence of buoyancy-driven flow on thermal distribution and the importance of *Ra* in characterizing convection-dominated heat transfer.

Figure 21.8 presents a plot illustrating the relationship between variables *x* and *y* under different conditions, labeled by *Q* and *t*. Each curve represents a unique combination of *Q* and *t*. As *x* increases, *y* rises from a negative value towards zero, showing a diminishing effect or stabilization. Higher values of *Q* and *t* appear to elevate the curves, indicating that increasing these parameters reduces the magnitude of *y*, which might imply a mitigating effect on a physical process (e.g., thermal or magnetic influence reducing resistance or drag). The close alignment of the curves with Kimura *et al.* [9] data indicates the model's accuracy in replicating established results. This

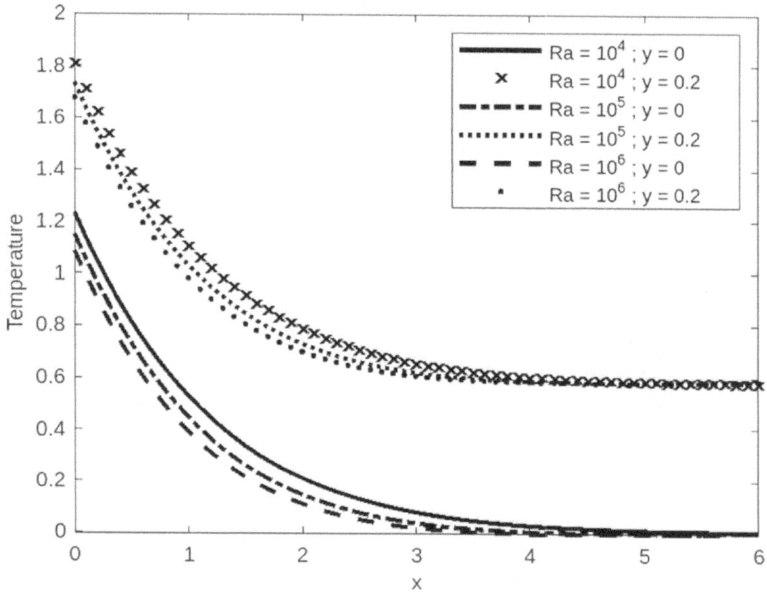

FIGURE 21.7 Temperature profile at $Q = 10$ and $H/L = 1$.

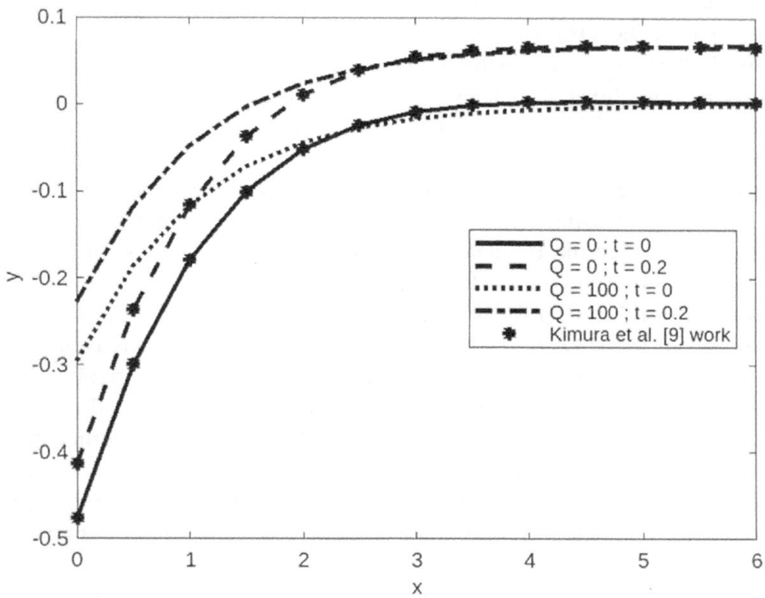

FIGURE 21.8 Isotherm at $H/L = 1$ and $Ra = 10^5$.

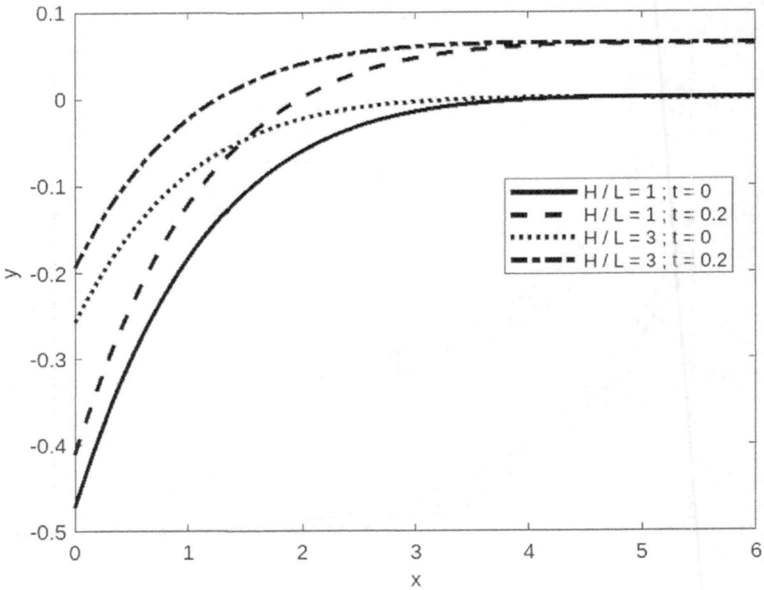

FIGURE 21.9 Isotherm at $Q = 10$ and $Ra = 10^5$.

plot likely demonstrates the influence of control parameters on a physical quantity in a fluid mechanics, heat transfer, or electromagnetism context.

Figure 21.9 shows a plot depicting the relationship between variables x and y under varying conditions of H/L and t Each curve corresponds to a specific combination of H/L and t, allowing a comparison of their effects on the behavior of y. As x increases, y gradually approaches zero from negative values, indicating a stabilizing or diminishing effect. Higher values of H/L cause the curves to shift upward, suggesting that increasing the geometric ratio reduces the magnitude of y. This behavior may indicate the influence of the aspect ratio on a physical phenomenon, such as flow behavior, thermal distribution, or magnetic effects in a system. Similarly, increasing t (from 0 to 0.2) further elevates the curves, indicating a transient effect or a response to the t. Overall, the figure demonstrates how increasing the aspect ratio and temperature mitigates the influence of y, suggesting a moderating effect in a physical process. This analysis could relate to heat transfer, fluid mechanics, or electromagnetic fields, where geometry and time-based changes significantly impact behavior. The thermal boundary layer in this case covers the viscous boundary layer. We infer that the boundary layer near the vertical walls, which is highly advantageous for controlling heat flow across the enclosure, is significantly influenced by the applied magnetic field.

Figure 21.10 illustrates a plot showing the relationship between variables x and y for several values of the Rayleigh number (Ra) and t. The Rayleigh number, Ra, is a dimensionless parameter often used in heat transfer and fluid dynamics to describe the influence of buoyancy-driven flow, indicating the balance between thermal driving

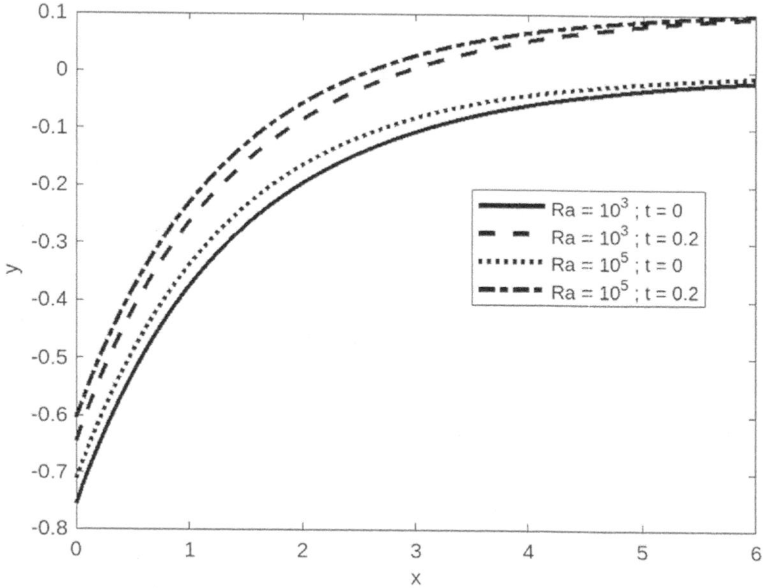

FIGURE 21.10 Isotherm at $Q = 10$ and $H/L = 1$.

forces and viscous dissipation. Higher values of *Ra* typically signify stronger convective currents. The curves represent combinations of *Ra* and *t*, with $Ra = 10^3$ and 10^5, while *t* ranges from 0 to 0.2. As *x* increases, *y* gradually approaches zero from a more negative value, indicating a stabilizing trend. Higher values of *Ra* shift the curves upward, suggesting that stronger buoyancy forces (larger *Ra*) reduce the effect captured by *y*. This behavior is consistent with intensified convection leading to a more uniform distribution or enhanced mixing. Similarly, increasing *t* elevates the curves, indicating a transient effect that further diminishes the influence of *y*. This plot likely depicts the impact of thermal or convective forces over time, highlighting how increased buoyancy (higher *Ra*) and *t* mitigate the observed effect. Such a graph would be relevant in studying heat transfer in fluids, natural convection, or thermal boundary layer behavior.

Figure 21.11 presents the relationship between the Nusselt number (*Nu*) and the Rayleigh number (*Ra*) for different values of *Q*, which likely represents a parameter related to an external influence, such as a magnetic field (in magnetohydrodynamics). The Nusselt number (*Nu*) is a dimensionless measure of convective heat transfer, while the Rayleigh number (*Ra*) quantifies the dominance of buoyancy-driven convection in a fluid system. The curves indicate that as *Ra* increases, *Nu* also increases, signifying enhanced convective heat transfer. Furthermore, higher values of *Q* reduce the heat transfer efficiency, as seen by the lower *Nu* values. This figure represents data from Kimura *et al.* [9], suggesting a comparison with previous work. This plot demonstrates the influence of *Q* on convective heat transfer, indicating that stronger external influences (e.g., magnetic fields) can suppress convective motion, leading to lower heat transfer efficiency.

Figure 21.12 illustrates the variation of the Nusselt number (*Nu*) as a function of the Rayleigh number (*Ra*) for different aspect ratios (*H*/*L*). The Nusselt number

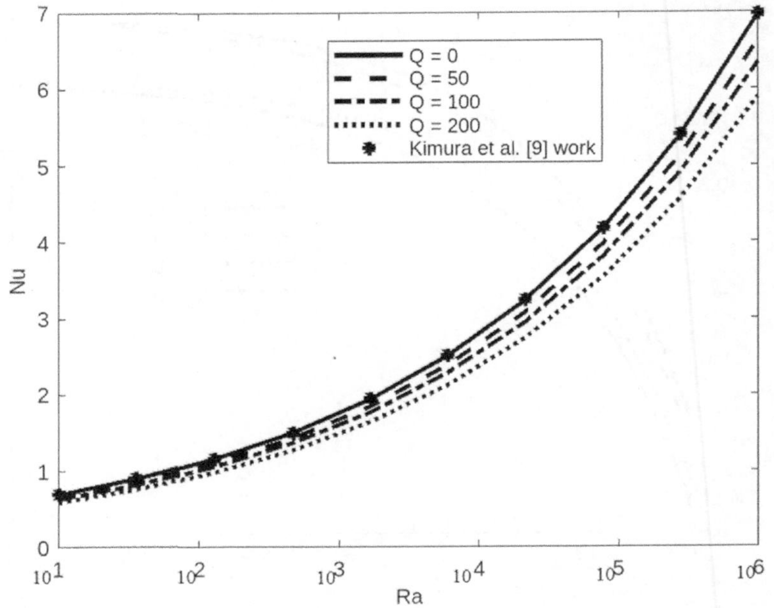

FIGURE 21.11 *Nu* vs *Ra* at $H/L = 1$ and $Ra = 10^5$.

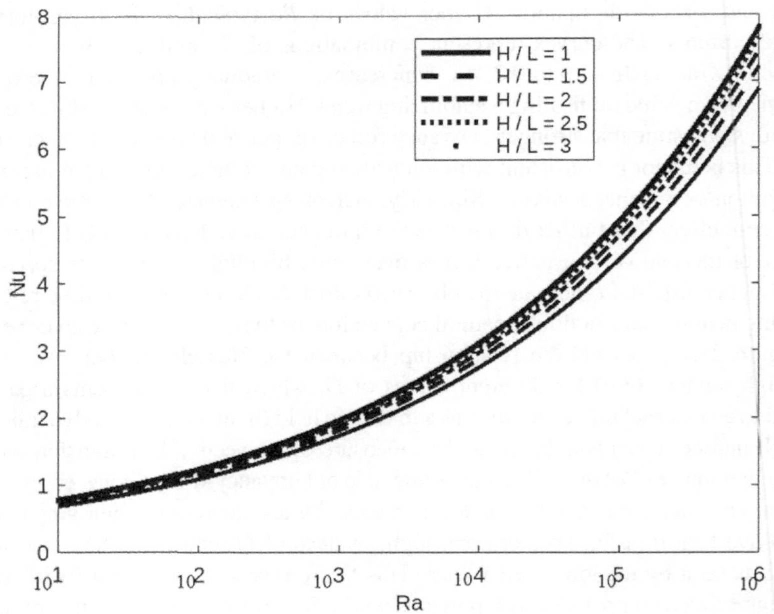

FIGURE 21.12 *Nu* vs *Ra* at $Q = 10$ and $Ra = 10^5$.

TABLE 21.1
The Effect of Chandrasekhar Number on Nusselt Number

H/L	Ra	Q	Nu	Kimura *et al.* [9]
1	10	0	0.567	0.566
	10^2	10	0.793	
	10^3	100	1.028	
1.5	10	0	0.594	0.594
	10^2	10	0.840	
	10^3	100	1.100	
2	10	0	0.613	0.612
	10^2	10	0.874	
	10^3	100	1.154	
2.5	10	0	0.628	0.628
	10^2	10	0.901	
	10^3	100	1.197	
3	10	0	0.641	0.640
	10^2	10	0.924	
	10^3	100	1.233	

quantifies the efficiency of convective heat transfer, while the Rayleigh number characterizes the intensity of buoyancy-driven convection. The different curves correspond to various aspect ratios, showing how changes in the domain geometry affect convective heat transport. As Ra increases, Nu also increases, indicating stronger convective heat transfer. Furthermore, larger aspect ratios (H/L) tend to enhance heat transfer slightly, as seen from the increasing Nu values. This suggests that taller enclosures favor more efficient convection by allowing the formation of stronger convective currents, leading to improved heat transport. The Nusselt number for different combinations of parameters is presented in Table 21.1.

21.4 CONCLUSION

A rectangular vertical enclosure with both side walls exposed to a uniform heat flux situation is examined in this research to document its flow and heat transfer characteristics. Understanding how magnetic fields affect convective heat transfer is the primary goal of researching magneto-Oberbeck convection in a rectangular cavity with vertical uniform heat flux limits. In technical applications including electronic cooling, metallurgy, and crystal formation, this study aids in the management and optimization of heat flow. By reducing turbulence or altering flow patterns, it helps stabilize or improve heat transmission. The results help to enhance thermal management and energy efficiency in geophysical and industrial systems. In the end, it promotes developments in the design of heat transfer systems with magnetic control. An analytical result establishes the solution framed for the boundary layer regime in a rectangular box with isothermal side walls along the lines of the subsequent demonstration. The major findings are as follows:

- The magnetic field has the effect of decreasing flow.
- The influence of the magnetic field is negligible very close to the boundary when the Rayleigh number, *Ra*, increases.
- The temperatures are falling exponentially. This is a feature of solutions that are based on Oseen linearization.
- Moving into the core region, or away from the outer layer, causes the temperature to stabilize. This means that the thermal field inside the core exhibits a linear vertical stratification, which is another feature of the core.
- The magnetic field's impact on isotherms is very consistent across all Rayleigh number levels. Since the improvement in buoyant force is characterized by an increase in *Ra*, the temperature field in the interior core exhibits a linear vertical stratification.
- They infer that the boundary layer near the vertical walls, which is highly advantageous for controlling heat flow across the enclosure, is significantly influenced by the applied magnetic field.
- When the magnetic field is relatively large, the steady-state Nusselt number gets closer to the conduction value. For a range of aspect ratios and Chandrasekhar number values, it provides the mean Nusselt number variation with the Rayleigh number. Thus, it is possible to use the magnetic field as an effective insulator.

21.5 FUTURE SCOPE

The future scope of this study is with significant implications in various scientific and engineering fields. Advanced computational modeling and high-fidelity simulations will enable a deeper understanding of heat transfer mechanisms under the influence of magnetic fields, improving turbulence modeling and predicting complex convection patterns. Industrial applications, such as cooling technologies for nuclear reactors, microelectronics, and spacecraft thermal management, will benefit from optimized MHD-based heat dissipation techniques. Metallurgical processes, including molten metal handling and crystal growth, can be enhanced through controlled magnetic convection. Energy systems, such as solar thermal collectors, magnetically controlled heat exchangers, and fusion reactors, will also see advancements in efficiency and performance. In material science, research into nanofluids and ferrofluids will improve thermal conductivity, while studies on variable viscosity and radiation effects will lead to smarter energy-efficient materials. Space and astrophysical applications, including planetary magneto-convection and thermal management in microgravity environments, will be key areas of exploration. Furthermore, biomedical applications such as MHD-controlled drug delivery and hyperthermia treatments for cancer therapy present new frontiers in medical technology. Environmental studies, including geothermal energy optimization and climate modeling, will also benefit from a deeper understanding of magneto-convective transport phenomena. As computational power and experimental techniques advance, magneto-Oberbeck convection research will continue to evolve, driving innovations in energy efficiency, space exploration, medical advancements, and industrial processing.

NOMENCLATURE

a_n	coefficient in equation	q''	heat flux from the side
c_p	specific heat	Ra	Rayleigh number = $g\beta q'' H^4 / (\alpha v k)$
T_0	reference temperature	ΔT	the average side-to-side temperature difference
v	kinematic viscosity	v, u	vertical and horizontal velocity components
g	gravitational acceleration	l	thermal boundary layer thickness
H	enclosure height	α	thermal diffusivity
k	thermal conductivity	β	thermal expansion coefficient
δ	boundary layer thickness	T	temperature
ρ	Density	x, y	horizontal and vertical coordinates
L	enclosure thickness	μ_f	viscosity of the fluid
Nu	Nusselt number	ρ_0	reference density
P	Pressure	ξ	vorticity function
Pr	Prandtl number	$p(y)$	even function
$\vec{q}$	velocity vector	$q(y)$	odd function
λ_n	function of y	Q	Chandrasekhar number = $\sigma\mu^2 H^2 H_0^2 / (\rho_0 v)$
Subscript: ∞ = indicating condition in the core			

REFERENCES

1. Cochrane, C. J., Vance, S. D., Castillo-Rogez, J. C., Styczinski, M. J., & Liuzzo, L. (2025). Stronger evidence of a subsurface ocean within Callisto from a multifrequency investigation of its induced magnetic field. *AGU Advances, 6*(1), e2024AV001237.
2. Deng, W., Dong, K., Ni, S., He, S., Zhao, P., Yu, T., & Zhao, J. (2025). Magnet gear-based triboelectric nanogenerator harvesting energy from boiling heat transfer environment. *Nano Energy, 137*, 110804.
3. Dhlamini, M., Zondo, K., Duve, P., Mondal, H., Mishra, S., Sibanda, P., ... & Motsa, S. (2024). Cattaneo-Christov heat flux-based micropolar nanofluid flow with relaxation, slip, and temperature jump effects. *Results in Engineering*, 102645.
4. Foroozan, F., Ianiro, A., Discetti, S., & Baars, W. J. (2025). Instantaneous convective heat transfer at the wall: A depiction of turbulent boundary layer structures. *arXiv preprint arXiv:2502.01494.*
5. Guo, B., Basem, A., Alizadeh, A. A., Najm, A. S., Kazemi-Varnamkhasti, H., Jasim, D. J., & Salahshour, S. (2025). Numerical analysis of turbulent natural convection in the presence of wire-induced non-uniform magnetic field inside a porous medium. *International Communications in Heat and Mass Transfer, 160*, 108398.
6. Ingason, H., Li, Y. Z., & Lönnermark, A. (2024). Heat flux and thermal resistance. In *Tunnel fire dynamics* (pp. 273–318). Cham: Springer International Publishing.

7. Kaddiri, M. (2024). Enhancing the convective heat transfer in vertical and horizontal rectangular enclosures using nanofluids: The crucial role of aspect ratio. *Physics of Fluids*, *36*(1), https://doi.org/10.1115/1.3246666.
8. Kianpour, E., Dehkordi, S. M. H. R., & Sidik, N. A. C. (2024). Numerical simulation of mhd nanofluid forced convection heat transfer in a rectangular enclosure with variable parametersters. *Innovative Mechanical Engineering, 3*(2), 43–55.
9. Kimura, S., & Bejan, A. (1984). The boundary layer natural convection regime in a rectangular cavity with uniform heat flux from the side. *ASME Journal of Heat Transfer*, *106*, 98–106.
10. Lin, W., Armfield, S. W., & Khatamifar, M. (2024). Scaling laws for natural convection boundary layer of a Pr > 1 fluid on a vertical solid surface subject to a sinusoidal temperature in a linearly-stratified ambient fluid. *Physics of Fluids*, *36*(1).
11. Luo, Y., Gui, N., Yang, X., Jiang, S., & Liu, Z. (2025). Analytical generalized thermal resistance model for conductive heat transfer in pebble beds based on heat flux weighted temperature difference. *International Journal of Heat and Mass Transfer, 236*, 126401. https://doi.org/10.1063/5.0191550.
12. Obalalu, A. M., Alfwzan, W. F., Memon, M. A., Darvesh, A., Adegbite, P., Hendy, A. S., & Ali, M. R. (2024). Energy optimization of quadratic thermal convection on two-phase boundary layer flow across a moving vertical flat plate. *Case Studies in Thermal Engineering, 55*, 104073.
13. Qi, S., Han, B., Zhu, X., Yang, B. W., Xing, T., Liu, A., & Liu, S. (2025). Machine learning in critical heat flux studies in nuclear systems: A detailed review. *Progress in Nuclear Energy*, *179*, 105535.
14. Saha, R., & Raju, B. H. S. (2025). Influence of inclined magnetic field on mixed convective cross flow across a cylinder with uniform heat flux condition. *International Journal of Heat and Fluid Flow, 111*, 109648.
15. Tegze, G., & Podmaniczky, F. (2024). Three-dimensional structure of the thermal boundary layer in turbulent Rayleigh-Bénard convection: A Lagrangian perspective. *Physical Review Fluids*, *9*(7), 074602.
16. Wakif, A., Alshehri, A., & Muhammad, T. (2024). Influences of blowing and internal heating processes on steady MHD mixed convective boundary layer flows of radiating titanium dioxide-ethylene glycol nanofluids. *ZAMM-Journal of Applied Mathematics and Mechanics/Zeitschrift für Angewandte Mathematik und Mechanik, 104*(4), e202300536.
17. Xu, H., Lu, K., Hu, F., Ding, Y., Li, B., & Zhou, K. (2024). Investigation on the restrain effect of air curtain to the spilled flame and its heat flux of a compartment-facade fire. *Applied Thermal Engineering, 256*, 124072.
18. Zhang, R., Zaydan, M., Alshehri, M., Raju, C. S. K., Wakif, A., & Shah, N. A. (2024). Further insights into mixed convective boundary layer flows of internally heated Jeffery nanofluids: Stefan's blowing case study with convective heating and thermal radiation impressions. *Case Studies in Thermal Engineering, 55*, 104121.
19. Zheng, H., He, P., Yang, S., Lu, Y., Guo, N., Li, Y., ... & Ding, G. (2024). Achieving ultra-high heat flux transfer in graphene films via tunable gas escape channels. *Advanced Science*, *12*, 2410913.
20. Zhou, Z. Y., Liao, B. K., Qi, K., Qiu, Y. B., Chen, Z. Y., & Guo, X. P. (2025). Effect of heat flux on the passive film and passive/active transition behavior of heat transfer surface of 304 SS in 0.5 M H2SO4 solution. *Surfaces and Interfaces*, *60*, 106021.

Index

P

Q

R

S

T

U

V

W